Geophysical Monograph Series

Including

IUGG Volumes
Maurice Ewing Volumes
Mineral Physics Volumes

Geophysical Monograph Series

121 **The History and Dynamics of Global Plate Motions** Mark A. Richards, Richard G. Gordon, and Rob D. van der Hilst (Eds.)

122 **Dynamics of Fluids in Fractured Rock** Boris Faybishenko, Paul A. Witherspoon, and Sally M. Benson (Eds.)

123 **Atmospheric Science Across the Stratopause** David E. Siskind, Stephen D. Eckerman, and Michael E. Summers (Eds.)

124 **Natural Gas Hydrates: Occurrence, Distribution, and Detection** Charles K. Paull and Willam P. Dillon (Eds.)

125 **Space Weather** Paul Song, Howard J. Singer, and George L. Siscoe (Eds.)

126 **The Oceans and Rapid Climate Change: Past, Present, and Future** Dan Seidov, Bernd J. Haupt, and Mark Maslin (Eds.)

127 **Gas Transfer at Water Surfaces** M. A. Donelan, W. M. Drennan, E. S. Saltzman, and R. Wanninkhof (Eds.)

128 **Hawaiian Volcanoes: Deep Underwater Perspectives** Eiichi Takahashi, Peter W. Lipman, Michael O. Garcia, Jiro Naka, and Shigeo Aramaki (Eds.)

129 **Environmental Mechanics: Water, Mass and Energy Transfer in the Biosphere** Peter A.C. Raats, David Smiles, and Arthur W. Warrick (Eds.)

130 **Atmospheres in the Solar System: Comparative Aeronomy** Michael Mendillo, Andrew Nagy, and J. H. Waite (Eds.)

131 **The Ostracoda: Applications in Quaternary Research** Jonathan A. Holmes and Allan R. Chivas (Eds.)

132 **Mountain Building in the Uralides Pangea to the Present** Dennis Brown, Christopher Juhlin, and Victor Puchkov (Eds.)

133 **Earth's Low-Latitude Boundary Layer** Patrick T. Newell and Terry Onsage (Eds.)

134 **The North Atlantic Oscillation: Climatic Significance and Environmental Impact** James W. Hurrell, Yochanan Kushnir, Geir Ottersen, and Martin Visbeck (Eds.)

135 **Prediction in Geomorphology** Peter R. Wilcock and Richard M. Iverson (Eds.)

136 **The Central Atlantic Magmatic Province: Insights from Fragments of Pangea** W. Hames, J. G. McHone, P. Renne, and C. Ruppel (Eds.)

137 **Earth's Climate and Orbital Eccentricity: The Marine Isotope Stage 11 Question** André W. Droxler, Richard Z. Poore, and Lloyd H. Burckle (Eds.)

138 **Inside the Subduction Factory** John Eiler (Ed.)

139 **Volcanism and the Earth's Atmosphere** Alan Robock and Clive Oppenheimer (Eds.)

140 **Explosive Subaqueous Volcanism** James D. L. White, John L. Smellie, and David A. Clague (Eds.)

141 **Solar Variability and Its Effects on Climate** Judit M. Pap and Peter Fox (Eds.)

142 **Disturbances in Geospace: The Storm-Substorm Relationship** A. Surjalal Sharma, Yohsuke Kamide, and Gurbax S. Lakhima (Eds.)

143 **Mt. Etna: Volcano Laboratory** Alessandro Bonaccorso, Sonia Calvari, Mauro Coltelli, Ciro Del Negro, and Susanna Falsaperla (Eds.)

144 **The Subseafloor Biosphere at Mid-Ocean Ridges** William S. D. Wilcock, Edward F. DeLong, Deborah S. Kelley, John A. Baross, and S. Craig Cary (Eds.)

145 **Timescales of the Paleomagnetic Field** James E. T. Channell, Dennis V. Kent, William Lowrie, and Joseph G. Meert (Eds.)

146 **The Extreme Proterozoic: Geology, Geochemistry, and Climate** Gregory S. Jenkins, Mark A. S. McMenamin, Christopher P. McKay, and Linda Sohl (Eds.)

147 **Earth's Climate: The Ocean–Atmosphere Interaction** Chuzai Wang, Shang-Ping Xie, and James A. Carton (Eds.)

148 **Mid-Ocean Ridges: Hydrothermal Interactions Between the Lithosphere and Oceans** Christopher R. German, Jian Lin, and Lindsay Parson (Eds.)

149 **Continent-Ocean Interactions Within East Asian Margina Seas** Peter Clift, Wolfgang Kuhnt, Pinxian Wang, and Dennis Hayes (Eds.)

150 **The State of the Planet: Frontiers and Challenges in Geophysics** Robert Stephen John Sparks and Christopher John Hawkesworth (Eds.)

151 **The Cenzoic Southern Ocean: Tectonics, Sedimentation, and Climate Change Between Australia and Antarctica** Neville Exon, James P. Kennett, and Mitchell Malone (Eds.)

152 **Sea Salt Aerosol Production: Mechanisms, Methods, Measurements, and Models** Ernie R. Lewis and Stephen E. Schwartz (Eds.)

153 **Ecosystems and Land Use Change** Ruth S. DeFries, Gregory P. Anser, and Richard A. Houghton (Eds.)

154 **The Rocky Mountain Region—An Evolving Lithosphere: Tectonics, Geochemistry, and Geophysics** Karl E. Karlstrom and G. Randy Keller (Eds.)

155 **The Inner Magnetosphere: Physics and Modeling** Tuija I. Pulkkinen, Nikolai A. Tsyganenko, and Reiner H. W. Friedel (Eds.)

156 **Particle Acceleration in Astrophysical Plasmas: Geospace and Beyond** Dennis Gallagher, James Horwitz, Joseph Perez, Robert Preece, and John Quenby (Eds.)

157 **Seismic Earth: Array Analysis of Broadband Seismograms** Alan Levander and Guust Nolet (Eds.)

158 **The Nordic Seas: An Integrated Perspective** Helge Drange, Trond Dokken, Tore Furevik, Rüdiger Gerdes, and Wolfgang Berger (Eds.)

Geophysical Monograph 159

Inner Magnetosphere Interactions: New Perspectives From Imaging

James Burch
Michael Schulz
Harlan Spence
Editors

American Geophysical Union
Washington, DC

Published under the aegis of the AGU Books Board

Jean-Louis Bougeret, Chair, Gray E. Bebout, Cari T. Friedrichs, James L. Horwitz, Lisa A. Levin, W. Berry Lyons, Kenneth R. Minschwaner, Andy Nyblade, Darrell Strobel, and William R. Young, members.

Library of Congress Cataloging-in-Publication Data

Inner magnetosphere interactions : new perspectives from imaging / James Burch, Michael Schulz, Harlan Spence, editors.
  p. cm. -- (Geophysical monograph, ISSN 0065-8448 ; 159)
 1. Magnetosphere--Remote-sensing images. 2. Imaging systems in astronomy. I. Burch, J. L., 1942- II. Schulz, Michael, 1960- III. Spence, H. (Harlan) IV. Series.

QC809.M35I477 2005
538'.766'0222--dc22   2005024704

   ISBN-10:0-87590-424-6 (hardcover)
   ISBN-13: 978-0-87590-424-5 (hardcover)

   ISSN 0065-8448

   Copyright 2005 by the American Geophysical Union
   2000 Florida Avenue, N.W.
   Washington, DC 20009

Figures, tables and short excerpts may be reprinted in scientific books and journals if the source is properly cited.

Authorization to photocopy items for internal or personal use, or the internal or personal use of specific clients, is granted by the American Geophyscial Union for libraries and other users registered with the Copyright Clearance Center (CCC) Transactional Reporting Service, provided that the base fee of $1.50 per copy plus $0.35 per page is paid directly to CCC, 222 Rosewood Dr., Danvers, MA 01923. 0065-8448/05/$01.50+0.35.

This consent does not extend to other kinds of copying, such as copying for creating new collective works or for resale. The reproduction of multiple copies and the use of full articles or the use of extracts, including figures and tables, for commercial purposes requires permission from the American Geophysical Union.

Printed in the United States of America.

CONTENTS

Preface
James Burch, Michael Schulz, and Harlan Spence vii

Plasmasphere Dynamics

The Global Pattern of Evolution of Plasmaspheric Drainage Plumes
J. Goldstein and B. R. Sandel 1

**Multipoint Observations of Ionic Structures in the Plasmasphere by CLUSTER—
CIS and Comparisons With IMAGE-EUV Observations and With Model Simulations**
*Iannis Dandouras, Viviane Pierrard, Jerry Goldstein, Claire Vallat, George K. Parks, Henri Rème,
Cécile Gouillart, Fréderic Sevestre, Michael McCarthy, Lynn M. Kistler, Berndt Klecker, Axel Korth,
Maria Bice Bavassano-Cattaneo, Philippe Escoubet, and Arnaud Masson*.. 23

Wave-Particle Interactions in the Plasmasphere and Ring Current

The Relationship Between Plasma Density Structure and EMIC Waves at Geosynchronous Orbit
B. J. Fraser, H. J. Singer, M. L. Adrian, D. L. Gallagher, and M. F. Thomsen.. 55

ULF Waves Associated with Enhanced Subauroral Proton Precipitation
Thomas J. Immel, S. B. Mende, H. U. Frey, J. Patel, J. W. Bonnell, M. J. Engebretson, and S. A. Fuselier............ 69

Afternoon Subauroral Proton Precipitation Resulting from Ring Current—Plasmasphere Interaction
M. Spasojević, M. F. Thomsen, P. J. Chi, and B. R. Sandel 85

The Influence of Wave-Particle Interactions on Relativistic Electron Dynamics During Storms
*Richard M. Thorne, Richard B. Horne, Sarah Glauert, Nigel P. Meredith, Yuri Y. Shprits, Danny Summers, and
Roger R. Anderson* 101

Distribution and Origin of Plasmaspheric Plasma Waves
James L. Green, Shing F. Fung, Scott Boardsen, and Hugh J. Christian... 113

Electric Fields and Currents in the Inner Magnetosphere

Direct Effects of the IMF on the Inner Magnetosphere
*R. A. Wolf, S. Sazykin, X. Xing, R. W. Spiro, F. R. Toffoletto, D. L. De Zeeuw,
T. I. Gombosi, and J. Goldstein*... 127

Non-Potential Electric Field Model of Magnetosphere-Ionosphere Coupling
Igor V. Sokolov, Tamas I. Gombosi, and Aaron J. Ridley 141

Pressure-Driven Currents Derived from Global ENA Images by IMAGE/HENA
Edmond C. Roelof 153

**On the Relation Between Electric Fields in the Inner Magnetosphere, Ring Current, Auroral Conductance,
and Plasmapause Motion**
*P. C. Brandt, J. Goldstein, B. J. Anderson, H. Korth, T. J. Immel, E. C. Roelof, R. DeMajistre,
D. G. Mitchell, and B. Sandel*... 159

Small-Scale Structure in the Stormtime Ring Current
Michael W. Liemohn and Pontus C. Brandt... 167

Solar and Ionospheric Plasmas in the Ring Current Region
T. E. Moore, M-C. Fok, S. P. Christon, S.-H. Chen, M. O. Chandler, D. C. Delcourt, J. Fedder, S. Slinker, and M. Liemohn 179

Statistical Properties of Dayside Subauroral Proton Flashes Observed With IMAGE-FUV
Benoît Hubert, Jean-Claude Gérard, Stephen B. Mende, and Stephen A. Fuselier 195

Ring Current and Radiation Belt Dynamics

Geospace Storm Processes Coupling the Ring Current, Radiation Belt and Plasmasphere
M.-C. Fok, Y. Ebihara, T. E. Moore, D. M. Ober, and K. A. Keller 207

Toward Understanding Radiation Belt Dynamics, Nuclear Explosion-Produced Artificial Belts, and Active Radiation Belt Remediation: Producing a Radiation Belt Data Assimilation Model
Geoffrey D. Reeves, Reiner H. W. Friedel, Sebastien Bourdarie, Michelle F. Thomsen, Sorin Zaharia, Michael G. Henderson, Yue Chen, Vania K. Jordanova, Brian J. Albright, and Dan Winske 221

Simulated Stormtime Ring-Current Magnetic Field Produced by Ions and Electrons
Margaret W. Chen, Michael Schulz, Shuxiang Liu, Gang Lu, Larry R. Lyons, Mostafa El-Alaoui, and Michelle Thomsen 237

Radiation Belt Responses to the Solar Events of October–November 2003
D.N. Baker, S.G. Kanekal, J.B. Blake, and J.H. Allen 251

Ionospheric Response to Solar-Wind Forcing

Hemispheric Daytime Ionospheric Response to Intense Solar Wind Forcing
A. J. Mannucci, Bruce T. Tsurutani, Byron Iijima, Attila Komjathy, Brian Wilson, Xiaoqing Pi, Lawrence Sparks, George Hajj, Lukas Mandrake, Walter D. Gonzalez, Janet Kozyra, K. Yumoto, M. Swisdak, J.D. Huba, and R. Skoug 261

Redistribution of the Stormtime Ionosphere and the Formation of a Plasmaspheric Bulge
John C. Foster, Anthea J. Coster, Philip J. Erickson, William Rideout, Frederick J. Rich, Thomas J. Immel, and Bill R. Sandel 277

Retrospective

Yosemite 2004—A Thirty Year Tradition
C. R. Chappell 291

PREFACE

We regard the *inner magnetosphere* as the region surrounded by a geomagnetic shell with an equatorial radius of approximately eight Earth radii. Of much concern to scientists and researchers, and the focus of this monograph, are the interactions that occur in this region of geospace. We include here hot-cold plasma interactions, plasmasphere dynamics, direct effects of the interplanetary magnetic field on the inner magnetosphere, ring-current/ionosphere interactions, ring-current effects on global electric fields, and ring-current/radiation-belt interactions. With new satellite measurements, especially those from IMAGE (Imager for Magnetopause-to-Aurora Global Exploration), and associated observations from ground-based facilities, our understanding of the interactions that take place in the inner magnetosphere has developed rapidly.

Indeed, the March 2000 launch of IMAGE has provided us with a much-needed global perspective on the dynamical interactions that take place in the inner magnetosphere during geomagnetic storms and magnetospheric substorms along with novel means of remotely sensing the ring current and plasmasphere. Previous imagers (e.g., on DMSP, Dynamics Explorer, UARS, and Polar) had focused on aurora and airglow at various wavelengths (ranging from infrared to X-ray) emitted from the atmosphere and ionosphere. These regions also were characterized rather well by spacecraft crossing through the inner magnetosphere, making continuous and detailed in-situ measurements of the resident cold and hot plasmas and energetic particles, and in some cases measuring ambient electric fields.

New imaging of the inner magnetosphere using extreme ultraviolet imagers (for global cold-He^+ detection) and energetic neutral-atom imagers (for global imaging of ions from low to high energies), however, has led to several surprises and insights about inner magnetosphere interactions. Among the surprises are (1) the direct influence of variations in the interplanetary magnetic field on the structure of the plasmapause, which resides deep inside the magnetosphere (far from its boundary with the solar wind); and (2) injection of the ring current into the post-midnight sector, whereas the magnetic signature of the main phase of a magnetic storm typically peaks in the dusk sector. As often happens after unexpected observations in space physics, a close examination of existing models showed that these features could have been predicted, or—as in case number (2)—the effect was predicted but the prediction was rejected because it seemed to violate the available data.

Among the new insights gained from imaging is the realization that newly-imaged plasmaspheric plumes in the equatorial plane map along magnetic field lines to the ionosphere, where they can be imaged by ground-based radars and detected by GPS receivers as widespread density enhancements. Furthermore, plasmaspheric plumes are shown to be loss sites for ring-current protons and thus agents for causing proton auroras. These observations provided some of the strongest evidence yet for close coupling of the magnetosphere to the ionosphere and upper atmosphere.

The current monograph thus features several recurring themes, including the following: (1) There are significant consequences of the regional overlap of plasmasphere with the partial and symmetric ring currents, as well as with the plasma sheet: Coulomb interactions, wave-particle interactions, and macroscopic interactions; (2) Electromagnetic waves, which are ubiquitous in the inner magnetosphere, provide the coupling processes for many of the interactions among the various plasma populations there; (3) Modeling and novel observations are confirming earlier theories and are leading to a new understanding of plasmasphere dynamics; (4) The inner magnetosphere responds both directly and indirectly to the solar wind and IMF in ways that are clear, as well as in ways that still need resolution; (5) The ring current and the ionosphere constitute a tightly coupled 3D system; (6) Ring-current and radiation-belt coupling is oftentimes subtle but critically important; and (7) Coupled models are needed to differentiate between competing physical ideas and possibly to reveal missing physics. They can elucidate the temporal and spatial evolution of global magnetic and electric field geometries, even as they reveal key macroscopic and microscopic parameters.

This monograph also has roots in the goal of a recent workshop: to stimulate the development of an integrated global view of the inner magnetosphere and to identify new research thrusts for the future. In the workshop, *Inner Magnetosphere Interactions,* held in February 2004 at Yosemite National Park, we explored these phenomena with modeling, along with data from imaging and in-situ spacecraft and from ground-based observations.

For the past thirty years now (beginning in February 1974) Yosemite has been the site of very nearly biennial workshops on various hot topics in space physics. A highlight of Yosemite 2004 was a presentation by Rick Chappell, one of the organizers of the first workshop, of videos of several of the talks given in February 1974. The videos, which appear on a CD-ROM in this monograph, provide fascinating glimpses into the past – both of the practitioners and of the ideas that laid the basis for much of the progress made in magnetosphere-ionosphere coupling (the topic of that workshop) since 1974. As you will find, many of the questions are the same, but most of the answers are quite different.

We would like to acknowledge and thank John Lynch of the Vanderbilt Television News Archive for digitizing the videotapes which are included on the CD-ROM which accompanies this volume. We thank the staff of Yosemite National Park, who have helped make this and previous Yosemite workshops both pleasant and productive. Thanks also go to the AGU for its willingness to publish this monograph.

James Burch
Southwest Research Institute

Michael Schulz
Lockheed Martin

Harlan Spence
Boston University

The Global Pattern of Evolution of Plasmaspheric Drainage Plumes

J. Goldstein

Space Science and Engineering Division, Southwest Research Institute, San Antonio, Texas

B. R. Sandel

Lunar and Planetary Laboratory, University of Arizona, Tucson, Arizona

We present observations of an 18 June 2001 erosion event obtained by the IMAGE extreme ultraviolet (EUV) imager. Following a 0304 UT southward turning of the interplanetary magnetic field (IMF), the plasmasphere on both nightside and dayside surged sunward, reducing the plasmasphere radius on the nightside and creating a broad drainage plume on the dayside. Over several hours this plume narrowed in magnetic local time (MLT), until shortly after a northward IMF turning between 1430 UT and 1500 UT, when the plume began corotating with the Earth. On a global scale, the 18 June EUV plasmasphere observations are consistent with the interpretation that dayside magnetopause reconnection (DMR) during southward IMF produced a sunward convection field in the inner magnetosphere. Using the Volland-Stern electric potential model normalized to the solar wind E-field, we performed a simple plasmapause test particle (PTP) simulation of the 18 June event and found good global agreement with EUV observations, but important sub-global differences as well. On a sub-global scale, proper treatment of plasmaspheric dynamics requires consideration of sub-auroral polarization streams (SAPS) and penetration electric field to explain narrow duskside plumes and preferential pre-dawn plasmapause motion, respectively. The 18 June 2001 EUV images contain evidence of a double plume (or bifurcation of a single plume) and dayside crenulations of the plasmapause, both of which remain unexplained. The observations suggest that strong convection suppresses or smooths plasmapause structure, which tends to increase during times of weak or absent convection. Analysis of the motion of the plasmapause on 18 June 2001 reveals some of the details of the initial erosion process, which apparently involves partial indentation of the plasmapause and subsequent widening of this indentation to other MLT sectors eastward and westward of the initial indentation, and produces 'rotated V' signatures in the electric field. Early erosion on 18 June was bursty, and modulated by the solar wind electric field; convection was turned on during southward IMF and turned off during northward IMF. Northward IMF apparently triggered overshielding, causing the formation of a midnight-to-dawn plasmapause bulge that subsequently corotated. It is clear that more detailed information about the inner magnetospheric E-field is required to fully understand plasmaspheric dynamics.

1. INTRODUCTION

1.1. Dayside Magnetopause Reconnection (DMR), Erosion, and Plume Formation

It is widely accepted that when the IMF turns southward, reconnecting field lines are dragged antisunward, driving magnetospheric convection in which the outer magnetospheric plasma moves tailward and inner magnetospheric plasma moves sunward [*Dungey*, 1961]. The strength of this dayside magnetopause reconnection (DMR) driven convection should fluctuate in time in accord with variations in the solar wind (SW) and interplanetary magnetic field (IMF). The main influence seems to be the polarity of the Z-component $B_{Z,IMF}$ of the IMF. During southward IMF (negative $B_{Z,IMF}$) DMR drives convection; during northward IMF $B_{Z,IMF} > 0$) DMR convection shuts off. Numerous studies (e.g., see *Carpenter et al.* [1993]; *Carpenter* [1995]; *Carpenter and Lemaire* [1997]; *Lemaire and Gringauz* [1998], and many observational and theoretical papers referenced therein) have shown that the strength of DMR-driven sunward convection is a primary influence on the dynamics and structure of the plasmasphere, the cold, rotating torus of plasma that surrounds the Earth and (on average) extends to equatorial distances of 4–6 earth radii (R_E). A strong change in DMR-driven convection can cause the outer boundary of the plasmasphere, the plasmapause, to move either radially inward (compression) or outward (rarefaction). A DMR convection increase may also produce an azimuthal plasma motion in which the outer layer of the plasmasphere is stripped away, a process known as plasmaspheric erosion. The hypothetical DMR-driven convection offers an explanation for why the plasmasphere shrinks during increased geomagnetic activity [*Chappell et al.*, 1970a], and why in situ observations imply the presence of a bulge near dusk [*Chappell et al.*, 1970b; *Higel and Wu*, 1984].

The details of the erosion process are not yet completely understood [*Carpenter and Lemaire*, 1997], but one known byproduct of erosion is the drainage plume. Plumes (also called 'tails') are regions of plasmaspheric plasma that are connected to the main body of the plasmasphere and extend outward into the surrounding tenuous plasma. Plumes were predicted on the basis of theoretical models of the effects of increases in DMR-driven convection [*Grebowsky*, 1970; *Chen and Wolf*, 1972; *Spiro et al.*, 1981; *Elphic et al.*, 1996; *Weiss et al.*, 1997; *Lambour et al.*, 1997]. In situ observations of outlying or 'detached' plasma, separated from the main plasmasphere [*Chappell*, 1974; *Carpenter and Anderson*, 1992], seemed consistent with the plume interpretation. An alternate explanation, that the detached plasma was due to 'blobs' completely separated from the plasmasphere, was offered by *Chappell* [1974], and the mechanism for creation of blobs by gravitational/centrifugal interchange instability was proposed by *Lemaire* [1975].

The existence of plumes of high-density plasmaspheric material has been conclusively demonstrated by global plasmaspheric images [*Sandel et al.*, 2001; *Burch et al.*, 2001; *Foster et al.*, 2002; *Goldstein et al.*, 2002, 2003a; *Spasojević et al.*, 2003; *Goldstein et al.*, 2003c, b, 2004b; *Sandel et al.*, 2003]. It should be noted that the global plasmaspheric images do not see the complete plasma distribution, but rather only the high-density portion, corresponding to total number densities above about 40 cm^{-3} [*Goldstein et al.*, 2003c; *Moldwin et al.*, 2003]. It is for this reason that we say 'plumes of high-density plasmaspheric material.' At densities below 40 cm^{-3}, completely detached blobs of plasma may indeed exist but still be invisible in plasmaspheric images. Plasmasphere images show there is a strong correlation between $B_{Z,IMF}$ polarity and the behavior of the plasmasphere during both southward [*Goldstein et al.*, 2003a; *Spasojević et al.*, 2003] and northward [*Goldstein et al.*, 2002, 2003d] IMF polarities. From plasmaspheric imaging, the formation and subsequent evolution of plasmaspheric plumes follows a predictable pattern that depends primarily on IMF polarity, as follows.

1.1.1. Sunward Surge. Following an increase in the magnitude of southward IMF, the plasmaspheric plasma surges sunward. On the nightside the plasmapause radius decreases (moves sunward/earthward), and on the dayside the plasmapause location increases (moves sunward). The increased extent of the dayside plasmasphere forms a plume that is broad in magnetic local time (MLT) extent, and which extends outward in the +X-direction.

1.1.2. Plume Narrowing. If the IMF polarity remains southward at its surge-time level for several hours, the plume formed during the initial sunward surge then undergoes a period of narrowing, in which the the dusk edge of the plume remains relatively stationary while the western edge of the plume slowly rotates eastward. Models provide some information about plume narrowing. The plume forms following an enhancement in convection, and concomitant inward motion of the corotation/convection boundary (CCB), and as time advances and the erosion progresses, less plasmaspheric material remains outside the CCB to 'feed' the plume, causing it to narrow. Also, the innermost western edge of the plume may lie within the CCB and thus tends to rotate with the Earth, while the dusk edge tends to line up with the CCB and thus is roughly stationary during steady convection. Models also suggest that if this plume narrowing phase continues indefinitely, the western edge eventually reaches the dusk edge, and the plume disappears/dissipates.

1.1.3. Plume Rotation/Wrapping Eventually, the IMF turns northward, and in EUV images the narrowed plume begins to rotate eastward and wrap about the main plasmasphere. The DMR convection hypothesis explains this plume rotation as follows. When the IMF turns northward, the CCB expands to larger radial distances, and the plume that was formerly in the convection zone is now in the corotation regime and thus begins to rotate. Inside the CCB, the rotation rate decreases with distance from the Earth on the dusk side, and this flow shear distorts the shape of the plume. The base of the plume (near the plasmasphere) moves faster than the end of the plume, so the plume lengthens as it rotates. If quiet conditions prevail long enough, the plume rotates until it encounters the new location of the CCB, and then it lengthens and wraps around the plasmasphere. *Spasojević et al.* [2003] showed a particularly dramatic example of this rotation/wrapping process that occurred on 10–11 June 2001.

The EUV-observed phases of plume evolution (sunward surge, plume narrowing, and plume rotation/wrapping) are entirely consistent with (and indeed were predicted by) model plasmaspheres subject to DMR-driven convection (e.g., *Grebowsky* [1970]; *Spiro et al.* [1981]), and also agree with prior in situ observations [*Elphic et al.*, 1996]. The eastward rotation of the plume during northward IMF is also in accord with in situ observations of the rotating duskside bulge [*Higel and Wu*, 1984; *Moldwin et al.*, 1994].

1.2. Details of the Erosion Process

Although the zero-order (i.e., global) active-time plasmaspheric dynamics are adequately described by the phases of plume evolution and the DMR-driven convection hypothesis, this simple picture is clearly incomplete.

There remain important questions about the mechanisms involved in transferring SW/IMF energy to the inner magnetosphere. The first question is, how is that energy transferred? *Goldstein et al.* [2003a] noted that there is a time delay (which they dubbed 'configuration delay' $\Delta\tau_C$) between the arrival of southward IMF at the magnetopause and the subsequent inward motion of the nightside plasmapause. This time delay $\Delta\tau_C$ has so far been consistently observed (in plasmasphere images) to be between 20 and 30 minutes when reasonably precise timing of the SW and IMF arrival at the magnetopause was available [*Spasojević et al.*, 2003; *Goldstein et al.*, 2003a, b, 2004b]. The cause of the delay $\Delta\tau_C$ is perhaps explainable as the time necessary for the entire magnetospheric DMR convection field to reconfigure following a southward IMF turning [*Coroniti and Kennel*, 1973]. This reconfiguration explanation has yet to be conclusively established, and the details of the reconfiguration process, surely involving coupled interactions of the ionosphere, plasmasheet, and ring current, remain unknown. Another question is, how much of the SW/IMF energy is transmitted to the inner magnetosphere? It is known that the inner magnetosphere inside the plasmasheet is to some (time-varying) degree shielded from DMR driven convection [*Jaggi and Wolf*, 1973]. However, effective shielding probably requires between 15 minutes and 1 hour to develop [*Kelley et al.*, 1979; *Senior and Blanc*, 1984; *Goldstein et al.*, 2003d], so that it probably cannot respond to more rapid changes in DMR-driven convection driven by the ever-present fluctuations in the SW and IMF. Thus, under quickly varying geomagnetic conditions the external convection field can 'penetrate' past the shielding layer. This so-called penetration E-field has been observed in ionospheric and equatorial in situ measurements [*Fejer et al.*, 1990; *Fejer and Scherliess*, 1995; *Scherliess and Fejer*, 1997; *Wygant et al.*, 1998], and there are indications that it can be 'focused' into the midnight-to-dawn MLT sector. From analysis and modeling of global plasmasphere images, *Goldstein et al.* [2003b, 2004c] estimated that between 12 and 25 percent of the solar wind E-field can be transmitted to the inner magnetosphere during plasmasphere erosion events.

A question of continuing interest (and at times, mild controversy) is: how and where does plasma redistribute itself to form a new plasmapause boundary, particularly during erosion? According to the DMR-driven convection hypothesis, when the plasmapause boundary moves inward (as it does on the nightside during erosion events), the plasma at the boundary moves both radially inward and azimuthally (either eastward or westward, depending on the MLT sector), and the net effect is a reduction of the plasmapause radius [*Grebowsky*, 1970; *Spiro et al.*, 1981]. However, the possible role of plasma instabilities in the erosion process is unknown. According to proponents of the gravitational/centrifugal interchange hypothesis [*Lemaire*, 1975; *Lemaire and Gringauz*, 1998], during erosion (i.e., inward radial plasmapause motion) the nightside plasma actually moves radially outward, forming detached blobs that might show up as fine-scale density structure outside the main plasmapause [*LeDocq et al.*, 1994; *Moldwin et al.*, 1995]. Possibly related to this topic is the unresolved issue of quiet-time plasmaspheric density structure. During or following extended quiet periods, the plasmasphere exhibits a great deal of as-yet unexplained meso-scale and fine-scale structure in the form of 'blobby' density regions, irregular plasmapause shapes, fingerlike density enhancements, and isolated high-density flux tubes found in the interior of the plasmasphere [*Moldwin et al.*, 1994, 1995, 2003; *Sandel et al.*, 2001; *Spasojević et al.*, 2003; *Dent et al.*, 2003; *Goldstein et al.*, 2004b]. What causes these density structures? One explanation is that the interchange instability, which during active times is suppressed by ring current pressure and/or high ionospheric conductivity [*Richmond*, 1973; *Huang et al.*, 1990], might during quiet times have a significant effect on plasmaspheric structure. It has also been suggested that during quiet times, in the absence of strong forcing by dayside reconnection, the inner magnetospheric electric field becomes disorganized and spatially structured, creating the observed quiet-time density characteristics [*Moldwin et al.*, 1994].

A significant modification of DMR-driven convection is the subauroral polarization stream (SAPS). SAPS—also known as subauroral ion drifts (SAID) or polarization jets—are a disturbance-time effect in which feedback between the ring current and ionosphere produces an intense, radially narrow, westward flow channel in the dusk-to-midnight MLT sector [*Foster and Burke*, 2002; *Foster et al.*, 2002; *Foster and Vo*, 2002; *Anderson et al.*, 2001; *Burke et al.*, 1998, 2000]. Ionospheric SAPS occur when the equatorward boundaries of the ion and electron precipitation separate, leading to a poleward Pedersen current in the subauroral ionosphere, connected to the the ion and electron plasmasheets via region 2 and region 1 field aligned currents, respectively. Due to the low conductivity at subauroral latitudes, the poleward Pedersen current generates intense poleward E-fields that are then mapped to the equatorial plane as radial E-fields confined between the inner edges of the ion and electron plasmasheets. Thus, SAPS form a radially-narrow (1 to 2 R_E) flow channel bordering or overlapping the dusk-to-midnight plasmasphere. Because of the ring-current/ionosphere feedback involved in SAPS generation, the magnetopause IMF polarity does not directly turn SAPS on and off as it does DMR convection; SAPS can persist even when DMR-driven convection has subsided following a northward IMF turning. SAPS have been demonstrated to modify plasmasphere dynamics in the dusk-to-midnight MLT sector by intensifying sunward convection, which sharpens the outer radial density gradient at the plasmapause boundary, smooths the MLT shape of the plasmapause, and at times creates narrow duskside plumes that are distinct from the broad dayside DMR-driven Grebowsky plumes [*Foster et al.*, 2002; *Goldstein et al.*, 2003b, 2004b, a].

1.3. 18 June 2001: Start-to-Finish Plume Evolution

In this paper we present global images of the plasmasphere obtained on 18 June 2001, when a plasmasphere erosion occurred following a southward IMF turning early in the day. We will examine these observations in the context of the hypothesis that dayside magnetopause reconnection (DMR) drives convection that exerts a primary global influence on the plasmasphere. The 18 June 2001 event exemplifies the pattern of plume evolution implied by the models of *Grebowsky* [1970] and others, and observed in part during other erosion events (e.g., *Elphic et al.* [1996], *Spasojevic' et al.* [2003], *Goldstein et al.* [2004b]). The appeal of this event is that global imaging observations were available to witness all the phases of plume evolution (sunward surge, plume narrowing, and plume rotation/wrapping), providing excellent coverage (with the exception of a data gap during the plume narrowing phase) of the plume formation and evolution from start to finish. By studying this single start-to-finish event, we can observe the creation and subsequent evolution of particular features of the plasma distribution, and more clearly identify ways in which the plasmasphere behaves both according to, and in disagreement with, the simple DMR-driven convection picture.

2. OBSERVATIONS 18 JUNE 2001

In this section we present solar wind and interplanetary magnetic field (IMF) data, and global plasmasphere observations, during a plasmasphere erosion event that occurred on 18 June 2001. On this day, the overall geomagnetic conditions were those of a weak-to-moderate magnetic storm. The storm sudden commencement occurred between 0300 UT and 0400 UT, and Dst reached a minimum of −61 nT between 0900 UT and 1000 UT.

2.1. Solar Wind and IMF

On 18 June 2001 the Advanced Composition Explorer (ACE) spacecraft [*Stone et al.*, 1998] was located approximately 244 R_E upstream of the Earth, and about 32 R_E duskward of the Earth-Sun line. Figure 1 shows data from the MAG [*Smith et al.*, 1998] and SWEPAM [*McComas et al.*, 1998] instruments. The ACE data have been propagated to the magnetopause by adding a time delay of 60 ± 10 minutes, calculated as V_{SW}/X. The uncertainty ±10 minutes in our propagation delay is for this event slightly larger than that of other published erosion events (e.g., *Goldstein et al.* [2003a]). The imprecise timing of the arrival of the solar wind at the magnetopause will affect the reliability of our estimate of $\Delta\tau_C$, the configuration delay for the 18 June 2001 event (see Section 2.2.2).

Figure 1a and Figure 1b plot the IMF polarity $B_{Z,IMF}$ and solar wind speed V_{SW}, respectively. At about 0300 UT a mild shock/transition arrived at the magnetopause, bringing a +60 km/s step-like increase in V_{SW} at 0257 UT and a somewhat noisy excursion from mild southward IMF ($B_{Z,IMF} \geq -3$ nT) to strong southward IMF ($B_{Z,IMF} \leq -10$ nT) at 0304 UT. According to the DMR-driven convection hypothesis, this southward IMF excursion should impose on the magnetosphere a duskward solar wind electric field (corresponding

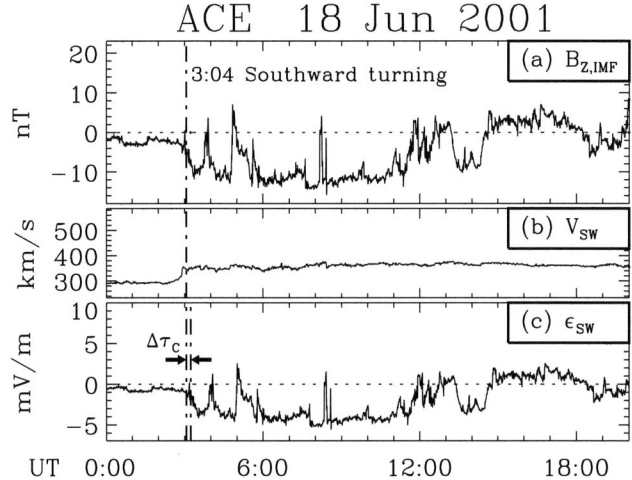

Figure 1. ACE MAG and SWEPAM data (courtesy N. Ness, C. Smith, D. McComas, and the ACE science center) on 18 June 2001, time-delayed to account for propagation to the magnetopause. (a) Z-component of the IMF $B_{Z,IMF}$; (b) solar wind speed V_{SW}; (c) solar wind electric field ϵ_{SW}, delayed by $\Delta\tau_C$ (described in text). Solar wind and IMF conditions changed at about 0300 UT, with a 0257 UT step-increase in V_{SW} and a 0304 UT southward excursion in $B_{Z,IMF}$. After 0314 UT ϵ_{SW} was negative, indicating a dawn-to-dusk global electric field.

to sunward $E \times B$ convection). Figure 1c plots the dawnward solar wind E-field ϵ_{SW}, defined as $\epsilon_{SW} \equiv V_{SW}B_{Z,IMF}$ so that ϵ_{SW} is negative when the IMF is southward. Therefore, in this paper we will use the terms 'negative ϵ_{SW}' and 'southward IMF' somewhat interchangeably. From the results of *Goldstein et al.* [2003a, b], we expect some delay $\Delta\tau_C$ between the IMF and the effects of SW on the plasmasphere, so we have delayed ϵ_{SW} by $\Delta\tau_C = 10$ minutes (as indicated in Figure 1c). (Determination of the value of $\Delta\tau_C$ for this event is discussed later, in Section 2.2.2). Because the DMR-driven E-field forms the conceptual framework of this paper, the plot of ϵ_{SW} in Figure 1c will be repeated to aid discussion of Plate 1 and Plate 3.

In the next section we present plasmasphere images obtained by the IMAGE EUV instrument.

2.2. Global Plasmasphere Images

The extreme ultraviolet (EUV) instrument on the IMAGE satellite is an imaging system composed of three cameras (with slightly overlapping fields of view) sensitive to the 30.4-nm ultraviolet light that is resonantly scattered by the He$^+$ ions in the plasmasphere [*Burch*, 2000; *Sandel et al.*, 2000, 2001]. IMAGE EUV sees the He$^+$ portion of the plasmasphere corresponding to total number densities above about 40 cm^{-3} [*Goldstein et al.*, 2003c; *Moldwin et al.*, 2003].

On 18 June 2001 the IMAGE satellite was in a roughly dawn-dusk orbit, with its 8.2 R_E apogee almost directly over the north magnetic pole. This apogee location provided an excellent viewing geometry, with minimal sunlight contamination and perspective distortion in the images, and a wide field of view that most of the time extended to or beyond geosynchronous orbit.

Plate 1 shows 12 panels of EUV image data, labeled a, b, c,..., j, k, l and arranged in 3 rows of 4 panels each, depicting a sequence of plasmapause images from 0010 UT through 1854 UT. In the figure, time increases from left to right in each row, and from top to bottom between rows, as indicated by the UT stamps at the bottom of the panels. Each panel shows the equatorial distribution of line-of-sight integrated He$^+$ column abundance. These equatorial maps were obtained using the procedure outlined in *Dent et al.* [2003] and *Goldstein et al.* [2004b]. Color indicates column abundance (in arbitrary units), increasing from black (zero) to white (very dense plasma). The plasmasphere is the green/white region surrounding the Earth. In each image, the plasmapause is the (often sharp) dropoff in signal intensity which occurs (on average in Plate 1) between $L = 2.5$ and $L = 4$. (For the reader unfamiliar with identification of the plasmapause in EUV images, see *Goldstein et al.* [2003c] and the plasmapause extractions of the bottom row of Plate 2.) Outside the plasmapause, the dark green speckled background represents plasma total number densities at or below the EUV lower noise floor (i.e., ≤ 40 cm^{-3}). The EUV field of view (FOV) edges vary as IMAGE progresses through its orbit. The FOV edges (labeled for demonstrative purpose in Plate 1d) are the black regions that may cut across the plasmasphere near the borders of the images.

To provide context for the 12 EUV images, the bottom panel of Plate 1 shows ϵ_{SW} (from Figure 1c). For reference, the times of the 12 EUV snapshots of Plate 1a through Plate 1l are shown as labeled vertical lines in the ϵ_{SW} plot. The solar wind E-field ϵ_{SW} is defined so that its sign (positive or negative) indicates IMF polarity (northward or southward). We plot ϵ_{SW} instead of $B_{Z,IMF}$ because (1) we wish to emphasize the finite delay $\Delta\tau_C$ between southward IMF at the magnetopause and its effect on the inner magnetosphere, and (2) it is the electric field that directly drives convection of cold $E \times B$-drifting plasma in the inner magnetosphere. Said another way, we wish to distinguish the effect (the E-field that drives plasma convection) from the cause (dayside magnetopause reconnection). Negative ϵ_{SW} means there is a dawn-to-dusk E-field, or sunward convection in the inner magnetosphere.

The 18 June 2001 EUV images of Plate 1 quite fortunately capture the plasmasphere for some finite time during each of the phases of plume evolution during a single erosion event. We will first discuss the overall global plasma behav-

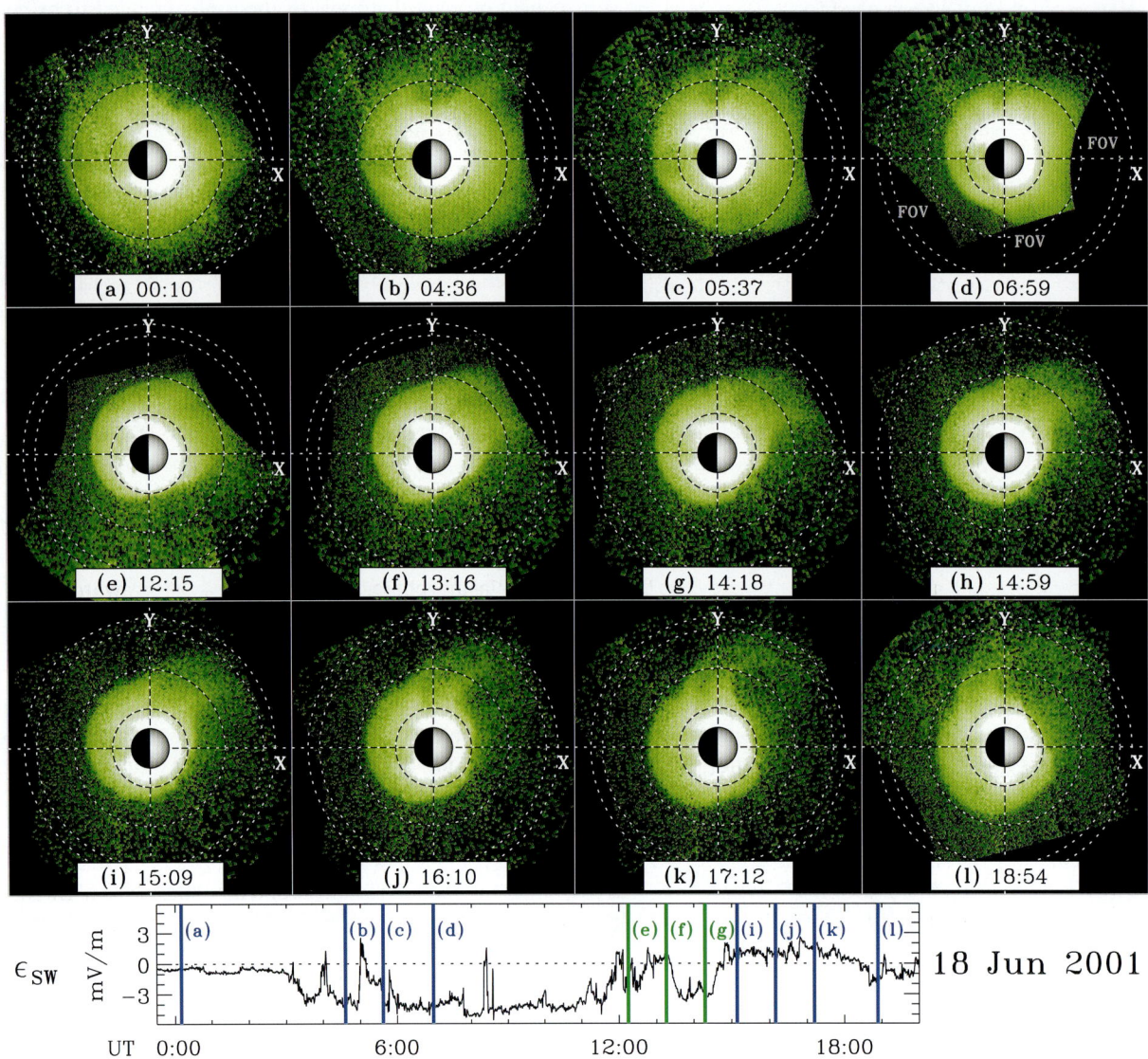

Plate 1. Panels a through l: IMAGE EUV global plasmasphere observations on 18 June 2001, depicting erosion of the plasmasphere and formation/evolution of a drainage plume. Each panel shows the equatorial plasmaspheric He$^+$ distribution versus X and Y (in SM coordinates). Color indicates column abundance (in arbitrary units), with black = zero. The Sun is to the right (positive X); the Earth is at the center. Circles are drawn at $L = 2, 4, 6$, and 6.62 (geosynchronous orbit). Bottom panel: ACE ϵ_{SW} from Figure 1c. Vertical lines labeled a through l indicate times of EUV snapshots above.

ior, and how it compares to the DMR-driven convection hypothesis.

2.2.1. Initial Plasmasphere. Plate 1a shows the plasmasphere at 0010 UT, at the start of the event. At that time, the plasmasphere was radially large and irregularly shaped, with a large ($L \approx 5$) bulge centered at noon MLT, and notches (i.e., regions of decreased plasmapause radius) at 1100 MLT and between 1600 MLT and 1800 MLT. The outer edge of the dayside plasmasphere was diffuse, representing gradual/ gentle outer density gradients. Outside these bulges and notches on the dayside, green speckling indicates some small amount of outlying plasma whose density was just above or at the EUV noise floor. In the image, the outer edge of the plasmasphere between 1800 MLT and 2100 MLT, and between 0600 MLT and 1100 MLT contains noisy brightness fluctuations that indicate spatial structure on scales below 0.5 R_E. Shape irregularity, diffuse boundaries, low-density outlying plasma, and spatial structure are characteristics typically found in quiet-time plasmasphere images [*Sandel et al.*, 2001; *Goldstein et al.*, 2003b]. On the other hand, in the nightside range 2100 MLT to 0600 MLT the plasmapause was relatively smooth and sharp, with little or no indication of outlying low-density plasma; this is typical of active time plasmasphere images. From the ϵ_{SW} plot in Plate 1, at 0010 UT the dawn-to-dusk solar wind E-field was well under 1 mV/m (which, according to the transmission factors of *Goldstein et al.* [2003b, 2004c] corresponds to 0.1–0.25 mV/m near the plasmasphere); i.e., at this time there was very weak but finite sunward convection and this weak convection had prevailed since 1600 UT of the previous day. So the dayside plasmasphere of Plate 1a reflects the relatively quiet conditions (presumably accompanied by dayside ionospheric filling of plasmaspheric flux tubes) that preceded 0010 UT, but the nightside reflects the mild convection that only had the ability to affect the local time range 2100 MLT to 0600 MLT. The DMR convection hypothesis has nothing to say about ionospheric filling of the plasmasphere, but the smooth plasmapause shape on the nightside is entirely consistent with the idea that mild sunward convection was in effect. After relatively steady convection for several hours, the DMR picture says the plasmapause location coincides approximately with the corotation/convection boundary (CCB), which was probably the case for the 0010 UT plasmapause between 2100 MLT and 0600 MLT. The 0010 UT plasmasphere illustrates how the plasmapause location and shape at different MLTs can arise due to a combination of accumulated effects and direct driving by the solar wind and IMF. Between 0010 UT and about 0314 UT the dayside bulges and notches corotated with the Earth, while the nightside smooth plasmapause remained almost perfectly stationary.

2.2.2. Sunward Surge: Plume Formation. On 18 June 2001 erosion commenced at some time between 0304 UT and 0324 UT. Unfortunately, the EUV images during the interval 0243–0324 UT contained an excessive amount of background noise, making interpretation very difficult, but not impossible. (Background noise in EUV images is believed to arise due to direct energetic particle excitation.) Our best guess is that the erosion started at 0314 UT (with an uncertainty of ±10 minutes due to the noise-related ambiguity in EUV images). In Plate 1, panels b, c, and d depict the initial phase of the 18 June erosion. The nightside plasmasphere contracted, moving 1–1.5 R_E inward in about 4 hours of UT. The dayside plasma surged sunward, forming a broad (in MLT) drainage plume. To illustrate the dayside sunward surge, compare Plate 1a (0010 UT) and Plate 1d (0659 UT). The noon MLT bulge of Plate 1a expanded sunward (in an apparent plasma rarefaction) to form the plume of Plate 1d, which extended outside the camera field of view (FOV). The simultaneous inward nightside motion and outward dayside motion is exactly the global behavior predicted by the DMR convection hypothesis. In this picture, the 0304 UT southward IMF excursion (Figure 1a) turned on dayside magnetopause reconnection, and at 0314 UT the effect of this DMR was felt as an enhanced dawn-to-dusk convection E-field at the plasmasphere, which initiated the sunward surge and plume formation. This suggests a configuration delay $\Delta \tau_C$ of 10 minutes, which is significantly shorter than the 20–30-minute delays reported in previous EUV-observed erosion events [Goldstein et al., 2003a, b]. However, ±10-minute uncertainty in both the ACE propagation delay (Section 2.1) and the EUV erosion onset timing (above) yields 200 percent uncertainty in $\Delta \tau_C$. Despite this timing ambiguity, the global evolution is consistent with the DMR convection picture. Note in Plate 1d how exceptionally sharp the plasmapause radial gradient is, and how smooth its MLT shape is on the nightside, especially east of 2100 MLT. This 0659 UT snapshot was obtained following a 1-hour interval of strong, steady solar wind E-field ϵ_{SW} (see bottom panel of Plate 1). The DMR picture says steady convection should produce a corotation/convection boundary (CCB) that is very stable in time, leading to a sharp nightside plasmapause gradient that coincides roughly with the CCB.

In Section 2.2.5 we will discuss some meso-scale features of the sunward surge plasmasphere (a narrow duskside plume and a predawn indentation) that suggest a more spatially structured inner magnetospheric flow field than is typically assumed in the simple global DMR-driven convection picture.

2.2.3. Plume Narrowing. Between 0709 UT and 1205 UT there are no EUV images available. (EUV turns off when close to the magnetic equator, and during perigee.) During this time, the dawnward solar wind E-field ϵ_{SW} was strong and fairly steady at an average value of $\epsilon_{SW} = -4$ mV/m (i.e., dawn-to-dusk E-field, sunward convection). Under DMR convection this should have resulted in a gradual narrowing of the broad dayside (sunward surge) plume, as the western edge of the plume rotated eastward toward the relatively stationary dusk edge of the plume. The plasmasphere of 0659 UT (Plate 1d) had a plume whose western edge was at about 0700 MLT, judging from the intersection of the plume with the FOV edge at about $L = 3.5$. Five hours later, at 1215 UT (Plate 1e), the western edge of the same plume crossed $L = 4$ at about 1200 MLT. This is clear evidence that the plume narrowing process did indeed occur during the five-hour gap in EUV data coverage.

Between 1215 UT and 1418 UT (the UT interval covered by the image sequence in panels e through g of Plate 1), ϵ_{SW} underwent slow, 3–4-mV/m peak-to-peak oscillations, presumably turning DMR convection on and off, and after about 1430 UT the IMF turned northward (ϵ_{SW} positive). Thus, the images in panels e, f, and g of Plate 1 were obtained at the tail end of the plume narrowing process; most of the plume narrowing had already occurred during the five-hour EUV data gap. Still, the western edge of the plume at 1418 UT was 1 to 1.5 MLT hours eastward of its location at 1215 UT, so this process continued until DMR convection turned off at 1430 UT.

As in the sunward surge phase, the EUV images of the plume narrowing phase contain sub-global density structures (a plume bifurcation and post-dawn crenulations) that will be discussed in Section 2.2.5.

2.2.4. Plume Rotating/Wrapping. At 1430 UT the IMF began a slow northward turning, as reflected in the ϵ_{SW} transition from −3 mV/m at 1430 UT to 2 mV/m at 1500 UT. Between 1500 UT and 1830 UT ϵ_{SW} had a mean value of about +1 mV/m, corresponding to northward IMF and much-reduced sunward convection. In response, the entire plume began rotating eastward, evident in the images of Plate 1, panels h through l. After 1712 UT (Plate 1k) the plume shape became distorted, lengthening as it just barely began the process of wrapping around the plasmasphere before EUV image coverage stopped.

The post-1500 UT interval of northward IMF saw outward radial motion of the plasmapause. Between 1509 UT (Plate 1i) and 1854 UT (Plate 1l) the post-midnight plasmapause moved outward by almost a full R_E. Part of this outward plasmapause motion may arise from rotation of the entire plasmasphere inside the CCB; since the nightside plasmapause radius increased in the westward direction, eastward rotation of the entire plasmasphere would result in an outward motion of the plasmapause at a fixed MLT value. However, analysis of the motion of the plasmapause during this time (see Section 2.3.3) suggests that this rotation scenario is an insufficient explanation.

Fine-scale and meso-scale structure in the EUV images of the plume rotation phase will be discussed in the next section.

2.2.5. Sub-Global Plasma Structures. In Sections 2.2.2 through 2.2.4 we showed to what extent the IMAGE EUV observations of the 18 June 2001 erosion event conformed to the global picture of plasmasphere evolution according to the simple DMR-driven convection picture. In this section we shall examine some sub-global (meso-scale and fine-scale) plasmaspheric density features from this event.

Plate 2 shows four selected snapshots from the event (with their corresponding panel letters from Plate 1), labeled according to the plume evolution phases: 'Initial' (Plate 2a), 'Sunward Surge' (Plate 2c), 'Plume Narrowing' (Plate 2g), and 'Plume Rotating' (Plate 2k). In each panel of the bottom row of Plate 2 is a plot of points (blue circles) that have been manually extracted from the EUV image directly above it. These extracted points do not necessarily follow contours of brightness, but rather are intended to highlight certain components of the EUV images that may not be obvious to the reader who is unfamiliar with EUV image interpretation.

The EUV images show some indirect evidence of a pre-dawn concentration of the effects of convection. During the sunward surge phase, the plasmapause is preferentially indented in pre-dawn MLT. Evidence of this can be seen in the 'flattening' of the 0537 UT plasmapause (Plate 1c, Plate 2c) between about 0300 MLT and 0500 MLT, and in the smooth indentation of the 0659 UT plasmapause (Plate 1d) between 0400 MLT and 0600 MLT. It is possible that this reflects a 'focusing' of the convection field in the pre-dawn MLT sector that has been identified in both models and observations [*Carpenter et al.*, 1972, 1993; *Carpenter and Smith*, 2000; *Senior and Blanc*, 1984; *Fejer and Scherliess*, 1995].

There is evidence of the presence of SAPS during the sunward surge phase, in the form of a narrow dusk-side plume (labeled in Plate 2c) that is separated from the main Grebowsky-type plume by a narrow density channel. In a comparison between EUV images and simulations, *Goldstein et al.* [2003b] showed how SAPS can create just such a distinct narrow plume. On 18 June, this duskside plume seemed to evolve from a sunward stretching of the duskside bulge whose western edge was at 1700–1800 MLT in the initial plasmasphere of 0010 UT (Plate 2a). Even though the initial erosion began at 0314 UT, the duskside bulge did not fully develop into a plume until about 0500 UT, when DMSP drift meter data show the early development of a mild SAPS

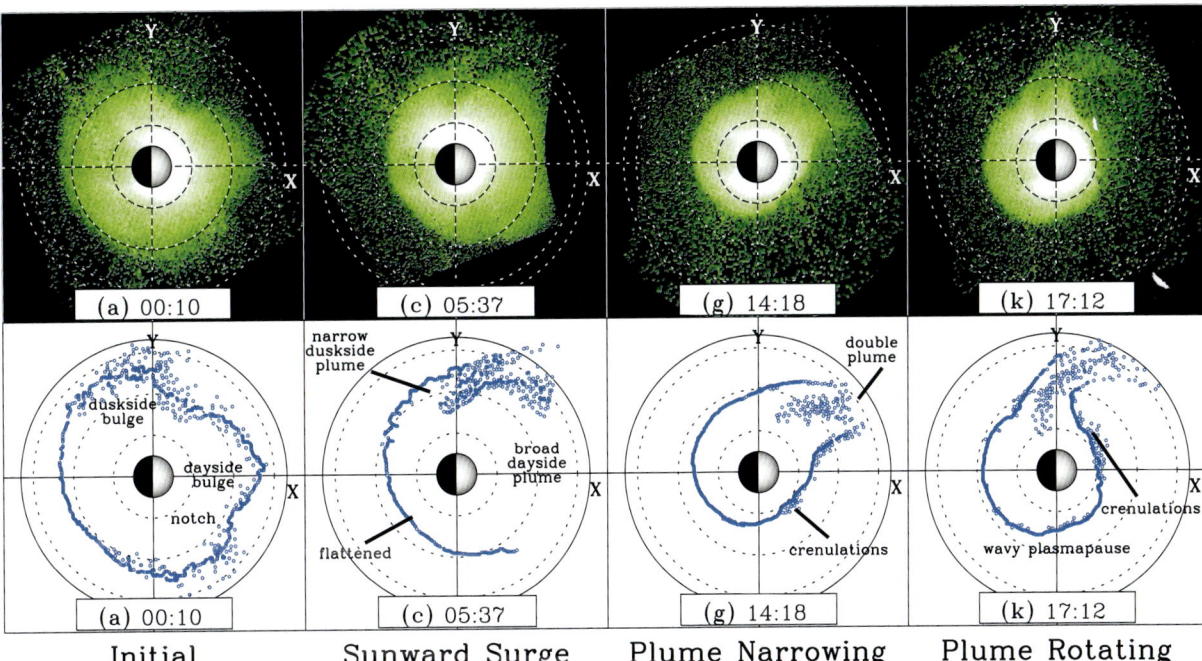

Plate 2. Top row: Four selected panels a, c, g, and k from Plate 1, each showing an EUV plasmasphere global image, as described in the caption for Plate 1. Bottom row: In each panel, the blue circles are manually extracted points from the EUV image above it, highlighting the features discussed in the text. (The apparently solid blue lines on the night side are actually composed of a very large number of blue circles.) At the bottom of each panel is a label indicating the phase of plume formation/evolution, as discussed in the text.

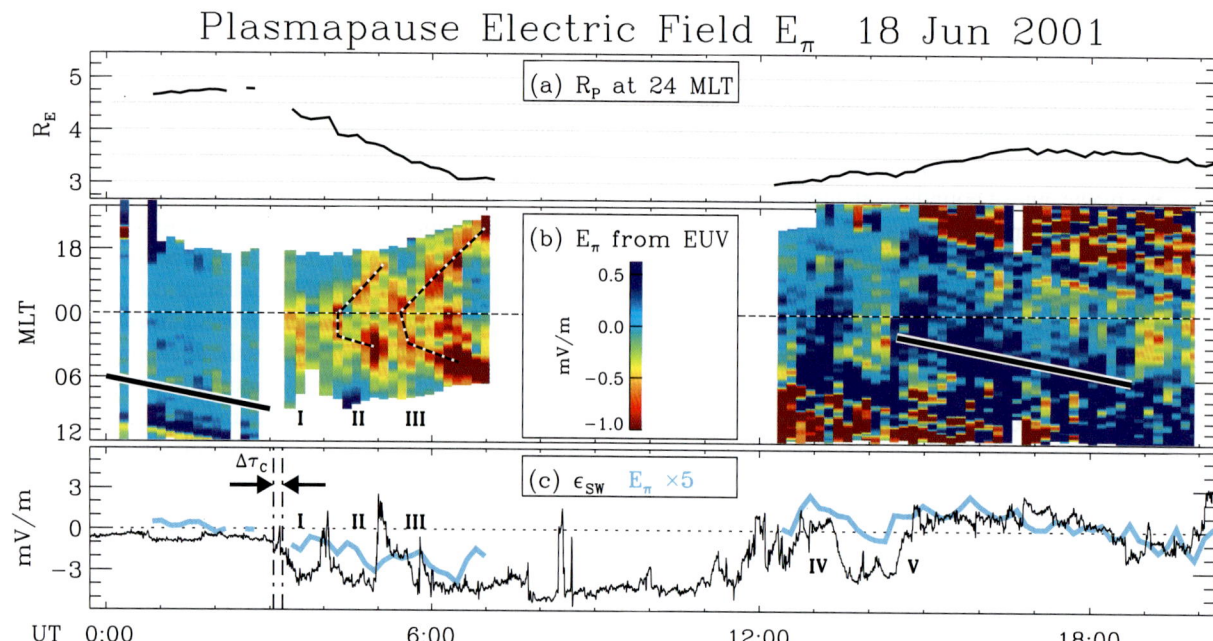

Plate 3. Electric field deduced from the motion of the plasmapause in IMAGE EUV images, 18 June 2001. (a) Plasmapause radius versus UT, at midnight MLT. (b) E-field component E_π tangential to the moving plasmapause, versus MLT and UT. Color indicates E_π strength in mV/m: red is $E_\pi < 0$ (inward motion), blue is $E_\pi > 0$ (outward motion), cyan is $E_\pi \approx 0$ (stationary plasmapause), and white is no data. Dotted line indicates midnight (2400) MLT. Diagonal lines (before 0300 UT, and between 1430 UT and 1830 UT) indicate strict corotation (1 MLT-hour per UT-hour). (c) Solar wind E-field ϵ_{SW} from Figure 1c (black) and $E_\pi \times 5$ (blue). Ratio of ϵ_{SW} to E_π indicates approximately 10 to 20 percent of dawn-to-dusk solar wind E-field transmitted to inner magnetosphere.

prior to the erosion event (as discussed in Section 2.2.1). On the dayside, the plasmapause shape corotated, mostly undistorted except for a mild outward motion of the dayside bulge (see Plate 2a). West of the dayside bulge was a shallow notch, which also corotated. The motion of the western edge of this notch left an observable feature in E_π, the diagonal blue feature between (0010 UT, 0900 MLT) and (0200 UT, 1100 MLT). The blue color of this feature corresponds to positive E_π, which is consistent with the notch's geometry and motion. At the western edge of the notch, positive E_π is by definition composed of a negative E_r component and a positive (eastward) E_φ component. As the notch rotated eastward past a given MLT, the plasmapause at this fixed MLT would appear to move eastward (negative E_r) and outward (positive E_φ). For reference, a bold line whose slope is that of strict corotation (1 MLT-hour/1 UT-hour) is drawn starting at 0600 MLT; by inspection, the slope of the blue diagonal feature is the same as the bold line. Thus, the post-dawn plasmapause was strictly corotating while the pre-dawn plasmapause was held stationary by steady mild convection.

2.3.2. Sunward Surge (Erosion Onset). The onset of the erosion first appears in Plate 3b at 0324 UT as a burst of negative E_π (red/yellow color) centered just east of midnight MLT. (Recall that the actual erosion probably began 10 minutes earlier, but the image quality at 0314 UT was too poor to infer E_π). In the period 0324 UT to 0659 UT the E_π 2D plot contains several bursts of red/yellow distributed in UT and MLT. These bursts reflect the fact that the plasmapause inward motion did not happen smoothly and uniformly. Instead, plasmapause motion was modulated by ϵ_{SW} (i.e., IMF polarity), and at any given UT was localized in MLT. Plate 3c contains a plot of ϵ_{SW}; after 0300 UT there were three distinct intervals of negative ϵ_{SW} (i.e., southward IMF), labeled 'I', 'II' and 'III'.

Interval I began the erosion at 0314 UT, and ended with a sharp upward turning of ϵ_{SW} at about 0400 UT that was preceded by a gradual increase in ϵ_{SW}. The E_π plot (Plate 3b) shows that the pre-midnight plasmapause was indented (red/yellow color) at the beginning of interval I, but this inward motion tapered off about 30 minutes after the onset.

Interval II initiated a second burst of inward motion at 0415 UT, also centered east of midnight MLT. The initial indentation of this second burst then apparently propagated both eastward (toward dawn) and westward (toward dusk) along the plasmapause, creating the signature that looks like the letter 'V' rotated 90 degrees clockwise. (This rotated V signature is emphasized in Plate 3b with black and white dotted lines.) Similarly, interval III also initiated an indentation, this time centered closer to midnight MLT, that then propagated both eastward and westward along the plasmapause.

Between 0618 UT and 0638 UT, as the interval III eastward-propagating indentation reached 0600 MLT another burst of inward motion occurred near midnight, just after another negative excursion in ϵ_{SW} (and presumably, another intensification of DMR-driven convection).

The V-shaped signatures indicate something about the process whereby the new plasmapause formed on 18 June 2001. Each interval of negative ϵ_{SW} (which corresponds to a distinct increase in DMR convection) initiated a new indentation of the nightside plasmapause. At the onset of erosion (0314 UT) the indentation process tapered off 30 minutes after the convection increase. For Intervals II and III the indentation widened across the nightside; at the edges of the widening indentation were eastward-moving and westward-moving ripples that create the V-shaped signature. Careful examination of the EUV plasmapause images during the erosion verifies that the E_π analysis brings out features that are actually in the image sequence (and not an artifact), but very hard to detect by visual inspection alone.

The bursts of erosion also may shed light on the process of shielding and penetration electric field. The fact that the initial erosion (interval I) tapered off after 30 minutes is consistent with estimates for the shielding time scale [*Kelley et al.*, 1979; *Senior and Blanc*, 1984; *Goldstein et al.*, 2003d]. The burstiness of the inward motion also fits with a shielding picture. The UT-width of any of the red bursts in Plate 3b, measured at a given MLT, is between 20 and 30 minutes, even though the intervals of negative ϵ_{SW} (I, II, and III) are between 40 and 60 minutes. The propagation of the indentation may indicate the finite speed of propagation of the sunward convective impulse, or it may in fact indicate that shielding does not develop over the entire inner magnetosphere at the same time. This is not unreasonable when one considers that shielding is accomplished via coupling between the ring current and ionosphere, and the distribution of ring current ions itself varies during a convection event. The bursts are generally more intense in the midnight-to-dawn MLT sector, consistent with statistical and theoretical models that show a concentration of the penetration E-field in this sector [*Carpenter et al.*, 1972, 1993; *Carpenter and Smith*, 2000; *Senior and Blanc*, 1984; *Fejer and Scherliess*, 1995].

The ratio of E_π to ϵ_{SW} can be taken as an estimate of how much of the dawn-to-dusk solar wind E-field was transmitted to the inner magnetosphere during the erosion. We calculated the average value of E_π versus UT over the entire nightside; in Plate 3c the average $E_\pi \times 5$ has been plotted on the same axes as ϵ_{SW}. The transmission factor was apparently between 10 and 20 percent, roughly consistent with the results of *Goldstein et al.* [2004c, 2003b].

2.3.3. Late-Stage Plasmasphere Evolution. We next discuss the evolution of the plasmasphere during the latter part of the plume narrowing phase, and the plume rotation (quieting) phase. This late-stage period is slightly more complex than the sunward surge phase, with different behavior on dayside and nightside.

2.3.3.1. Nightside plasmasphere. After 1200 UT there were two positive ϵ_{SW} excursions (i.e., northward IMF turnings), labeled 'IV' and 'V' in Plate 3c. Following each of these positive excursions, the nightside plasmapause moved outward for about 1 UT hour, as indicated by the roughly vertical blue bands that coincide with intervals IV and V. In Section 2.2.4 it was suggested that this outward motion may have been caused by rotation of the larger duskside plasmapause into the post-midnight sector. However, such a rotation would produce a visible diagonal blue signature prior to intervals IV and V. The clear absence of such a diagonal signature before interval IV in Plate 3b means that the outward plasmapause motion was due to a positive radial flow of nightside plasma. *Carpenter et al.* [1972]; *Carpenter and Smith* [2000] observed that after temporally isolated substorms, nightside plasmaspheric plasma flowed antisunward, and speculated that this was due to the overshielding effect [*Kelley et al.*, 1979]. Overshielding occurs following a convection decrease that occurs faster than the shielding time scale; upon the lessening of convection, the residual shielding field (which has not yet had time to dissipate) imposes antisunward convection upon the inner magnetosphere. It has been demonstrated that overshielding can cause a bulging out of the midnight-to-dawn plasmapause, creating plasmaspheric shoulders [*Goldstein et al.*, 2002, 2003d]. It was apparently the case that overshielding, or some form of 'reverse' (i.e., antisunward) convection, caused the outward plasmapause motion in interval IV.

After interval IV the IMF turned southward again, producing the negative ϵ_{SW} excursion between 1330 UT and 1430 UT, and enhancing DMR convection. During this post-interval-IV convection enhancement, the nightside plasmapause ceased moving outward (E_π close to zero, cyan color), and moved slightly inward (E_π slightly negative, yellow/red color). The small amount of inward motion was localized to the pre-dawn MLT sector, again suggesting a concentration of penetration E-field. Although ϵ_{SW} during 1330–1430 UT (after interval IV) was comparable to that during intervals I and II, the effect (as reflected in E_π) of this later convection increase was much smaller. According to the DMR convection picture, after several hours of strong convection (which had occurred during 0600–1200 MLT), further strong convection has a lessened effect. This argument is strengthened by the fact that the magnitude of ϵ_{SW} during 1330–1430 UT was smaller than that of the strongest convection at earlier times.

After 1430 UT (interval V), the effects of northward IMF (positive ϵ_{SW}) dominated the nightside. The DMR convection reduction at the start of interval V caused a second outward plasmapause motion, also apparently related to overshielding. This outward motion created a nightside plasmapause bulge that proceeded to rotate eastward at a rate commensurate with strict corotation (as indicated by the bold diagonal line in Plate 3b). This nightside corotation continued until about 1830 UT, when ϵ_{SW} again became negative. This increase in DMR convection coincided with a more pronounced inward plasmapause motion than that between intervals IV and V. About four hours of reduced convection before the 1830 UT negative ϵ_{SW} excursion increased the nightside plasmapause radius by almost 1 R_E, as discussed in Section 2.2.4, and it may have been the presence of this larger nightside plasmasphere that most likely increased the effectiveness of the post-1830 UT convection. But the diffuse yellow/red diagonal band between (1600 UT, 2100 MLT) and (2000 UT, 0100 MLT) suggests that a region of reduced plasmapause radius (i.e., a shallow notch) rotated into the post-midnight region about the same time as the convection enhancement, and thus contributed to inward plasmapause motion there.

2.3.3.2. Dayside plasmasphere. In contrast to the nightside plasmapause behavior, which was very much driven by changes in ϵ_{SW}, the dayside plasmasphere (from 0600 MLT to 1800 MLT) corotated after 1200 UT. We infer the rotation rate from E_π by tracking the azimuthal (MLT) motion of distinctive dayside features such as the plume and crenulations, which have recognizable diagonal E_π signatures (i.e., red and blue diagonal bands at the top and bottom of Plate 3b). Judging from the slope of the diagonal signatures as compared to the slope of the line indicating strict corotation, the plasmasphere east of 0600 MLT and west of noon MLT (at the bottom part of the plot in Plate 3b) strictly corotated with the Earth. There is evidence of some slight subcorotation (shallower MLT/UT slope) west of 1800 MLT and east of noon MLT (at the top of Plate 3b). This subcorotation could be attributed to the presence of duskside convection (both DMR-driven and SAPS-driven) which is directed opposite to eastward corotational flows.

3. SIMULATION 18 JUNE 2001

In this section we simulate the response of the plasmasphere to a simple global convection field driven by dayside magnetopause reconnection. Plasmaspheric dynamics can be modeled by assuming that the plasmapause boundary is composed of cold test particles subject only to $E \times B$ drift. In a time-varying electric field such as is expected in response to the variable rate of DMR, plasmapause evolution is simulated by the changing shape of the curve defined by the aggregate of these test particles. This approach, used by *Grebowsky* [1970], *Chen and Wolf* [1972], and others, will hereinafter be called the plasmapause test particle (PTP) simulation. The PTP method is best applied to represent steep outer plasmaspheric density gradients, because a boundary with an indistinct edge (i.e., a gradually dropping density) is not well represented by a single plasmapause contour. This method is adaptive; as the PTP plasmapause curve evolves in time, test particles are added or removed as necessary to resolve its structure. Thus, the PTP simulation preserves structure without numerical diffusion; the shape of the evolving PTP plasmapause depends entirely on the initial conditions, and the details of the time-varying E-field used to drive the simulation. For initial conditions, we used a 40-term Fourier expansion of the extracted plasmapause of 0010 UT (Plate 2a).

To drive our simulation of 18 June 2001, we chose the simple and popular model of *Volland* [1973] and *Stern* [1975]. This model is not necessarily the most realistic, but if properly normalized to the solar wind electric field it is a good representation of the DMR-driven convection paradigm. The Volland-Stern (VS) model potential is $\Phi_{VS}(r,\varphi) = -A_0 r^2 \sin \varphi$. We normalized this function so that $A_0 = 0.2 |\epsilon_{SW}| (6.6 R_E)^{-1}$, which is equivalent to 20 percent of the solar wind electric field applied across the inner magnetosphere inside geosynchronous orbit. We chose 20 percent from the upper limit of the ratio E_π/ϵ_{SW} that we found from Plate 3c. To include a finite viscous interaction between the solar wind and magnetosphere during northward IMF, $|\epsilon_{SW}|$ in the VS model is constrained to be ≥ 0.5 mV/m.

Kp-based normalizations (e.g., *Maynard and Chen* [1975]) parameterize the VS model according to all the different geomagnetic phenomena that might contribute to Kp. These phenomena would include not only DMR-driven convection, but also internal magnetospheric processes like substorms and SAPS. Our ϵ_{SW}-based normalization allows us to parameterize only the DMR-driven portion of convection. The goal of our simulation is to present the response of the plasmasphere to a simplified global DMR-driven convection E-field. We will compare this DMR-driven response to the real response of the plasmasphere as seen by EUV, in order to study the limits of the validity of the DMR convection picture.

The results of the 18 June 2001 PTP simulation are presented in Plate 4 and Plate 5, which are formatted similarly to Plate 1 and Plate 2, respectively. The panel labels (a through l) and time stamps are the same for the EUV and PTP figures. Comparison of the EUV images and PTP simulated plasmapause curves reveals both similarites and differences.

The most obvious agreement is the global behavior of the plasmasphere during active times. The PTP simulated plasmasphere evolves according to the same phases of plume formation and evolution as the EUV imaged plasmasphere. The erosion begins when ϵ_{SW} turns negative (i.e., the IMF turns southward) and DMR-driven convection becomes strong. Sunward surging on both nightside and dayside produces a reduced nightside plasmapause radius, and a broad dayside plume. Over the course of several hours, the initial surge plume then narrows in local time, and the edges of the plume (both dusk and western) are in reasonable agreement with the EUV data, although the PTP model dusk edge is at a slightly larger Y value than that of EUV. With reference to the discussion of the 3–4-mV/m ϵ_{SW} oscillations in Plate 3, note that between 1215 UT (Plate 5e) and 1316 UT (Plate 5f) the dusk edge of the plume (at the dusk terminator, 1800 MLT) moves outward in Y by about 0.4 R_E in response to the positive ϵ_{SW} excursion between these two times. Plume rotation and wrapping commences after 1500 UT in the model, although during this quieting phase the PTP-EUV differences become more severe (as we will discuss below).

In the PTP model, some of the structure (especially in the early sunward surge phase) depends on the initial conditions; e.g., the duskside bulge at 0010 UT (Plate 4a) evolves into a spiky structure near dusk at 0436 UT (Plate 4b). This is consistent with the long-standing idea that at a given instant of time, the plasmaspheric configuration reflects the time-integrated effects of a few or several previous hours of geomagnetic conditions. Thus, plasmapause models often depended not on instantaneous Kp, but rather some representation (e.g., maximum or average) of a few or several previous hours of Kp (e.g., *Carpenter and Anderson* [1992]). However, in the PTP simulation most of the original spatial structure is eventually washed away during the erosion, and the final state of the plasmasphere is dominated by the spatial form of the VS electric field model, normalized to the time-varying solar wind E-field. Said another way, the global properties of the plasmasphere (which is all one can hope to capture when using the VS model which does not contain any sub-global spatial variation of the inner magnetospheric E-field) are indeed directly driven by the state of dayside magnetopause reconnection-driven convection. The sub-global scale features of the plasmasphere must then depend on the prior history of the plasmasphere and/or the sub-global spatial structure of the convection field.

It is on the sub-global scale that the PTP simulation differs from the EUV plasmasphere evolution of 18 June 2001. The absence of SAPS in the Volland-Stern potential means that the location of the duskside edge of the PTP plume is eastward of the EUV-observed location. Also, the narrow duskside plume of Plate 2c fails to develop in the PTP simulation, again presumably due to the lack of SAPS to strengthen sunward convection near dusk. Instead of the narrow duskside plume, the PTP simulated plasmapause of Plate 5c has a spiky bulge (mentioned in the previous paragraph) near 2100

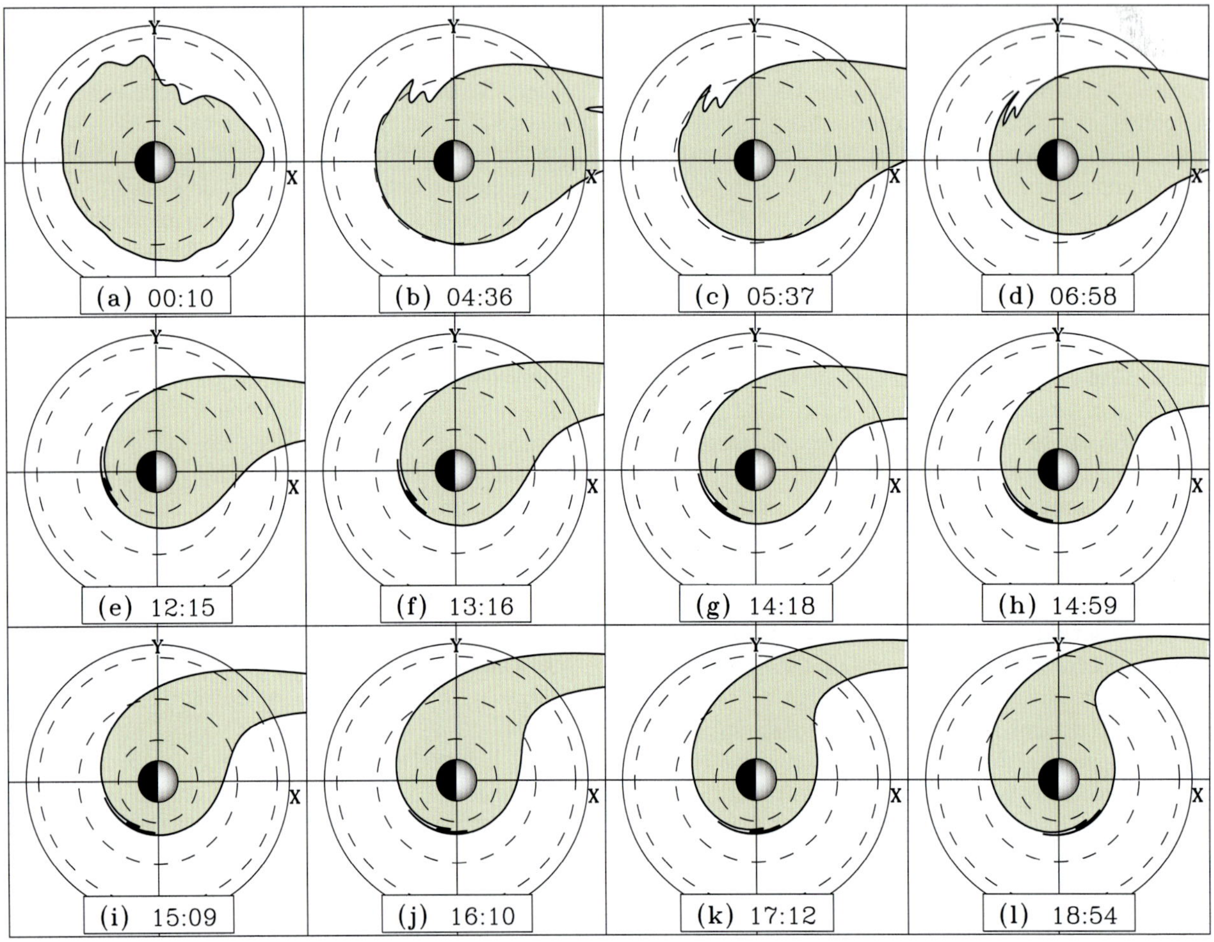

16 PLUME EVOLUTION

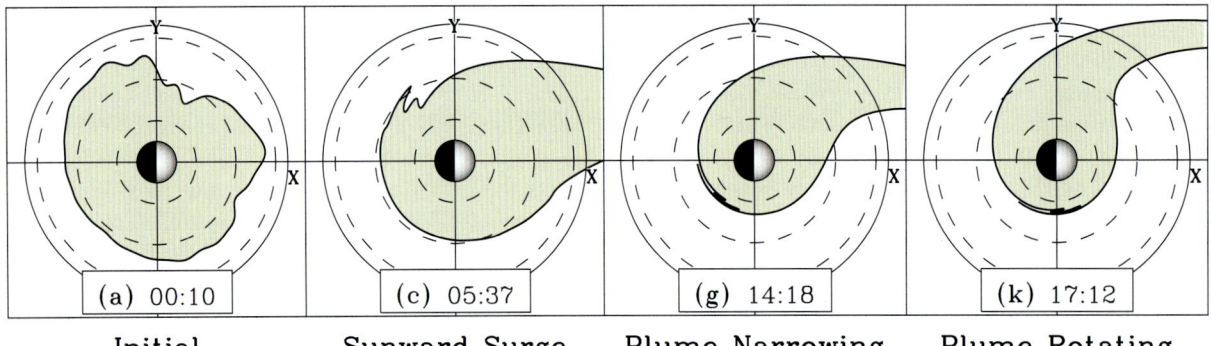

Plate 4. Plasmapause test particle (PTP) simulation of 18 June 2001 event, plotted in a format similar to that of Plate 1, with snapshots at the same times as panels a through l of Plate 1. In each plot, the plasmasphere is indicated by the green region surrounding the Earth. The simulation results indicate an erosion and plume development sequence that agrees with the EUV images on a global level, but there are important meso-scale and fine-scale differences, as discussed in the text.

Plate 5. Four selected panels a, c, g, and k from Plate 4, each showing a PTP simulated plasmasphere. This figure should be directly compared with Plate 2. The phases of plume evolution from Plate 2 are evident in the simulation results (see text).

MLT that is a remnant of the large dusk-side bulge of Plate 5a. This spiky feature then gets compressed by nightside convection and rotates across the nightside, reaching 0300 MLT by 1418 UT (Plate 5g). In the EUV images there is no such rotating spiky feature because in the EUV plasmasphere the initial duskside bulge is elongated by strong duskside (SAPS) convection into a narrow duskside plume (see Plate 2c and *Goldstein et al.* [2003b]).

The compression of this spiky feature is interesting because it shows that under the right circumstances, sub-global spatial structure (in the plasmaspheric density distribution) can survive the trip across the night-side even during strong sunward convection. If such a structure were in fact present on the nightside plasmapause in EUV images, it would not be resolved by the 0.1-R_E EUV pixels. When convection is relaxed, this structure would be free to expand again inside the corotation/convection boundary (CCB). This could be a partial explanation for why the plasmasphere seems to develop increased spatial structure as soon as convection drops off; the structure is compressed by the contraction of the CCB that accompanies strong convection, but it survives and gets elongated or distorted once free to evolve inside an expanded CCB. In this scenario, plasmapause crenulations form in the post-dawn MLT sector where the plasmapause lies inside the CCB, and wavy variations of the plasmapause develop during the quieting (plume rotation/wrapping) phase. (This hypothetical scenario was mentioned earlier, at the end of the section titled 'Sub-Global Plasma Structures.')

This brings up the next point: if the PTP simulation preserves structure, and if that structure can survive strong convection intervals, why doesn't the double plume develop in the PTP model, and why is the PTP model so much less structured during the plume rotation phase? One explanation might be that the initial conditions failed to capture all of the (perhaps sub-EUV-pixel) structure of the 0010 UT plasmasphere, and some of this uncaptured structure becomes important in later stages of the evolution. Even if this is the case (and it quite probably is), it is also undeniably true that the Volland-Stern model is too simple to properly capture anything but global plasmasphere evolution, and so it is not surprising that the double plume, the crenulations, and other meso-scale and fine-scale features are not reproduced. Besides SAPS, another important known effect not included in the VS model is the concentration of electric field in the midnight-to-dawn MLT sector. This effect is quite evident in the E_π 2D plot of Plate 3b. To highlight this difference more quantitatively, Plate 6 shows R_p versus MLT plots of both EUV and PTP plasmapauses taken from the four panels (a, c, g, and k) of Plate 2 and and Plate 5. By visual inspection it is clear that during the sunward surge phase (Plate 6c), the inward plasmapause motion between midnight MLT and 0600 MLT is more pronounced for the EUV plasmapause (blue circles) than the PTP model plasmapause (bold line). It is midnight-to-dawn convection concentration that apparently produces the flattening and indentation evident in panels b, c, and d of Plate 1, and in Plate 2b. This flattening is not apparent in the PTP model snapshots of Plate 4, panel b, c, and d, Similarly, the outward bulging evident in the EUV midnight-to-dawn plasmapauses of Plate 6g and Plate 6k is absent in the PTP model plasmapauses of the same MLT range. This demonstrates the necessity for a more sophisticated (i.e., MLT-dependent) treatment of the penetration E-field than that of the Volland-Stern model.

The observed quiet-time meso-scale and fine-scale complexity of the plasmaspheric density distribution deserves much more attention. The EUV images suggest a level of fine-scale structure that is unresolved by the 0.1 R_E EUV pixels, and this is in agreement with the very structured density profiles that have been observed in situ [*Moldwin et al.*, 1994, 1995; *Carpenter and Lemaire*, 1997]. Plasma instabilities and other subtle (non $E \times B$-driven) physics may very well be involved in the quiet-time plasmasphere evolution, when DMR-driven convection is mild or completely absent.

4. CONCLUDING REMARKS

4.1. Alternate Plume Formation Mechanism

We have shown that the global pattern of plume evolution observed on 18 June 2001 fits with the DMR-driven convection picture. An alternate plume formation scenario was proposed by *Lemaire* [2000], in which a plasmapause bulge forms on the dayside and subsequently evolves into a duskside plume. Because rotation speed decreases with radial distance, the bulge experiences a shear in the eastward convection speed, and it gets stretched/distorted into a plume. This scenario was apparently verified by EUV observations on 10 June 2001 [*Spasojevic' et al.*, 2003].

4.2. Comments on Interchange Driven Erosion

The validity of the DMR-driven convection picture relies on the assumption that plasmaspheric plasma is subject only to $E \times B$-drift in a global convection field, and lacking complete knowledge of the inner magnetospheric E-field, it is not known precisely how the new plasmapause boundary forms, especially during an erosion. As mentioned in Section 1, *Lemaire* [1975] proposed that the plasmapause forms under the influence of small-scale electric fields that arise due to the gravitational/centrifugal interchange instability. One of the predictions of this hypothesis is that during an erosion, the nightside plasma at the boundary moves radially outward,

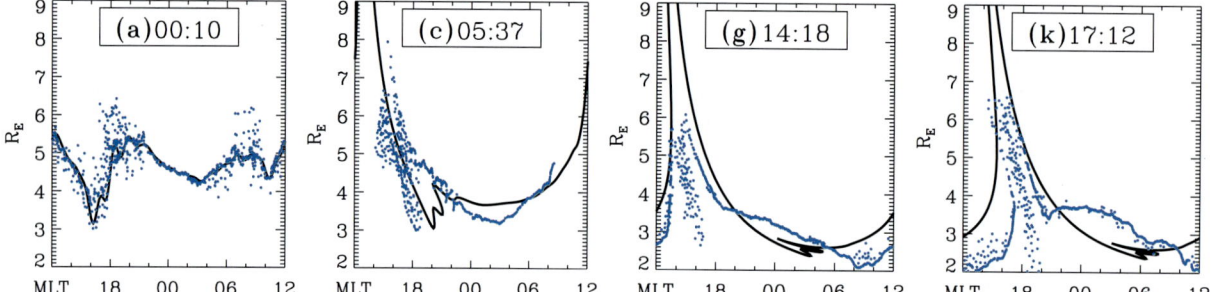

Plate 6. A more quantitative comparison between the EUV data of Plate 2 and the PTP simulation of Plate 5. In each of panels a, c, g, and k: blue circles plot the manually extracted plasmapause from EUV, and the solid black curve is the PTP simulated plasmapause. Although the zero-order global features are reproduced by the model, there are key differences, as discussed in the text.

detaching from the main plasmasphere as blobs. The gradual removal of a large number of blobs produces a net inward motion of the plasmapause [*Lemaire and Gringauz*, 1998]. The maximum allowable speed of the interchange-driven blob was estimated by *Lemaire and Gringauz* [1998] to be about V_{max}=0.03 R_E/hour.

To date, EUV images acquired during erosion have revealed no evidence of this proposed outward motion of detached parcels of nightside plasma. In fact, all of the evidence in EUV images suggests that plasma motion is sunward; both the dayside and nightside boundaries move sunward during the initial stage of the erosion. Because EUV cannot see low-density plasma below 40 cm^{-3}, it is conceivable that the nightside blobs are invisible to the EUV cameras because as they move radially outward, their density quickly drops below the lower density threshold. Let us examine this premise in the context of the 18 June 2001 event. In Plate 3a is a plot of R_p versus UT, showing the initial plasmapause location at $L = 4.4$ at 0325 UT. Suppose that the plasmapause plasma moves outward, and as it does, its density drops according to the inverse of the flux tube volume, i.e., L^{-4}. If we assume the plasmaspheric density of *Carpenter and Anderson* [1992], this outermost plasma parcel has electron density of about 295 cm^{-3}. In order for the plasma parcel to become invisible to EUV, its final density must be ≤40 cm^{-3}. Then its final L-shell is $L_f \geq L_i\,(295/40)^{-\frac{1}{4}}$, or $L_f \geq 7.2$. For this outward motion to avoid being captured by EUV, it must travel to this final location L_f in 10 minutes or less (the time cadence of EUV images); the plasma parcel must have speed $V_f \geq 2.8\ R_E$/10 min., or 17 R_E/hr. The lower limit of the required speed is 570 times faster than V_{max}, the maximum interchange speed. Even if the proposed blobs are of a size that is below the resolution of the EUV imager, to account for the large amount of nightside cold plasma that is removed during erosion, a large number of blobs would be required. One would still expect to see the collective motion of such a large number of tiny blobs with a poorly-resolved image, just as a person with poor vision can still see the collective motion of a large number of individual drops of rain, or individual grains of sand on a windy day.

Thus, the invisible outward-moving blob scenario seems unlikely, for the following reasons. First, the dayside plasmasphere expands/rarefies during erosion, and still remains visible to EUV at geosynchronous orbit and beyond [*Goldstein et al.*, 2004b]. Second, it is not probable that the blobs could move fast enough to avoid leaving some detectable signature in EUV images. Other than a pixelated noise background with no evidence of systematically outward-moving blobs, the nightside in EUV images typically appears to be evacuated of plasmaspheric plasma exterior to the inward-moving plasmapause during erosions (e.g., see Plate 1). Third, if one supposes that the blobs have density below the EUV threshold when they first detach from the surrounding dense nightside plasmapause, or that they are of a spatial size too small to be resolved by EUV (and thus they are invisible to EUV at all times), then it is difficult to account for the large amount of dense nightside plasmaspheric plasma removed during the erosion by such tenuous blobs. With its slow growth rate, the interchange instability probably does not play a strong role during geomagnetically active times. On the other hand, the role of interchange during extended quiet periods ought to be investigated further, because during such periods slower processes may have enough time to act effectively.

Summary

The 18 June 2001 EUV observed erosion event serves as an example of the global-scale pattern of plume evolution that is repeatedly found in plasmasphere images during geomagnetically active times. Given an initial plasmaspheric configuration that is subject to an increase in the strength of sunward convection, plume formation and evolution follows three main phases: sunward surge, plume narrowing, and plume rotating/wrapping. The excellent EUV image coverage of the 18 June event contains examples of all of these phases from a single erosion event. On a global scale, the 18 June plasmasphere observations are consistent with the DMR-driven convection interpretation, as represented by both prior modeling work by *Grebowsky* [1970] and others, and by our own simulation specifically tailored for the 18 June event. The EUV observations of this event are also consistent with prior in situ observations of shrinking plasmaspheres, detached plasma regions, and duskside bulge rotation [*Moldwin et al.*, 2003].

On a sub-global scale, proper treatment of plasmaspheric dynamics requires a more sophisticated treatment than offered by the global DMR-driven convection picture. A handful of interesting sub-global plasmaspheric features were observed on 18 June; some of these features have plausible explanations. The narrow duskside plume (Plate 2c) and sharpening of the duskside plasmapause (e.g., Plate 1d) both indicate the presence of SAPS, a coupling phenomenon not directly driven by dayside reconnection. The pre-dawn indentation (e.g., Plate 1d) or flattening (Plate 2c) suggests the presence of pre-dawn concentration of the penetration electric field [*Carpenter and Smith*, 2000]. Similarly, the outward excursion of the midnight-to-dawn plasmapause during northward IMF (i.e., positive ϵ_{SW}) probably reflects the presence of overshielding [*Carpenter and Smith*, 2000].

Other features defy immediate explanation. The plume bifurcation (or double plume) and crenulations (both in Plate 2g) might arise due to spatial structure in the initial (pre-erosion) plasmasphere, or might be created by spatial structure in the dayside convection field. The increased spatial structure and complexity in the plasmaspheric distribution during the quieting phase (plume rotating/wrapping) is similarly unexplained. Does this spatial structure arise due to density fluctuations in the initial plasmasphere that grow and change shape? Is the structure due to spatially structured quiet-time electric fields? Or is it due to plasma instabilities and non-$E \times B$ motion? Whatever the cause of meso-scale and fine-scale density variations in the plasmasphere, general EUV observations, and the 18 June images in particular, suggest that convection suppresses or smooths the structure, while the absence or lessening of strong convection seems to encourage its growth.

Our electric field (E_π) analysis yielded some insight into the process of erosion. The rotated-V signatures (Plate 3b) suggest that during the early phase of the erosion, there is partial indentation of the plasmapause near or east of midnight MLT, with subsequent eastward and westward spreading/widening of the indentation, as discussed by *Carpenter and Lemaire* [1997] and similarly observed by *Goldstein et al.* [2004c, a]. We observed bursts of erosion whose intensity was modulated by the sign of ϵ_{SW} (a time-delayed proxy for the IMF polarity). Some of the bursty behavior suggests a shielding time scale of 20–30 minutes, and the midnight-to-dawn concentration of the bursts probably reflects pre-dawn concentration of the penetration E-field. From our E_π analysis we estimated two quantities. The time delay $\Delta\tau_C$ between IMF polarity reversal at the magnetopause and the resulting motion of the plasmapause was found to be 10±20 minutes. The inner magnetospheric E-field was found to be about 10–20 percent of the solar wind E-field. During the later phase of the erosion (once the sunward surge was over), we observed different behavior on the dayside and nightside. The dayside tended to corotate, independent of ϵ_{SW} polarity. The nightside behavior was modulated by ϵ_{SW}. When $\epsilon_{SW} < 0$, mild sunward motion occurred, less pronounced than in the early stages of the erosion. When $\epsilon_{SW} > 0$, apparent overshielding caused an outward bulging of the midnight-to-dawn plasmapause, and this bulge subsequently corotated during extended positive ϵ_{SW}. From this event it is clear that more information about the detailed structure of the inner magnetospheric E-field is needed. There is hope that future comparisons between models and global EUV images, and EUV E analysis of other events, will yield insight into this electric field and its eect on the plasmasphere.

Acknowledgments. We wish to thank N. Ness, C. Smith, D. McComas, and the ACE science center for the easy availability of the excellent ACE data set. The IMAGE mission under NASA contract NAS5-96020 supported work at Southwest Research Institute (JG) and the University of Arizona. Additional funding for research at Southwest Research Institute was provided by the NASA SEC Guest Investigator program under NAG5-12787. The first author is most grateful to M. Spasojevic´, J. Burch. D. Carpenter, and R. Wolf for intelligent and enjoyable discussions about the plasmasphere during the era of the IMAGE mission.

REFERENCES

Anderson, P. C., D. L. Carpenter, K. Tsuruda, T. Mukai, and F. J. Rich (2001), Multisatellite observations of rapid subauroral ion drifts (SAID), *J. Geophys. Res., 106,* 29,585.

Burch, J. L. (2000), IMAGE mission overview, *Space Sci. Rev., 91,* 1.

Burch, J. L., D. G. Mitchell, B. R. Sandel, P. C. Brandt, and M. W¨uest (2001), Global dynamics of the plasmasphere and ring current during magnetic storms, *Geophys. Res. Lett., 28,* 1159.

Burke, W. J., A. G. Rubin, N. C. Maynard, L. C. Gentile, P. J. Sultan, F. J. Rich, O. de La Beujardière, C. Y. Huang, and G. R. Wilson (2000), Ionospheric disturbances observed by DMSP at middle to low latitudes during the magnetic storm of June 4–6, 1991, *J. Geophys. Res., 105,* 18,391.

Burke, W. J., et al. (1998), Electrodynamics of the inner magnetosphere observed in the dusk sector by CRRES and DMSP during the magnetic storm of June 4–6, 1991, *J. Geophys. Res., 103,* 29,399.

Carpenter, D. L. (1995), Earth's plasmasphere awaits rediscovery, *EOS Trans. AGU, 76,* 89.

Carpenter, D. L., and R. R. Anderson (1992), An ISEE/Whistler model of equatorial electron density in the magnetosphere, *J. Geophys. Res., 97,* 1097.

Carpenter, D. L., and J. Lemaire (1997), Erosion and recovery of the plasmasphere in the plasmapause region, *Space Sci. Rev., 80,* 153.

Carpenter, D. L., and A. J. Smith (2000), The study of bulk plasma motions and associated electric fields in the plasmasphere by means of whistler-mode signals, *J. Atmos. Solar-Terr. Phys.,* p. 1.

Carpenter, D. L., K. Stone, J. C. Siren, and T. L. Crystal (1972), Magnetospheric electric fields deduced from drifting whistler paths, *J. Geophys. Res., 77,* 2819.

Carpenter, D. L., B. L. Giles, C. R. Chappell, P. M. E. Decreau, R. R. Anderson, A. M. Persoon, A. J. Smith, Y. Corcuff, and P. Canu (1993), Plasmasphere dynamics in the dusk-side bulge region: a new look at an old topic, *J. Geophys. Res., 98,* 19,243.

Chappell, C. R. (1974), Detached plasma regions in the magnetosphere, *J. Geophys. Res., 79,* 1861.

Chappell, C. R., K. K. Harris, and G. W. Sharp (1970a), A study of the influence of magnetic activity on the location of the plasmapause as measured by OGO5, *J. Geophys. Res., 75,* 50.

Chappell, C. R., K. K. Harris, and G. W. Sharp (1970b), The morphology of the bulge region of the plasmasphere, *J. Geophys. Res., 75,* 3848.

Chen, A. J., and R. A. Wolf (1972), Effects on the plasmasphere of a time-varying convection electric field, *Planet. Space Sci., 20,* 483.

Coroniti, F. V., and C. F. Kennel (1973), Can the ionosphere regulate magnetospheric convection?, *J. Geophys. Res., 78,* 2837.

Dent, Z. C., I. R. Mann, F. W. Menk, J. Goldstein, C. R. Wilford, M. A. Clilverd, and L. G. Ozeke (2003), A coordinated ground-based and IMAGE satellite study of quiet-time plasmaspheric density profiles, *Geophys. Res. Lett., 30*(12), 1600, doi:10.1029/2003GL016,946.

Dungey, J. W. (1961), Interplanetary magnetic field and the auroral zones, *Phys. Rev. Lett., 6,* 47.

Elphic, R. C., L. A. Weiss, M. F. Thomsen, D. J. McComas, and M. B. Moldwin (1996), Evolution of plasmaspheric ions at geosynchronous orbit during times of high geomagnetic activity, *Geophys. Res. Lett., 23,* 2189.

Fejer, B. G., and L. Scherliess (1995), Time dependent response of equatorial ionospheric electric fields in magnetospheric disturbances, *Geophys. Res. Lett., 22,* 851.

Fejer, B. G., R. W. Spiro, R. A. Wolf, and J. C. Foster (1990), Latitudinal variation of perturbation electric fields during magnetically disturbed periods: 1986 SUNDIAL observations and model results, *Ann. Geophys., 8,* 441.

Foster, J. C., and W. J. Burke (2002), SAPS: A new categorization for sub-auroral electric fields, *EOS Trans. AGU, 83,* 393.

Foster, J. C., and H. B. Vo (2002), Average characteristics and activity dependence of the subauroral polarization stream, *J. Geophys. Res., 107*(A12), 1475, doi:10.1029/2002JA009409.

Foster, J. C., P. J. Erickson, A. J. Coster, and J. Goldstein (2002), Ionospheric signatures of plasmaspheric tails, *Geophys. Res. Lett., 29*(13), 1623, doi:10.1029/2002GL015067.

Goldstein, J., R. W. Spiro, P. H. Reiff, R. A. Wolf, B. R. Sandel, J. W. Freeman, and R. L. Lambour (2002), IMF-driven overshielding electric field and the origin of the plasmaspheric shoulder of May 24, 2000, *Geophys. Res. Lett., 29*(16), doi:10.1029/2001GL014534.

Goldstein, J., B. R. Sandel, W. T. Forrester, and P. H. Reiff (2003a), IMF-driven plasmasphere erosion of 10 July 2000, *Geophys. Res. Lett., 30*(3), doi:10.1029/2002GL016478.

Goldstein, J., B. R. Sandel, P. H. Reiff, and M. R. Hairston (2003b), Control of plasmaspheric dynamics by both convection and sub-auroral polarization stream, *Geophys. Res. Lett., 30*(24), 2243, doi:10.1029/2003GL018390.

Goldstein, J., M. Spasojević, P. H. Reiff, B. R. Sandel, W. T. Forrester, D. L. Gallagher, and B. W. Reinisch (2003c), Identifying the plasmapause in IMAGE EUV data using IMAGE RPI in situ steep density gradients, *J. Geophys. Res., 108*(A4), 1147, doi:10.1029/2002JA009475.

Goldstein, J., R. W. Spiro, B. R. Sandel, R. A. Wolf, S.-Y. Su, and P. H. Reiff (2003d), Overshielding event of 28–29 July 2000, *Geophys. Res. Lett., 30*(8), 1421, doi:10.1029/2002GL016644.

Goldstein, J., B. R. Sandel, M. R. Hairston, and S. B. Mende (2004a), Plasmapause undulation of 17 April 2002, *Geophys. Res. Lett., 31,* L15,801, doi:10.1029/2004GL019959.

Goldstein, J., B. R. Sandel, M. F. Thomsen, M. Spasojević, and P. H. Reiff (2004b), Simultaneous remote-sensing and in situ observations of plasmaspheric drainage plumes, *J. Geophys. Res., 109,* A03,202, doi:10.1029/2003JA010,281.

Goldstein, J., R. A. Wolf, B. R. Sandel, and P. H. Reiff (2004c), Electric fields deduced from plasmapause motion in IMAGE EUV images, *Geophys. Res. Lett., 31*(1), L01,801, doi:10.1029/2003GL018797.

Grebowsky, J. M. (1970), Model study of plasmapause motion, *J. Geophys. Res., 75,* 4329.

Higel, B., and L. Wu (1984), electron density and plasmapause characteristics at 6.6 RE: A statistical study of the GEOS 2 relaxation sounder data, *J. Geophys. Res., 89,* 1583.

Huang, T. S., R. A. Wolf, and T. W. Hill (1990), Interchange instability of the Earth's plasmapause, *J. Geophys. Res., 95,* 17,187.

Jaggi, R. K., and R. A. Wolf (1973), Self-consistent calculation of the motion of a sheet of ions in the magnetosphere, *J. Geophys. Res., 78,* 2852.

Kelley, M. C., B. G. Fejer, and C. A. Gonzales (1979), An explanation for anomalous ionospheric electric fields associated with a northward turning of the interplanetary magnetic field, *Geophys. Res. Lett., 6,* 301.

Lambour, R. L., L. A. Weiss, R. C. Elphic, and M. F. Thomsen (1997), Global modeling of the plasmasphere following storm sudden commencements, *J. Geophys. Res., 102,* 24,351.

LeDocq, M. J., D. A. Gurnett, and R. R. Anderson (1994), Electron number density fluctuations near the plasmapause observed by the CRRES spacecraft, *J. Geophys. Res., 99,* 23,661.

Lemaire, J. (1975), The mechanisms of formation of the plasmapause, *Ann. Geophys., 31,* 175.

Lemaire, J. (2000), The formation of plasmaspheric tails, *Phys. Chem. Earth (C), 25,* 9.

Lemaire, J. F., and K. I. Gringauz (1998), *The Earth's Plasmasphere*, Cambridge University Press, Cambridge.

Maynard, N. C., and A. J. Chen (1975), Isolated cold plasma regions: Observations and their relation to possible production mechanisms, *J. Geophys. Res., 80,* 1009.

McComas, D. J., S. J. Bame, P. Barker, W. C. Feldman, J. L. Phillips, P. Riley, and J. W. Griffee (1998), Solar wind electron proton alpha monitor (SWEPAM) for the Advanced Composition Explorer, *Space Sci. Rev., 86,* 563.

Moldwin, M. B., M. F. Thomsen, S. J. Bame, D. J. McComas, and K. R. Moore (1994), An examination of the structure and dynamics of the outer plasmasphere using multiple geosynchronous satellites, *J. Geophys. Res., 99,* 11,475.

Moldwin, M. B., M. F. Thomsen, S. J. Bame, D. McComas, and G. D. Reeves (1995), The fine-scale structure of the outer plasmasphere, *J. Geophys. Res., 100,* 8021.

Moldwin, M. B., B. R. Sandel, M. Thomsen, and R. Elphic (2003), Quantifying global plasmaspheric images with in situ observations, *Space Sci. Rev., 109,* 47.

Richmond, A. D. (1973), Self-induced motions of thermal plasma in the magnetosphere and stability of the plasmapause, *Rad. Sci., 8,* 1019.

Sandel, B. R., R. A. King, W. T. Forrester, D. L. Gallagher, A. L. Broadfoot, and C. C. Curtis (2001), Initial results from the IMAGE extreme ultraviolet imager, *Geophys. Res. Lett., 28,* 1439.

Sandel, B. R., J. Goldstein, D. L. Gallagher, and M. Spasojević (2003), Extreme ultraviolet imager observations of the structure and dynamics of the plasmasphere, *Space Sci. Rev., 109,* 25.

global images, and the complementarity of the two approaches; local measurements giving the "ground truth" (including plasma composition, distribution functions etc.) and global images allowing to put local measurements into a global context, and to deconvolve spatial from temporal effects.

1. INTRODUCTION

The plasmasphere is the torus of cold (~1 eV) dense plasma that encircles the Earth occupying the inner magnetosphere out to a boundary known as the plasmapause, where the density can drop by 1 to 2 orders of magnitude. The configuration and dynamics of the plasmasphere are highly sensitive to geomagnetic disturbances. During extended periods of relatively quiet geomagnetic conditions the outer plasmasphere can become diffuse, with a gradual fall-off of plasma density. During increasing magnetospheric activity, however, the plasmasphere is eroded and plasmaspheric ions can be peeled off and escape toward the outer magnetosphere.

The outer plasmasphere region is located at the interface between the expanded ionosphere, corotating with the Earth, and the internal magnetosphere, dominated by sunward convection [e.g. *Lemaire and Gringauz*, 1998]. In contrast with the inner plasmasphere, where the density repartition is smooth, the outer plasmasphere is characterized by complex plasma structures, formed by fluctuations of the convective large-scale electric field governed by solar wind conditions. Observations and modelling efforts have demonstrated that, for instance, plasma tongues can be wrapped around the plasmasphere, shoulders can be formed, or that plasma irregularities can be detached from the main body of the plasmasphere [*Lemaire*, 2001; *Goldstein et al.*, 2003a; *Sandel et al.*, 2003]. The in situ observations of the outer plasmasphere obtained by the Cluster constellation provide some novel views of this region.

In this study we use data provided by the Cluster Ion Spectrometry (CIS) experiment [*Rème et al.*, 2001] to analyze the ionic structures observed locally during the Cluster spacecraft crossings of the plasmasphere. The perigee of the four Cluster spacecraft, at ~4 R_E, allows cuts through the outer plasmasphere. The CIS observations of the plasmapause position are then compared to the simulation results using an interchange instability numerical model for the plasmapause deformations [*Pierrard and Lemaire*, 2004]. The CIS local ion measurements have also been correlated with global images of the plasmasphere, obtained by the EUV instrument onboard Image [*Sandel et al.*, 2000], for an event where the Cluster spacecraft were within the field-of-view of EUV (Plate 1).

2. CLUSTER ORBIT AND INSTRUMENTATION

The Cluster mission is based on four identical spacecraft launched on similar elliptical polar orbits with a perigee at about 4 R_E and an apogee at 19.6 R_E [*Escoubet et al.*, 2001]. This allows Cluster to cross the ring current region, the radiation belts and the outer plasmasphere, from South to North, during every perigee pass. Orbital manoeuvres that change the inter-spacecraft separation take place once or twice per year, allowing the study of different characteristic scales in the various plasma regions in the magnetosphere and in the solar wind. The tetrahedron formed by the four spacecraft can thus have characteristic sizes ranging between 100 km and greater than 10 000 km.

The Cluster Ion Spectrometry (CIS) experiment on board Cluster consists of the two complementary spectrometers CODIF (or CIS-1) and HIA (or CIS-2), and provides three-dimensional ion distributions (from about 0 to 40 keV/q) with one spacecraft spin (4 seconds) time resolution [*Rème et al.*, 2001]. Furthermore, the mass-resolving spectrometer CODIF provides the ionic composition of the plasma for the major magnetospheric species (H^+, He^+, He^{++} and O^+), from the thermal energy to about 40 keV/q. In addition CODIF is equipped with a Retarding Potential Analyzer (RPA), which allows more accurate measurements in the about 0.7–25 eV/q energy range, covering the plasmasphere energy domain. The operation on CODIF of the RPA mode and of the normal magnetospheric modes (which provide a 25 eV/q to 40 keV/q energy range) is mutually exclusive. The RPA mode is thus operated on one out of 10 orbits, on the average, and not always on all of the spacecraft.

The magnetic field data, used here to calculate ion pitch angle distributions, come from the FGM (Fluxgate Magnetometer) experiment on board Cluster [*Balogh et al.*, 2001].

The Image spacecraft was launched in March 2000 into a highly inclined elliptical orbit with an apogee altitude of 7.2 R_E and a perigee altitude of 1000 km [*Burch*, 2000]. On board Image, the Extreme Ultraviolet Imager (EUV) provides global images of the plasmasphere by imaging the distribution of He^+ in its 30.4 nm resonance line [*Sandel et al.*, 2000].

3. OBSERVATIONS AND ANALYSIS

3.1 Plasmasphere Cut: 4 July 2001 Example

Plate 2 shows an example of a Cluster crossing of the plasmasphere, in the post-noon sector (15:30 MLT), during quiet magnetospheric conditions (4 July 2001 event: Kp = 1+). Plate 3 shows the corresponding orbit plot, which highlights the 10-14 UT interval. During this event CODIF

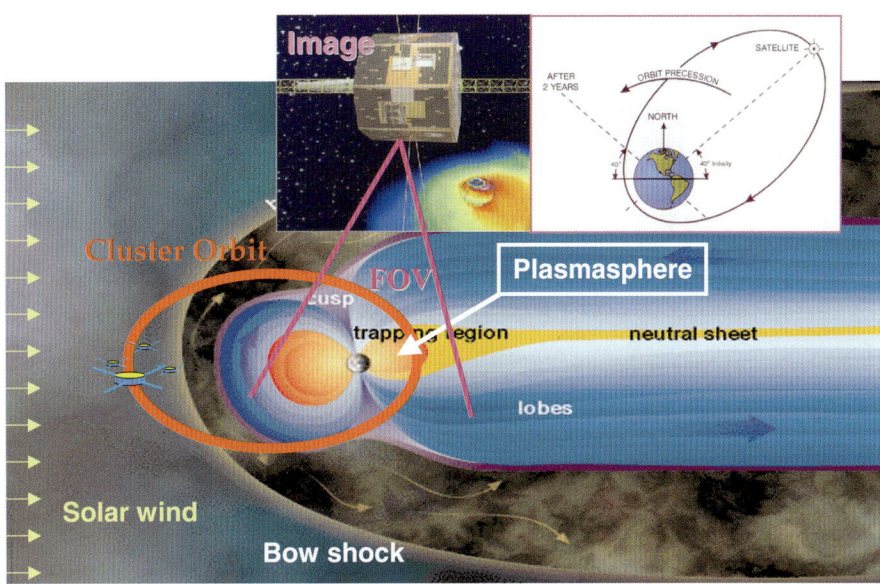

Plate 1. CIS (Cluster) and EUV (Image) measurements comparison principle: the CIS local ion measurements are correlated with global images of the plasmasphere, obtained by the EUV instrument onboard Image, for an event where the Cluster spacecraft were within the field-of-view of EUV.

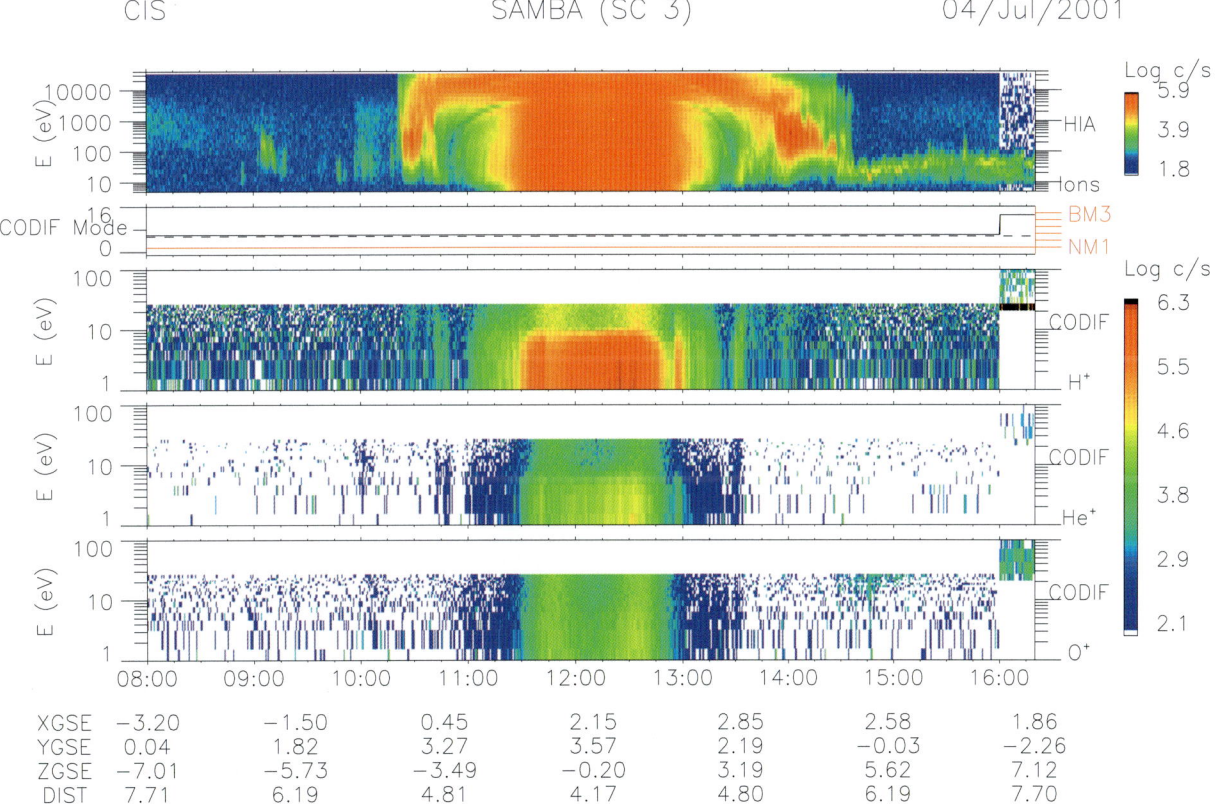

Plate 2. Cluster spacecraft 3 ion data for July 4, 2001. From top to bottom: HIA energy-time ion spectrogram (normal magnetospheric mode: 5 eV/q–32 keV/q), in corrected-for-detection-efficiency counts per sec. (c/s) ; CODIF mode (in black: RPA mode until 16:00 UT); CODIF Energy-time ion spectrograms, separately for H^+, He^+, and O^+; spacecraft coordinates (GSE system) and geocentric distance, in R_E.

26 IONIC STRUCTURES IN THE PLASMASPHERE

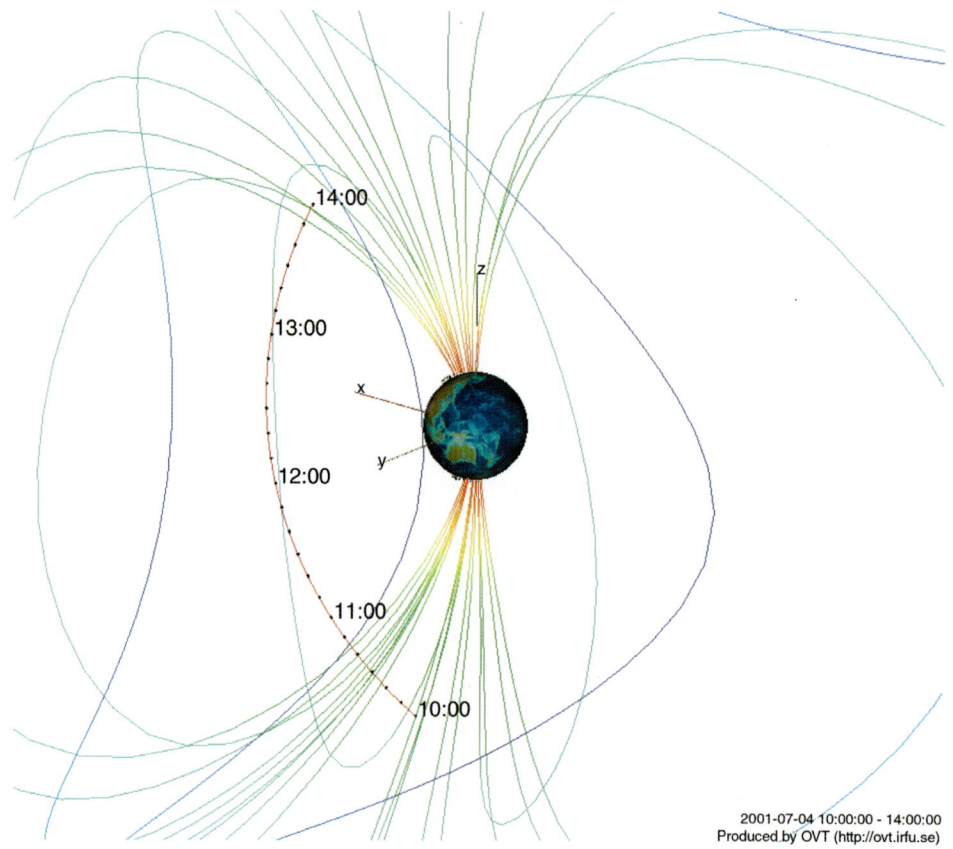

Plate 3. Cluster spacecraft 3 orbit (in red) for July 4, 2001, projected on the Tsyganenko 89 magnetic field model. Orbit Visualization Tool plot, courtesy of the OVT Team.

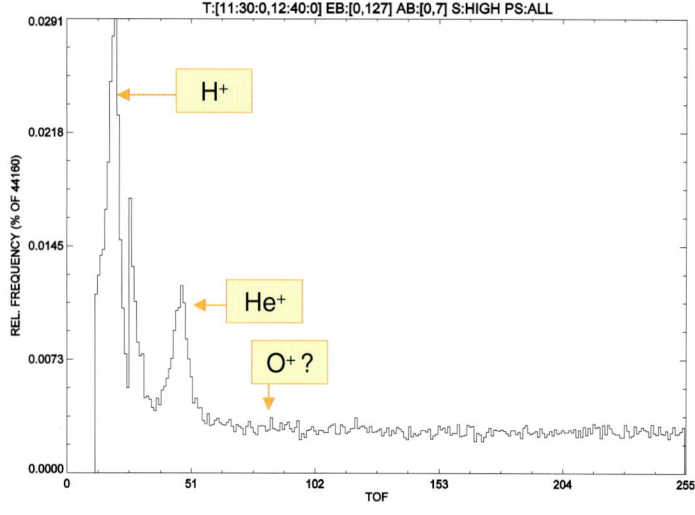

Plate 4. Time-of-flight spectrum for the ions detected by CODIF between 11:30 and 12:40 UT (Cluster sc 3, July 4, 2001). The abscissa axis is the time-of-flight channel number (inversely proportional to the ion velocity), and the ordinate axis is the number of particles in a given channel (with two different sampling laws above and below channel 26).

(bottom 3 spectrograms in Plate 2) was in the RPA mode (~0.7–25 eV/q) until 16:00 UT, when it switched back to a normal magnetospheric mode (full energy coverage). HIA (top spectrogram in Plate 2) was continuously in a normal magnetospheric mode. Cluster was in the southern lobe until ~10:20 UT, when it crossed a first boundary, entering into the southern plasma sheet (cf. HIA data). At ~10:45 UT the spacecraft entered into the ring current, where it remained until ~13:40 UT, characterized in these data by intense particle fluxes at energies above 7 keV, showing the presence of high-energy ions subject to gradient and curvature drift [*Vallat et al.*, 2004]. The spacecraft then traversed through the northern plasma sheet and entered into the northern lobe, at the outbound leg of its trajectory, at ~14:38 UT.

Between 11:30 and 12:55 UT HIA suffered a strong background due to penetrating particles from the radiation belts, appearing as a high counting rate at all energies. This background presents two maxima centered on L-shell values around 4.5, one at the inbound leg and the other at the outbound leg of the orbit.

CODIF, which during this orbit interval was in the RPA mode, first detected the presence of a diffuse low-energy ion population at 11:00 UT, and then entered the main plasmasphere at 11:30 UT. This is characterized by high ion fluxes at energies below 7.7 eV/q, with respect to the spacecraft potential. As will be shown later, the Cluster spacecraft potential in the plasmasphere was of the order of 1-2 V. The plasmaspheric ion data shown here cover thus an energy domain of ~2 to ~9 eV, and they correspond to the tail of the distribution function, which in the plasmasphere has typical temperatures of the order of 1 eV [*Comfort*, 1996; *Bezrukikh et al.*, 2001]. The plasmasphere is detected until about 13:00 UT, in the outbound leg, corresponding to geocentric distances less than 4.6 R_E.

Ionic composition is provided by CODIF thanks to the time-of-flight technique, where ions, after being accelerated through a 15 kV potential, have their velocity determined by measuring the time-of-flight of each ion through a given length. The spectrograms plotted in Plate 2, for the three main ion species, correspond each to the time-of-flight interval of the given species. In order to verify the mass separation and the eventual contamination by background, we plotted, in Plate 4, a time-of-flight spectrum of the ions detected between 11:30 and 12:40 UT. The characteristic peaks of H^+ and He^+ are clearly present. He^{++}, if present, would be almost "washed-out" by the tail of the H^+ distribution (spillover). Note that for H^+ the height of the peak is not proportional to the relative abundance, because a different sampling law was used, in the spectrum, for H^+ and for the other ion species. Plate 4, however, does not show the presence of O^+ ions (although it cannot be completely excluded).

A persistent background is present over all energy channels, due to penetrating particles from the radiation belts, and it produces the two faint yellow strips shown in the O^+ spectrogram in Plate 2. O^+, if present, should have an extremely low signal-to-noise ratio. Note also that the radiation belt background is relatively low, due to the time-of-flight technique that eliminates a large number of counts from penetrating particles, and it allows the clear identification, in this event, of dominant species as H^+ and He^+. This is not the case with HIA, which does not use the time-of-flight technique, and where the radiation belt background can become overwhelming (cf. upper panel of Plate 2).

The moments of the ion distribution functions, calculated in the ~0.7 eV/q to 25 eV/q energy range (with respect to spacecraft potential), are shown in Plate 5a for H^+ and in Plate 5b for He^+. The densities profiles are similar for these two ion species, with the He^+ densities being lower by a factor of ~15. This is consistent with the measurements made by DE 1 for the same geocentric distance [*Craven et al.*, 1997]. Note the characteristic density drop at the plasmapause, by about one to two orders of magnitude. The absolute values of the densities measured here are comparable, but slightly lower, to those typically measured onboard Cluster by the Whisper resonance sounder experiment [*Décréau et al.*, 2001; *Moullard et al.*, 2002; *Darrouzet et al.*, 2003], because CODIF-RPA measures particles only in a finite energy range.

The H^+ and He^+ velocities measured here show a clear Vx < 0 and a Vy > 0 component, consistent with a corotating plasma, given the spacecraft position at 15:30 MLT (theoretical corotation velocity: 2.1 km s^{-1}). The systematic bias to negative values, shown by the Vz component, is the artefact of the instrument particle detection efficiency being inhomogeneous across the anodes looking at different elevations, and not completely compensated by the calibration values.

The H^+ and He^+ temperatures within the plasmasphere show typical values of 1 eV. Outside the plasmaspheric dense plasma the temperature and velocity calculations suffer from reduced counting statistics.

3.2 Detached Plasmasphere Observations: 31 October 2001 Event

Plate 6 shows a crossing of the plasmasphere by Cluster spacecraft (sc) 1, 3 and 4, in the morning sector (08:45 MLT), during initially quiet magnetospheric conditions (31 October 2001 event: Kp = 0+ during the 9-12 UT interval). The onset of a negative auroral bay is however observed in the AE index at ~12:30 UT, i.e. close to the outbound plasmapause crossing by the Cluster spacecraft, and the Kp index jumped from 0+ to 3 in the 12-15 UT interval.

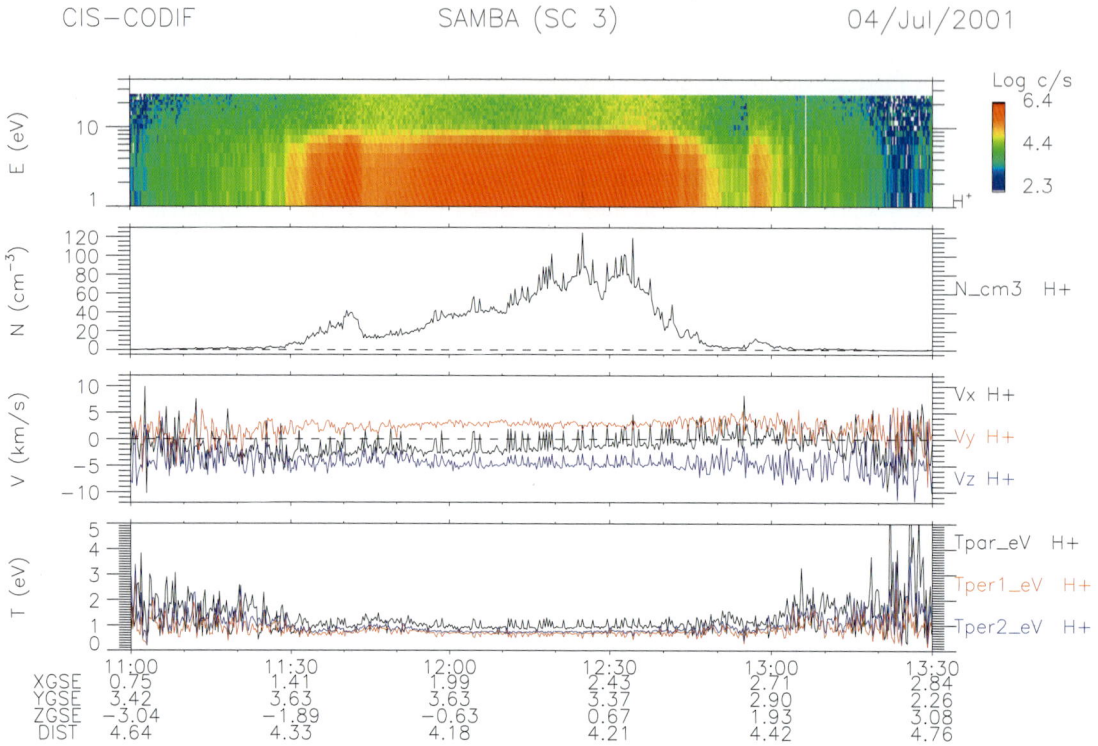

Plate 5a. Cluster sc 3 CODIF H$^+$ data for July 4, 2001. From top to bottom: Energy-time ion spectrogram and moments of the distribution functions (~0.7 eV/q to 25 eV/q energy range): density, velocity (in GSE coordinates), and parallel and perpendicular temperatures.

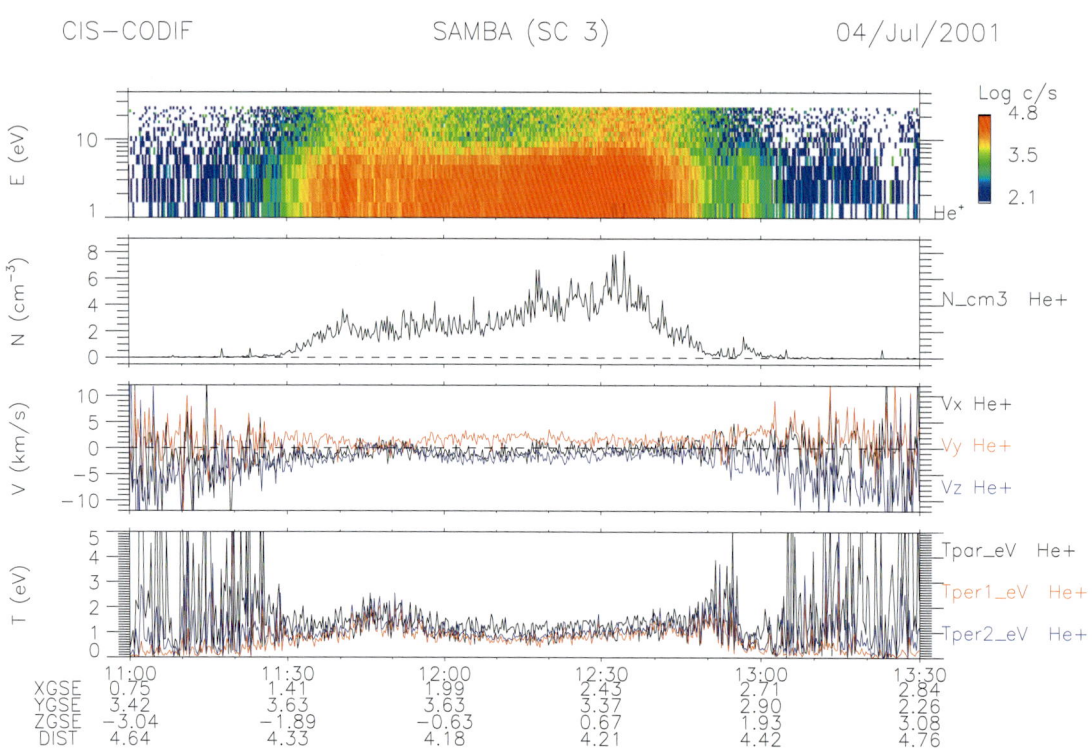

Plate 5b. Same as Plate 5a, but for He$^+$ ions.

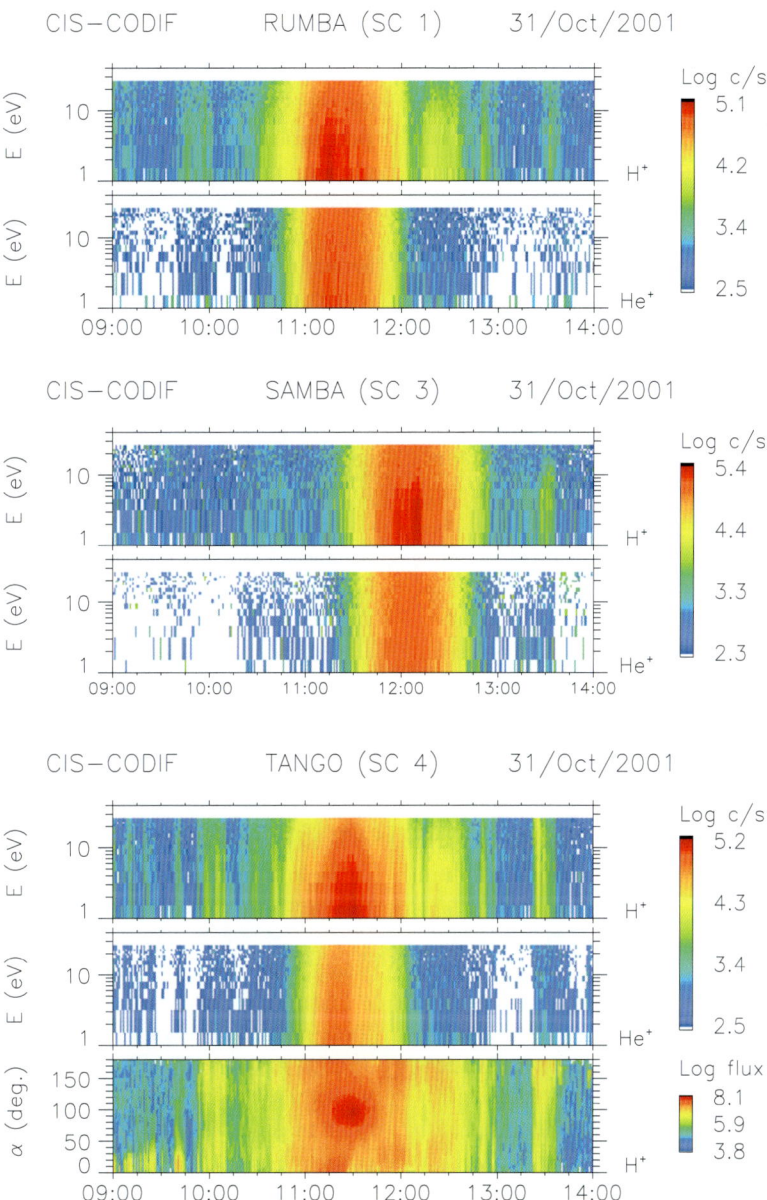

Plate 6. Cluster sc 1, 3 and 4 H$^+$ and He$^+$ energy-time ion spectrograms, in corrected-for-detection-efficiency counts per sec., for October 31, 2001. All spacecraft in RPA mode. Bottom panel: sc 4 pitch-angle distributions for H$^+$ ions (~0.7 eV/q to 25 eV/q energy range), in particle flux units (cm^{-2} sr^{-1} s^{-1} keV^{-1}).

All spacecraft were in the RPA mode. Cluster sc 3 was lagging on its orbit, with respect to the other spacecraft, which explains why it crossed the plasmasphere about 30 minutes later. Sc 1 and 4 crossed the plasmospheric main ion population between ~10:47 UT (L ≈ 5.1) and ~12:04 UT (L ≈ 5.8). However, detached plasma of lower density was also observed before the entry into the main plasmasphere, at around 10:00 UT (L ≈ 11), and after the exit from the plasmasphere, at around 12:25 UT (L ≈ 8) and at around 13:30 UT (L ≈ 55).

The bottom panel of Plate 6 shows the H^+ pitch-angle distributions (pad) for sc 4. The distributions in the main plasmasphere are relatively isotropic, while around the crossing of the equatorial plane the major part of the proton population is centred at 90° (pancake distributions). On the detached plasmasphere observations, however, the distributions are still trapped (distributions symmetric with respect to the magnetic field), but are of the butterfly type, presenting a deficiency of particles perpendicular to the magnetic field direction. Plate 7, which shows the H^+ distribution cuts for three times (main plasmasphere and detached plasmasphere), confirms the bi-directional character of these distributions, showing that the particles in these detached plasma observations are trapped.

Plate 8 shows the time-of-flight spectra for three intervals, corresponding to these distribution cuts. As for the July 4, 2001 event, a persistent background due to penetrating particles from the radiation belts is present in the first spectrum, obtained in the main plasmasphere. For the two following spectra, however, corresponding to the detached plasmasphere observations in the outbound leg of the orbit, the background disappears. These spectra show clearly the absence of O^+ ions. All spectra show the characteristic peaks of H^+ and He^+.

The moments of the ion distribution functions of H^+, calculated in the ~0.7 eV/q to 25 eV/q energy range (with respect to spacecraft potential), are shown in Plate 9. The density values measured during the detached plasmasphere observations are by about an order of magnitude lower than the ones measured in the main plasmasphere, consistently with the substantial density reduction for the detached plasma shells predicted by the peeling-off models [*Lemaire*, 2001]. The H^+ velocities measured here show a clear $V_x > 0$ and a $V_y > 0$ component in the main plasmasphere (~11–12 UT), corresponding to a corotating plasma (08:45 MLT). For the detached plasmasphere observations, however, around 13:30 UT, the measured velocities are dominated by $V_y < 0$ and $V_z > 0$ showing a strong outward expansion of the plasma tube. This expansion velocity increases as a function of the L-shell value: the spacecraft gets from L ≈ 50 at 13:22 UT to L ≈ 110 at 13:36 UT, on high-latitude field lines, and the measured expansion velocity goes from ~3 km s^{-1} to ~10 km s^{-1} in that interval.

3.3 Detached Plasmasphere and Upwelling Ions Observations: 12 November 2001 Event

Plate 10 shows a crossing of the plasmasphere by Cluster sc 1 and 3, in the early morning sector (07:50 MLT), during quiet magnetospheric conditions (Kp = 0+, very quiet AE). Both spacecraft were in the RPA mode. Cluster sc 3 was lagging on its orbit, with respect to the other spacecraft, which explains why it crossed the plasmasphere about 30 minutes later. Sc 1 crossed the plasmospheric main ion population between ~08:22 UT (L ≈ 5.3) and ~09:42 UT (L ≈ 6.3).

Around 07:00 UT (L ≈ 20), before the entry of sc 1 in the main plasmasphere, there appears in the energy-time spectrograms what looks as detached plasma of lower density. A similar observation appears also after the exit from the plasmasphere, at around 10:20 UT (L ≈ 11.7). An examination of the pitch-angle distributions (two middle panels in Plate 10) however shows that these two observations correspond to particle populations of very different characteristics. The observation around 07:00 UT, in the southern hemisphere (inbound leg), shows a very strong anisotropy, and is dominated by particles with pitch angles close to 0°. These are upwelling H^+ and O^+ ions, escaping from the ionosphere along the magnetic field lines [*Moore et al.*, 1986; *Chappell et al.*, 1987; *Sauvaud et al.*, 2004]. Although they are observed at low-energies, it is clear from the spectrograms that the population should also extend at energies above the RPA upper limit (25 eV/q).

The observation around 10:20 UT, on the contrary, at the outbound leg, is symmetric with respect to the magnetic field and corresponds to a trapped H^+ population. Contrary to the first one, this second observation corresponds thus to a detached plasmasphere.

Plate 11, which shows the H^+ and O^+ distribution cuts for three instances, confirms the observation of escaping H^+ and O^+ ions at ~07 UT. Their strong anisotropy contrasts with the bi-directional distributions detected later in the main plasmasphere and in the detached plasma event at the outbound leg.

Plate 12 shows the time-of-flight spectra for three intervals, corresponding to these distribution cuts. As for the July 4, 2001 event, a persistent background due to penetrating particles from the radiation belts is present in the second spectrum, obtained in the main plasmasphere. In the other spectra, however, corresponding to the upwelling ion observation (upper panel) and to the detached plasmasphere observation (lower panel), there is no background. The first spectrum shows clearly the presence of O^+ ions, and even of

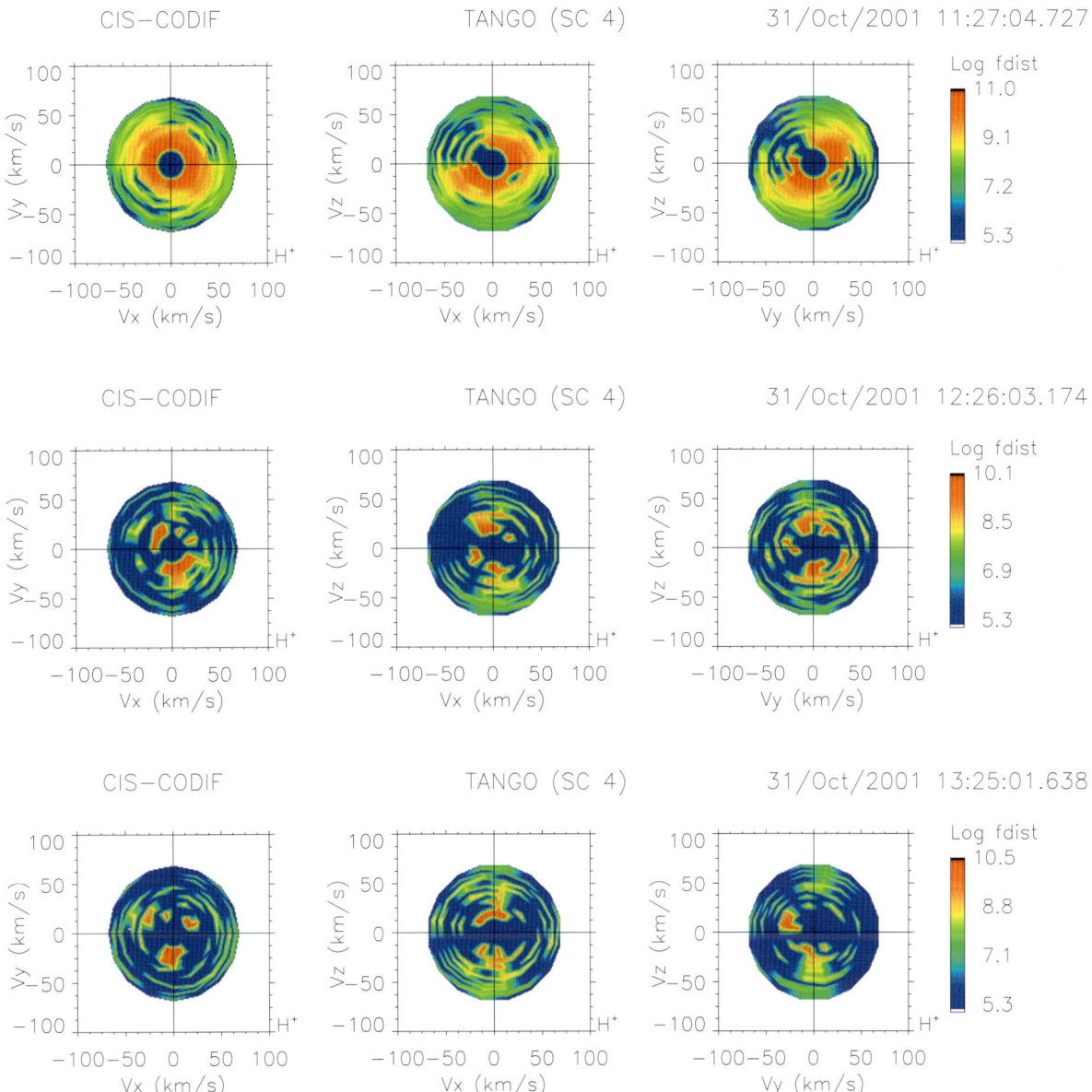

Plate 7. H^+ distribution cuts in three perpendicular planes each, for the October 31, 2001 event, in GSE coordinates and in particle phase-space density units (sec^3 km^{-6}), for three instants: 11:27:04 UT (main plasmasphere), 12:26:03 UT (detached plasmasphere), and 13:25:01 UT (detached plasmasphere).

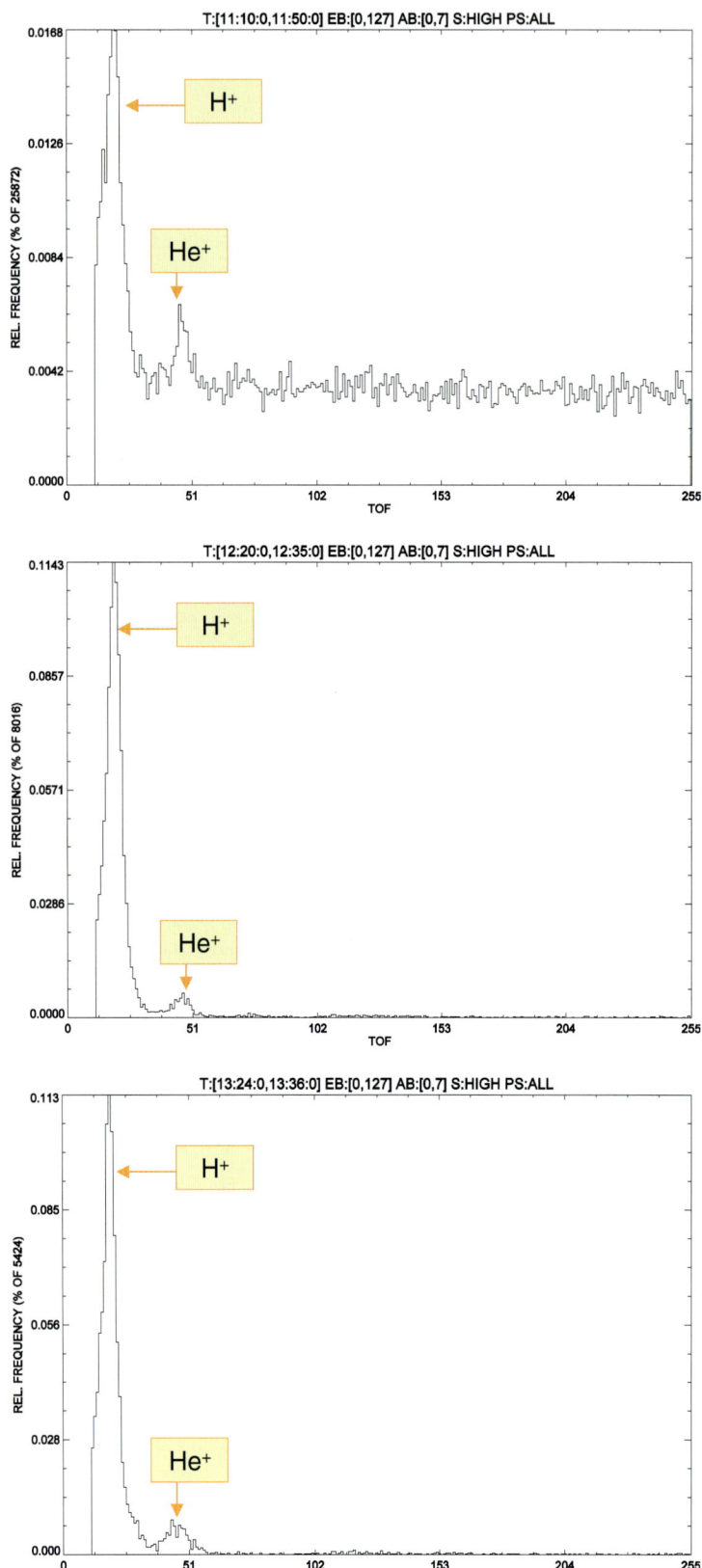

Plate 8. Time-of-flight spectra, same format as in Plate 4, for the ions detected by CODIF on sc 4 during the October 31, 2001 event, for three intervals: 11:10–11:50 UT (main plasmasphere), 12:20–12:35 UT (detached plasmasphere), and 13:24–13:36 UT (detached plasmasphere). These three intervals correspond to the distributions shown in Plate 7.

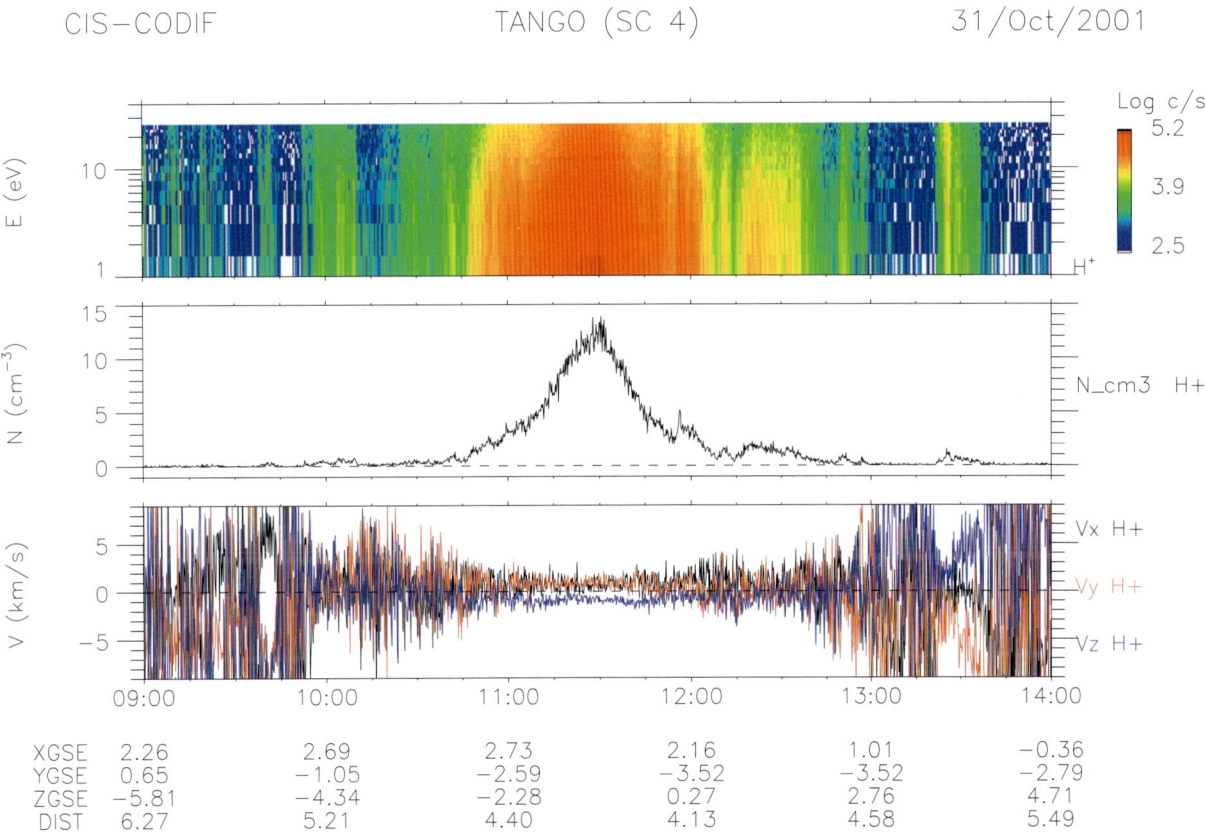

Plate 9. Cluster sc 4 CODIF H$^+$ data for October 31, 2001. From top to bottom: Energy-time ion spectrogram and moments of the distribution functions (~0.7 eV/q to 25 eV/q energy range): density, velocity (in GSE coordinates).

O^{++} ions in this escaping population [*Truhlík*, 1997; *Wilber et al.*, 2003], in addition to the H^+ and He^+ ions. As shown in Plate 13, these upwelling ions are observed on auroral field lines. The spectrum in the bottom panel of Plate 12, however, shows the absence of ions heavier than He^+ in the detached plasmasphere.

3.4 Cluster Multi-Spacecraft Plasmasphere Observations and Modeling: 17 February 2002 Event

In the first six months of 2002 the Cluster inter-spacecraft separation was reduced so as to obtain a regular tetrahedron of a 100 km characteristic size when traversing the cusp. This resulted in an elongated tetrahedron, of ~70 km width and ~240 km height mainly along the GSE z-axis, when crossing perigee.

Plate 14 shows a crossing of the plasmasphere by Cluster sc 1, 3 and 4, in the night sector (00:45 MLT), during moderately disturbed magnetospheric conditions (Kp = 2, AE ≈ 200 nT). Sc 1 and 3 were in the main plasmasphere between ~21:32 UT (L ≈ 4.3) and ~22:02 UT (L ≈ 4.2). Sc 4, however, did not at all detect the main plasmasphere, but only a low-density cold plasma, revealing the close presence of the plasmapause and the associated density gradient. The separation between sc 3 and sc 4, projected in the equatorial plane, was only 65 km. This gives a measure of the density gradients in the vicinity of the plasmapause, the density measured by sc 4 being lower by a factor of the order of 50-100 with respect to the one measured by sc 3. Note that the H^+ gyroradius at this region, for 2 eV plasmaspheric ions, is 0.6 km, i.e. consistent with the density gradients observed.

This localisation of the plasmapause, between the closely spaced Cluster spacecraft, allows us to compare the observation with model predictions. The interchange instability numerical model for the plasmapause deformations [*Pierrard and Lemaire*, 2004] was used here to simulate the February 17, 2002 event. This model uses as input an empirical Kp-dependent equatorial electric field model [*McIlwain*, 1986], and it determines, using kinetic simulations, the plasmapause position as the location where plasma interchange peels off the plasmasphere, i.e., where and when the magnetospheric convection velocity is enhanced at the onset of substorms [*Lemaire*, 1974, 2001]. According to this physical mechanism, the plasmapause is formed in the post-midnight MLT sector where and when the field-aligned component of the centrifugal pseudo-force overcomes that of the gravitational force.

The simulation results appear in Plate 15, bottom panel (equatorial plane). The blue dot corresponds to the Cluster spacecraft position (sc 1, 3 and 4), which appear as a single dot due to their close spacing: less than 100 km. It appears clearly that the spacecraft are almost at the edge of a plasmapause bulge, formed by plasma brought by the interchange instability. This explains why only some of the spacecraft (1 and 3) entered the plasmasphere.

The February 17, 2002 plasmasphere observation included also an eclipse which, for example for sc 3, was between 21:46 UT and 22:09 UT (cf. Plate 16). The suppression of photoelectron production allows us to evaluate the change in the spacecraft potential. The entrance in the plasmasphere (for sc1 and 3) is well before the start of the eclipse, and the exit from the plasmasphere (for sc1 and 3) is during the eclipse. The depth of the eclipse is almost the same for all sc. The only effect of the eclipse on the low-energy ions, detected by CIS-RPA, is a slight enhancement of the detected ion fluxes, and an increase in their energy by 1-2 eV, due to a more negative spacecraft potential by 1-2 V. This is compatible with the spacecraft potential measurements by the EFW electric field experiment onboard Cluster [*Gustafsson et al.*, 2001]. In the low-density magnetospheric lobes, however, the spacecraft potential can become positive by several Volts or even tens of Volts, unless the ASPOC ion emitter [*Torkar et al.*, 2001], used for the spacecraft active potential control, is operating.

At the outbound leg, around 01:00 UT (February 18, 2002) sc 4 detected three successive spikes of low-energy H^+ ions. As the pitch-angle distributions of these ions show (bottom panel of Plate 14), these are upwelling ions escaping from the ionosphere along the magnetic field lines. Their composition included only H^+ and He^+, and no O^+, as can be seen in the time-of-flight spectrum of Plate 17. It should be noted that the ASPOC ion emitter was operating, on sc 4, during this interval.

3.5 Cluster and Image Correlated Plasmasphere Observations: 9 August 2001 Event

On August 9, 2001, the Cluster spacecraft went through perigee in the noon sector (13:30 MLT), during the onset of a negative magnetic bay in the auroral zone (max AE = 500 nT, cf. Plate 18, Kp = 2), following a long period of quiet conditions (Kp = 1 for several hours). Sc 3 was in the plasmasphere between ~04:50 UT (L ≈ 4.4) and ~05:50 UT (L ≈ 5.3). Sc 1, however, did not at all detect the main plasmasphere, and the only signature in the particle data is increased background from the radiation belts, also seen by sc 3 (Plate 19). It should be noted that sc 1 was leading on the orbit by 45 minutes: sc 1 went through perigee at 04:18 UT and sc 3 at 05:03 UT. The sc 3 orbit got also deeper into the inner magnetosphere, with a minimum L-shell value of 4.2, versus 4.3 minimum L for sc 1.

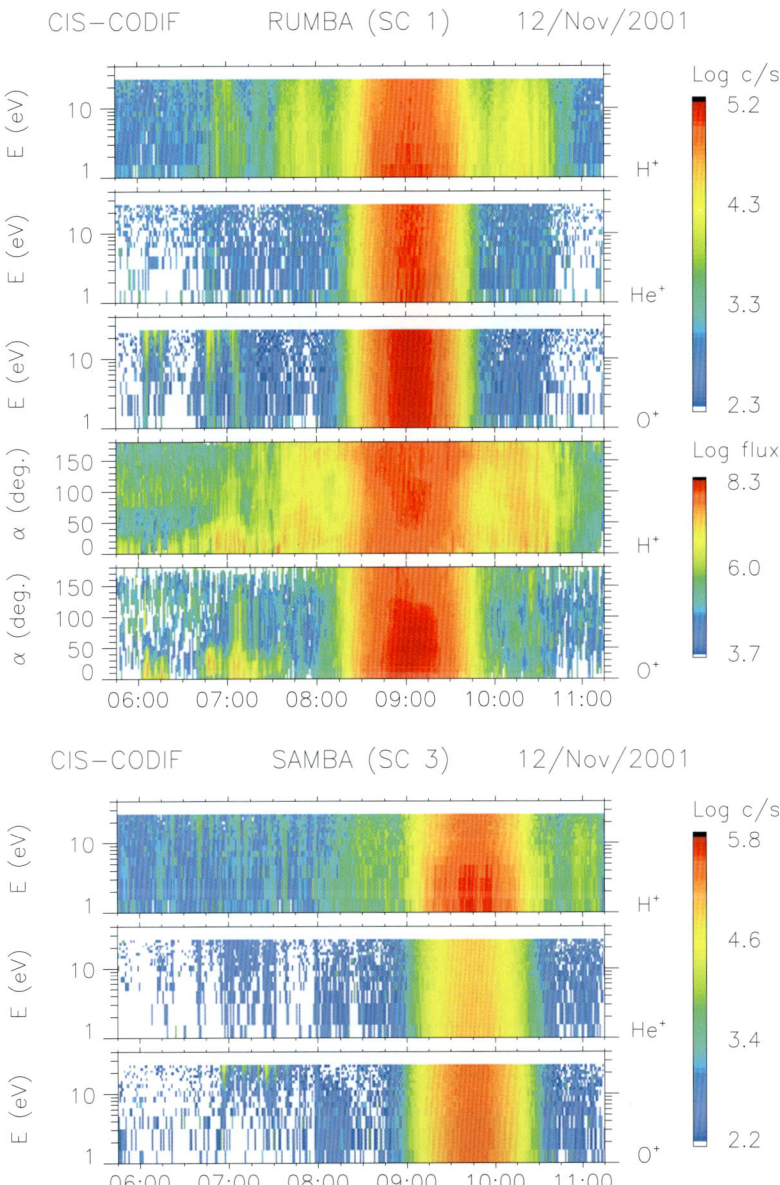

Plate 10. Cluster sc 1 and 3 H$^+$, He$^+$ and O$^+$ energy-time ion spectrograms, in corrected-for-detection-efficiency counts per sec., for November 12, 2001. Both spacecraft in RPA mode. Middle two panels show sc 1 pitch-angle distributions for H$^+$ and O$^+$ ions (~0.7 eV/q to 25 eV/q energy range), in particle flux units (cm^{-2} sr^{-1} s^{-1} keV^{-1}).

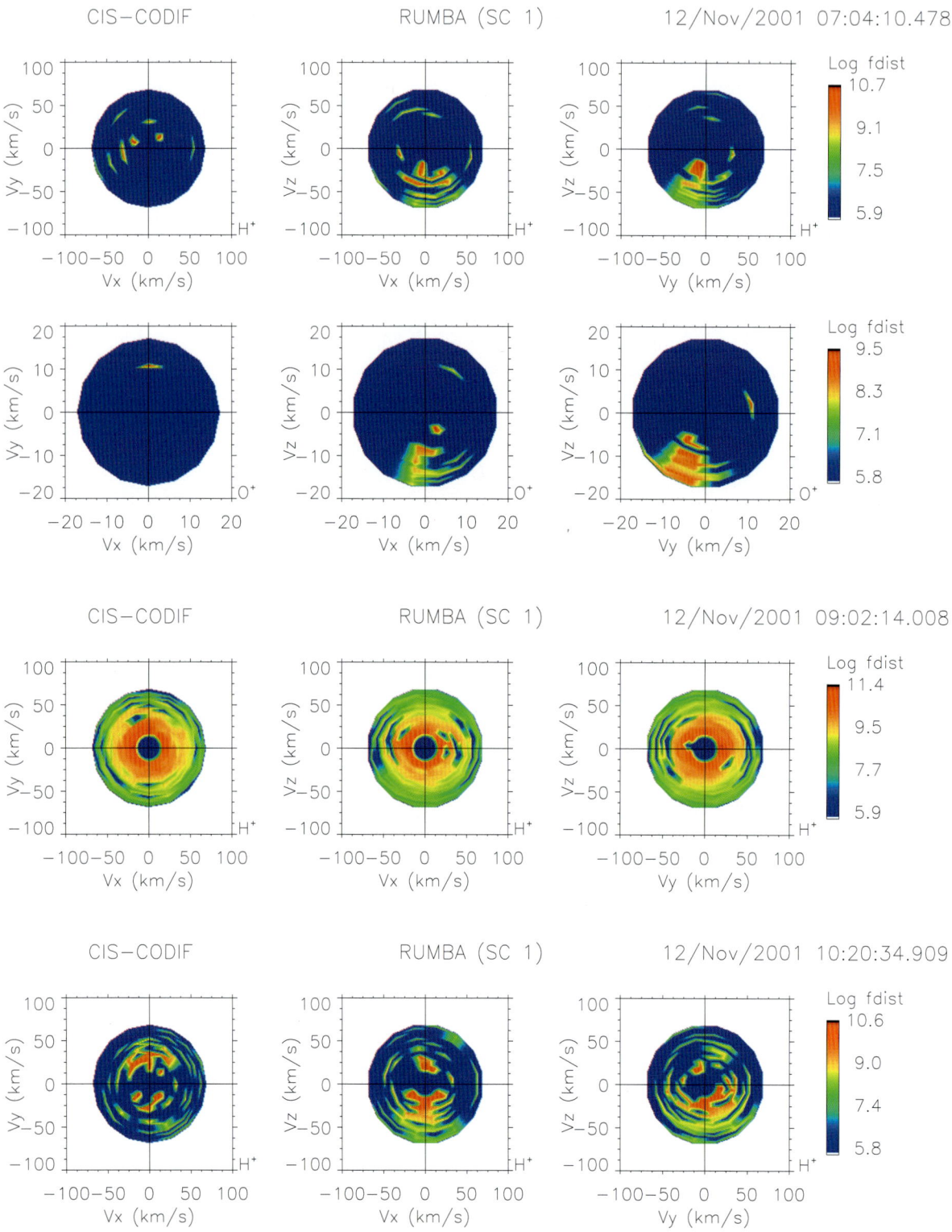

Plate 11. H$^+$ and O$^+$ distribution cuts in three perpendicular planes each, for the November 12, 2001 event, in GSE coordinates and in particle phase-space density units (sec^3 km^{-6}), for three instants: 07:04:10 UT (upwelling H$^+$ and O$^+$ ions), 09:02:14 UT (main plasmasphere), and 10:20:34 UT (detached plasmasphere).

In order to interpret the difference in observations between these two Cluster spacecraft (the only ones operating in the RPA mode during this event), and to deduce whether it was due to spatial effects (plasmapause situated between the trajectories of sc 1 and sc 3), or to temporal effects (boundary motion), we examined the plasmaspheric images provided by the EUV experiment onboard the Image spacecraft [*Sandel et al.; 2000*].

Plate 20 shows a time series of three EUV images of the plasmasphere, projected in the magnetic equatorial plane. The reddish haze around the Earth is EUV-observed 30.4 nm emissions from He^+ (see plate caption for details). In each image, the Cluster sc 1 and sc 3 orbits (in blue and green respectively) are also mapped. The bottom panel of each image shows radial slices of the normalized EUV intensity. The first of the images in this time series is close to the Cluster sc 1 perigee pass (no plasmasphere detected in-situ by this spacecraft) and the second one is during the plasmaspheric observation by sc 3.

What appears in these images is a very diffuse plasmasphere, with a gradual fall-off of the plasma density and no clear plasmapause boundary, and also a lot of azimuthal variation. This resulted from the extended period of relatively quiet geomagnetic conditions, preceding the observations. It is possible, however, to define an ad-hoc plasmapause as where there is a gradient in intensity that passes through 100 or so [*Goldstein et al., 2003b*].

The above definition of an ad-hoc plasmapause has been used to produce the white dotted line plots in the two panels of Plate 21, which show the intensity of the EUV 30.4 nm emissions at 13.5 MLT, as a function of UT and L. The projected sc 1 or sc 3 orbit is indicated with the blue or green curve, respectively, in the two panels.

These panels show a temporal dependence of the radial extent of the plasmasphere at 13.5 MLT. This is most clearly seen by looking for the bright (orange-yellow) region. The edge of this region starts at about L = 2.2 at 03:00 UT, and moves upward to about L = 2.8 by 05:45 UT. This bright orange-yellow region is the densest part of the plasmasphere. It is possible also to follow the red-orange (less dense) parts and see that they too move outward. The ad-hoc plasmapause moves outward between 04:30 and 05:45 UT as well. Sc 1 goes through 13.5 MLT when the ad-hoc plasmapause is just at about L = 4.2, just skimming above it. Sc 3 passes through 13.5 MLT when the ad-hoc plasmapause (at this MLT) has moved outward to L ≈ 4.8, passing well inside of it. This is consistent with the in-situ observation of the plasmasphere by sc 3, and the absence of plasmasphere detection by sc 1 (cf. Plate 19).

The outward motion of the plasmapause at 13.5 MLT, between the perigees of sc 1 and sc 3, is not a result of dynamic expansion of the global plasmapause. It results rather from the fact that the radial density profile of the plasmasphere varies with MLT, and a more extended radial profile "rotated" into 13.5 MLT in between the sc 1 and sc 3 perigees, due to the plasmasphere co-rotation with the Earth. Because there is so much azimuthal structure in the plasma density, this caused a slightly larger radial extent of the dense plasma to rotate into 13.5 MLT.

The interchange instability numerical model for the plasmapause deformations [*Pierrard and Lemaire, 2004*] was also used here to simulate this correlated Cluster – Image observation of the plasmasphere. Plate 22 shows the simulation results for the plasmapause geometry, for 04:21 UT (close to the Cluster sc 1 perigee pass) and for 05:43 UT (during the plasmaspheric observation by sc 3). The azimuthal structure in the "plasmapause" position is very clear. Since this is an event following an extended period of low – Kp, the model gives also a plasmapause position extending between 4 and 5 R_E. The blue dot, superposed on these simulations, corresponds to the Cluster sc 1 position, and the green dot to sc 3.

In the 04:21 UT simulation results sc 3 is well outside the plasmasphere. Sc 1, however, appears as being situated within the plasmasphere, which is contrary to the Cluster and to the Image observations. To explain this we should take into account the fact that the only input parameter to the simulation is Kp, which is a 3-hour index. There is thus an uncertainty in the rotation phase of the plasmapause by about 1–2 hours. Introducing a phase delay of 2 hours in the plasmapause rotation is equivalent to rotating the projected spacecraft position, on the simulation results, by adding 2 hours in its MLT. This shifted azimuthal spacecraft position of sc 1 appears in the 04:21 UT simulation, Plate 22, as a red dot which now is very close to the boundary. Given the diffuse nature of the "plasmapause", for this event, this simulation result is consistent with the Cluster and Image observations, showing sc 1 skimming just above it.

The same 2 hours phase delay in the plasmapause rotation was also applied to the 05:43 UT simulation results. The azimuthal position of sc 3, which was observing the plasmasphere, was shifted by adding 2 hours in its MLT (red dot). Sc 3 appears then well inside a bulge in the plasmasphere, again consistent with the Cluster and Image observations.

3.6 Plasmapause Position: Statistical Analysis

Using the CIS-RPA data, a statistical analysis of the observed plasmapause positions was performed, for the period July 2001–March 2003. The results are shown in Plate 23, in an L–MLT azimuthal plot. Black crosses cor-

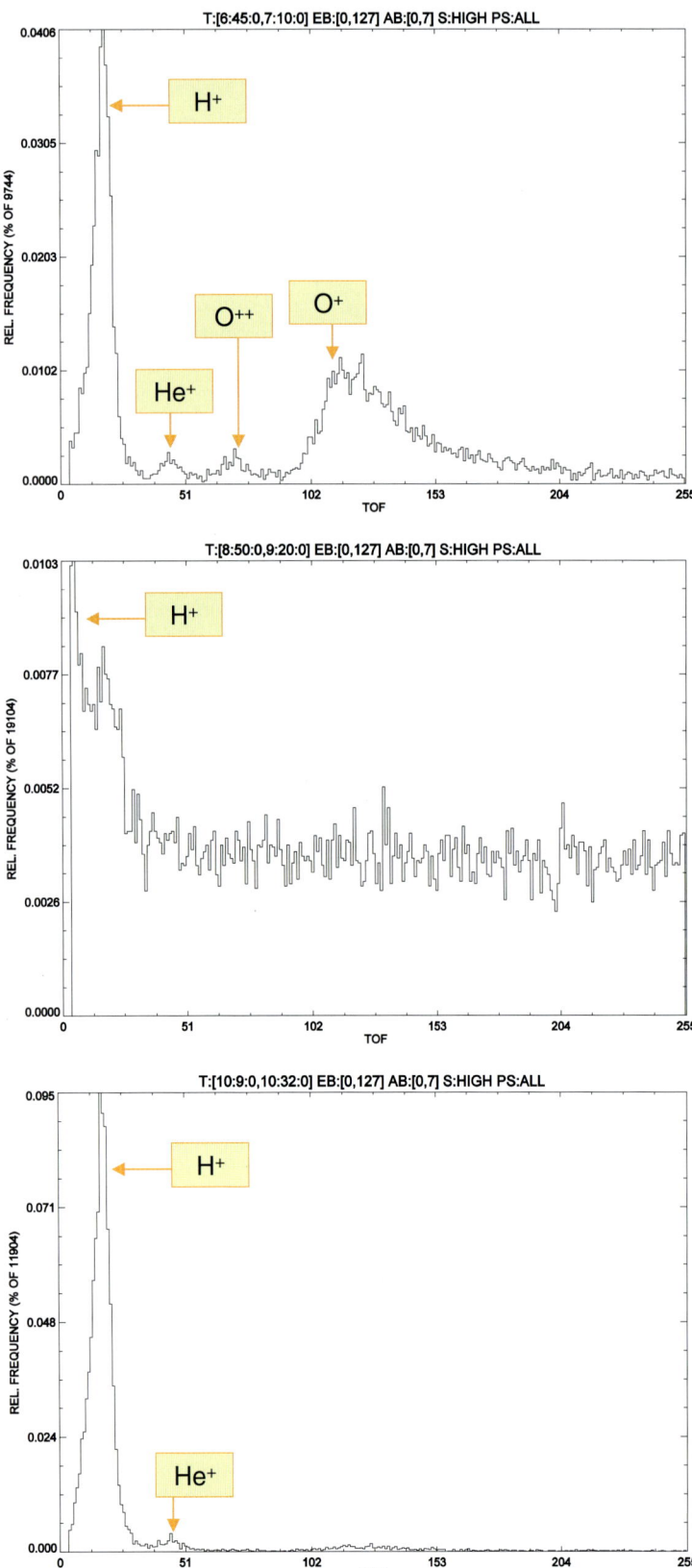

Plate 12. Time-of-flight spectra, same format as in Plates 4 and 8, for the ions detected by CODIF on sc 1 during the November 12, 2001 event, for three intervals: 06:45–07:10 UT (upwelling ions), 08:50–09:20 UT (main plasmasphere), and 10:09–10:32 UT (detached plasmasphere). These three intervals correspond to the distributions shown in Plate 11.

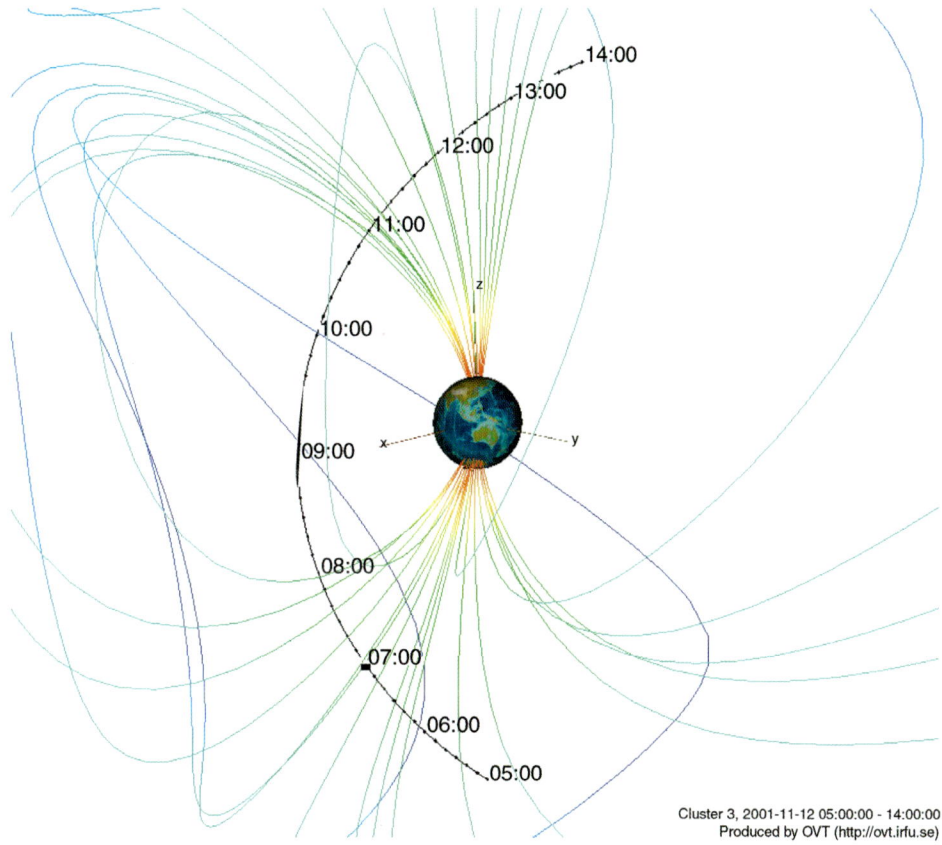

Plate 13. Cluster sc 1 orbit (in black) for November 12, 2001, projected on the Tsyganenko 89 magnetic field model. Orbit Visualization Tool plot, courtesy of the OVT Team.

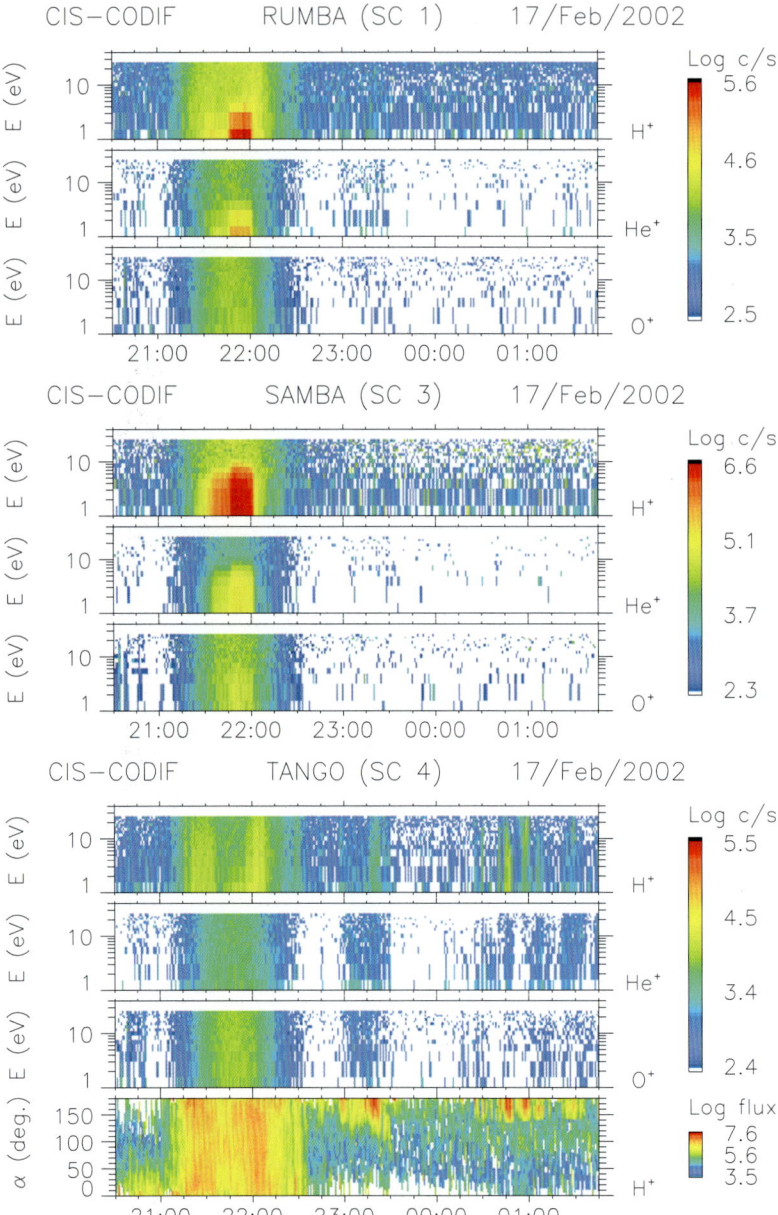

Plate 14. Cluster sc 1, 3 and 4 H^+, He^+ and O^+ energy-time ion spectrograms, in corrected-for-detection-efficiency counts per sec., for February 17, 2002. All spacecraft in RPA mode. Bottom panel: sc 4 pitch-angle distributions for H^+ ions (~0.7 eV/q to 25 eV/q energy range), in particle flux units ($cm^{-2}\ sr^{-1}\ s^{-1}\ keV^{-1}$).

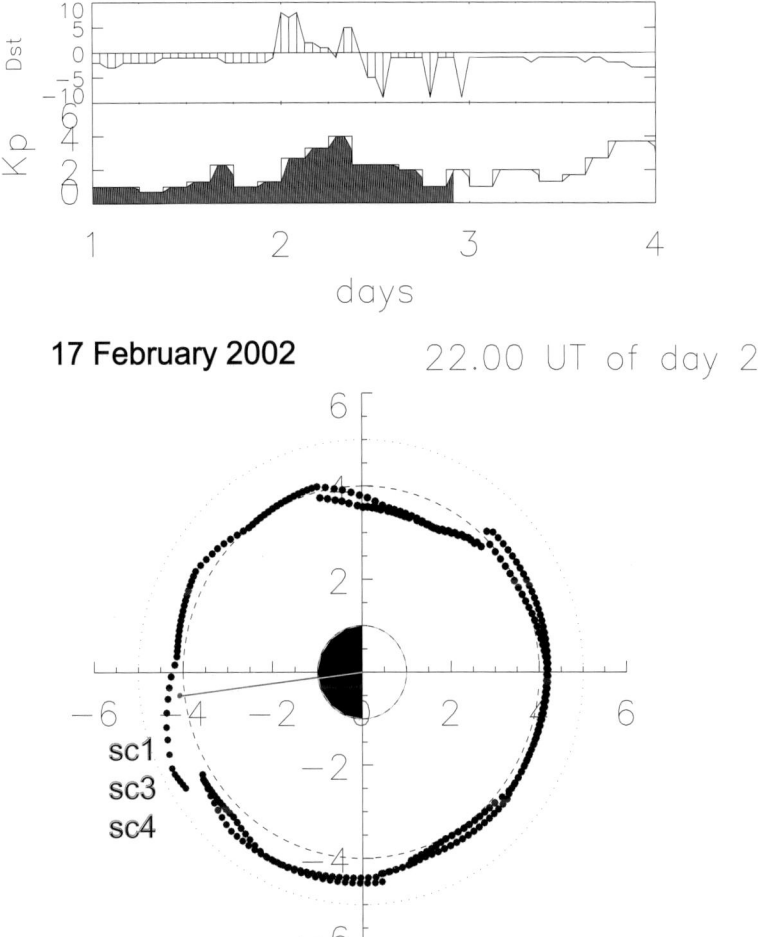

Plate 15. Numerical simulation results of the plasmapause deformations, for February 17, 2002, using the interchange instability model. Upper panel: Kp index time history, used as input parameter for the simulation. Bottom panel: simulation results of the plasmapause deformations, in the equatorial plane. The blue dot corresponds to the Cluster spacecraft positions (sc 1, 3 and 4), which appear as a single dot due to their close spacing.

42 IONIC STRUCTURES IN THE PLASMASPHERE

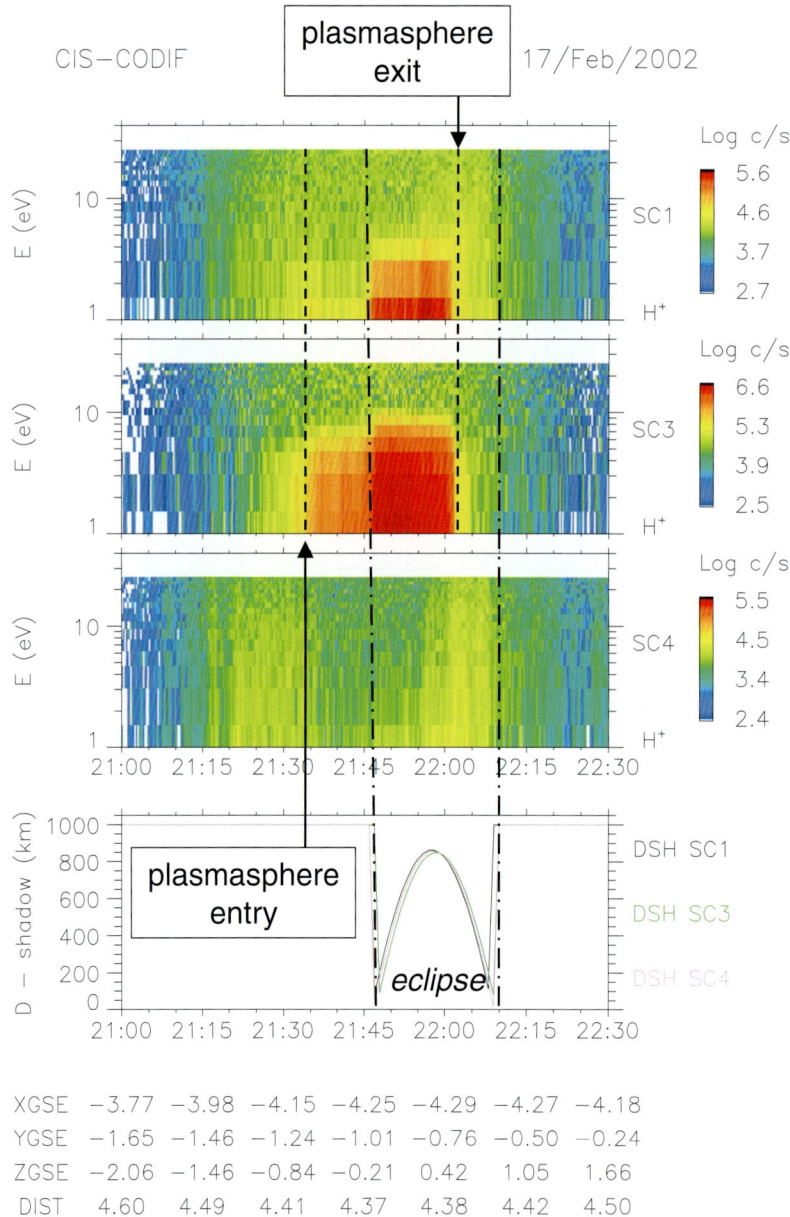

Plate 16. Cluster sc 1, 3 and 4 H$^+$ energy-time ion spectrograms, for February 17, 2002. The bottom panel shows an indicator of the spacecraft penetration within the eclipse shadow cone.

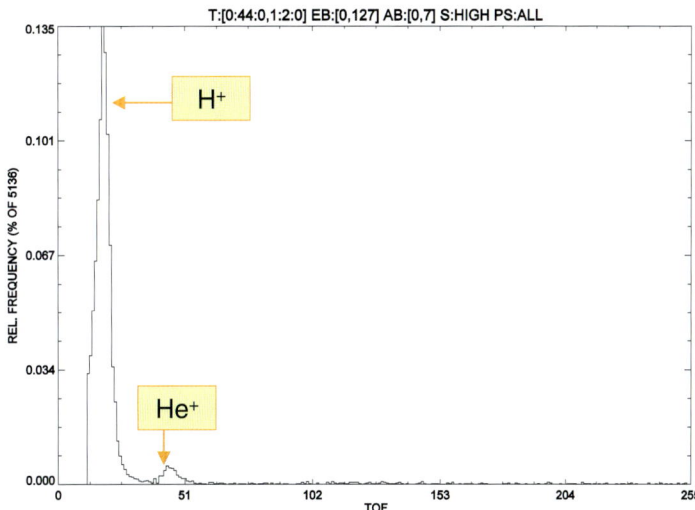

Plate 17. Time-of-flight spectrum, same format as in Plate 4, for the upwelling ions detected by CODIF on sc 4 during the interval 00:44–01:02 UT on February 18, 2002.

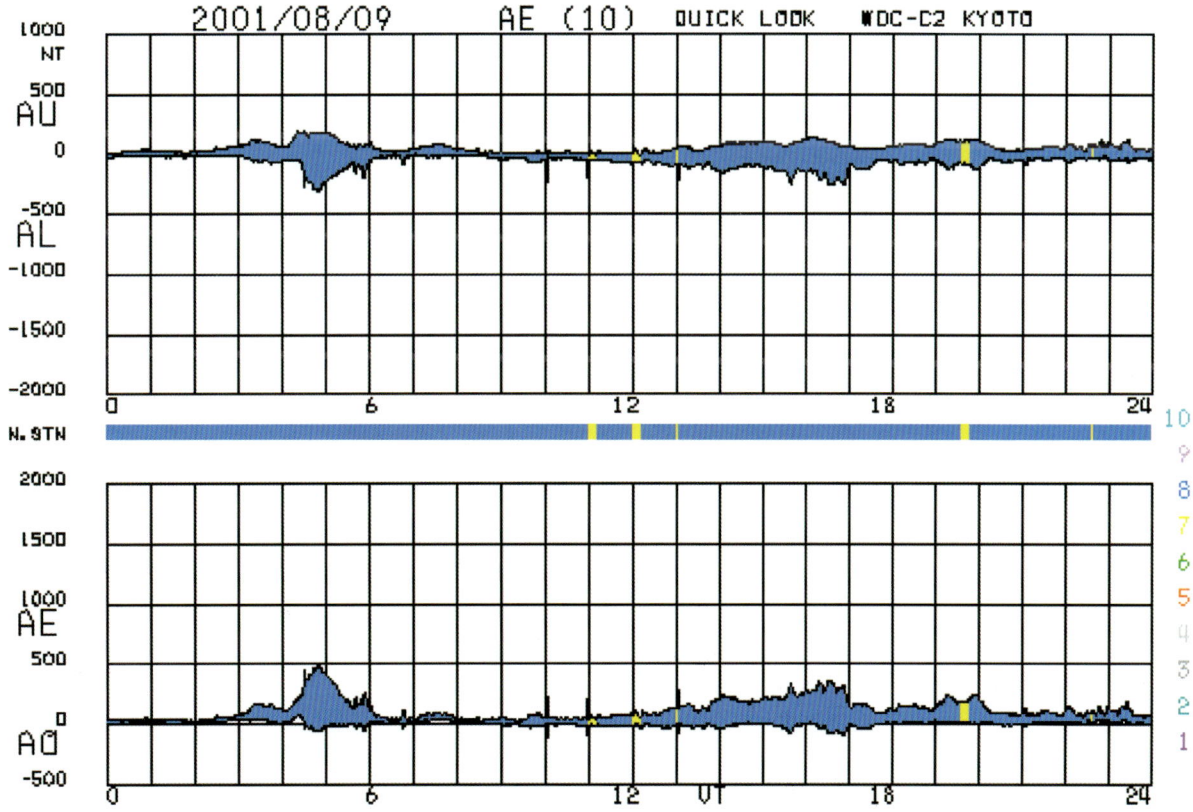

Plate 18. AE index for August 9, 2001.

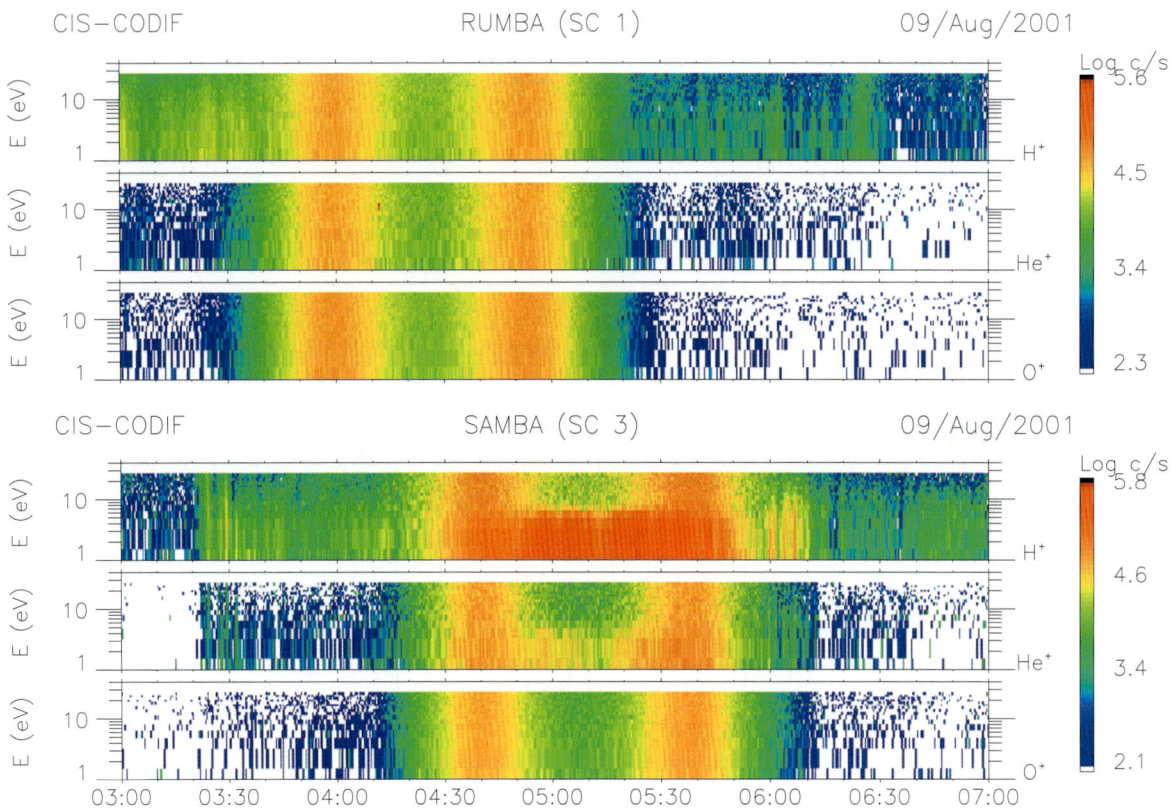

Plate 19. Cluster sc 1 and 3 H^+, He^+ and O^+ energy-time ion spectrograms, in corrected-for-detection-efficiency counts per sec., for August 9, 2001. Both spacecraft in RPA mode.

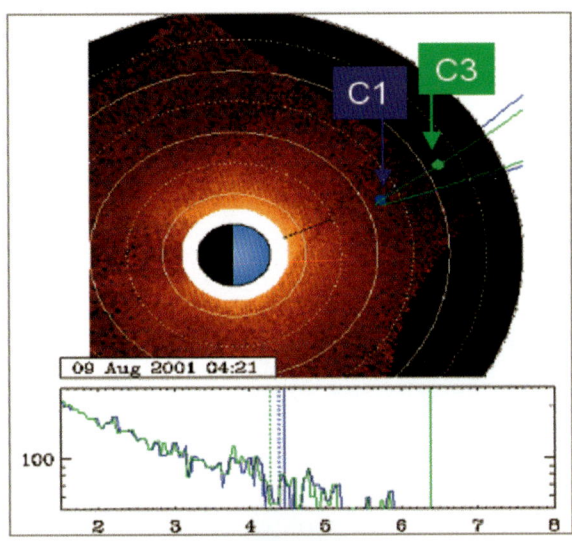
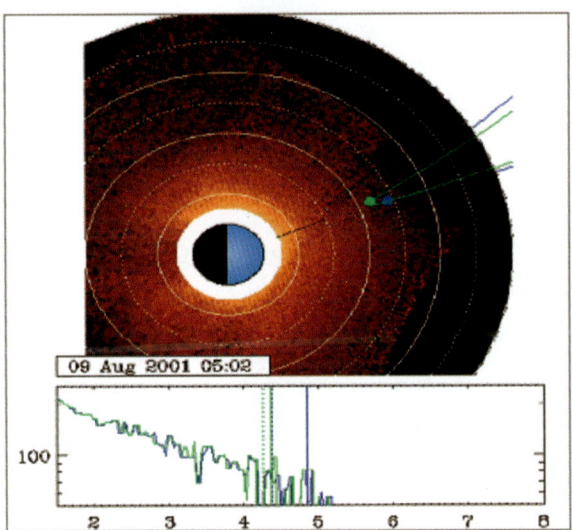
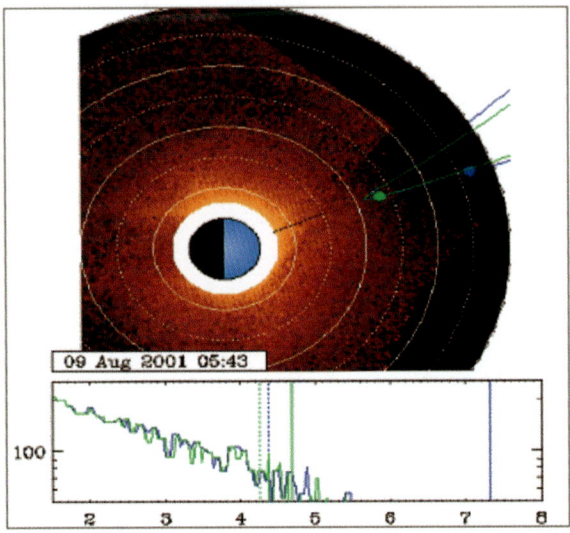

Plate 20. Series of three EUV images of the plasmasphere, for August 9, 2001. In each image the top panel shows the projection in the magnetic equatorial plane. The Earth is in the center, and the Sun direction is to the right. The reddish haze around the Earth is EUV-observed 30.4 nm emissions from He^+. The black regions mark the edges of the EUV field of view, distorted because the edges have been mapped to the equator. The blue and green lines are the Cluster sc 1 and 3 orbits (respectively), mapped to the equator. The dot shows where the satellite is at the given time. The bottom panel of each image shows radial slices through the EUV equatorial image, with the vertical axis being intensity (from 0–255) and the horizontal axis being L-shell. The blue curve is a radial intensity slice at the Cluster sc 1 perigee MLT and the green curve is for sc 3. (Sc 1 and sc 3 are at slightly different MLT, although rounded off to one decimal place they are both at 13.5 MLT). The vertical dotted lines mark the perigee L-shell of each spacecraft (blue for sc 1 and green for sc 3). The vertical solid lines show the instantaneous L-positions of each spacecraft.

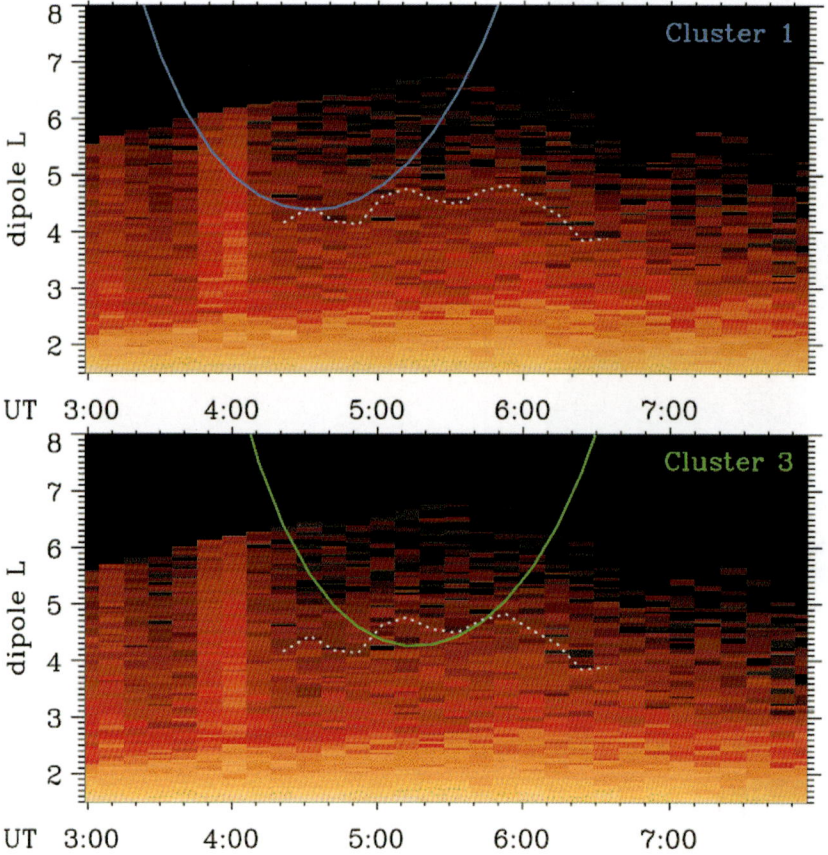

Plate 21. Two panels of Image-EUV data (upper and lower, one for each Cluster spacecraft sc1 and sc3). In each panel the color corresponds to the intensity of the EUV 30.4 nm emissions at 13.5 MLT, as a function of UT and L. The projected sc 1 or sc 3 orbit is indicated with the blue or green curve, respectively. The ad-hoc plasmapause L-value (see text) is the white dotted line. The intervals [2:59, 3:10, 3:51, 4:01, 4:11] correspond to high background noise.

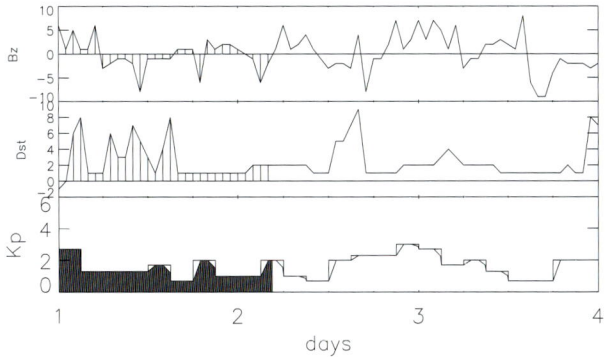

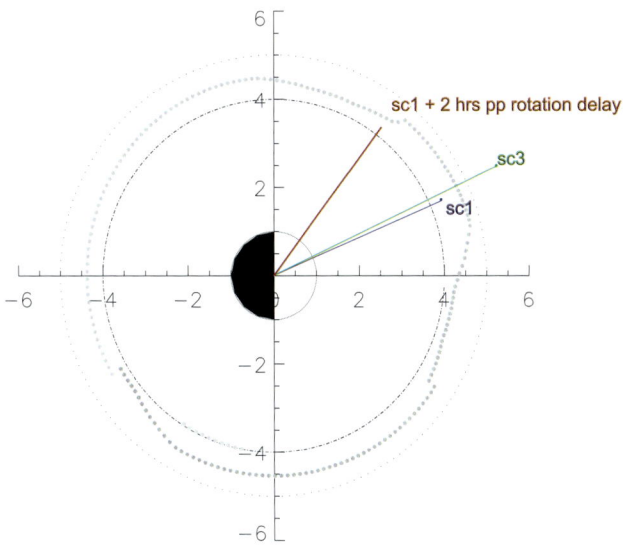

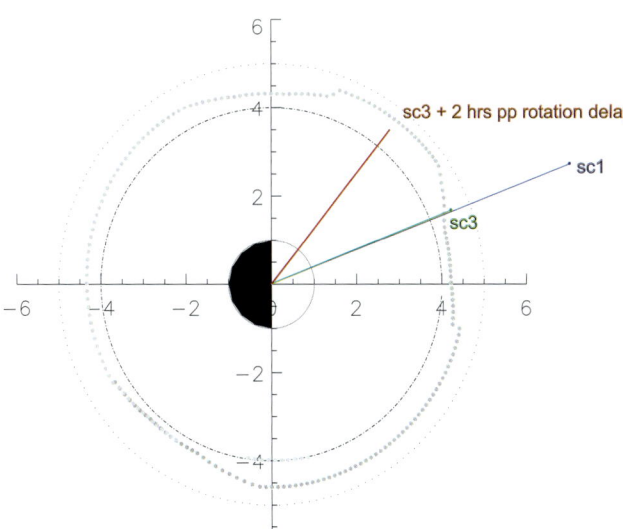

Plate 22. Numerical simulation results of the plasmapause deformations for August 9, 2001, using the interchange instability model. Upper panel: Kp index time history, used as input parameter for the simulation. Middle and bottom panels: simulation results of the plasmapause deformations, in the equatorial plane, at 04:21 UT and at 05:43 UT. The blue dot corresponds to the Cluster sc 1 spacecraft position and the green to sc 3. The red corresponds to the relative position of sc 1 (middle panel) and sc 3 (bottom panel), if a 2 hours delay is introduced in the plasmasphere rotation (see text).

48 IONIC STRUCTURES IN THE PLASMASPHERE

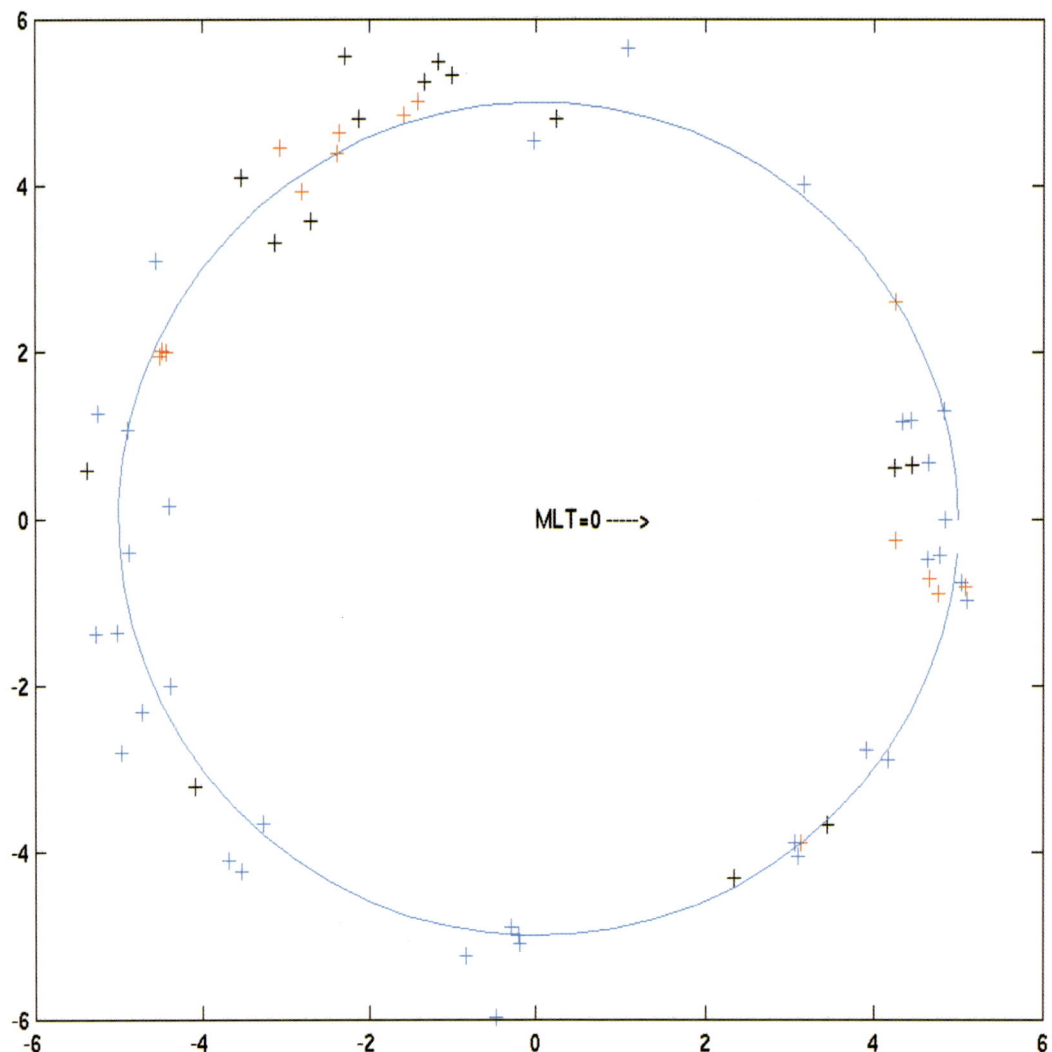

Plate 23. Observed plasmapause positions by the CIS experiment onboard Cluster, for the period July 2001–March 2003, in an L–MLT azimuthal plot (local noon at left). Black crosses correspond to Kp ≤ 1 events, blue to 1 < Kp < 3, and red to Kp ≥ 3. The blue circle corresponds to L = 5.

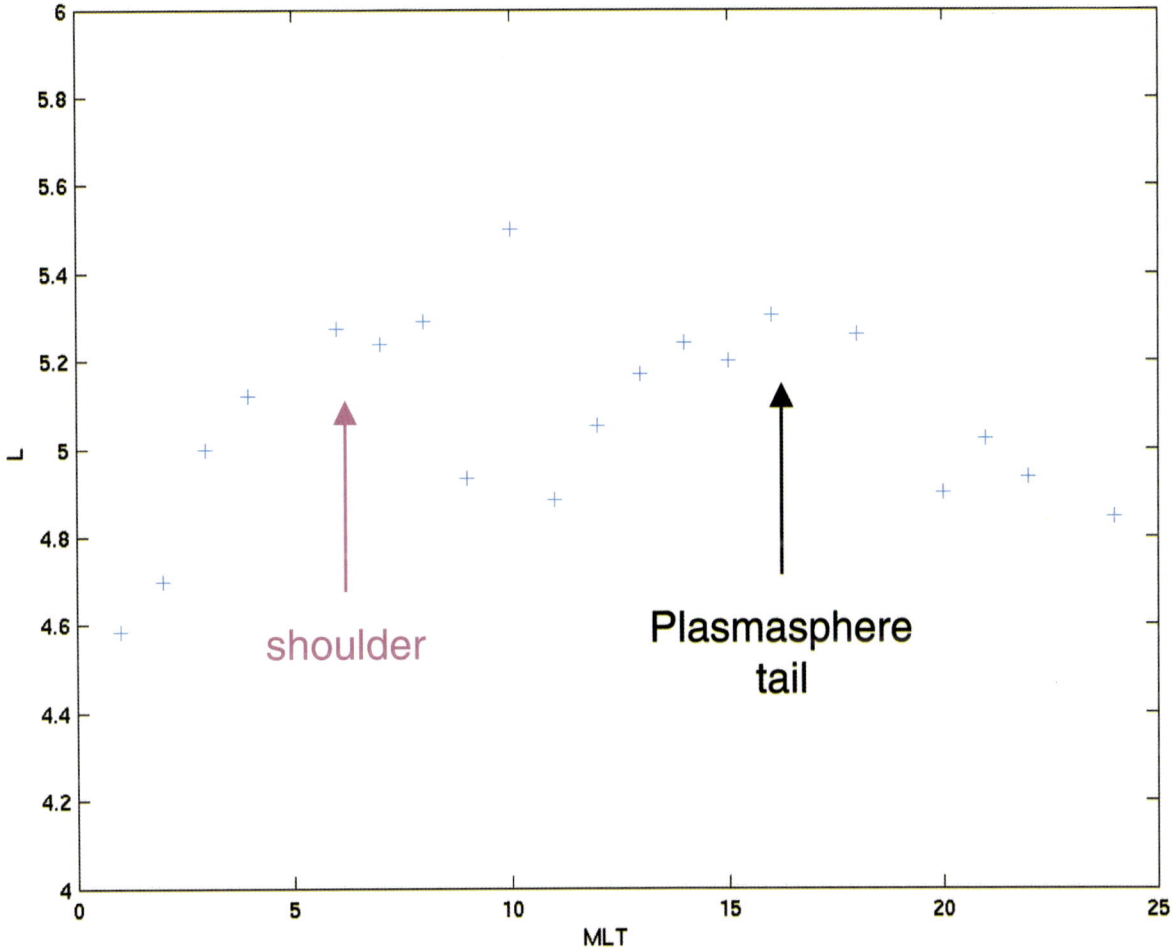

Plate 24. Average plasmapause L-shell positions (all Kp) observed by CIS, as a function of MLT.

50 IONIC STRUCTURES IN THE PLASMASPHERE

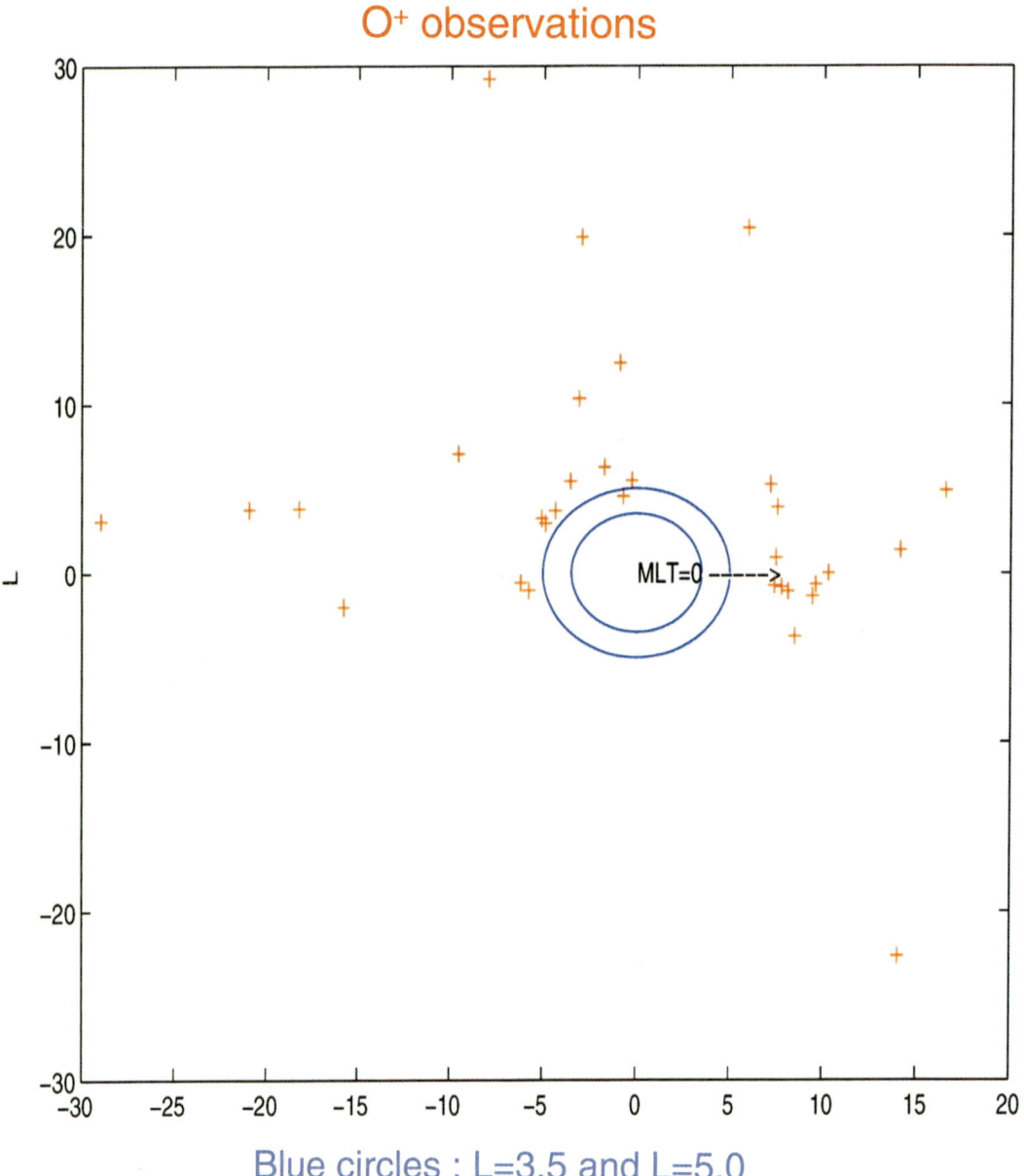

Plate 25. Observed low-energy O^+ ions by CIS-RPA, for the period July 2001–March 2003, in an L – MLT azimuthal plot (local noon at left). The blue circles correspond to L = 3.5 and to L = 5.0.

respond to the Kp ≤ 1 events, blue to 1 < Kp < 3, and red to Kp ≥ 3. These results show the plasmasphere slightly more compressed in the night sector and extended outwards in the noon sector, consistently with the Image-EUV observations [*Sandel et al., 2003*], and with the plasmapause observations by the Whisper resonance sounder experiment onboard Cluster [*Darrouzet et al., 2003*]. The plasmasphere erosion during high Kp activity appears as reduced L-shell values of the observed plasmapause positions.

Our results also show the characteristic "shoulder" in the 6–8 MLT sector (observations in the upper/upper-left part of Plate 23), which is an asymmetric bulge in the plasmapause and which was seen for the first time in the EUV images [*Burch et al., 2001*]. This is more clearly seen in Plate 24, which shows the average L-shell values of the plasmapause as a function of MLT (one average per MLT-hour). The formation in the afternoon-dusk sector of the plasmaspheric tail, or drainage plume [*Carpenter, 1983; Lemaire, 2000; Sandel et al., 2003*], is another feature that appears also statistically in Plate 24. It should be noted, however, that our observations do not allow us to evaluate the full spatial extent of these structures, due to the limitations introduced by the Cluster spacecraft orbit (sampling effect).

The observations by CIS-RPA of low-energy O^+ ions, for the period July 2001–March 2003, in an L – MLT azimuthal plot, are shown in Plate 25. All these observations are outside the main plasmasphere, and most of them correspond to upwelling ions, escaping from the ionosphere along the magnetic field lines (cf. November 12, 2001 event). For few of these events, however, the O^+ distributions are bi-directional, indicating detached plasma, originating from deeper in the plasmasphere, and having an outwards expansion velocity. This was for example the case during the October 2, 2001 event, around 23:20 UT, when sc 4 observed in the morning sector (09:45 MLT, L ≈ 6) detached plasma, including O^+ ions, presenting symmetric bi-directional pitch-angle distributions and having an outwards expansion velocity of ~3 km/s (not shown). We should notice, however, that O^+ ions were never observed in the main plasmasphere, at the Cluster altitudes (perigee at about 4 R_E).

4. DISCUSSION AND CONCLUSIONS

We have analyzed some typical events, representative of the ionic structures observed in or close to the outer plasmasphere by the CIS experiment onboard the Cluster spacecraft [*Rème et al., 2001*]. A statistical study has been also performed on these observations. Our data allow us to reconstruct statistically the plasmapause morphology and dynamics, but they also reveal new and interesting features. From our analysis, we can conclude that:

- The H^+ and He^+ ions show mostly similar density profiles, with the He^+ densities being lower by a factor of ~15.
- O^+ ions, however, are not observed as part of the main plasmaspheric population at the Cluster orbit altitudes ($R \geq 4\ R_E$).
- Detached plasmasphere events, that are observed by CIS during some of the passes at about 0.5 to 1 R_E outside of the plasmapause, are also present. The symmetric bi-directional pitch-angle distribution functions of these detached plasmaspheric populations allow us to distinguish them from upwelling ion populations.
- The density values measured in the detached plasmasphere observations are by about an order of magnitude lower than the ones measured in the main plasmasphere, consistently with the substantial density reduction for the detached plasma shells predicted by the peeling-off models [*Lemaire, 2001*].
- The plasmasphere co-rotation with the Earth is observed in the ion distribution functions, acquired within the main plasmasphere. In the detached plasmasphere observations, however, the plasma is not corotating, but has a strong outward expansion velocity, which is increasing as a function of the L-shell value.
- The pitch-angle distributions in the main plasmasphere are relatively isotropic, while around the crossing of the equatorial plane the major part of the proton population is centred at 90° (pancake distributions). At higher latitudes and in the detached plasmasphere observations, however, the distributions are of the butterfly type, presenting a deficiency of particles perpendicular to the magnetic field direction.
- Low-energy ($E < 25$ eV) O^+ is observed only as upwelling ions, escaping from the ionosphere along auroral field lines. O^{++} can be also observed in these upwelling ion populations. Furthermore, in few cases low-energy O^+ ions have been observed as detached plasma, showing symmetric bi-directional pitch-angle distributions and having an outwards expansion velocity.
- The CIS-RPA observations of the plasmapause position have been simulated with an interchange instability numerical model for the plasmapause deformations [*Pierrard and Lemaire, 2004*], that uses as input an empirical Kp-dependent equatorial electric field model [*McIlwain, 1986*]. The numerical model reproduces in a very satisfactory way the CIS observations.
- The CIS local ion measurements have been also correlated with global images of the plasmasphere, obtained by the EUV instrument onboard Image [*Sandel et al., 2000*], for an event where the Cluster spacecraft were within the field-of-view of EUV. The EUV images show, for this event, that the difference observed between two Cluster

spacecraft was temporal (boundary motion): the radial density profile of the plasmasphere varies with MLT, and a more extended radial profile "rotated" in between the two Cluster spacecraft perigee passes. They thus show the necessity for correlating local measurements with global images, and the complementarity of the two approaches; local measurements giving the "ground truth" (including plasma composition, distribution functions etc.) and global images allowing to put local measurements into a global context, and to deconvolve spatial from temporal effects.

Acknowledgments. V. Pierrard thanks the Belgian Science Policy SPP for the grant Action 1 MO/35/010. Support at UC Berkeley is from a NASA Cluster grant. The authors are grateful to the PI of the EUV experiment, Bill R. Sandel. The Kp index was provided by the GeoForschungsZentrum, Potsdam, and the AE index by the World Data Center for Geomagnetism, Kyoto.

REFERENCES

Balogh, A., C. M. Carr, M. H. Acuña, M. W. Dunlop, T. J. Beek, P. Brown, K.-H. Fornaçon, E. Georgescu, K.-H. Glassmeier, J. Harris, G. Musmann, T. Oddy, K. Schwingenschuh, The Cluster Magnetic Field Investigation: overview of in-flight performance and initial results, *Ann. Geophys., 19,* 1207, 2001.

Bezrukikh, V. V., M. I. Verigin, G. A. Kotova, L. A. Lezhen, Yu. I. Venediktov, J. Lemaire, Dynamics of the plasmasphere and plasmapause under the action of geomagnetic storms, *J. Atmosph. Sol.-Terr. Phys., 63,* 1179–1184, 2001.

Burch, J. L., (Ed), *The IMAGE Mission*, Kluwer Acad., Norwell, Mass., 2000 (Reprinted from Space Sci. Rev., 91), 2000.

Burch, J. L., D. G. Mitchell, B. R. Sandel, P. C. Brandt, M. Wüest, Global dynamics of the plasmasphere and ring current during magnetic storms, *Geophys. Res. Lett., 28,* 6, 1159–1162, 2001.

Carpenter, D. L., Some aspects of plasmapause probing by whistlers, *Radio Sci., 18,* 917–925, 1983.

Chappell, C. R., T. E. Moore, and J. H. Waite, Jr., The ionosphere as a fully adequate source of plasma for the earth's magnetosphere, *J. Geophys. Res., 92,* 5896–5910, 1987.

Comfort, R. H., Thermal structure of the plasmasphere, *Adv. Space Res., 17,* (10) 175–(10) 184, 1996.

Craven, P. D., D. L. Gallagher, and R. H. Comfort, Relative concentration of He^+ in the inner magnetosphere as observed by the DE1 retarding ion mass spectrometer, *J. Geophys. Res., 102,* 2279–2289, 1997.

Darrouzet, F., P. M. E. Décréau, J. Lemaire, A. Masson, E. Le Guirriec, J. G. Trotignon, J. L. Rauch, P. Canu, F. Sedgemore, and M. André, Density irregularities at the plasmapause: Cluster observations, *EGS - AGU - EUG Joint Assembly,* Nice, France, 6–11 April 2003.

Décréau, P. M. E., P. Fergeau, V. Krasnoselskikh, E. LeGuirriec, M. Lévêque, Ph. Martin, O. Randriamboarison, J. L. Rauch, F. X. Sené, H. C. Séran, J. G. Trotignon, P. Canu, N. Cornilleau, H. de Féraudy, H. Alleyne, K. Yearby, P. B. Mögensen, G. Gustafsson, M. André, D. C. Gurnett, F. Darrouzet, J. Lemaire, C. C. Harvey, P. Travnicek, and Whisper experimenters, Early results from the Whisper instrument on Cluster: an overview, *Ann. Geophys., 19,* 1241–1258, 2001.

Escoubet, C. P., M. Fehringer, and M. Goldstein, The Cluster mission, *Ann. Geophys., 19,* 1197–1200, 2001.

Goldstein, J., B. R. Sandel, M. R. Hairston, and P. H. Reiff, Control of plasmaspheric dynamics by both convection and sub-auroral polarization stream, *Geophys. Res. Lett., 30,* 24, 2243, doi: 10.1029/2003GL018390, 2003a.

Goldstein J., M. Spasojevic', P. H. Reiff, B. R. Sandel, W. T. Forrester, D. L. Gallagher, and B. W. Reinisch, Identifying the plasmapause in IMAGE EUV data using IMAGE RPI in situ steep density gradients, *J. Geophys. Res., 108,* 1147, doi: 10.1029/2002JA009475, 2003b.

Gustafsson, G., M. André, T. Carozzi, A. I. Eriksson, C.-G. Fälthammar, R. Grard, G. Holmgren, J. A. Holtet, N. Ivchenko, T. Karlsson, Y. Khotyaintsev, S. Klimov, H. Laakso, P.-A. Lindqvist, B. Lybekk, G. Marklund, F. Mozer, K. Mursula, A. Pedersen, B. Popielawska, S. Savin, K. Stasiewicz, P. Tanskanen, A. Vaivads, and J-E. Wahlund, First results of electric field and density observations by Cluster EFW based on initial months of operation, *Ann. Geophys., 19,* 1219–1240, 2001.

Lemaire, J., The "Roche-limit" of ionospheric plasma and the formation of the plasmapause, *Planet. Space Sci., 22,* 757–766, 1974.

Lemaire, J. F., The formation of plasmaspheric tails, *Phys. Chem. Earth, 25,* 9–17, 2000.

Lemaire, J. F., The formation of the light-ion trough and peeling off the plasmasphere, *J. Atmosph. Sol.-Terr. Phys., 63,* 1285–1291, 2001.

Lemaire, J. F., and K. I. Gringauz, with contributions from D. L. Carpenter and V. Bassolo, *The Earth's Plasmasphere,* Cambridge University Press, Cambridge, 1998.

McIlwain, C. E., A Kp dependent equatorial electric field model, The Physics of Thermal plasma in the magnetosphere, *Adv. Space Res., 6(3),* 187–197, 1986.

Moore, T. E., M. Lockwood, M. O. Chandler, J. H. Waite, Jr., C. R. Chappell, A. Persoon, and M. Sugiura, Upwelling O^+ ion source characteristics, *J. Geophys. Res., 91,* 7019–7031, 1986.

Moullard, O., A. Masson, H. Laakso, M. Parrot, P. Décréau, O. Santolik, and M. André, Density modulated whistler mode emissions observed near the plasmapause, *Geophys. Res. Lett., 29,* 20, 1975, doi: 10.1029/2002GL015101, 2002.

Pierrard, V., and J. F. Lemaire, Development of shoulders and plumes in the frame of the interchange instability mechanism for plasmapause formation, *Geophys. Res. Lett., 31,* L05809, doi: 10.1029/2003GL018919, 2004.

Rème, H., C. Aoustin, J. M. Bosqued, I. Dandouras, B. Lavraud, J. A. Sauvaud, A. Barthe, J. Bouyssou, Th. Camus, O. Coeur-Joly, A. Cros, J. Cuvilo, F. Ducay, Y. Garbarowitz, J. L. Medale, E. Penou, H. Perrier, D. Romefort, J. Rouzaud, C. Vallat, D. Alcaydé, C. Jacquey, C. Mazelle, C. d'Uston, E. Möbius, L.M. Kistler, K. Crocker, M. Granoff, C. Mouikis, M. Popecki, M. Vosbury, B. Klecker, D. Hovestadt, H. Kucharek, E. Kuenneth, G. Paschmann,

M. Scholer, N. Sckopke, E. Seidenschwang, C. W. Carlson, D. W. Curtis, C. Ingraham, R. P. Lin, J. P. McFadden, G. K. Parks, T. Phan, V. Formisano, E. Amata, M. B. Bavassano-Cattaneo, P. Baldetti, R. Bruno, G. Chionchio, A. Di Lellis, M.F. Marcucci, G. Pallocchia, A. Korth, P. W. Daly, B. Graeve, H. Rosenbauer, V. Vasyliunas, M. McCarthy, M. Wilber, L. Eliasson, R. Lundin, S. Olsen, E. G. Shelley, S. Fuselier, A. G. Ghielmetti, W. Lennartsson, C. P. Escoubet, H. Balsiger, R. Friedel, J-B. Cao, R. A. Kovrazhkin, I. Papamastorakis, R. Pellat, J. Scudder, and B. Sonnerup, First multispacecraft ion measurements in and near the earth's magnetosphere with the identical CLUSTER Ion Spectrometry (CIS) Experiment, *Ann. Geophys., 19,* 1303–1354, 2001.

Sandel, B. R., A. L Broadfoot, C. C Curtis, R. A. King, T. C. Stone, R. H. Hill, J. Chen, O. H. W. Siegmund, R. Raffanti, D. D. Allred, R.S. Turley, and D.L. Gallagher, The extreme ultraviolet imager investigation for the IMAGE mission, *Space Science Reviews, 91,* 197–242, 2000.

Sandel, B. R., J. Goldstein, D. L. Gallagher, and M. Spasojević, Extreme ultraviolet imager observations of the structure and dynamics of the plasmasphere, *Space Sci. Rev., 109,* 25–46, 2003.

Sauvaud, J. A., P. Louarn, G. Fruit, H. Stenuit, C. Vallat, J. Dandouras, H. Rème, M. André, A. Balogh, M. Dunlop, L. Kistler, E. Möbius, C. Mouikis, B. Klecker, G. K. Parks, J. McFadden, C. Carlson, F. Marcucci, G. Pallocchia, R. Lundin, A. Korth, and M. McCarthy, Cases studies of the dynamics of ionospheric ions in the Earth's magnetotail, *J. Geophys. Res., 109,* A01212, 10.1029/2003JA009996, 2004.

Torkar, K., W. Riedler, C. P. Escoubet, M. Fehringer, R. Schmidt, R. J. L. Grard, H. Arends, F. Rüdenauer, W. Steiger, B. T. Narheim, K. Svenes, R. Torbert, M. André, A. Fazakerley, R. Goldstein, R. C. Olsen, A. Pedersen, E. Whipple, and H. Zhao, Active spacecraft potential control for Cluster—implementation and first results, *Ann. Geophys., 19,* 1289–1302, 2001.

Truhlík, V., Theoretical interpretation of thermal O^{++} ions density enhancement in the midlatitude outer ionosphere, *Adv. Space Res., 20,* 3, 419–423, 1997.

Vallat, C., I. Dandouras, P. C:son Brandt, D. G. Mitchell, E. C. Roelof, R. deMajistre, H. Rème, J.-A. Sauvaud, L. Kistler, C. Mouikis, M. Dunlop, and A. Balogh, First comparisons of local ion measurements in the inner magnetosphere with ENA magnetospheric image inversions: Cluster-CIS and IMAGE-HENA observations, *J. Geophys. Res., 109,* A04213, 10.1029/2003JA010224, 2004.

Wilber, M., G. Parks, H. Rème, L. Kistler, I. Dandouras, M. McCarthy, C. Carlson, J. McFadden, and J. A. Sauvaud, Results from initial Cluster/CIS survey of O^{++}, *EGS - AGU - EUG Joint Assembly,* Nice, France, 6–11 April 2003.

M. B. Bavassano-Cattaneo, CNR-IFSI, Via del Fosso del Cavaliere, 00133 Roma, Italy. (bice@ifsi.rm.cnr.it)

I. Dandouras, C. Gouillart, H. Rème, F. Sevestre and C. Vallat, Centre d'Etude Spatiale des Rayonnements, 9 Ave. du Colonel Roche, B.P. 4346, F-31028 Toulouse Cedex 4, France. (Iannis.Dandouras@cesr.fr; Claire.Vallat@cesr.fr; Henri.Reme@cesr.fr; Cecile.Gouillart@cesr.fr; Frederic.Sevestre@cesr.fr)

P. Escoubet and A. Masson, ESA/ESTEC (SCI-RSSD), Postbus 299, Keplerlaan, 1, 2200 AG Noordwijk, The Netherlands. (Philippe.Escoubet@esa.int; Arnaud.Masson@esa.int)

J. Goldstein, Space Science and Engineering Division, Southwest Research Institute, San Antonio, TX 78238, USA. (jgoldstein@swri.edu)

L. M. Kistler, Space Science Center, University of New Hampshire, Durham, NH 03824, USA. (lynn.kistler@unh.edu)

B. Klecker, Max-Planck-Institut für Extraterrestrische Physik, Giessenbachstrasse, D-85748 Garching, Germany. (berndt.klecker@mpe.mpg.de)

A. Korth, Max-Planck-Institut für Sonnensystemforschung, Max-Planck-Str. 2, D-37191 Katlenburg-Lindau, Germany. (korth@linmpi.mpg.de)

M. McCarthy, Geophysics, Box 351650, University of Washington, Seattle, WA 98195-1650, USA. (mccarthy@geophys.washington.edu)

G. K. Parks, Space Sciences Laboratory, University of California, Berkeley, CA 94720-7450, USA. (parks@ssl.berkeley.edu)

V. Pierrard, Belgian Institute for Space Aeronomy, Ringlaan 3, B-1180 Brussels, Belgium. (Viviane.Pierrard@oma.be)

The Relationship Between Plasma Density Structure and EMIC Waves at Geosynchronous Orbit

B. J. Fraser[1], H. J. Singer[2], M. L. Adrian[3], D. L. Gallagher[4], and M. F. Thomsen[5]

Recent IMAGE satellite EUV helium density observations of plasmaspheric plumes extending beyond the plasmapause into the plasma trough region of the magnetosphere have been associated with sub-auroral proton arcs observed by the IMAGE FUV instrument. Also proton precipitation has been associated with electromagnetic ion cyclotron (EMIC) waves seen on the ground as Pc1-2 ultra-low frequency (ULF) waves. This evidence suggests a relationship between plasma plumes, proton precipitation and EMIC waves, and supports the EMIC wave-particle interaction with ring current ions as a possible ring current loss mechanism. Using high-resolution (0.5 s) fluxgate magnetometer data from the GOES-8 and GOES-10 geosynchronous satellites we show two case studies on 9–10 and 26–27 June 2001, where EMIC waves in the 0.1–0.8Hz frequency range are observed within plasma plumes extending to geosynchronous orbit. These plumes are also seen in LANL geosynchronous satellite MPA data. The results suggest that EMIC waves may be preferentially generated in enhanced plasma density created by the plasma plume. The EMIC waves are unstructured and have the properties of the well-known intervals of pulsations with diminishing period (IPDP) seen on the ground and in space in the afternoon sector. They are classical EMIC transverse waves showing left-hand circular and elliptical polarization below the helium cyclotron frequency. The observation of a slot in the wave spectrum suggests the presence of He^+ ions with relative concentrations in the range 6–16%, in a predominantly H^+ plasma. This is consistent with the IMAGE-EUV He^+ observations of plasma plumes.

---

[1]School of Mathematical and Physical Sciences, CRC for Satellite Systems, University of Newcastle, Callaghan, New South Wales, Australia.
[2]NOAA Space Environment Center, Boulder, Colorado, USA.
[3]Laboratory for Extraterrestrial Physics, Interplanetary Physics Branch, NASA Goddard Space Flight Center, Greenbelt, Maryland, USA.
[4]Science Directorate, NASA Marshall Space Flight Center, Huntsville, Alabama, USA.
[5]Los Alamos National Laboratories, Los Alamos, New Mexico, USA

Inner Magnetosphere Interactions: New Perspectives from Imaging
Geophysical Monograph Series 159
Copyright 2005 by the American Geophysical Union.
10.1029/159GM04

1. INTRODUCTION

Plasma waves with frequencies below about 5 Hz are ubiquitous in the Earth's magnetosphere and plasmasphere. In common with most magnetospheric processes the wave energy is either directly or indirectly transmitted from the solar wind. Internal sources of energy include the radiation belts and ring current where instability is associated with cyclotron, bounce and drift motion of particles whose distributions are anisotropic. These energy sources may be pressure gradients, velocity shears or rapid changes in magnetospheric geometry associated with storm and substorm processes (see reviews by *Samson*, 1991; *Kangas et al.*, 1998). In this paper we report on one genre of plasma waves that pervades the magnetosphere and plasmasphere.

These are the electromagnetic ion cyclotron (EMIC) waves generated in the middle magnetosphere by ring current wave-particle interaction and observed on the ground as Pc1-2 (0.1–5 Hz) ultra-low frequency (ULF) waves. These waves are somewhat localised in the magnetosphere and their morphology is described in papers by *Anderson et al.* (1992) and *Fraser and Nguyen* (2001). The motivation for the study of EMIC waves relates to the long realized need to obtain experimental evidence of the enhancement of EMIC wave activity in association with enhanced plasma density regions in the plasmatrough. This has become possible using high-resolution (0.5 s) magnetometer data from the GOES geosynchronous satellites and the associated IMAGE satellite extreme ultraviolet (EUV) imager remote observations of He^+ plasma density. These data are supplemented with thermal ion plasma density data, predominantly H^+, from the LANL geostationary satellites.

The relationship between the occurrence of EMIC waves in the magnetosphere, which are seen on the ground as Pc1-2 pulsations (0.1–5 Hz), and cold plasma populations has been a continuing topic of interest since what were previously conjectured to be detached plasma regions were found to occur in the mid-afternoon–midnight sector by OGO 5 (*Chappell et al.*, 1971). These are times when Pc1–2 pulsation activity on the ground maximizes at middle and high latitudes (*Fraser*, 1968). However, the availability of simultaneous EMIC wave and plasma density observations has been somewhat limited in the past. An exception is the CRRES spacecraft where EMIC waves observed by the onboard fluxgate magnetometer can be related to the cold electron density data from the Iowa plasma wave instrument. The results from one particular unpublished event are shown in Figure 1 (N. Meredith, private communication, 2003). EMIC waves are seen to occur in association with an enhanced electron plasma region on 30 April 1991 (DOY 120) over a radial range L = 4.4 to 6.2 in the 2200–0000 UT (1805–1949 MLT) afternoon sector. This plasma profile probably results from CRRES passing through an enhanced plasma density plume, although this cannot be confirmed from a single radial pass. This event is considered in more detail by *Fraser et al.* (1996).

EMIC waves are considered to be generated by wave-particle interaction, involving 10–100 keV protons in the magnetosphere over L ~ 4–9. They propagate as field aligned left-hand mode EMIC wave packets, away from the equatorial generation region down to the topside ionosphere where they are reflected. This process leads to a bouncing wave packet phenomenon which explains the repetitive Pc1–2 fine structure seen in ground signatures (e.g. *Tepley*, 1964). The theoretical basis for increased EMIC wave activity seen in association with enhanced plasmasphere or plasmatrough density has been noted in a number of parameter modelling studies. For example, using linear theory *Kozyra et al.* (1984) noted that an increase in H^+ density from 10 to 500 cm^{-3} in a multi-component plasma provided a three-fold increase in EMIC wave temporal growth rates. It is generally considered that the presence of cold heavy ions (He^+, O^+) in the background magnetospheric plasma may also contribute to an increased growth rate of the EMIC instability. *Hu and Fraser* (1994) found growth rates increased outside the plasmapause in the lower density trough region when He^+ concentrations increased. Consequently, the presence of heavy ions, in addition to an H^+ plasma density increase, may also enhance the generation of EMIC waves seen at synchronous orbit. Typical energies involved here are illustrated in the top panel in Figure 1 and given in more detail by *Meredith et al.* (2003). A necessary condition for EMIC instability is a positive temperature anisotropy $T_{perp}/T_{||} \geq 1$ of the free energy ring current source, namely 10–100 keV protons (*Kennel and Petschek*, 1966). Positive temperature anisotropies may result from substorm injection (*Ishida et al.*, 1987), impulsive magnetospheric compression (*Anderson and Hamilton*, 1993) and polar cusp injections (*Morris and Cole*, 1991). These studies show that considerable experimental and theoretical support exists for enhanced EMIC wave growth associated with increased cold/thermal plasma density.

A brief description of instrumentation and data analysis procedures will be presented, followed by results illustrating the relationship between EMIC waves seen by GOES, and IMAGE-EUV and LANL satellite plasma data emphasizing the role of cold/thermal ions in plasma plumes on the observation and properties of the waves.

2. GOES, IMAGE AND LANL DATA

The triaxial fluxgate magnetometers onboard the GOES series of geosynchronous satellites have over the last decade provided a rich source of data to study ultra-low frequency (ULF) waves. The GOES 8–12 satellites are 3-axis stabilised spacecraft and fluxgate magnetometer data are typically analyzed in the spacecraft coordinate frame. In the spacecraft coordinate system (Hp, He, Hn), Hp is parallel to the Earth's spin axis for zero inclination orbit, or approximately parallel to the field. The He component is defined perpendicular to Hp and directed Earthward. The Hn component completes the orthogonal system and directed eastward. Data from two GOES satellites in 2001 are used in this study, with GOES-8 located at 75°W (geographic) and GOES-10 at 135°W (geographic). The high-resolution data, sampled at 0.5 s are able to provide information on the low end of the electromagnetic ion cyclotron (EMIC) wave spectrum, from 0.1 Hz up to the 1 Hz Nyquist frequency. The response between 0.5–1.0 Hz is

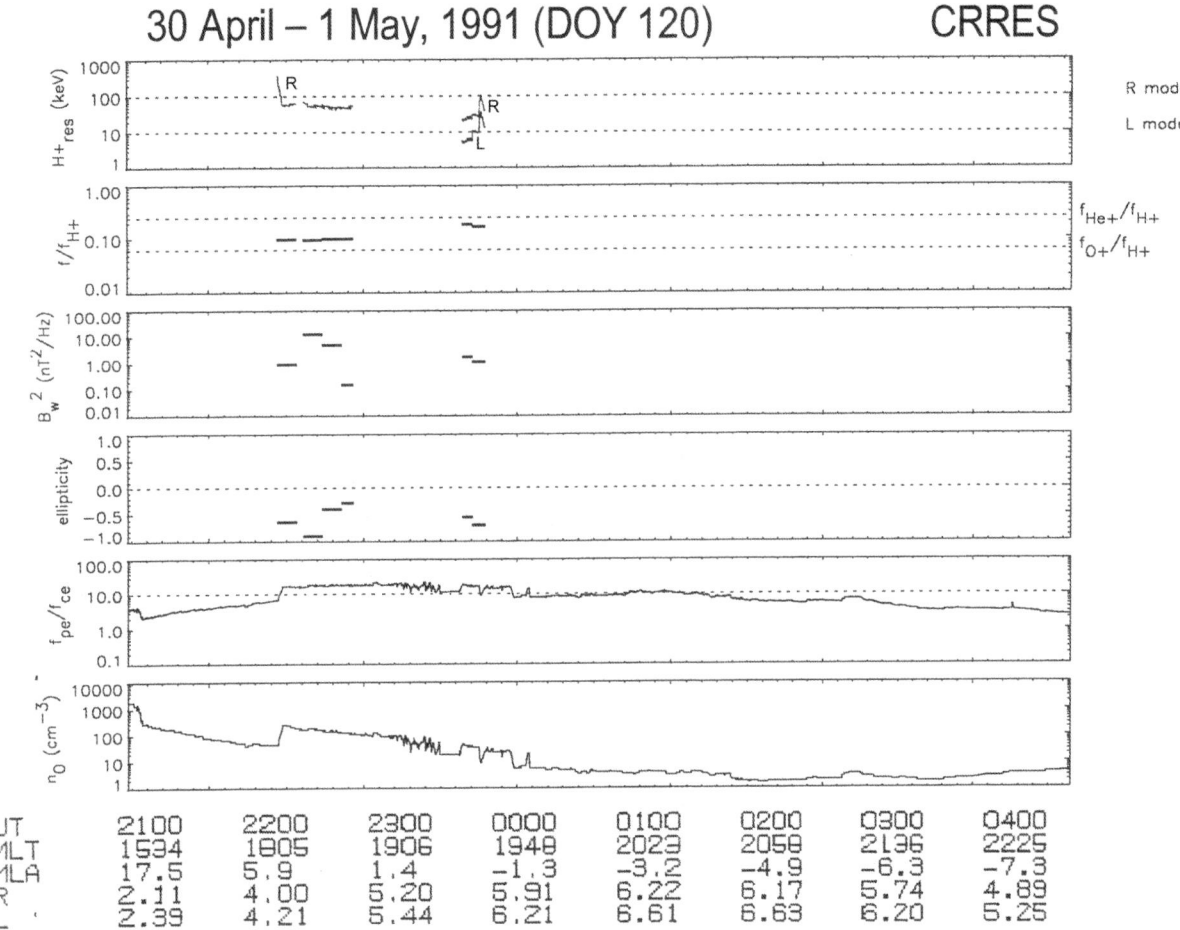

Figure 1. CRRES observations of EMIC waves shown by horizontal lines in the panels 2–4 from the top. Here the ratio of the wave frequency to the local proton cyclotron frequency for times where EMIC waves are observed is plotted in the second panel. The third panel shows transverse EMIC wave energy is between 0.1–10nT^2/Hz and the fourth panel the wave polarization which is left-hand. The fifth panel shows the ratio of the electron plasma frequency to the electron cyclotron frequency. The enhanced density region over L = 4.4–6.2 (2200–0000 UT) is seen in the bottom panel. The top panel shows the parallel resonant energy of the protons generating the EMIC (L) waves assuming a multi-ion plasma with concentrations of 70% H$^+$, 20% He$^+$ and 10% O$^+$. (From N. P. Meredith, private communication).

reduced by a 5-pole Butterworth low-pass anti-aliasing filter. The GOES-9 triaxial fluxgate magnetometer response which is typical for the GOES-8 to 12 satellites is shown in Figure 2. Although the filter 3 dB point is 0.5 Hz, the slow drop off in filter response allows EMIC waves with amplitude ≥ 1 nT to be seen up to ~ 0.8 Hz. Below 0.5 Hz the noise level is ~ 0.1 nT. This frequency response provides an opportunity to observe the lower end of the EMIC wave spectrum which is generally below ~ 1 Hz in the local afternoon and evening sector (*Fraser*, 1968; *Anderson et al.*, 1992). The noon–midnight sector is also the region of plasma plume formation and evolution (*Burch et al.*, 2001).

The IMAGE global plasmasphere imaging satellite provided the first global images of plasmaspheric plumes in snapshots taken by the extreme ultraviolet (EUV) instrument. This instrument measures He$^+$ resonance scattering of solar 30.4 nm radiation from the plasmasphere. The integrated intensity of the 30.4 nm emissions provides the column density with a spatial resolution of ~ 0.1 Re and temporal resolution ~ 10 min when seen from apogee at ~ 8 Re (*Burch*, 2000; *Sandel et al.*, 2001). The deduced He$^+$ density threshold is typically equivalent to 30–50 electrons cm^{-3} (*Goldstein et al.*, 2003; *Moldwin et al.*, 2004) or 6–10 He$^+$ cm^{-3}, depending on the He$^+$/H$^+$ ratio. The outer limit of the

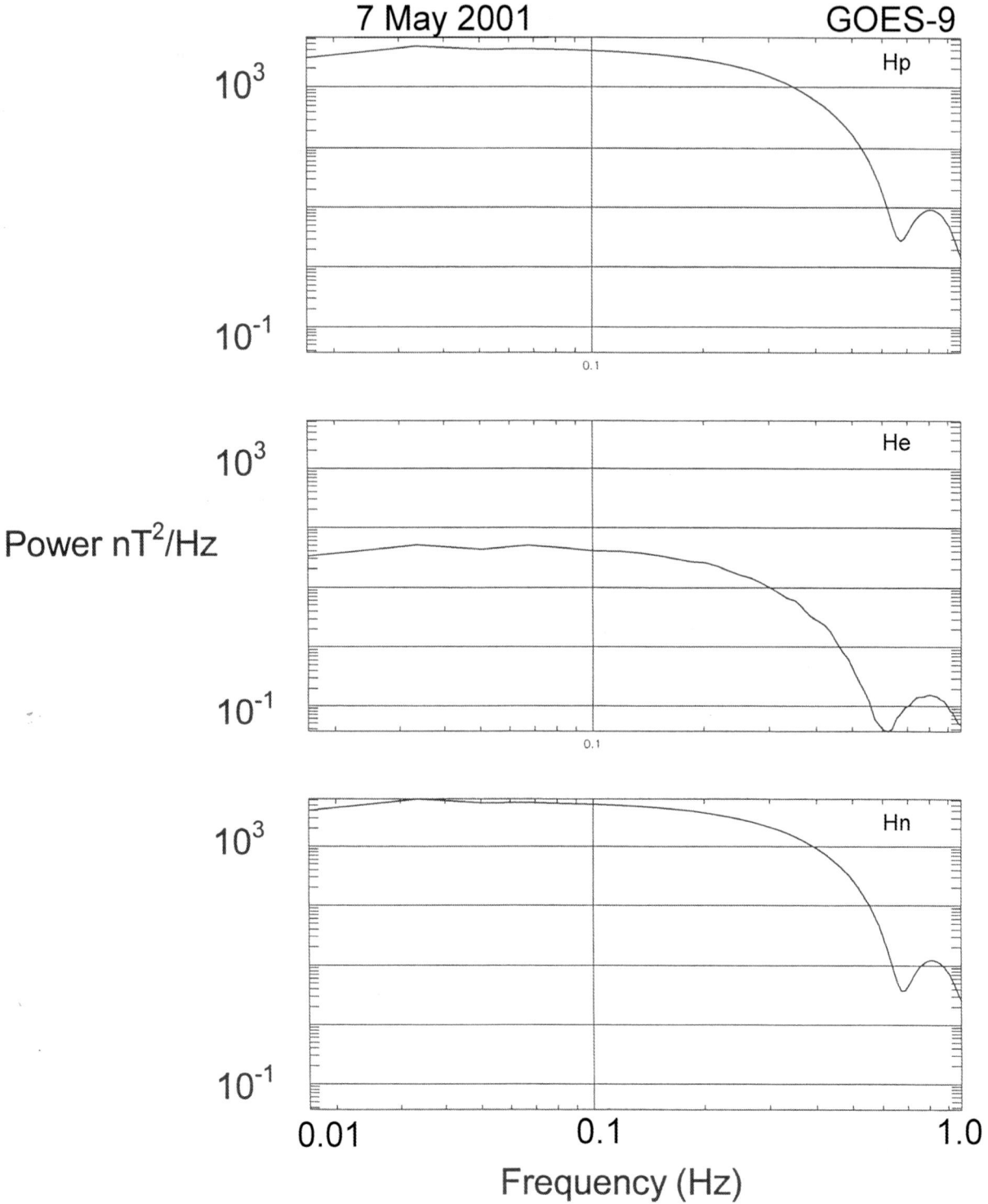

Figure 2. The frequency response of the 3-components of the GOES-9 fluxgate magnetometer in spacecraft coordinates (Hp, He, Hn). GOES-8 and -10 magnetometers have similar responses. The different spectral density in each panel is related to the different input signal amplitudes. All components have equal sensitivity.

EUV field of view varies between 4–8 Re and is frequently at or mostly within geosynchronous orbit.

In situ geostationary plasma density observations were obtained from the Magnetospheric Plasma Analyser (MPA) instruments on board the Los Almos National Laboratories satellites (*McComas et al.,* 1993). Here various data from the MPAs on four of the six LANL satellites, 1990-095 (located at 38.6° W geographic) 1991-080 (164.7°W), 1994-084 (145.5°W), and LANL-01A (8.4°E) have been used. Ion density measurements at energies < 100 eV are of interest and presumed to be predominantly protons, although heavy He^+ and O^+ may be included. The MPA can measure densities < 1 cm^{-3} with 2–3 times better spatial resolution than the EUV He^+ measurements.

3. GOES SPECTRAL ANALYSIS

Specific days relating to plasma plumes observed by IMAGE-EUV were analyzed in the GOES 0.5 s high-resolution magnetometer data. Daily dynamic spectral survey plots in UT of the Hp, He and Hn components over a logarithmic frequency range of 0–1 Hz were used to identify the presence of EMIC (Pc1–2) waves in the 0.1–1 Hz band. They also show Pc3–4 waves in the 10–100 mHz band and Pc5 waves in the 1–10 mHz band. A typical dynamic spectrum illustrating the waves seen in the Hn component at GOES-10 is shown in Plate 1. Here a Hanning window is applied to the time series and FFT's are computed over 20 minute (2400 points) segments of data. With 0.5 s sampled data over a 6 hr data interval this provides a frequency resolution of 0.8 mHz. A burst of EMIC wave activity is seen at 0.4–0.6 Hz around 0700 UT while broad band impulse Pi2 substorm signatures are observed after local midnight at 1000 UT. Pc5 wave activity centered on 5 mHz commenced near dawn and persisted for the remainder of the UT day. Coincidentally, Pc3 harmonic structure showing 7 harmonics in the 10–100 mHz band also commenced near dawn. For more detailed analysis EMIC wave dynamic spectra are displayed in linear frequency format with an FFT window of 2–5 minutes. This provides a spectral resolution of 0.027 Hz with 72 degrees of freedom. Examples of these are included in Plates 2 and 3.

4. EMIC WAVE EVENT STUDIES

The study of the relationship between IMAGE-EUV plasma density and the associated observations of EMIC waves by GOES-8 and 10 was partly motivated by the EUV data and plasmasphere response recently published for two geomagnetically disturbed periods in 2001 (*Spasojević et al.,* 2003). The results from these two intervals are presented below.

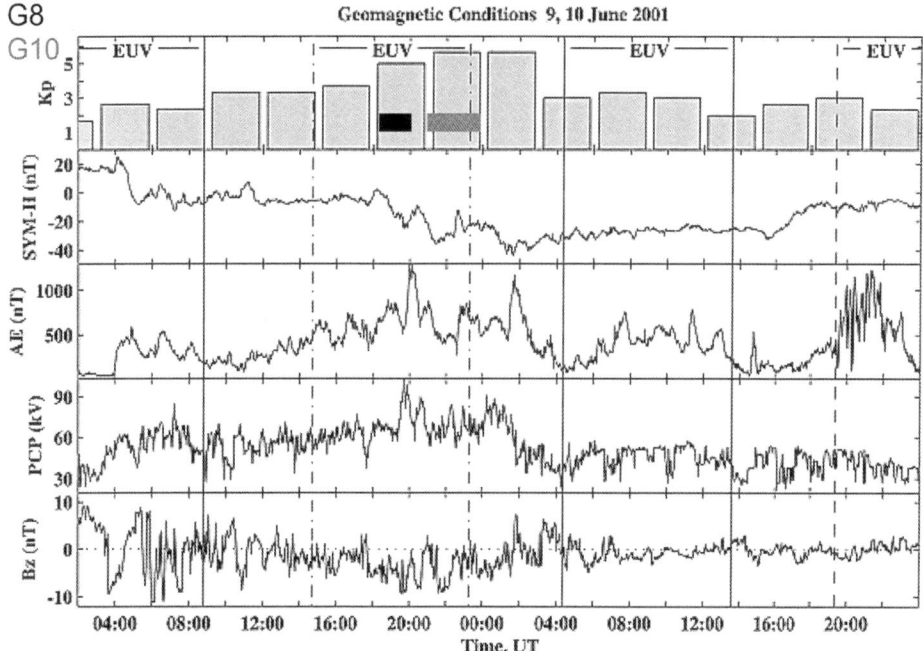

Figure 3. Solar wind and geomagnetic conditions for a 46 hour period over 9–10 June, 2001. (Figure 10 from *Spasojević et al.,* 2003). EMIC waves observed by GOES-8 (1820–2000 UT) and GOES-10 (2100–2400 UT) are indicated by the horizontal bars in the top Kp panel.

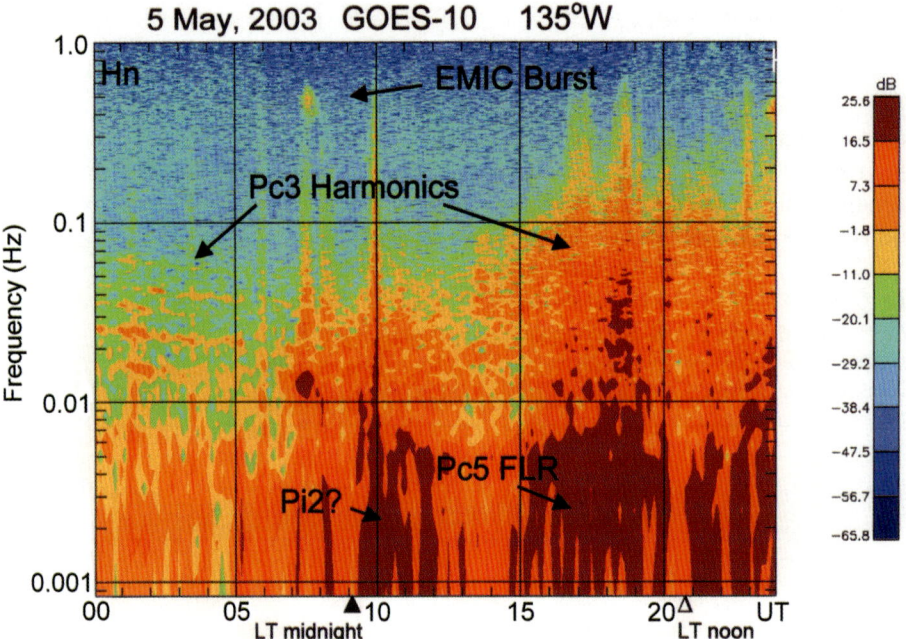

Plate 1. A typical GOES daily dynamic spectrum of the azimuthal (Hn) component showing the various types of waves observed.

9–10 June 2001

Figure 3 reproduces Figure 10 from *Spasojević et al.* (2003) and shows solar wind and geomagnetic conditions for a 46 hour disturbed period over 9–10 June 2001. During this time Kp reached a maximum of 5+ at 21–24 UT on 9 June 2001. The AE index indicates multiple onset substorm activity commencing at ~ 0340 UT on 9 June just following a southward Bz turning. A decrease in SYM-H begins ~ 20 minutes later. SYM-H remains reasonably steady, although fluctuating until the second southward Bz turning around 1800 UT. Four EUV image intervals were described by *Spasojević et al.* and these are indicated in the top of Figure 3. Of particular interest here is the second interval from ~ 1430–2330 UT, where the EUV determined plasmapause locations at 1705 UT and 2121 UT are shown in Figure 4(b) and EMIC waves were seen at GOES. *Spasojević et al.* (2003) showed that while the pre-midday boundary remained reasonably stationary for the previous 9 hours, by 1705 UT in Figure 4(b) the daytime boundary had moved significantly inwards and formed a radially extended plume over 1430–1630 MLT. A sudden surge of plasma sunward on the dayside following the second southward turning at ~ 1740 UT by 2121 UT had moved the western edge of the plume against corotation to an earlier near noon local magnetic time and extended its width to ~ 3 hours over 12–15 MLT. Between the time the two images in Figure 4(b) were taken, GOES-8 (G8) and GOES-10 (G10) passed through the extended plasma plume at geostationary orbit. At ~ 1820 UT EMIC waves appeared at GOES-8 and almost three hours later at GOES-10, in association with SYM-H minima. Over the intervals 1820–2000 UT and 2100–2400 UT GOES-8 and GOES-10 respectively observed EMIC waves in the plume. Dynamic frequency-time spectra of these EMIC waves are shown in Plate 2. GOES-10 over 12–15 LT observed five EMIC wave emissions in the 0.2–0.4 Hz band, at 2110–2118 UT, 2140–2150 UT, 2240–2300 UT, 2315–2325 UT and 2345–2350 UT. GOES-8 observed an emission over 0.1–0.35 Hz at 1850–1920 UT.

The wave properties of the EMIC wave spectral segments illustrated in Plate 2 are shown in Figure 5 where the cross-spectra, coherence and phase between the He (radial) and Hn (azimuthal) magnetic components are plotted. In Figure 5 GOES-10 data over the 2110–2120 UT interval shows a distinctly coherent peak between 0.25–0.42 Hz and a phase difference of −100° to −160°, indicating a left-hand (LH) circular/elliptical polarized wave. The second interval, 2140–2150 UT shows two spectral peaks at 0.20–0.35 Hz and 0.45–0.55 Hz, both with high coherency. The corresponding wave polarisation is LH for the lower frequency peak and a mix of linear and LH for the higher frequency peak. A broad double peak spectral structure is also observed in the third interval, 2235–2305 UT and both bands are LH polarised. Over 2335–2345 UT two narrow peaks are seen over 0.25–0.35 Hz and 0.50–0.55 Hz. They both show LH polarisation where coherence is high. Deductions from the two GOES-8 EMIC events (not shown) indicate for the interval 1850–1920 UT, a broad spectral peak over 0.10–0.35 Hz with high coherency over 0.1–0.6 Hz and almost linear polarisation over this broad band. The very narrow band emission centred on 0.33 Hz in

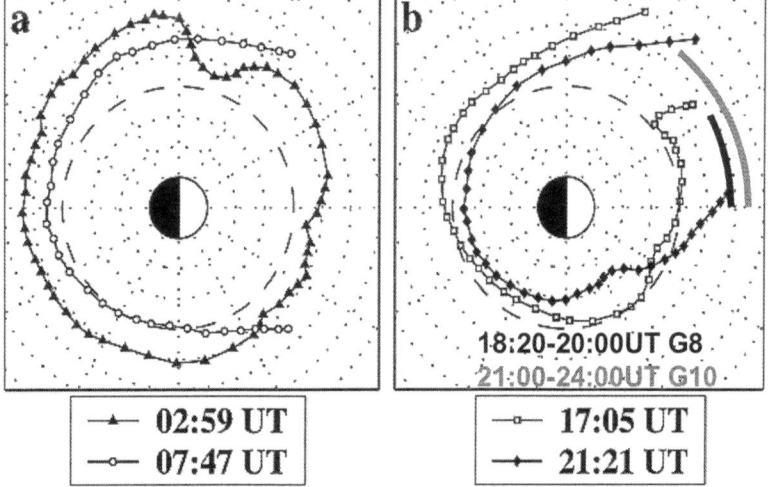

Figure 4. Panels (a) and (b) showing the plasmapause locations during plasma plume development (from Figure 11, *Spasojević et al.*, 2003). The local times of GOES-8 and GOES-10 observations of EMIC waves are associated with the plume development are shown in panel (b).

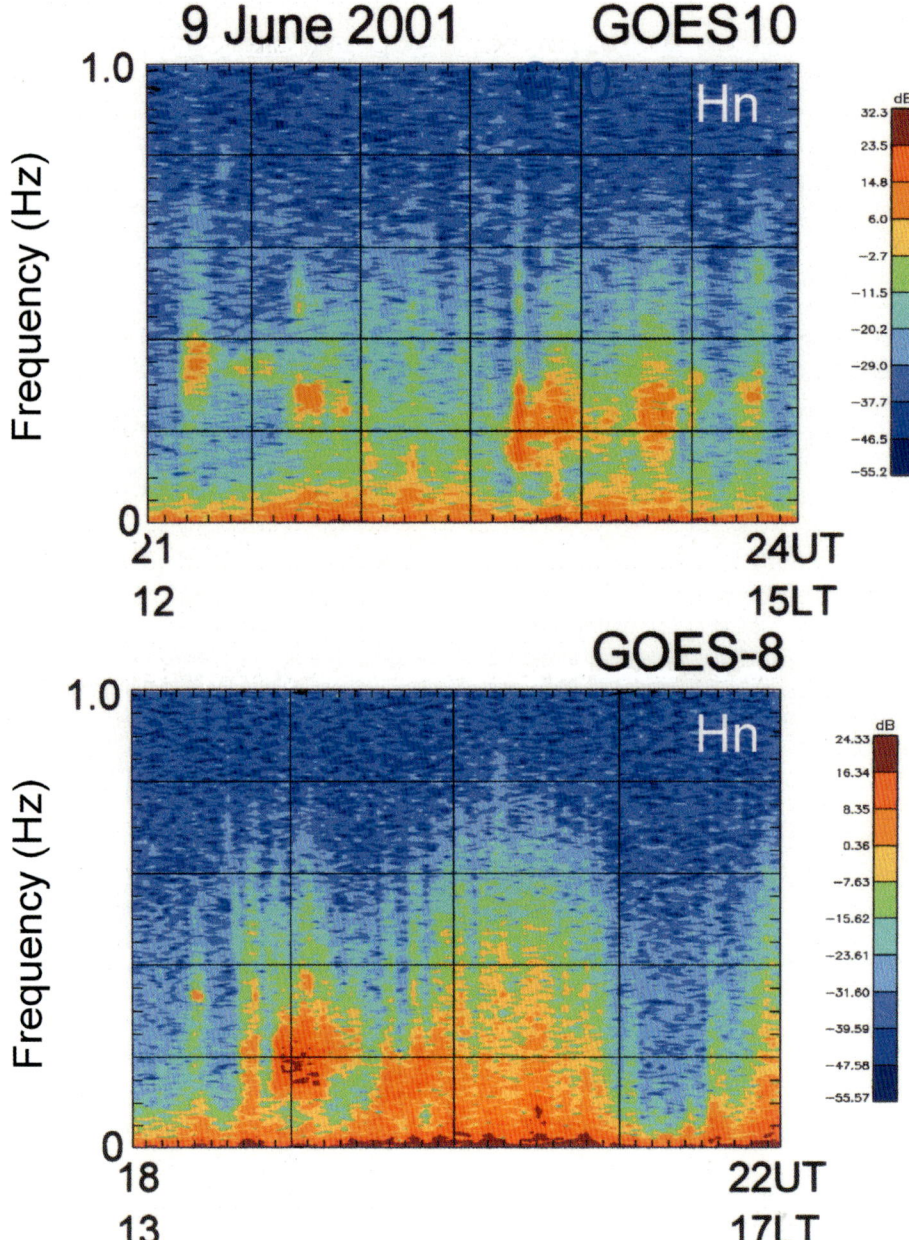

Plate 2. Dynamic spectra showing EMIC waves observed in association with the EUV plasma plume shown in Figure 4. Top Panel: Five bursts of EMIC waves are seen by GOES-10 in the 0.2–0.4 Hz band over 2110–2118 UT, 2140–2150 UT, 2240–2300 UT, 2315–2325 UT and 2345–2350 UT. Bottom Panel: EMIC waves seen by GOES-8 in the 0.1–0.35 Hz over 1850–1920UT.

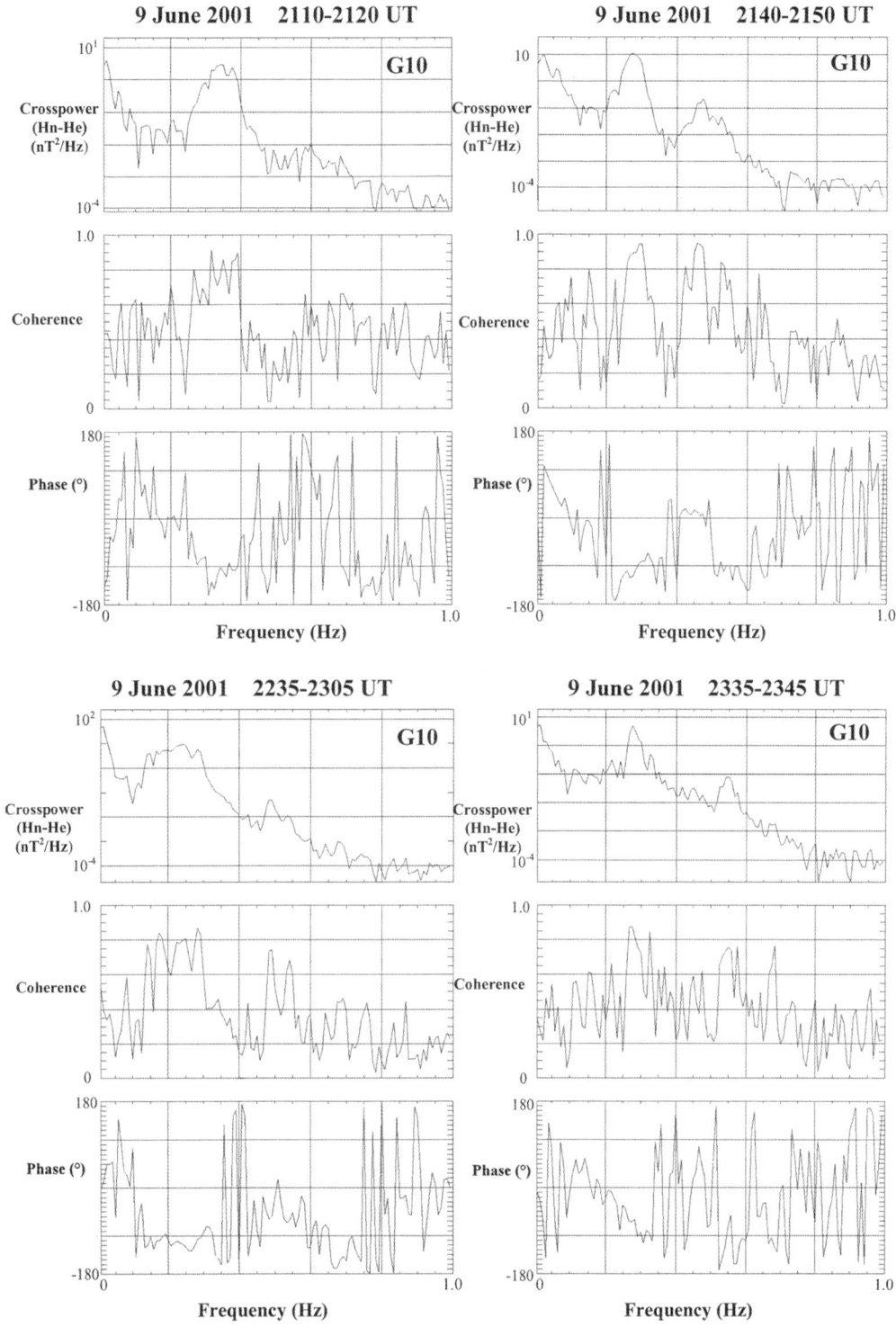

Figure 5. EMIC wave properties for four of the bursts seen by GOES-10 over the intervals shown in Plate 2. The top plot in each of the four intervals shows the transverse wave power spectrum (Hn-He); the middle plot in each panel shows the transverse (Hn-He) coherence and the bottom plot the (Hn-He) phase where negative phase indicates left-hand polarisation.

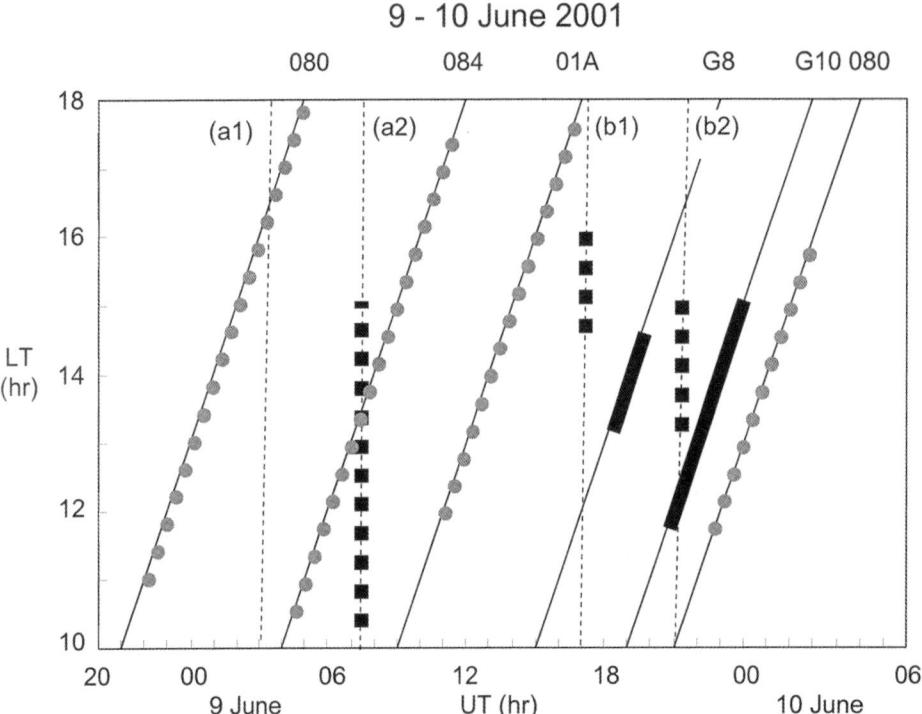

Figure 6. UT-LT plots of geostationary satellite GOES-8 and 10 EMIC waves (solid lines) and LANL 080, 084 and 01A plasma density enhancements (light dot) and IMAGE-EUV enhanced density plumes (dashed squares). Geostationary orbits appear as sloping lines and EUV data as vertical lines at the UT image snapshot time.

the 1820–1830 UT interval (Plate 2) also shows LH polarisation. In general LH polarisation persists over all spectral segments with the one exception of linear polarisation. This LH polarisation is typical for EMIC waves which propagate in a plasma where H^+ ions dominate and heavy ion concentrations including He^+ are minimal (*Mauk et al.,* 1982; *Fraser,* 1985). Also, they can probably be associated with the unstructured IPDP (intervals of pulsations with diminishing period) type of events seen in the afternoon–evening sector, on the ground and in the outer magnetosphere (*Anderson et al.,* 1996). In contrast, the observation of linear polarisation and the double peaked spectra may indicate the presence of a significant population of heavy ions (He^+, O^+) (*Young et al.,* 1981; *Fraser and McPherron,* 1982). This will be discussed later.

A summary of the relationship between EMIC waves observed by GOES, plasma observations from the IMAGE-EUV spatial snapshots and LANL geostationary satellite data is presented in a UT-LT plot in Figure 6. Here the GOES-8 and 10 and LANL-080, 084 and 01A geostationary satellite orbits showing EMIC wave occurrence and MPA cold ion plasma enhancements ($N_i > 10 cm^{-3}$; 1–130eV/q) respectively are plotted on diagonal orbit lines with the snapshot IMAGE plasma density enhancements, extrapolated visually to geostationary orbit, plotted on vertical lines. A broad daytime region of enhanced plasma density is seen by LANL-080, 084 and 01A over 8–9 June prior to the southward turning of Bz which occurred at about 1730 UT 9 June. At 0259 UT 9 June IMAGE-EUV places the plasmapause inside geostationary orbit at L < 6 at all MLT (Figure 4). The greater sensitivity of the LANL satellites with a minimum detectable ion density of ~ 1 cm^{-3} compared with the IMAGE-EUV density at ~ 40 cm^{-3}, shows in the LANL-080 and 084 data which straddle at this time a broad dayside plasma enhancement extending to geostationary orbit. In Figure 6 this broad density region is seen by LANL-080, 084 and 01A satellites while IMAGE at 0747 UT observes the increased density of the region over 1000–1500 LT out to the radial limits of measurement (Figure 4). After 1700 UT both IMAGE-EUV and LANL-080 densities show a broad plume-like structure restricted in MLT in the afternoon sector. EMIC waves occur at times when Bz is southward at -8 nT (Figure 3) and in a well defined plume seen by IMAGE-EUV and a broader plasma structure seen by LANL-01A and 080.

26 June 2001

Another example showing EMIC waves occurring at GOES-10 in the afternoon over 2305–2340 UT (1405–1440

LT) and in association with a narrowing plasma structure, tending towards a plume structure, is shown in Plate 3. Waves were also seen in the same location by GOES-8 at 1855–1940 UT (1355–1440 LT). Again this is an IMAGE-EUV case study discussed by *Spasojević et al.* (2003) from the point of view of global dynamics and plasma density enhancements. Figure 5 in *Spasojević et al.*, reproduced in the two rows of panels in Plate 3(b), shows the evolution of the plasmapause location and the formation of a plume in the afternoon. During a period of sustained southward IMF commencing at about 1220 UT the dayside plasmasphere increased in radial extent with a broad local time coverage (Plate 3(b), panel a). The plume then undergoes a period of narrowing under the southward IMF with the western edge slowly corotating eastwards. Earlier, the LANL-01A ion data in Figure 7 also shows a broad enhanced density region commencing prior to 1200 MLT and extending beyond 18 MLT. In Figure 7 the narrow plume is seen by LANL-080 over 1400–1500 MLT, in agreement with the IMAGE-EUV observations at 0110 UT and 0222 UT. This evolutionary process is seen in panels (b) to (h) in Plate 3(b) and is described in detail by *Spasojević et al.* (2003) and *Goldstein et al.* (2004). EMIC waves are seen by both GOES-8 and GOES-10 as they pass through the broad and narrowing plume some 4 hours apart.

5. MULTI-ION PLASMA EFFECTS

The spectral characteristics of the EMIC wave seen by GOES-10 on 26 June 2001 over 2305–2340 UT in the plume are shown in Figure 8. There are two emission bands, one over 0.22–0.33 Hz and a possible harmonic over 0.43–0.66 Hz. They are LH polarised in both bands and the local He^+ cyclotron frequency at 0.37 Hz measured from the main field falls within the spectral slot. The presence of a harmonic may suggest a situation similar to that seen for EMIC waves observed by CRRES where these waves were often seen to be propagating in plasma slots or biteouts (*Fraser et al.*, 1994). However, there is a more probable explanation here. If a significant concentration of He^+ ions is present in the cold plasma in addition to H^+ then the spectral slot may represent the non-propagation stopband which exists between the He^+ cyclotron frequency (f_{He+}) and the He^+ cutoff frequency (f_{co}). The IMAGE-EUV He^+ observations support this argument. From the relationship $f_{co} = (1 + 3 ß) f_{He+}$ with $f_{co} = 0.45$ Hz from Figure 8 we can estimate the relative He^+ concentration (ß) at ~ 7%, assuming a H^+–He^+ plasma. If f_{He+} was taken as the upper edge of the lower frequency band at 0.32 Hz in Figure 8, rather than the main field calculated value of 0.37 Hz then this places an upper limit of 14% on the relative He^+ concentration. The double peaks observed in

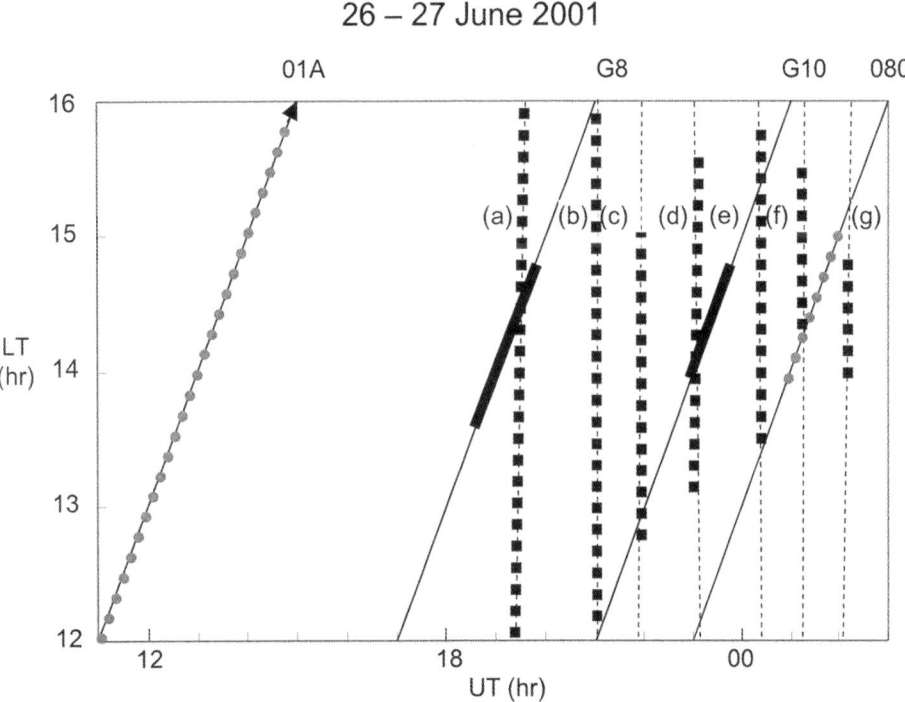

Figure 7. UT-LT plot for 26–27 June 2001 showing GOES EMIC wave observations and LANL satellite enhanced plasma density observations. EUV plumes from panels (a) to (g) are included. See Figure 6 caption for details.

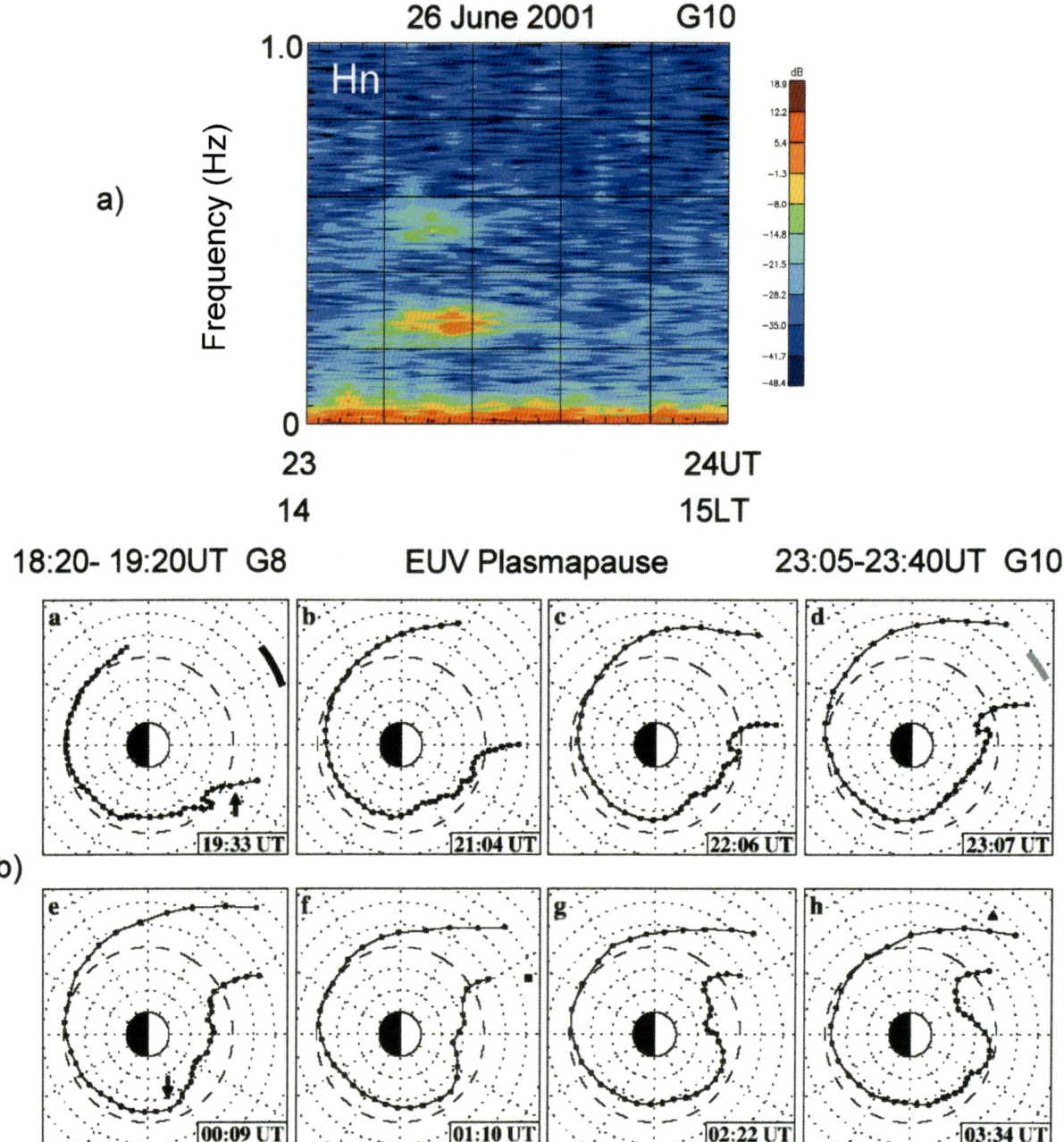

Plate 3. (a) Dynamic Spectrum of the EMIC wave event at 2305–2335 UT, showing two bands separated by a spectral slot. (b) EUV snapshot images showing the plasmapause location and the evolution of a plume over 8 hours [from Figure 5 *Spasojević et al., 2003*]. The local times EMIC waves that were seen by GOES-8 and -10 are also shown in panels (a) and (d).

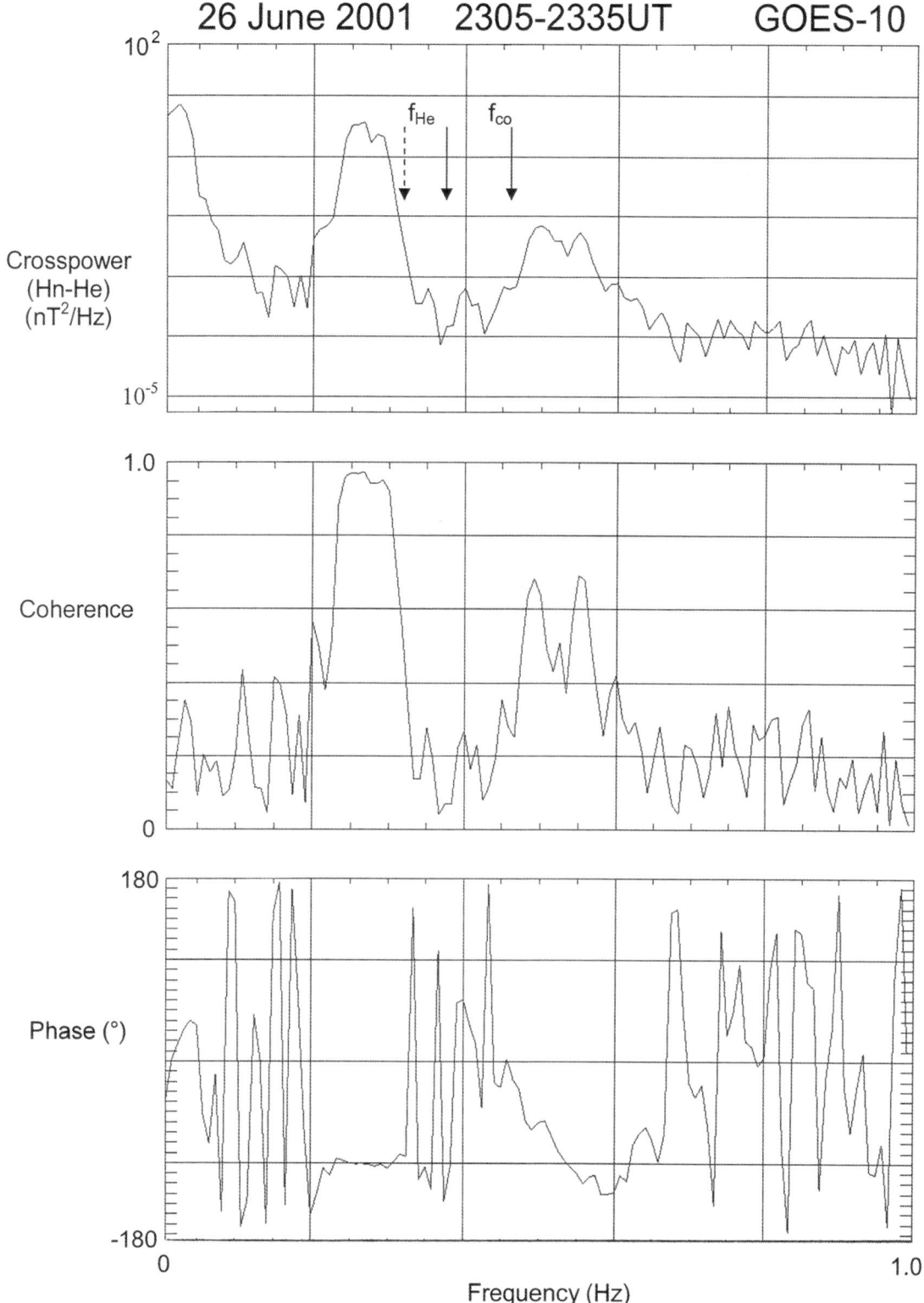

Figure 8. Transverse wave (Hn-He) characteristics for the GOES-10 EMIC event shown in Plate 3(a). See Figure 5 caption for details.

Figure 5 in the GOES-10 spectra on 9 June 2001 may also indicate the non-propagation stopband between f_{He+} and f_{co} as described above. The He$^+$ ion concentration required to explain the spectral slots in the 2140–2150, 2235–2305 and 2335–2345 UT intervals are in the ranges of 5–13%, 4–14% and 6–23%, respectively. The averages of these upper and lower limits of the concentrations of He$^+$ at GOES-10 are 6–16%, in agreement with earlier observations of *Fraser* (1985). For more details on the propagation characteristics of EMIC waves see *Young et al.* (1981), *Fraser* (1985) and *Rauch and Roux* (1982).

6. DISCUSSION AND CONCLUSIONS

This study resulted from an investigation of high-resolution GOES geosynchronous magnetometer data on 9–10 June and 25–26 June 2001 during which the evolution of plasma plumes over two geomagnetically disturbed intervals were studied by *Spasojević et al.* (2003). In both intervals EMIC waves in the 0.1–0.8 Hz band were seen at GOES. The interplay between solar wind induced convection and the corotation electric field produces a consistent plasma plume response (*Goldstein et al., 2004*). This pattern can be seen in the UT-LT plots in Figures 6 and 7 where broad dayside inflated plasma regions, extending over 6–8 hours in LT, narrow and corotate to a width of ~2 hours in LT over ~10 hours in UT. These He$^+$ observations are supported by the higher resolution ion density observations (predominantly H$^+$ ions) from the LANL geosynchronous satellites, which show similar trends in plume evolution. The stability of the plumes in LT over ~4 hours of UT has allowed us to relax the LT-UT coincidence between EMIC wave observations as GOES transits in LT through the UT snapshot IMAGE-EUV images of plumes. For example, GOES-8 in Plate 3(b) sees EMIC waves at 1933 UT while GOES-10 sees waves at 2307 UT at the same LT.

The EMIC wave properties illustrated in Figures 5 and 8 are typical of LH polarized mode transverse waves propagating along the ambient field direction and seen at geosynchronous orbit (e.g. *Fraser,* 1985). The slot between the two spectral peaks has been explained by the presence of a non-propagation stopband introduced by the presence of He$^+$ heavy ions with concentrations in the range 6–16%. The observation of a plasma plume by the IMAGE-EUV instrument confirms the presence of He$^+$ ions in association with the dominant H$^+$ ions seen by the LANL satellites.

The generation and occurrence of EMIC waves in the magnetosphere is primarily controlled by internal magnetosphere processes, in particular the proton cyclotron instability through temperature anisotropy, ring current energy and ambient cold/thermal plasma density. In this context it is important to identify the parameters that increase EMIC wave growth in the magnetosphere in the presence of enhanced plasma density. Early theoretical work confined amplification to the region near the plasmapause (e.g. *Criswell,* 1969). The interpretation of this result in a single ion plasma is that the Alfven velocity ($V_A = B(\mu_0\rho)^{-1/2}$) has a minimum just inside the plasmapause because of the low magnetic field (B) and high cold/thermal plasma density (ρ) (*Fraser et al.,* 1988). Consequently the parallel resonant energy of the interacting protons is lowest and the ring current protons may resonate with the waves (*Hu et al.,* 1990). Also, a low Alfven velocity in the equatorial region allows the waves more time for amplification in the interaction region. *Kozyra et al.* (1984) showed that increasing the cold plasma density in a single ion plasma increased the convective growth rate. For a multi-ion plasma (e.g. H$^+$, He$^+$), *Hu and Fraser* (1994) showed that minimum proton energy can occur on more than one L shell for the same wave frequency. Furthermore, ion cyclotron instability growth rates integrated along the field aligned propagation path maximise over a wide range of L values, both inside and outside the plasmapause. *Hu and Fraser* also showed that EMIC waves in the He$^+$ and H$^+$ wave branches (f < f_{He+} and f_{He+} < f < f_{H+} respectively) are amplified over the outer magnetosphere.

There are a number of points for consideration from these theoretical results with respect to the GOES observations at geosynchronous orbit. Firstly, EMIC waves have been observed in association with enhanced cold/thermal plasma densities in plume structures. As well as affecting the wave growth, increased cold plasma density may shift the unstable frequency range (*Gendrin et al.,* 1984). This could contribute to the low EMIC wave frequencies seen in the GOES data (0.1 < f < 0.8). In contrast to this, the integrated growth rate increases with increasing cold plasma density over all frequencies (f < f_{H+}), while the inclusion of He$^+$ cold ions shows that the integrated growth rate increases with increasing He$^+$ concentration (*Hu and Fraser,* 1994). In this situation He$^+$ ions may contribute to increased EMIC wave instability in plumes.

EMIC wave growth is also influenced by other factors including the ring current ion free energy source, typically in the 10–100 keV range, the anisotropy (A = $T_{perp}/T_{||}$–1) and the background geomagnetic field (*Kozyra et al.,* 1984; *Hu and Fraser,* 1994). We have no measure of these parameters in this study and their consequences will not be discussed further. However, they are expected to influence the amplitude and frequency of EMIC waves seen at GOES and this may explain why the waves are not seen at all times in plasma plumes, although a thorough study of the occurrence of waves in plumes has not yet been carried out.

Another important aspect of the wave-particle interaction is its contribution to ring current decay. It has been postulated for a number of years, and supported by modeling (*Kozyra et al., 1997; Jordanova et al., 2001*) that this mechanism is a significant contributor to the loss of ring current protons during the storm recovery phase. In this scenario the instability produces pitch angle diffusion which leads to the partial refilling of the loss cone, resulting in the precipitation of KeV particles. It is only recently that experimental support for this mechanism has been presented. For example, recent work by *Yahnina et al.* (2003) has shown a close relationship between the morphological features of EMIC waves observed on the ground with localized energetic proton precipitation (~ 30–80 KeV) observed by the NOAA-POES low altitude satellites. Also, *Spasojević et al.* (2004) has shown an association between the occurrence of a plume and a subauroral proton arc where modeling suggested that the arc could have resulted from proton precipitation induced by EMIC wave interaction with ring current ions.

This study supports the earlier elliptically orbiting and geosynchronous satellite results where EMIC waves were seen in plasma islands or detached plasma regions, but has also shown that these regions are now identified as attached plasma plumes in the spatial topology seen by the IMAGE EUV instrument. There are important implications for these results with respect to ring current loss processes where cold plasma convection and electric fields may contribute to these losses through EMIC wave-particle interaction. It is however important to note that not all EMIC wave events observed are associated with plumes or radial plasma structures in the noon–midnight sector. None of the so-called "pearl" pulsation EMIC waves showing a bouncing wave packet fine structure signature have been seen at GOES in association with plumes although they have been seen earlier in the day on a number of occasions.

Future research should concentrate on the relationship between EMIC waves, plasma plumes and other types of radial plasma structures, proton arcs and precipitation in order to further understand the role of EMIC waves in ring current loss mechanisms. It is also important to differentiate between instability parameters relating to EMIC waves observed within plasma plumes compared with those observed at other local times.

Acknowledgments. This research was carried out with financial support from the Commonwealth of Australia through the Cooperative Research Centres Program and the University of Newcastle. BJF undertook part of this work under a NRC Senior Research Associateship at NOAA/SEC Boulder Colorado. N. P. Meredith MSSL, University College London is thanked for providing the CRRES plot.

REFERENCES

Anderson, B. J., R. E. Erlandson, and L. J. Zanetti, A statistical study of Pc1-2 magnetic pulsations in the equatorial magnetosphere 1. Equatorial occurrence distributions, *J. Geophys. Res.*, *97*, 3075, 1992.

Anderson, B. J., and D. C. Hamilton, Electromagnetic ion cyclotron waves stimulated by modest magnetospheric compressions, *J. Geophys. Res.*, *98*, 11369, 1993.

Anderson, B. J., R. E. Erlandson, M. J. Engebretson, J. Alford, and R. L. Arnoldy, Source region of 0.2 to 1.0 Hz geomagnetic pulsation bursts, *Geophys. Res. Lett.*, *23(7)*, 769, 1996.

Burch, J. L., IMAGE mission overview, *Space Sci. Rev.*, *91* (1/2), 1–14, 2000.

Burch, J. L., D. G. Mitchell, B. R. Sandel, P. C. Brandt, and M. Wuest, Global dynamics of the plasmasphere and ring current during magnetic storms, *Geophys. Res. Lett.*, *28* (6), 1159–1162. 2001.

Chappell, C. R., K. K. Harris, and G. W. Sharp, The dayside of the plasmasphere, *J. Geophys Res.*, *76*, 7632–7647, 1971.

Criswell, D., Pc1 micropulsation activity and magnetospheric amplification of 0.2 to 5.0 Hz hydromagnetic waves, *J. Geophys. Res.*, *74*, 205, 1969.

Fraser, B. J., Temporal variations in Pc1 geomagnetic micropulsations, *Planet. Space Sci.*, *16*, 111, 1968.

Fraser, B. J., Observations of ion cyclotron waves near synchronous orbit and on the ground, *Space Sci. Rev.*, *42*, 357, 1985.

Fraser, B. J., and R. L. McPherron, Pc1-2 magnetic pulsation spectra and heavy ion effects at synchronous orbit: ATS 6 results, *J. Geophys. Res.*, *87*, 4560, 1982.

Fraser, B. J., R. L. McPherron, and C. T. Russell, Radial Alfven velocity profiles in the magnetosphere and their relation to ULF wave field line resonance, *Adv. Space Res.*, *8*, 49, 1988.

Fraser, B. J., H. J. Singer, W. J. Hughes, R. R. Anderson, and J. R. Wygant, Electromagnetic ion cyclotron harmonic waves near the plasmapause, *Adv. Space Res.*, *5*, 255–258, 1994.

Fraser, B. J., H. J. Singer, W. J. Hughes, J. R. Wygant, R. R. Anderson, and Y. D. Hu, CRRES Poynting vector observations of electromagnetic ion-cyclotron waves near the plasmapause, *J. Geophys. Res.*, *101*, 15331, 1996.

Fraser, B. J., and T. S. Nguyen, Is the Plasmapause a Preferred Source Region of Electromagnetic Ion Cyclotron Waves in the Magnetosphere, *J. Atmos. Solar Terrestr. Phys.*, *63*, 1225–1247, 2001.

Gendrin, R. M. A. Ashour-Abdalla, Y. Omura, and K. Quest, Linear analysis of ion cyclotron interaction in a multicomponent plasma, *J Geophys. Res.* 89, 9119, 1984.

Goldstein J., M. Spasojević, P. H. Reiff, B. R. Sandel, W. T. Forrestor, D. L. Gallagher and B. W. Reinisch, Identifying the plasmapause in IMAGE-EUV data using IMAGE-RPI in-situ steep density gradients, *J. Geophys. Res.*, *108*, 1147, doi.1029/2002JA004975, 2003.

Goldstein, J., B. R. Sandel, M. F. Thomsen, M. Spasojević, and P. H. Reiff, Simultaneous remote sensing and in situ observations of plasmaspheric drainage plumes, *J. Geophys. Res.*, *109*, A03202, doi:10.1029/2003JA010281, 2004.

Hu, Y.D., and B. J. Fraser, EMIC wave amplification and source regions in the magnetosphere, *J. Geophys. Res., 99*, 263–272, 1994.

Hu, Y. D., B. J. Fraser, and J. V. Olson, Amplification of electromagnetic ion cyclotron waves along a wave path in the Earth's multicomponent magnetosphere, *Geophys. Res Lett. 17*, 1053, 1990.

Ishida, J., S. Kokubun, and R. L. McPherron, Substorm effects on spectral structures of Pc1-2 waves at synchronous orbit, *J. Geophys. Res., 92*, 143, 1987.

Jordanova, V. K., C. J. Farrugia, R. M. Thorne, G. V. Khazanov, G. D. Reeves, and M. F. Thomsen, Modeling ring current proton precipitation by electromagnetic ion cyclotron waves during the May 14–16, 1997, storm, *J. Geophys. Res., 106*, 7, 2001.

Kangas, J., A. V. Guglielmi, and O. A. Pokhotelov, Morphology and physics of short-period magnetic pulsations: A review, *Space Sci. Rev., 83*, 435, 1998.

Kennel, C. F., and H. E. Petschek, Limit on instability trapped particle fluxes, *J. Geophys. Res., 71*, 1, 1966.

Kozyra, J. U., V. K. Jordanova, R. B. Horne, and R. M. Thorne, Modelling of the contribution of electromagnetic ion cyclotron (EMIC) waves to stormtime ring current erosion, in *Magnetic Storms, Geophys. Monogr. Ser.*, vol 98, edited by B. T. Tsurutani et al., p187, AGU, Washington DC, 1997.

Kozyra, J. U., T. E. Cravens, A. F. Nagy, E. G. Fontheim, and R. S B. Ong, Effects of energetic heavy ions on electromagnetic ion cyclotron wave generation in the plasmapause region, *J. Geophys. Res., 89*, 2217, 1984.

Mauk, B. H., Helium resonance and dispersion effects on geostationary Alfven/ion cyclotron waves, *J. Geophys. Res., 87*, 9107, 1982.

McComas, D. J., S. J. Bame, B. L. Barraclough, J. R. Donart, R. C. Elphic, J. T. Gosling, M. B. Moldwin, K. R. Moore, and M. F. Thomsen, Magnetospheric plasma analyzer (MPA): Initial three-spacecraft observations from geosynchronous orbit, *J. Geophys. Res., 98*, 13453, 1993.

Meredith, N. P., R. M. Thorne, R. B. Horne, D. Summers, B. J. Fraser, and R. R. Anderson, Statistical analysis of relativistic electron energies for cyclotron resonance with EMIC waves observed on CRRES, *J. Geophys. Res., 108*(A6), 1250, doi:10.1029/2002JA009700, 2003.

Moldwin M. B., B. R. Sandel, M. F. Thomsen and R. C. Elphic, Quantifying global plasmaspheric images with in-situ observations, *Space Sci. Rev.*, in press, 2004.

Morris, R. J., and K. D. Cole, High latitude day-time Pc1-2 continuous magnetic pulsations: A ground signature of the polar cusp and cleft projection, *Planet. Space Sci., 39*, 1473–1491, 1991.

Rauch, J. L., and A. Roux, Ray tracing of ULF waves in a multicomponent magnetospheric plasma: Consequences for the generation mechanism of ion cyclotron waves, *J. Geophys. Res., 87*, 8191, 1982.

Samson, J. C., Geomagnetic pulsations and plasma waves in the Earth's magnetosphere, *Geomagnetism, 4*, 481, 1991.

Sandel, B. R., R. A. King, W. T. Forrester, D. L. Gallagher, A. L. Broadfoot, and C. C. Curtis, Initial results from the IMAGE extreme ultraviolet imager, *Geophys. Res. Lett., 28*, 1439, 2001.

Spasojević, M., J. Goldstein, D. L. Carpenter, U. S. Inan, B. R. Sandel, M. B. Moldwin and B. W. Reinisch, Global response of the plasmasphere to geomagnetic disturbance, *J. Geophys. Res. Vol. 108*,A9, 1340, doi 10.1029/2003JA009987, 2003.

Spasojević, M., H. U. Frey, M. F. Thomsen, S. A. Fuselier, S. P. Gary, B. R. Sandel, and U. S. Inan, The link between a detached subauroral proton arc and a plasmaspheric plume, *Geophys. Res. Lett., 31*, L04803, doi:10.1029/2003GL018389, 2004.

Tepley, L., Low-Latitude Observations of Fine-Structured Hydromagnetic Emissions, *J. Geophys. Res., 69*, 2273, 1964.

Yahnina, T. A., A. G. Yahnin, J. Kangas, J. Manninen, D. S. Evans, A. G. Demekhov, V. Yu. Trakhtengerts, M. F. Thomsen, G. D. Reeves, and B. B. Gvozdevsky, Energetic particle counterparts for pulsations of Pc1 and IPDP types, *Ann. Geophysicae, 21*, 2281, 2003.

Young, D. T., S. Perraut, A. Roux, C. d. Villedary, R. Gendrin, A. Korth, G. Kremser, and D. Jones, Wave-particle interactions near Ω_{He^+} observed on GEOS 1 and 2, 1, Propagation of ion cyclotron waves in He^+ - rich plasma, *J. Geophys. Res., 86*, 6755, 1981.

M. L. Adrian, Laboratory for Extraterrestrial Physics, Interplanetary Physics Branch, NASA Goddard Space Flight Center, Code 130, Greenbelt, Maryland, 20771, USA, (e-mail: mark.l.adrian@nasa.gov)

B. J. Fraser, School of Mathematical and Physical Sciences, CRC for Satellite Systems, University of Newcastle, Callaghan, NSW, 2308, Australia, (e-mail: brian.fraser@newcastle.edu.au)

D. L. Gallagher, Science Directorate, NASA Marshall Space Flight Center, NSSTC/SDSO, 320 Sparkman Drive, Huntsville, Alabama, 35805, USA, (e-mail: dennis.l.gallagher@nasa.gov)

H. J. Singer, NOAA Space Environment Center, 325 Broadway, Boulder, Colorado, 80305-3328, USA, (e-mail: howard.singer@noaa.gov

M. F. Thomsen, Los Alamos National Laboratories, NIS-1, Los Alamos, New Mexico, 87545, USA, (e-mail: mthomsen@lanl.gov)

ULF Waves Associated With Enhanced Subauroral Proton Precipitation

Thomas J. Immel, S. B. Mende, H. U. Frey, J. Patel, and J. W. Bonnell

Space Sciences Laboratory, University of California, Berkeley

M. J. Engebretson

Department of Physics, Augsburg College

S. A. Fuselier

Lockheed-Martin Advanced Technology Center

Several types of sub-auroral proton precipitation events have been identified using the Spectrographic Imager (SI) onboard the NASA-IMAGE satellite, including dayside subauroral proton flashes and detached proton arcs in the dusk sector. These have been observed at various levels of geomagnetic activity and solar wind conditions and the mechanism driving the precipitation has often been assumed to be scattering of protons into the loss cone by enhancement of ion-cyclotron waves in the interaction of the thermal plasmaspheric populations and more energetic ring current particles. Indeed, recent investigation of the detached arcs using the MPA instruments aboard the LANL geosynchronous satellites has shown there are nearly always heightened densities of cold plasma on high-altitude field lines which map down directly to the sub-auroral precipitation. If the ion-cyclotron instability is a causative mechanism, the enhancement of wave activity at ion-cyclotron frequencies should be measurable. It is here reported that magnetic pulsations in the Pc1 range occur in the vicinity of each of 4 detached arcs observed in 2000–2002, though with widely varying signatures. Additionally, longer period pulsations in the Pc5 ranges are also observed in the vicinity of the arcs, leading to the conclusion that a bounce-resonance of ring-current protons with the azimuthal Pc5 wave structure may also contribute to the detached precipitation.

1. INTRODUCTION

The studies of magnetospheric processes operative across a range of time scales, involving plasmas across a broad range of mean energies, are facilitated by the monitoring the thermospheric airglow signatures of precipitating energetic particles originating in the magnetosphere. The glow of the polar aurora is the most obvious thermospheric emission that reflects an enormous variety of processes whose end result is the conversion of energy contained in magnetospheric plasmas to heat and light. Identifying the individual process that produce particular aurorae, discrete or diffuse, dynamic or steady, and related to the precipitation of electrons or protons (or both), remains a top priority for the fields of magnetospheric and auroral physics.

The polar aurora is most simply described as a wide ring of high-altitude atmospheric emissions which encircles the magnetic poles of the earth. The aurora expands equatorward with increasing magnetospheric activity, and its brightness and width vary in punctuated steps of magnetospheric energy releases known as substorms [*Akasofu et al.*, 1963]. Embedded in the auroral oval are numerous forms and concentrations of light in arcs, bands, patches, and diffuse glows broadly distributed in longitude and latitude. These manifestations can be attributed to precipitation of electrons, protons, or a combination thereof whose proportion very generally depends on the local time of the observation. The average energy of the electrons and protons has been separately quantified using large sets of in-situ observations [*Hardy et al.*, 1985; *Hardy et al.*, 1989, 1991], while numerous case studies have focused on the global distribution of precipitating particle species and energies, the phases of auroral morphology and the evolution of auroral forms during storms [e.g., *Sharber et al.*, 1998] or other more localized events [e.g., *Sandahl et al.*, 1980; *Sears and Vondrak*, 1981; *Sandholt and Newell*, 1996]. Numerous studies have treated and modeled the auroral emissions expected from aurorae of particular energies and species [e.g., *Rees and Luckey*, 1974; *Strickland and Anderson*, 1983; *Lummerzheim and Lilensten*, 1994; *Hubert et al.*, 2001].

In addition to substorms, other magnetospheric conditions can lead to significant departures of the auroral morphology from a basic oval. Precipitation of solar wind protons into the magnetospheric cusp can produce stronger localized emissions [e.g., *Reiff et al.*, 1977] near noon, and extending to higher latitudes than the main oval [*Frey et al.*, 2002]. Large gaps in the brightness of the nightside oval can appear after substorms [*Chua et al.*, 1998]. A localized enhancement of electron precipitation is often seen in the afternoon sector [*Meng and Lundin*, 1986; *Lui et al.*, 1989] as electron populations drift eastward into the compressed dipole field of the dayside magnetosphere. Processes such as these indicate the existence of particular physical processes which are localized to particular latitudes or local times of the aurora, often favoring the precipitation of a particular species.

The properties of the magnetospheric plasma vary considerably with radial distance, particularly at the plasmaspheric boundary where closed single-particle drift paths dominated by corotational electric fields cede to open paths driven by magnetospheric convection. Though the magnetic field lines in the region of the auroral oval generally map to the dayside boundary layer or nightside plasma sheet, there are times when the lower latitude boundary of the auroral oval can clearly map to locations closer to the plasmaspheric boundary. Prime examples of this mapping are the large Stable Auroral Red (SAR) arcs often seen during large geomagnetic storms at mid-latitudes. These arcs extend well away from the oval, as far south as Florida, for instance, for many hours, originating at what are normally considered plasmaspheric L-shells. Though these are generally associated with significant fluxes of superthermal electrons from the outer plasmasphere [*Kozyra et al.*, 1997; *McEwen and Huang*, 1995], heightened proton precipitation has been observed as a significant component [*Mendillo et al.*, 1989] of the total precipitating energy flux, sometimes in localized patches.

With the launch and operation of a Far-Ultraviolet global imager on the NASA-IMAGE satellite specifically designed to observe the proton aurora, several instances and types of proton precipitation at latitudes equatorward of the main auroral oval have been observed. For instance, protons with mean energies greater than 30 keV have recently been observed to precipitate in sub-corotating patches at latitudes that map out to the plasmaspheric boundary [*Frey et al.*, 2004]. Precipitation dominated by protons is observed at sub-auroral latitudes on the dayside in short-lived bursts (<15 minutes) which are well-correlated with the arrival of shocks in the solar wind at the magnetopause [*Hubert et al.*, 2003; *Zhang et al.*, 2003; *Fuselier et al.*, 2004].

Another striking feature that FUV images of the proton aurora can reveal is the dynamic development of detached proton arcs in the afternoon sector. These phenomena are somewhat rare, but over 20 significant events have been observed during the first two years of the IMAGE mission. Detached arcs were first observed during the ISIS 2 mission, described in reports by *Anger et al.* [1978] and *Moshupi et al.* [1979]. From those early studies, it was concluded that electron precipitation was the main component of the detached arcs, and that electron cyclotron resonance was the means by which the normally stably trapped electron population was scattered into the loss cone [*Wallis et al.*, 1979]. Detached arcs and patches have also been seen in ground based optical and radar observations [*Mendillo et al.*, 1989, and R. Greenwald, private communication]. Recent observations give strong indications that proton precipitation is the dominant component of the detached arcs observed by IMAGE [*Immel et al.*, 2002]. Calculations of the mean energy and number flux of the detached proton precipitation, using the ratio of the proton-induced FUV emissions of N_2 and O I, show mean energies of 20–30 keV per proton, and energy fluxes up to 1 mW/m^2. Correlative data from the NASA-FAST and DMSP satellites confirms these energy calculations [see *Immel et al.*, 2002 and *Burch et al.*, 2002].

The source of the detached arcs must be reconsidered in light of the fact that the greatest portion of precipitating energy in the arcs seen is carried by protons. The source

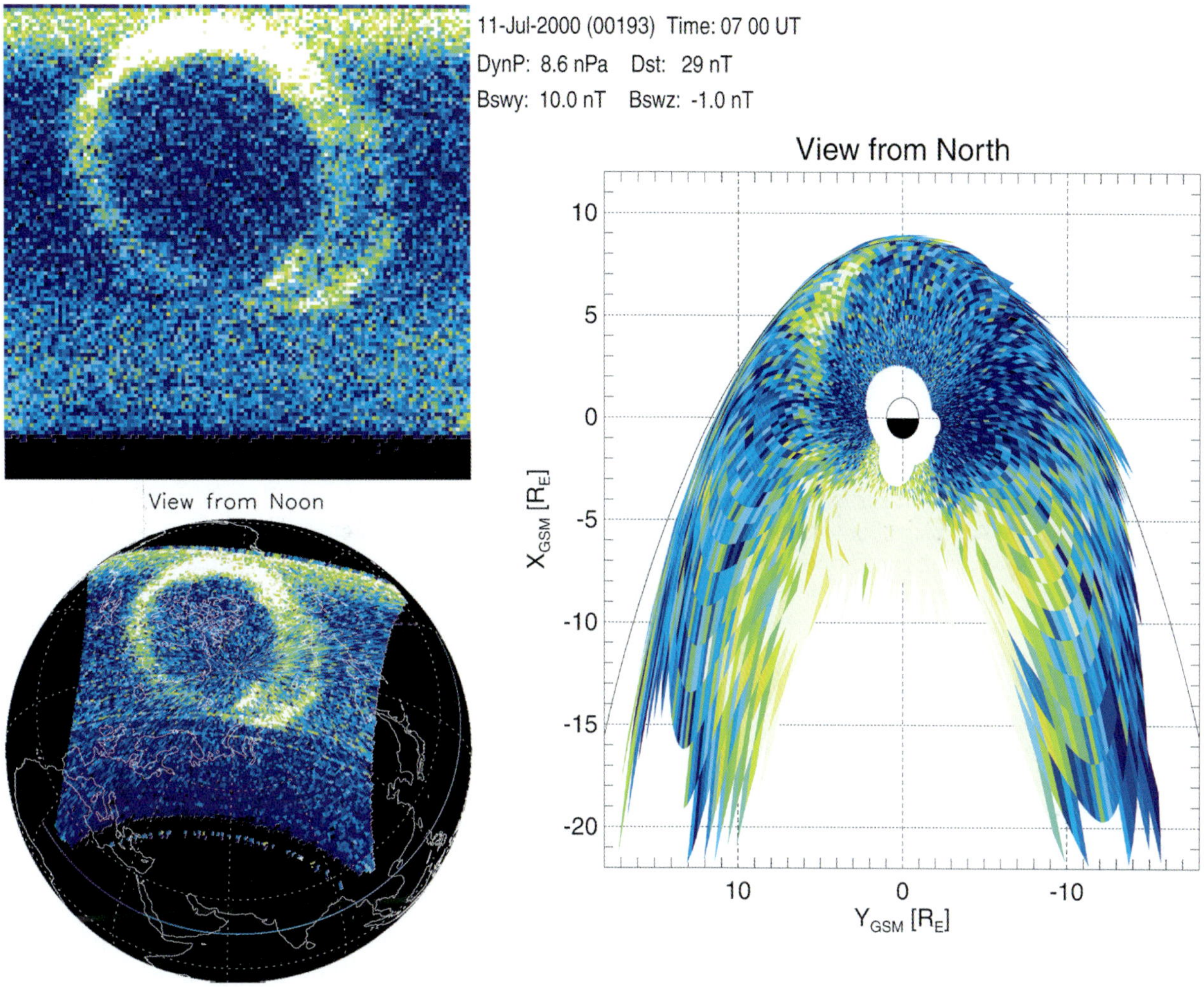

Plate 1. Mapping of a detached proton arc to the GSM x-y plane. The three panels show the original image, the image minus dayglow mapped to Earth, and the proton auroral emissions projected out to the GSM x-y plane. Note the appearance of the detached arc in the afternoon sector of the magnetospheric mapping. This example occurred over the northern coast of Siberia.

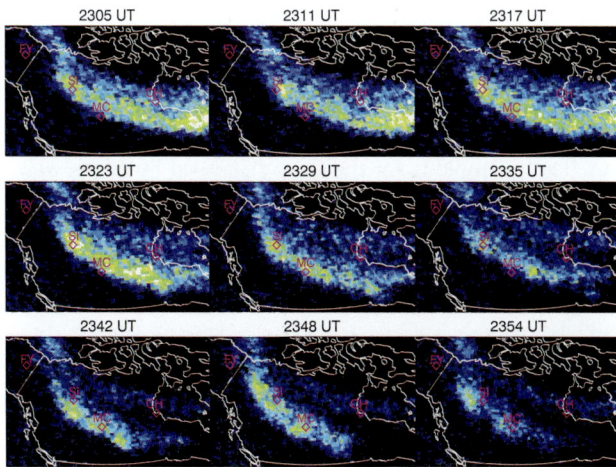

Plate 2. Series of SI-12 images of a detached proton arc appearing on January 23, 2001 mapped to a fixed sprojection of geographic coordinates. The magnetometer stations from which come thepower spectra in Plates 3 and 4 are indicated with FY and SI abbreviations, respectively.

population is almost certainly ring current protons. In the observed energy range of 20-30 keV, these protons will ∇B-drift in the westward direction after being transported to L < 7 Re near the midnight sector by substorm electric fields [*Roederer*, 1970]. After moving through the evening and afternoon sectors, they can either drift out of the magnetosphere on open paths or continue to circle the earth many times, depending on the configuration of magnetospheric convection electric fields. When mapped out to the GSM x-y plane using the *Tsyganenko* [1995] magnetospheric magnetic field model, the emissions of the detached proton aurora mark a path through the magnetosphere that is remarkably similar to open drift paths of 20–30 keV particles. This is shown with a single image in Plate 1, where the 3 panels show the original image (upper left), the image with airglow removal applied and mapped to the earth (lower left), and the full mapping of the original image in the x-y plane.

By what mechanism the ring current protons are caused to depart from their bounce-drift motion to precipitate into the atmosphere is the pertinent question. A prime candidate is the ion-cyclotron instability, which under particular conditions can grow and work to divert cyclotron energy into translational energy parallel to the field, thus adding particles to the loss cone [*Brice*, 1964]. However, there are other means for diverting particles into the loss cone, as are discussed in this report.

2. OBSERVATION TECHNIQUES

2.1 Magnetometer Measurements

Important supporting data will be obtained from three magnetometer networks: The Geophysical Institute Magnetometer Array (GIMA), Canadian Auroral Network for the Open Program Unified Study (CANOPUS), and a network of stations deployed across the Antarctic continent. GIMA is distributed across several sites in Alaska. CANOPUS provides sites in Canada covering latitudes between approximately 55 to 70°. Several magnetometers are located at manned and autonomous stations across Antarctica.

The CANOPUS and GIMA observation sites are all equipped with a 3-component fluxgate ring-core type magnetometers. The sampling rate for these instruments is 8 samples per second, though final data are provided at 5 and 1-second intervals.

Of the stations in Antarctica, we use data from three Autonomous Geophysical Observatory stations operated by the British Antarctic Survey (A80, A81, and A84). Another magnetometer on the continent is supported by the United States National Science Foundation at South Pole Station.

The magnetometer at South Pole provides vector samples of dB/dt at local geomagnetic coordinates with X northward at a rate of 10 samples per second [*Taylor et al.*, 1975; *Engebretson, M. J., et al.*, 1997]. A search coil magnetometer is also included at each of the three multi-instrument AGO stations [*Arnoldy et al.*, 1998]. These search coils provide data at a rate of 2 samples per second. A fluxgate magnetometer is also installed at all four of these sites.

2.2 IMAGE-FUV Measurements

The NASA-IMAGE satellite was launched in March, 2000, on a mission to study magnetospheric phenomena through their emissions of FUV and EUV photons, radio waves and neutral atom fluxes [*Burch*, 2000]. The Spectrographic Imaging component of the FUV instrument [*Mende et al.*, 2000] obtains simultaneous 2D images of terrestrial FUV emissions at wavelengths of 121.8 and 135.6-nm. Most important of these is the former of the two channels as it detects Doppler-shifted emissions of hydrogen produced in the recombination of protons and electrons in the proton aurora.

3. COMBINED MAGNETOMETER AND IMAGING OBSERVATIONS

The work for this study began with a visual examination of every third SI-12 image obtained in the 2000–2002 period. More than 20 significant detached arcs were observed in this process, all >20 minutes in duration and appearing in the dusk-sector, with attachment to the main oval often observed at the end of the arc closest to noon. From these characteristics, it is believed that these all represent similar events as those described in earlier studies by *Immel et al.* [2002], *Burch et al.* [2002] and *Spasojević et al.* [2004].

3.1 Comparisons Between Si-12 and Alaskan/Canadian Stations : Pc5 Pulsations

The first magnetometer pulsations observed to coincide with detached proton arcs were large Pc5 pulsations. An excellent example is for the time discussed by *Immel et al.* [2002], where the proton auroral oval was seen to produce an afternoon detached arc several times in a 4-hour imaging period. The strongest arcs were observed during the 23–24 UT period on that day, selected images of which are shown in Plate 2.

A magnetic spectrogram for one of the ground-based magnetometers that falls within the field of view of the SI-12 imager and close to the detached arc is shown in Plate 3. The spectrogram shows strong enhancements in magnetic wave

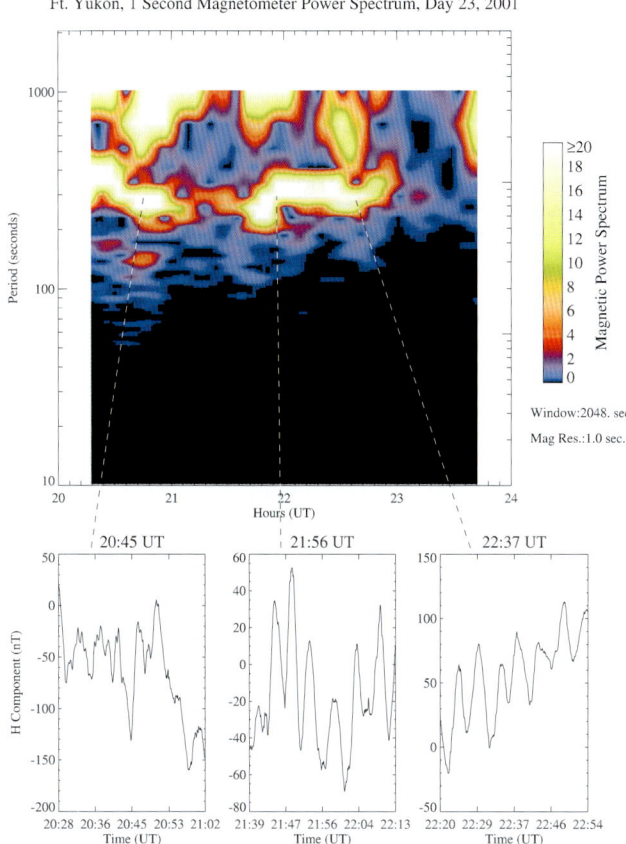

Plate 3. Magnetic power spectrum for a 4-hour period including the time of imaging shown in 2, from the Fort Yukon magnetometer station.

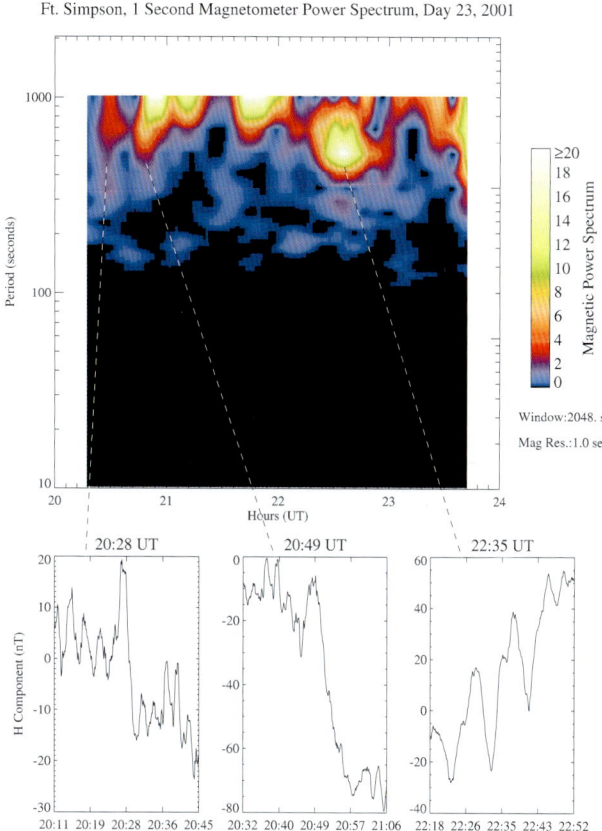

Plate 4. Magnetic power spectrum for a 4-hour period including the time of imaging shown in 2, from the Fort Simpson magnetometer station.

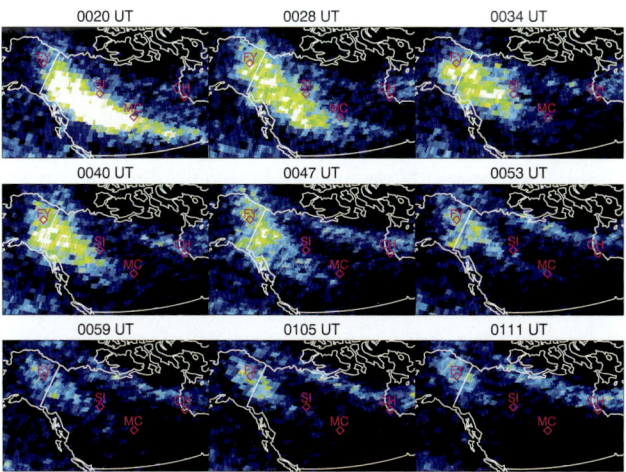

Plate 5. Series of SI-12 images showing a detached proton arc appearing on November 10, 2001, presented in a manner similar that used in Plate 2

power in the $\tau \sim 300$ second band at Fort Yukon, Alaska. Individual time series of the horizontal (H) component of the magnetic field are shown at 3 selected times during the 4 hour time series where Pc5 waves are particularly strong.

The Alaskan station is about an hour earlier in local time than the western end of the detached arc. A magnetic spectrogram from the Canadian CANOPUS magnetometer station at Fort Simpson, that is basically colocated with the bright emissions of the arc, is shown in Plate 4. At this station, the pulsations have a markedly different signature, with a short-lived peak in Pc5 power around 2230 UT, but significant waves with $\tau \sim 180$ sec earlier at 2038 UT. The power at this frequency is at the bottom of the color table, but the pulsations can be seen in the 30 minute window plot (lower left panel).

Another example of the correspondence of detached proton arcs and large Pc5 pulsations is found in the case described by *Burch et al.* [2002], where another detached proton arc was observed over Canada on November 10, 2000. The development of the arc is shown in Plate 5, with the same important ground magnetometer sites including Forts Yukon and Simpson indicated therein. This arc has a different morphology than the previous example, with a greater latitudinal extent at early local times. It also is the predominant signature of the entire proton aurora in the hemisphere, whereas on January 23, 2001, the proton auroral oval was generally bright.

The power spectrum of the H-component of B is shown for the Forts Yukon and Simpson ground based stations in Plates 6 and 7, respectively. The overall level of power in the Pc5 range is higher than in the previous case, though an isolated clear tone such as that seen in the Fort Yukon magnetometer at 2230 UT on Jan.23 is not evident here. Long period waves continue to be observed an hour after the detached arc has faded (see lower right panels of Plates 6 and 7).

In both cases, a recent magnetospheric substorm is the likely source for the enhanced ion fluxes in the ring current that make up the detached arc. In each of the cases shown in Plates 2 and 5, a substorm onset preceeds the first appearance of the detached arc (about 23:23 and 00:20, respectively) by ~ 40 minutes. The Auroral Electrojet (Ae) indices determined in the 6 hours leading up to the detached arc observation are shown in Plate 8. The relationship between arc observation and substorm onset and peak is apparent. The occurrence of a magnetospheric substorm was described by *Immel et al.* [2002] an important and necessary condition for the development of large-scale detached proton arcs in the afternoon sector. The short lifetime of the detached arcs shown here (less than one hour) suggests a similarly short-lived source population of energetic particles. The magnetospheric substorm provides just such an impulsive and initially highly-localized source population of ions that will move from the nightside into the afternoon sector, and then beyond to earlier local times or completely out of the magnetosphere (cf. Plate 1) on a time scale of tens of minutes.

In summary, Pc5 pulsations are seen in both of these events, though they are usually stronger at locations away from the precipitation than directly underneath. In each case, it is the Alaskan station, closer to noon, that shows the stronger Pc5 signature. Recent studies of Pc5 pulsations have found evidence that modulation of the solar wind pressure at similar frequencies can drive the Pc5 pulsations, and this is possibly a result of that effect [*Kepko et al.*, 2002]. Previous research had generally found that Pc5 pulsations were a natural oscillation mode of the magnetosphere [e.g., *Crowley et al.*, 1987]. In either case, the structure of the magnetic and electric fields in the magnetosphere which is subject to Pc5 oscillations can interact with ring-current particles at particular particle energies, adding to the parallel energy with each mirroring cycle, which will eventually enhance precipitation of energetic particles [*Southwood et al.*, 1969; *Glassmeier et al.*, 1999]. It has also been found that magnetospheric substorms that would enhance the fluxes of 20-40 keV ions in the afternoon sector of the magnetosphere soon after their occurrence precede the appearance of afternoon detached proton arcs.

3.2 Comparisons Between Si-12 and Conjugate AGO Stations : Pc1 Pulsations

For comparison to the Antarctic magnetometers, events were selected for further analysis if any portion of the arc extended into the ocean gap between Greenland and Canada, where the magnetically conjugate points of the AGO magnetometer stations A80, A81, and A84 fall. Of all the events, four met this criterion. All of the images for each event were processed by mapping each into a fixed $0.5°\times 0.5°$ grid of geographic coordinates and averaging each grid location in a three-image (six-minute) running average. The resulting averaged images have a better signal-to-noise ratio than individual images. Nine successive images that contain the (qualititively determined) best defined detached arc are shown in the following plates. The cadence of the imaging is adjusted to fully cover the development of the arc.

The locations magnetically conjugate to four of the observatories on the Antarctic continent are identified with a diamond and denoted with a simple abbreviation (e.g., SP=South Pole). The conjugate mapping is accomplished simply by maintaining the magnetic longitude and reversing the sign of the magnetic latitude of each station. Use of a more sophisticated magnetic field model, in this case the Tsyganenko 1996 field model [*Tsyganenko*, 1995], can show

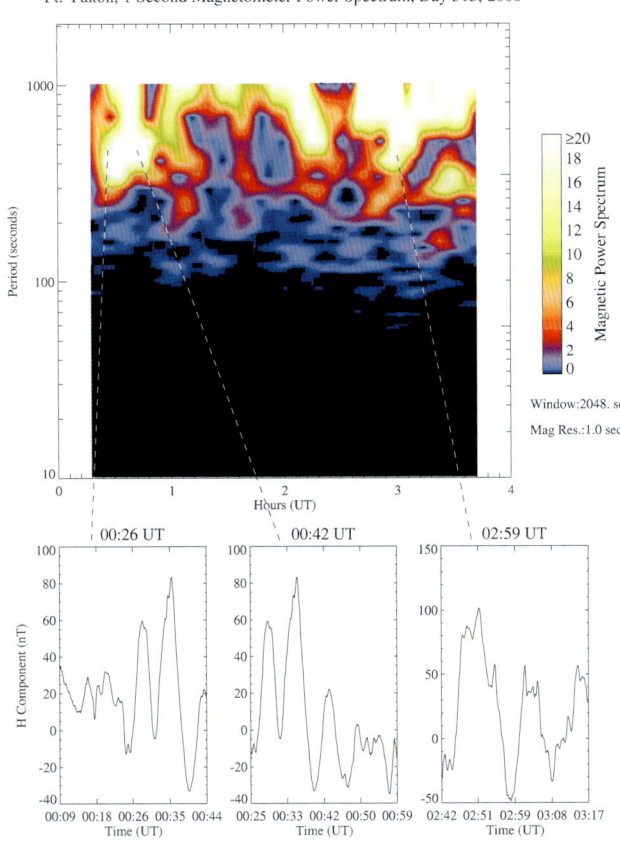

Plate 6. Magnetic power spectrum for a 4-hour period including the time of imaging shown in Plate 5, from the Fort Yukon GIMA magnetometer station.

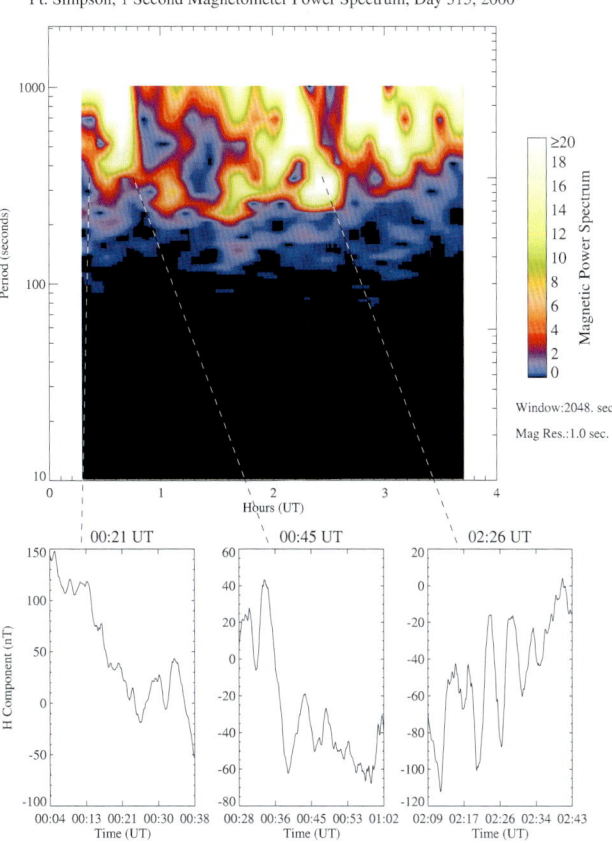

Plate 7. Magnetic power spectrum for a 4-hour period including the time of imaging shown in Plate 5, from the Fort Simpson CANOPUS magnetometer station.

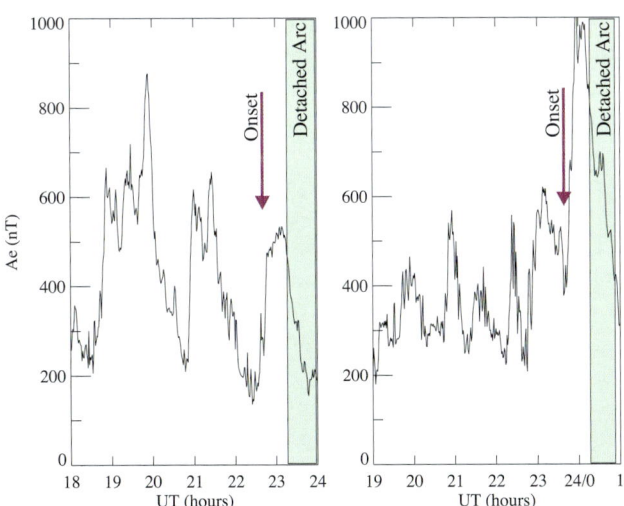

Plate 8. Ae indices for the six hours ending in the detached arc observations shown in Plates 2 and 5. The onset times of the substorms which likely provide the source population of ions are indicated. The ranges of time when detached arcs are observed and shown in the earlier plates are indicated by the green shaded regions.

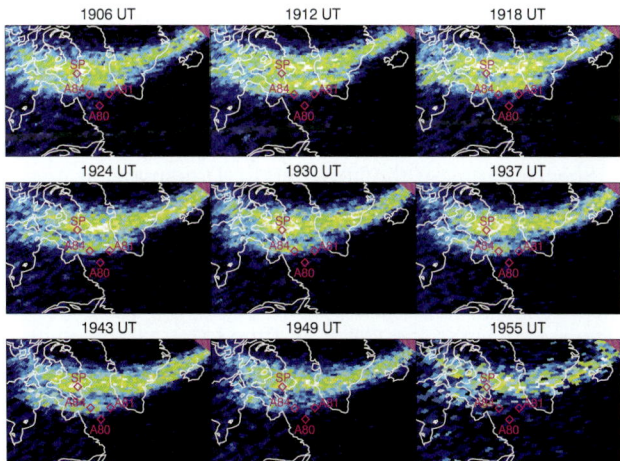

Plate 9. Series of detached arc images for May 22 (day 142), 2001

a 100–200 km change in the the predicted conjugate footpoint of the A81–A84 stations, and an often highly unpredictable mapping of the South Pole station. A preliminary study shows, however, that no gain in understanding of the pulsation signatures is made by using the Tsyganenko field model. It is decided for the sake of continuity that the simplest conjugate mapping will be used in all of the four cases of detached proton arcs that follow.

Accompanying each case is the magnetic power spectrum (0.05–10 Hz) of the X-component of the available antarctic magnetometer data. These are prepared using a 128-second sliding sampling window. South Pole station is usually fairly far from the detached arc, but since it is not now known how broadly any Pc1 signature may appear to extend, that spectrum is also shown when available. Represented in these plots are all of the stations that returned data during the detached arc events.

First case: May 22, 2001, 1930 UT The first detached arc in the sector conjugate to the stations in Antarctica was observed on May 22, 2001. A series of nine 6-minute averaged images is shown in Plate 9. The signature of the bright detached arc begins as a especially wide proton auroral oval over the Canadian mainland, evident at 1856 UT, which extends separated from the main oval to later local times in the following 30 minutes. The detached morphology is quite obvious from 1914 through 1945 UT, in which time it extends to cover the conjugate points of A84 and A81, while A80 remains further south. The conjugate point of South Pole station remains on the equatorward boundary of the main oval throughout the time of observation. This is in contrast to other cases such as that the November 10, 2000, where the main oval retreats to higher latitude while the detached arc remains equatorward, often shifting even further equatorward with time. Plate 9.

Spectrograms of the horizontal (H) component of the magnetic field at the stations in Antarctica that provided data during the time of the detached arc observations are shown in Plate 10. As opposed to the spectrograms generated to show the power in the $\tau=100$–1000 sec range, the AGO data are shown in frequency vs. time spectrograms, with a useful range of .1-1 Hz. The magnetic spectrum is analyzed between 1600 and 2200 UT, with several hours preceeding the ~1 hour of imaging data shown in Plate 9. The two useful stations are South Pole (top panel) and A84 (center panel). Each shows a variety of similar Pc1 signatures, beginning first with a series of dispersed Pc1 "pearls" [*Mende et al.*, 1980] at 0.3–0.5 Hz that continue to slightly larger frequencies over the next 3 hours. These pearls are periodic with τ ~300 seconds, in the Pc5 frequency range. As time proceeds at South Pole, a weaker group of pulsations around .15 Hz fades to below the detection threshold by 2000 UT, while the same pulsations are even weaker at A84 until ~1930 UT where they strengthen significantly, and independently of the South Pole observations.

This interesting coincidence of the development of a detached proton arc with a corresponding enhancement in wave power in the Pc1 range gives an indication that the processes are related. Furthermore, a brief examination of the fluxgate magnetometer readings at A81 and A84 show significant Pc5 pulsations with amplitudes of 20–40 nT for several hours about the time of the detached arc and Pc1 observations. Further study of the correspondence of the Pc1 pearls and Pc5 waveform will be important, but will be addressed in a future study. The study at hand focuses on three more known cases of detached arcs occurring conjugate to the Antarctic stations, as described in the following sections.

Second case: June 18, 2001, 1605 UT Another case of a detached proton arc developing conjugate to the Antarctic stations was found on June 16 (day 169), 2001. A series of images which show the development of the arc can be seen in Plate 11. This is obviously a significant detached arc event, with brightnesses in the detached arc at the same level as the main oval. The first appearance of the arc is at 1500 UT, and unlike rapidly developing arcs such as that seen on Nov 10, 2001, the detached proton arc maintains its form for more than an hour. For that reason, the images are shown in 10 minute increments covering a 90 minute interval. Plate 11.

At 1500 UT when the detached arc begins to develop, the conjugate Antarcticstations A80-A84 are not within the proton auroral boundaries, though South Pole is right in the oval. Within 20 minutes, though, the edge of the detached arc approaches the stations, and by 1630 UT, the higher latitude stations A84 and A81 are enveloped by the arc. South Pole is in an interesting location at 1602 UT, just at the location where the detached arc connects to the main oval. These locations relative to the detached arc are markedly different than the last case, mainly because the local time at the stations is 4 hours earlier at the beginning of the detached arc development than in the previous example. The stations are just crossing into the afternoon sector to which the detached arcs are almost always confined.

The magnetic power spectrograms for this detached arc event are shown in Plate 12, and the differences from the previous case are striking. There is an enhancement in power in the 0.1-0.8 Hz range, which is most significant at A84 between 1435 and 1450 UT. The timeframe of the signal is earlier at South Pole (1420–1440 UT), and later at

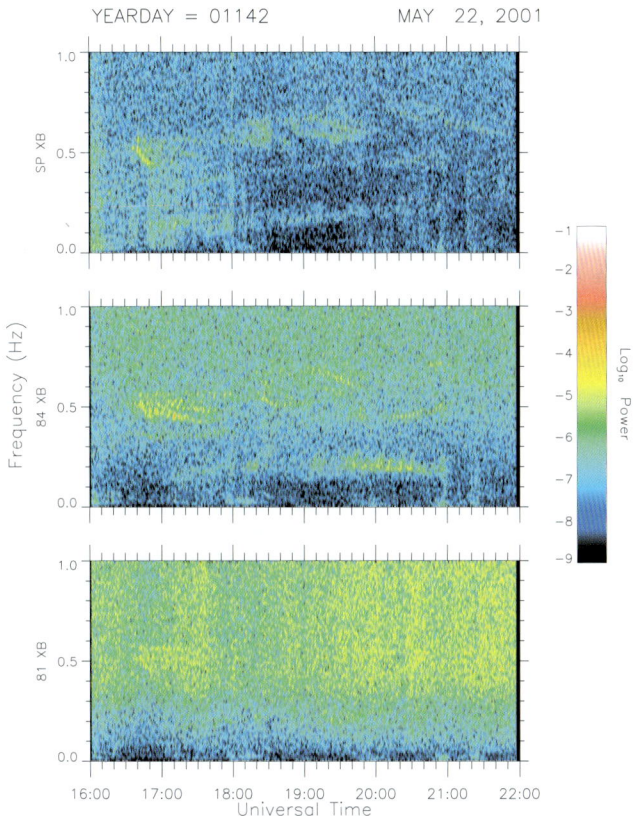

Plate 10. Frequency-time magnetic power spectra from AGO stations conjugate to the detached arc observed on May 22 (day 142), 2001

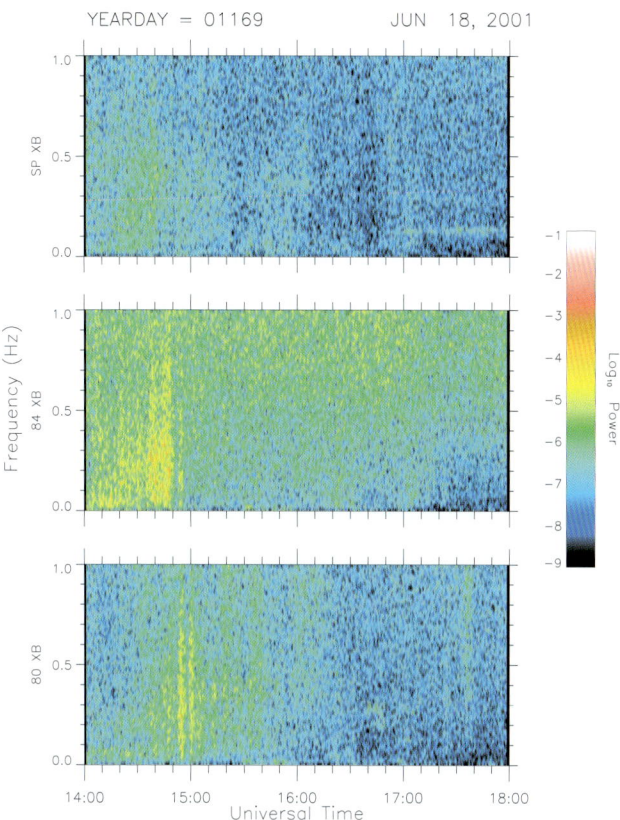

Plate 12. Frequency-time magnetic power spectra from Antarctic stations conjugate to the detached arc observed on June 18 (day 169), 2001

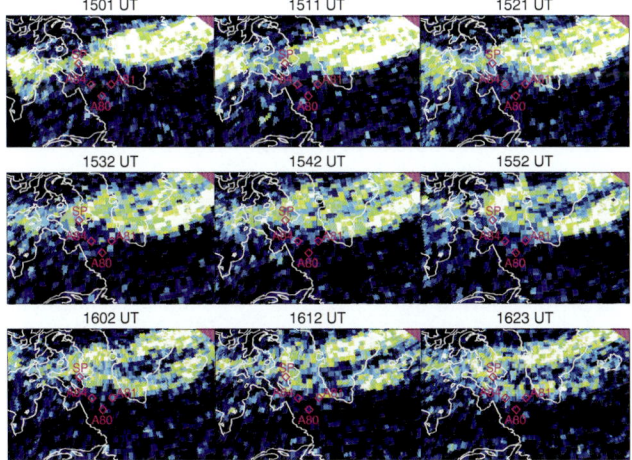

Plate 11. Series of detached arc images for June 18 (day 169), 2001

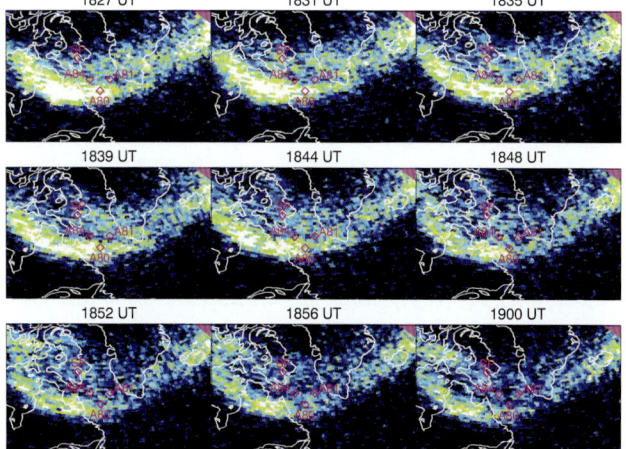

Plate 13. Series of detached arc images for July 15 (day 196), 2001

A80 (1455-1540 UT). Each of these signals, like the previous example, is modulated at Pc5 frequencies (e.g. $\tau \sim 400$ seconds at A80). It is interesting to note that the strongest signal seems to migrate to lower latitudes with time. What is clear is that there is no correspondence between the location of the detached arc over the station's conjugate point and the occurrence of Pc1 as measured at that station. This non-correspondence provides a counterexample to the argument that EMIC waves are related to detached proton arcs, though further discussion later in this report will address this question again.

Third case: July 15, 2001, 1845 UT The next detached proton arc manifested itself as a strong dusk emission that gradually detached from the main auroral oval, which itself appeared to recede overall to higher latitudes. This is evident in the imaging series of Plate 13. The morphology of this case is initially similar to that observed on Nov 10, 2000, as discussed in Section 3.1, with a broad extent in latitude at early local times, narrowing rapidly in the eastward direction. The arc, however, narrows significantly in the 1827-1900 UT imaging time shown in 13. This arc is shorter lived than that presented in Plate 11, so the imaging cadence is back down to 4 minutes between images.

The magnetic power spectra from all 4 Antarctic stations are shown in Plate 14. Here is found the most powerful Pc1 pulsation of all the cases discussed in this report. The signals at A84 (top-center panel) and A80 (bottom panel) are similar, though much stronger in the 0.15–0.25 Hz range at the lower latitude station (A80) in the 1715–1900 UT time frame. This strong enhancement in Pc1 power coincides with the development of the detached arc. A80 is conjugate to the detached arc, while A84 is conjugate to the main oval. This powerful signal is not seen at A81 at all, though the later enhancement in Pc1 pulsation power at 0.1 Hz, seen after 2000 UT in all three stations. This demonstrates what appears to be extreme localization in the penetration of Pc1 power generated in the magnetosphere through the ionosphere. Plate 13.

Fourth case: June 16, 2002, 1845 UT The last case is an example of a relatively weak detached proton aurora seen on July 15 (day 167), 2002. It has, however, a significant corresponding signature in the Pc1 range. The imaging sequence shown in Plate 15 begins at 1905 UT and shows strong proton precipitation in the dusk sector with A84 and A81 on the equatorward edge of the proton arc. The gap in emissions over central Greenland is an artifact of the slightly changed flatfield from the previous year's levels. The main oval actually extends from the South Pole footpoint eastward through central Greenland. The detached arc extends eastward to the edge of Iceland. The next image is ~ 8 min later at 1912 UT, though all successive images are at 4 minute intervals.

In this time frame, a detached arc is seen to fade and become more narrow, but also to become more distinct from the main oval, extending roughly along the initial proton auroral boundary seen at 1905 UT while proceeding 100-200 km south of this initial location. The arc is clearly detached by 1938 UT, which happens mainly by the narrowing of the initial bright arc, leaving the main oval over central Greenland. These observations are near the threshold of the proton imager's capability, but clearly indicate emissions separate from the main proton auroral oval which lie in the vicinity of the conjugate magnetic footpoints of the AGO stations.

Two of the AGO stations returned data for this event, A84 and South Pole. Compared to previous observations, the Pc1 signatures are remarkably similar. Each shows a rapid enhancement in the 0.35-0.6 Hz range at 1840 UT, which continues until about 2000 UT at each station. Each Pc1 signature is marked with slightly dispersed periodic variability in power at the upper end of Pc5 frequencies ($\tau \sim 240$ sec). This enhancement in power is relatively weak compared to several examples shown earlier (e.g. Plate 14), but shows very clear features that are not obscured by additional wave sources. This indeed may be the best example of Pc1 waves which correspond directly to the development of a detached proton arc.

4. DISCUSSION AND CONCLUSIONS

This research effort investigates the connection between magnetic pulsations observed at ground-based stations and proton aurorae which detach from the main oval and advance equatorward, mapping to regions much deeper in the magnetosphere than usual for auroral proton energy fluxes of 1 mW/m^2. It is found that magnetic pulsations on the order of 1 Hz and 0.003 Hz are often observed in apparent association with the occurrence of detached proton arcs. The Pc5 waves are observed at Alaskan and Canadian stations in the vicinity of the arcs, while the Pc1 waves are observed at Antarctic stations conjugate to the observed arcs.

The long-period waves are in the Pc5 range and can occur at the location of the arc, but in the examples shown here (see Section 3.1), are stronger at earlier local times than at the location of the arcs. The short-period pulsations are usually in the 0.2-0.5 Hz Pc1 range and can be strongest in the immediate vicinity of the detached arc. In contrast to the two Pc5 cases discussed, the Pc1 waves are generally present when the detached arc is coincident. In the case where the Antarctic magnetometer stations are at earlier local times during the initial development of the detached proton arc (June 18 (day

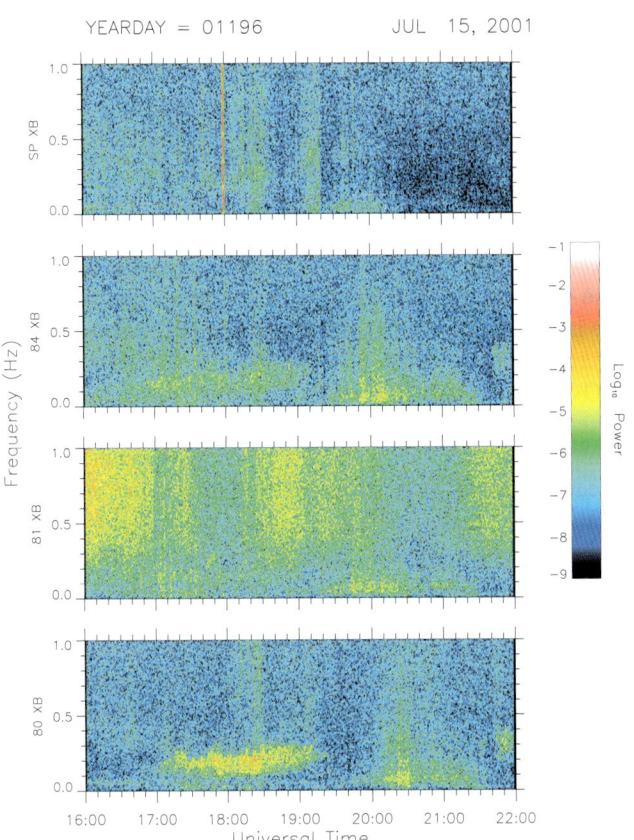

Plate 14. Frequency-time magnetic power spectra from AGO stations conjugate to the detached arc observed on July 15 (day 196), 2001

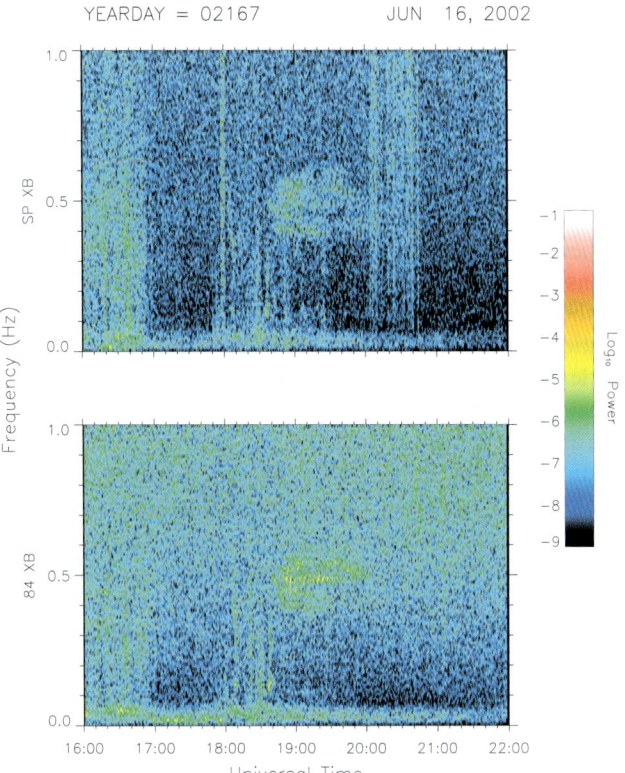

Plate 16. Frequency-time magnetic power spectra from AGO stations conjugate to the detached arc observed on June 16 (day 167), 2002

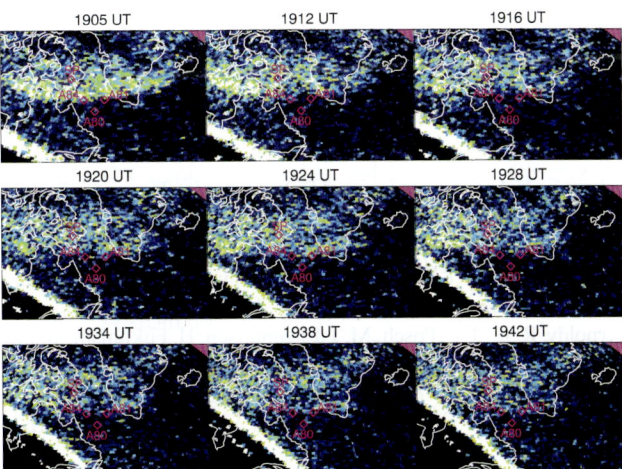

Plate 15. Series of detached arc images for June 16 (day 167), 2002

Spasojević, M., H. U. Frey, M. F. Thomsen, S. A. Fuselier, S. P. Gary, B. R. Sandel, and U. S. Inan, The link between a detached subauroral proton arc and a plasmaspheric plume, *Geophys. Res. Lett., 31*, 4803–4806, 2004.

Strickland, D. J., and D. E. J. Anderson, Dependence of auroral FUV emissions on the incident electron spectrum and neutral atmosphere, *J. Geophys. Res., 88*, 8051, 1983.

Taylor, W. W. L., B. K. Parady, P. B. Lewis, R. L. Arnoldy, and L. J. C. Jr., Initial results from the search coil magnetometer at Siple, Antarctica, *J. Geophys. Res.*, pp. 4762–4769, 1975.

Tsyganenko, N. A., Modelling the Earth's magnetospheric magnetic field confined within a realistic magnetopause, *J. Geophys. Res., 100*, 5599–5612, 1995.

Wallis, D. D., J. R. Burrows, M. C. Moshupi, C. D. Anger, and J. S. Murphree, Observations of particles precipitating into detached arcs and patches equatorward of the auroral oval, *J. Geophys. Res., 84*, 1347–1360, 1979.

Zhang, Y., L. J. Paxton, T. J. Immel, H. U. Frey, and S. B. Mende, Sudden solar wind dynamic pressure enhancements and dayside detached auroras: IMAGE and DMSP observations, *Journal of Geophysical Research (Space Physics), 108*, 2–1, 2003.

T. J. Immel, S. B. Mende, H. U. Frey, Space Sciences Laboratory, University of California, Berkeley, CA, 94720 (email: immel@ssl.berkeley.edu)

Afternoon Subauroral Proton Precipitation Resulting From Ring Current—Plasmasphere Interaction

M. Spasojević[1,2], M. F. Thomsen[3], P. J. Chi[4], B. R. Sandel[5]

We investigate the occurrence of arcs of precipitating protons equatorward of and detached from the afternoon proton auroral oval and their relationship with the plasmasphere and electromagnetic ion cyclotron waves. In a four month study interval including sixteen events, we find that the detached proton arcs are more likely to occur during geomagnetically disturbed periods and specifically at times when enhanced energetic ion densities and temperature anisotropies are observed in the equatorial magnetosphere. The disturbance-time arcs tend to be located at lower magnetic latitudes and are consistently associated with plasmaspheric plumes. Conversely, arcs which occur during quiet times tend to be located at higher latitudes, and their relationship with regions of enhanced cold plasma density remains unclear. Wave data available for two of the detached arc events indicate the presence of strong ion cyclotron waves near the equator in the vicinity of the proton precipitation region.

INTRODUCTION

Although the dominant loss processes for terrestrial ring current ions are collisional, including charge exchange with the neutral geocorona and Coulomb collisions within the plasmasphere, wave-particle interactions are also believed to play an important role as they provide a mechanism for the rapid decay of the ring current during the early recovery phase of geomagnetic storms [*Kozyra et al.*, 1997]. Resonant interaction between energetic ring current ions and electromagnetic ion cyclotron (EMIC) waves results in pitch angle scattering and subsequent precipitation of the energetic ions into the upper atmosphere. Considerable attention has been given to regions of spatial overlap between energetic, anisotropic ring current ions and cold, dense plasmaspheric material that should be particularly conducive to the growth of EMIC waves [*Cornwall et al.*, 1970; *Lyons and Thorne*, 1972].

In terms of magnetospheric dynamics, global magnetospheric convection and substorms act to energize and transport plasma sheet particles into the inner magnetosphere, building up the ring current [e.g. *Fok et al.*, this volume]. Convection also acts to erode the plasmasphere, and plasmaspheric material is transported sunward [e.g., *Goldstein and Sandel*, this volume] and into the path of westward-drifting ring current ions. Previously stable energetic ions distributions encounter the cold, dense plasma, become unstable to the growth of ion cyclotron waves, and some fraction of the energetic ions are scattered into the loss cone.

The free energy for wave growth in the electromagnetic ion cyclotron instability is provided by the temperature anisotropy of energetic ring current ions ($T_\perp > T_\parallel$). Energy and momentum exchange can occur when the Doppler shifted wave frequency matches the cyclotron frequency of the individual resonant particles. According to linear Vlasov dispersion theory, the kinetic energy of protons that

---

[1] Space Sciences Laboratory, University of California, Berkeley, California.
[2] now at Space, Telecommunications, and Radioscience Laboratory, Stanford University, Stanford, California.
[3] Los Alamos National Laboratory, Los Alamos, New Mexico.
[4] Institute of Geophysics and Planetary Physics, University of California, Los Angeles, California.
[5] Lunar and Planetary Laboratory, University of Arizona, Tucson, Arizona.

can resonate with a given frequency wave decreases with increasing cold plasma density [*Kennel and Petchek*, 1966]. The wave growth rate is then proportional to the temperature anisotropy of the energetic protons and the fractional number of protons near resonance. For example, at geosynchronous orbit on the dayside, the introduction of moderate densities of cold plasma (10 to 50 cm^{-3}) can reduce the resonant proton energy so that it falls within the range of the bulk of the ring current (<200 keV), and thus, more energetic protons are available for resonance. If the energetic proton temperature anisotropy is sufficiently large, waves will grow and scatter the protons until the distribution has stabilized. The maximum amplification of the ion cyclotron waves occurs near the equatorial plane, where magnetic field values are low, and for wave normal vectors parallel to the magnetic field direction [*Thorne and Horne*, 1992]. Also, heavy ions (He$^+$, O$^+$) which are present in both the hot and cold plasma distributions significantly modify the ion cyclotron wave frequencies and growth rates including the formation of stop bands above the heavy ion gyrofrequencies [*Young et al.*, 1981; *Kozyra et al.*, 1984].

The effects of wave scattering have been including in global ring current models [e.g. *Kozyra et al.*, 1997, *Jordanova et al.*, 2001, *Khazanov et al.*, 2003]. *Jordanova et al.* [2001] developed a time-dependent global EMIC wave model to study the spatial and temporal evolution of precipitating proton fluxes during different phases of a geomagnetic storm. The most intense fluxes of precipitating protons are found along the duskside plasmapause during the storm main and early recovery phases. The global precipitation patterns move to lower L shells during the main phase as the plasmasphere is eroded and recede to larger L shells during the storm recovery as the plasmasphere refills.

Although the global impact of wave-particle interactions has not been experimentally verified, a large body of supporting observational evidence does indicate that they may at times have an important influence on the evolution of the ring current. The inner edge of the ion ring current can at times penetrate the duskside plasmasphere by 0.5 to 2 R_E [*Frank*, 1971; *LaBelle et al.*, 1988; *Burch et al.*, 2001], and numerous event studies have reported ion cyclotron waves and changes in pitch angle distributions consistent with expectations from wave scattering in this overlap region [*Williams and Lyons*, 1974; *Taylor and Lyons*, 1976; *Kintner and Gurnett*, 1977; *Mauk and McPherron*, 1980; *Young et al.*, 1981]. More recently, *Erlandson and Ukhorskiy* [2001] found a direct correlation between EMIC wave spectral density and the flux of energetic protons in the loss cone. Similarly, *Yahnina et al.* [2000] reported a close association between ground based observations of Pc1 pulsations and precipitating energetic protons.

Statistically, EMIC waves (in the range 0.1–5 Hz) have been found to be primarily a phenomenon of the outer dayside magnetosphere (L>7) [*Anderson et al.*, 1992]. However, at lower L values (L=4 to 7) the occurance as well as wave amplitude is strongly peaked in the afternoon local time sector [*Anderson et al.*, 1992; *Fraser and Nguyen*, 2000; *Erlandson and Ukhorskiy*, 2001] in the region where sunward extensions of cold plasma have been frequently observed [e.g. *Chappell*, 1974]. Although *Fraser and Nyguyen* [2000] showed that the plasmapause itself is not necessarily the primary source region for waves in this L region, the vast majority of the wave events did occur where the cold plasma density exceeded 10 cm^{-3}.

2. AFTERNOON DETACHED SUBAURORAL PROTON ARCS

Global imaging of the proton aurora by the Far Ultraviolet (FUV) Spectrographic Imager (SI) [*Mende et al.*, 2000] onboard the IMAGE satellite [*Burch*, 2000] has led to the identification of arcs of precipitating protons at latitudes equatorward of and separated from the main proton oval. The detached subauroral proton arcs appear over several hours of local time in the afternoon sector, and satellite observations magnetically connected to the detached arc confirm the presence of precipitating protons and an absence of precipitating electrons [*Immel et al.*, 2002; *Burch et al.* 2002]. The afternoon detached subauroral arcs can persist for about thirty minutes up to several hours. Thus, they are distinct from other recent observations of so-called subauroral dayside proton flashes which last only for tens of minutes and are triggered by sudden increases in solar wind dynamic pressure [*Zhang et al.* 2002; *Hubert et al.* 2003].

Wave-particle interactions within the plasmasphere have been suggested as a precipitation mechanism for the afternoon detached arcs, and in one of the reported events, 10 Nov 2000, the Magnetospheric Plasma Analyzer (MPA) [*Bame et al.*, 1993] onboard the geosynchronously orbiting 1989-046 spacecraft observed enhanced fluxes of plasmaspheric ions in the region where the equatorial extension of the subauroral arc was expected to map [*Burch et al.*, 2002]. Another event study by *Spasojević et al.* [2004] indicated that the detached arc on 18 June 2001 was directly associated with a globally observed plasmaspheric plume. Predicted by numerical modeling for many years [e.g., *Grebowsky*, 1970; *Chen and Wolf*, 1972], plasmaspheric plumes, also referred to as plasma tails, are regions of cold plasma which extend sunward from the plasmasphere and are formed during periods of enhanced magnetospheric convection. They were first observed globally by the IMAGE Extreme Ultraviolet (EUV) imager [*Burch et al.*, 2001; *Sandel et al.*, 2001]. The EUV instrument [*Sandel et al.*, 2000] images the plasmasphere by detecting 30.4-nm solar radiation resonantly scattered by plasmaspheric He$^+$ ions.

For the 18 June 2001 event, the assertion that the proton arc was a result of wave induced scattering was supported by two main arguments [*Spasojević et al.*, 2004]. First, the auroral arc when mapped to the magnetic equatorial plane overlapped with the plasmaspheric plume as defined by a broad region of enhanced cold plasma density observed globally by IMAGE EUV and *in situ* at geosynchronous altitude by LANL-01a MPA. Second, wave growth calculations based on MPA observations of the hot and cold plasma parameters [e.g. *Gary et al.*, 1995] indicated positive growth of the proton cyclotron instability within the plume region whereas the energetic proton distributions were stable outside the plume. The wave growth calculations suggest that subauroral proton precipitation may occur at times when the hot and cold plasma distributions are similar to those observed on 18 June 2001.

Previous studies have shown that plasmaspheric plumes form readily in the afternoon local time sector during periods of enhanced convection with cold ion densities (in the range of ~1–130 eV/q) at geosynchronous orbit commonly exceeding 40 cm^{-3} [e.g., *Moldwin et al.*, 1995; *Spasojević et al.* 2003]. However, at the time of the detached arc on 18 June 2001, the hot ion density (in the range of 0.13–45 keV/q) observed by LANL-MPA was about 1.5 to 2 times larger than the average value at the same local time under similar geomagnetic conditions ($Kp \approx 5$) [*Korth et al.*, 1999]. Hence in order to further explore the relationship between detached proton arcs and plasmaspheric plumes, we examine how often the plasma parameters at geosynchronous orbit are similar to those of 18 June 2001, and more specifically, whether subauroral detached proton arcs are also observed at those times.

The association of detached arcs with plasmaspheric plumes and corresponding wave growth calculations provide only indirect evidence that the precipitation is due to EMIC wave scattering. Thus, we also investigate observations of ion cyclotron waves by the Polar Magnetic Field Experiment (MFE) [*Russell et al.*, 1995] near the geomagnetic equator in association with detached arc events.

3. IDENTIFYING DETACHED ARC EVENTS BASED ON GEOSYNCHRONOUS PARTICLE SIGNATURES

In order to identify intervals when subauroral proton precipitation may be expected, we searched the MPA data set to find times when the hot ion moments were similar to those reported for the 18 June 2001 event. Specifically, we identified intervals during which a LANL satellite was located on the dayside, 06 > MLT > 18, and for at least 30 minutes or more MPA observed a hot ion density, n_{ih}, greater than 1 cm^{-3} and a hot ion temperature anisotropy, $A \equiv T_\perp/T_\parallel - 1$, greater than 0.25. No restrictions were placed on the cold ion density as we wanted to independently determine whether the presence of cold plasma was a necessary condition for the observation of the detached arcs. The intervals were then further restricted to times when IMAGE FUV was imaging in the northern hemisphere. We performed the search over four months of data using the two LANL satellites with the best data coverage for each month. The intervals were all from 2001, in the months March (using satellites 1989–046 and 1994–084), May (1991–080 and 1994–084), June (1991–080 and 1994–084), and July (1991–080 and 1994–084).

The months of May through July 2001 were chosen because at that time the IMAGE orbit geometry was particularly favorable for EUV imaging of the plasmasphere and long intervals of high quality images are available during each orbit. During March 2001, the Polar spacecraft crossed the magnetic equator just outside of geosynchronous orbit in the post-noon sector and thus, for portions of its orbit, was in a suitable location for monitoring wave activity associated with detached arcs. Some EUV data are also available during March 2001.

Once the intervals meeting the MPA hot ion criteria ($n_{ih} > 1$ and $A > 0.25$) were identified, we surveyed the FUV S12 images for several hours around each MPA interval in search of subauroral detached proton arcs. Figure 1 gives an overview of the arc events identified using this technique over the four month study period. Overall, 13 intervals were identified when the MPA hot ion criteria were met. Of those, 11 had detached proton arcs observed by FUV, indicated by the solid vertical lines in Figure 1. The proton arcs were not necessarily observed at the same universal time or local time sector as the MPA hot ion observations. For the remaining 2 intervals, indicated by the dotted vertical lines in Figure 1, no evidence of detached subauroral precipitation was observed, and we will show that in these two cases there was no cold plasma extending to large radial distances in the afternoon sector.

3.1 Arc Events with MPA Criteria Met

We will now explore the relationship between the subauroral proton arcs and regions of enhanced cold plasma density. For each of the 11 arc events identified as a result of the MPA search criteria ($n_{ih} > 1$ and $A > 0.25$), a characteristic FUV image was selected and the subauroral proton arc was mapped to the equatorial plane so that its location could be compared with global EUV observations of the plasmasphere as well as *in situ* cold ion density measurements along geosynchronous orbit.

88 RING CURRENT–PLASMASPHERE INTERACTION

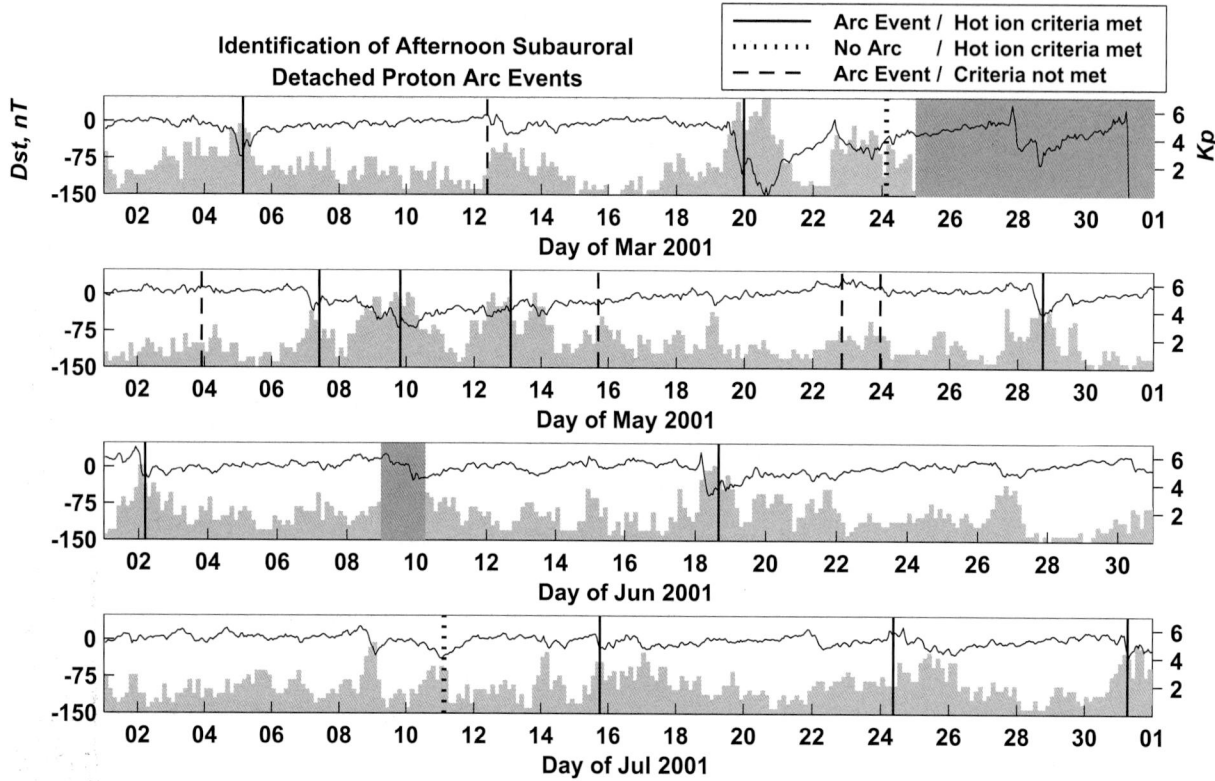

Figure 1. Geomagnetic indices *Dst* (black line) and *Kp* (light gray bars) for the months of March, May, June and July 2001. The solid vertical lines indicate times when subauroral detached proton arcs where observed and the *in situ* hot ion criteria were met (see text). The dotted lines indicate times when the hot ion criteria were met but no detached subauroral precipitation was observed. The dashed lines indicate arc events found when the hot ion criteria were not met. The regions covered by dark gray boxes are times when FUV data are unavailable (late Mar) or of poor quality (9–10 Jun).

3.1.1 Proton Arc Event: 05 Mar 2001 An example event from 05 Mar 2001 is shown in Figures 2–4. On that day LANL 1994-084 observed elevated hot ion densities and ion temperature anisotropies which exceeded the hot ion criteria for about an hour from 02:35 to 03:45 UT in the local time sector from 9.2 to 10.4 (segment between the two gray diamonds in Figures 2 and 4). The FUV imager observed subauroral proton precipitation in the afternoon sector for about a 3.5 hour period from about 02:45 to 05:15 UT around the time of minimum *Dst* for that disturbance interval (Figure 1). The subauroral precipitation appears to fade and rebrighten several times within the 3.5 hour period. At 03:47 UT the proton precipitation consisted of several bright spots in the afternoon sector generally extending from a magnetic latitude of ~68° at 14 MLT to ~60° at 17 MLT (Figure 3).

There appears to be a strong association between the mapped precipitation region and a plasmaspheric plume, as illustrated in Figure 4. Points bounding the subauroral arc were selected and mapped to the Solar Magnetic (SM) equatorial plane using the T96 magnetic field model [*Tsyganenko and Stern*, 1996] and prevailing solar wind conditions, (open squares in Figure 4). This is the same region indicated by the white dashed line in Figure 3. Overlayed are the plasmapause locations (black dots) extracted from an EUV image at 03:54 UT (center of the 10-minute integration window). The plasmapause locations were determined by selecting points along the sharp brightness gradient in the EUV image, finding the field line with the minimum apex along the line of sight to each point using the T96 magnetic field model, and tracing that field line to the SM equatorial plane. A plasmapause location was selected only at local times for which a sharp brightness gradient could be reliably identified in the EUV image [*Goldstein et al.*, 2003]. A distinct plasmaspheric plume can be seen extending sunward in the afternoon sector having been formed as result of a prolonged period of enhanced magnetospheric convection. The proton precipitation region maps within the region of the plasmaspheric plume, just inside the duskside plasmapause as determined by EUV.

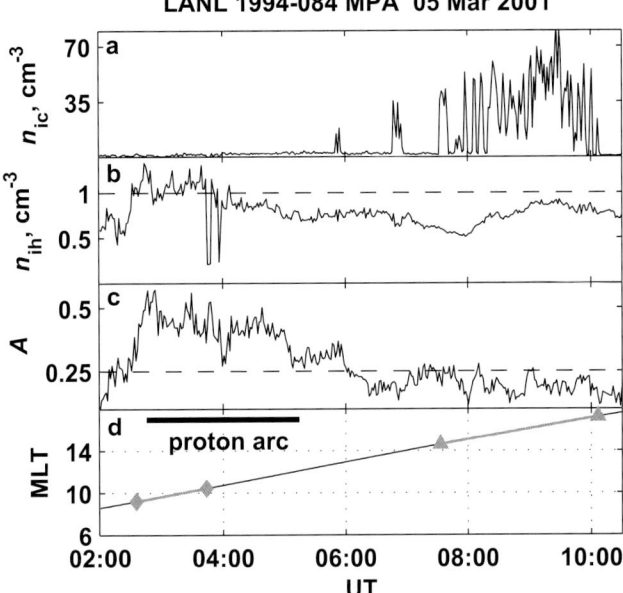

Figure 2. LANL 1994-0084 MPA observations on 05 Mar 2001 of (a) cold ion density (b) hot ion density, (c) hot ion temperature anisotropy along with (d) the satellite magnetic local time. The MPA hot ion criteria threshold is indicated by the dashed horizontal lines in (b) and (c). The MPA observations exceeded the threshold over the segment between the two gray diamonds in (d), and later that day MPA observed a plasmaspheric plume in the segment between the two gray triangles.

Unfortunately, there were no simultaneous *in situ* measurements of the hot and cold plasma parameters directly within the mapped precipitation region. However, several hours later LANL 1994-084 transversed the plume in afternoon sector as indicated by the region between the two gray triangles in Figures 2d and 4. Convection had weakened in the intervening time such that the eastern edge of the plume as measured by MPA at 10:06 UT was located at a later local time as compared to the EUV image at 03:47 UT. The average ion density across the plume was 31 cm^{-3}, and as is typically seen at geosynchronous altitude, the plume contained highly irregular density structure (Figure 2a) [*Moldwin et al.*, 1995, *Spasojević et al.*, 2003, *Goldstein et al.*, 2004].

All 11 arc events identified as a result of the MPA search criteria appear to be associated with regions of enhanced cold plasma density as determined by either global EUV images of the plasmasphere or *in situ* cold plasma density measurements at geosynchronous orbit. In characterizing association between the proton arc and regions of cold plasma, we found that the events could be broken down into one of two categories:

1. All or a significant part of the arc directly maps inside the EUV field of view and within regions of enhanced density. Events of this type include the 05 Mar 2001 event described above as well as 19 Mar 2001, 09 May 2001, 18 June 2001, 24 July 2001, and 31 July 2001.
2. The majority of the arc maps outside the EUV field of view, but EUV images indicate that a plasmaspheric plume likely extends to the mapped precipitation region. Events of this type include 07 May 2001, 13 May 2001, 28 May 2001, 02 June 2001, and 15 July 2001.

For events of both types, there are supporting MPA observations of enhanced cold ion densities in the vicinity of the mapped precipitation region.

Of the 11 proton arc events identified as a result of the MPA hot ion criteria (solid lines in Figure 1), 10 occurred during periods of at least moderately enhanced geomagnetic activity. Correspondingly, each of those 10 arcs was associated with plasmaspheric plume type extensions in the afternoon sector resulting from prolonged periods of enhanced convection. The one exception is the arc event of 24 July 2001 which occurred during a period of relatively weak geomagnetic activity. The proton arc in this event is associated with a quiet time "duskside bulge" type feature [e.g. *Carpenter*, 1970] within which the MPA instrument observed hot ion densities and temperature anisotropies in excess of the selected threshold criteria.

3.1.2 Proton Arc Event: 09 May 2001 Another example of the first type of event can be seen in Figures 5 and 6. Like

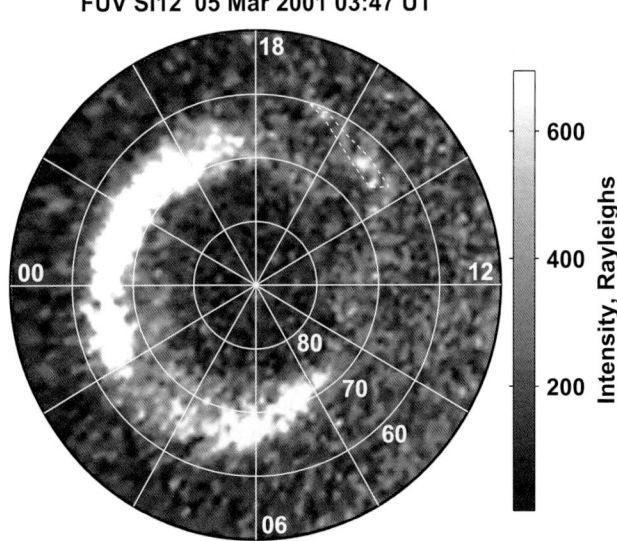

Figure 3. FUV SI12 image of the proton aurora on 05 Mar 2001 at 03:47 UT mapped onto the magnetic APEX coordinates with noon to the right. A subauroral detached proton arc can be seen in the afternoon sector between ~60° to 68° magnetic latitude, as indicated by the dashed white line.

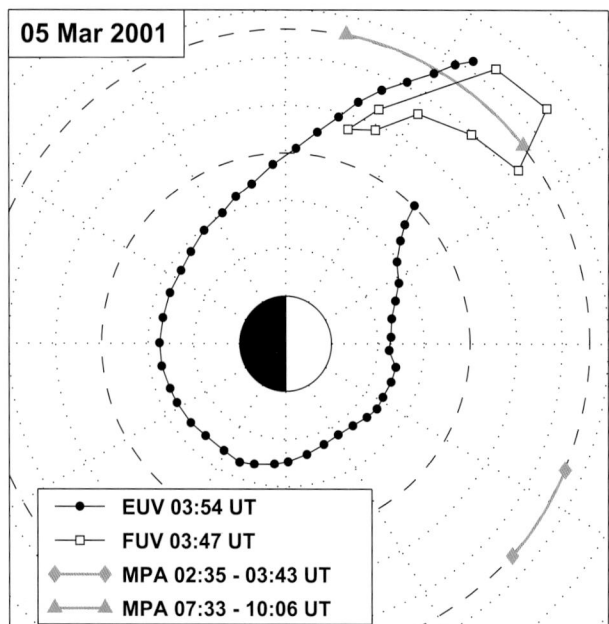

Figure 4. EUV plasmapause locations (black dots) from 05 Mar 2001 at 03:54 UT are shown along with the mapped proton precipitation region (open squares) from the FUV SI12 image at 03:47 UT. The region over which the MPA hot ion criteria were met is indicated by the gray diamonds, and MPA observations of the plasmaspheric plume were made several hours later in the region indicated by the gray triangles (as also shown in Figure 2). The sun is to the right, dotted circles are spaced 1 R_E apart, and dashed circles are at 4 and 6.6 R_E.

the previous event, on 09 May 2001 the LANL 1991-080 MPA instrument observed hot ion densities and temperature anisotropies (not shown) in excess of the selected threshold criteria in the morning sector (region between the gray diamonds in Figure 6), and there were no observations of the hot ion parameters in the afternoon sector at the time of the detached arc. Although the subauroral proton arc on 09 May 2001 was nearly twice as bright (Figure 5) as the 05 Mar 2001 event, FUV observed the arc for only about a half hour, from 19:45 to 20:15 UT. The arc was located at a later local time (17–19 MLT) and lower magnetic latitude (56°–60°) than any other event identified thus far. In addition, it is the only subauroral arc in this study which mapped completely within the EUV field of view (Figure 6) lying primarily within 4 R_E just inside the duskside plasmapause.

3.1.3 Proton Arc Event: 02 June 2001 The 02 June 2001 arc event is an example of the second type described above, where the majority of the arc maps outside the EUV field of view. The proton arc persisted from about 04:30 to 05:00 UT and maps to the equatorial plane outside of a radial distance of ∼ 6 R_E (Figure 7). In addition, the arc was closer to noon than the previous examples. The EUV images at that time show a broad region of enhanced density extending from prenoon across the afternoon sector, corresponding to early stage plume development [*Spasojević et al.*, 2003]. A period of enhanced magnetospheric convection began near 00:20 UT on 02 June 2001 as determined by the the first evidence of inward motion of the nightside plasmapause and also corresponding to a strong southward turning of the interplanetary magnetic field (IMF) (indicated by upstream solar wind monitors) [*Goldstein et al.*, 2004]. The proton arc was observed after only about 4 hours of enhanced convection whereas for the two previous examples (05 Mar 2001 and 09 May 2001) the period of enhanced convection began 12 or more hours before the proton arc was observed. Thus, the observed plumes in those events (Figure 4 and 6) were narrower in local time extent as a result of the long duration of combined sunward convection and eastward corotation on the dayside.

The proton precipitation region on 02 June 2001 maps outside the EUV field of view but likely within the sunward extension of the plasmaspheric plume. At the time of the mapped images in Figure 7, the LANL 1994-084 was located near the western edge of the plasmaspheric plume (gray star in Figures 7 and 8) where the MPA instrument observed a cold ion density of ∼50 cm^{-3}. MPA continued to observe plume material as it traversed the afternoon sector (region between the gray triangles). The plume began to

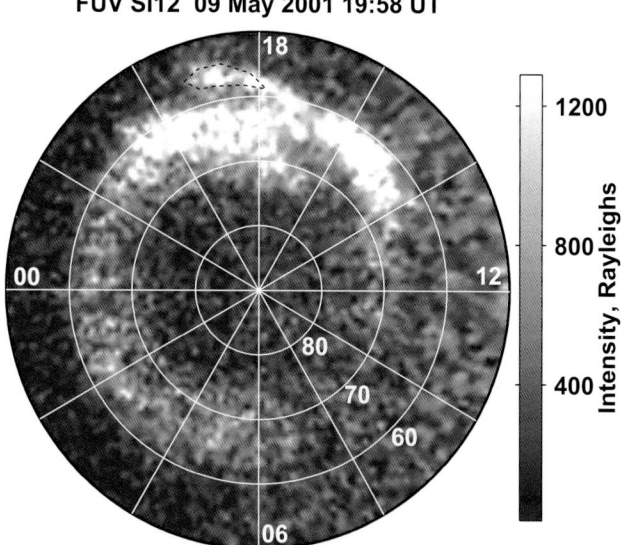

Figure 5. FUV SI12 image of the proton aurora on 09 Mar 2001 at 19:58 UT. A subauroral detached proton arc can be seen near dusk extending from ∼60° near 17 MLT to ∼56° near 19 MLT, as indicated by the dashed black line.

rotate eastward when convection decreased around 06:00 UT and thus LANL 1994-084 did not exit the plume until 11:53 UT. In addition, LANL 1994-084 observed enhanced hot ion densities and temperature anisotropies across much of the dayside (Figure 8b–c).

3.1.4 Proton Arc Event: 28 May 2001 A second example of a proton arc which maps outside the EUV field of view is the event of 28 May 2001 (Figure 9). Again in this event, EUV observed a well defined plume in the afternoon sector. The proton precipitation region observed by FUV maps completely outside of the EUV field of view but to a narrow local time region in what appears to be the radial extension of the plume.

3.2 MPA Criteria Met Without Subauroral Precipitation

For two of the 13 intervals which met the MPA hot ion criteria, there were no indications of subauroral proton precipitation in the FUV images. EUV images for both of these events indicate the absence of plasmaspheric plumes or bulges that extend to large radial distances in the afternoon sector.

The extracted plasmapause locations on 24 Mar 2001 from an EUV image during the interval over which MPA observed

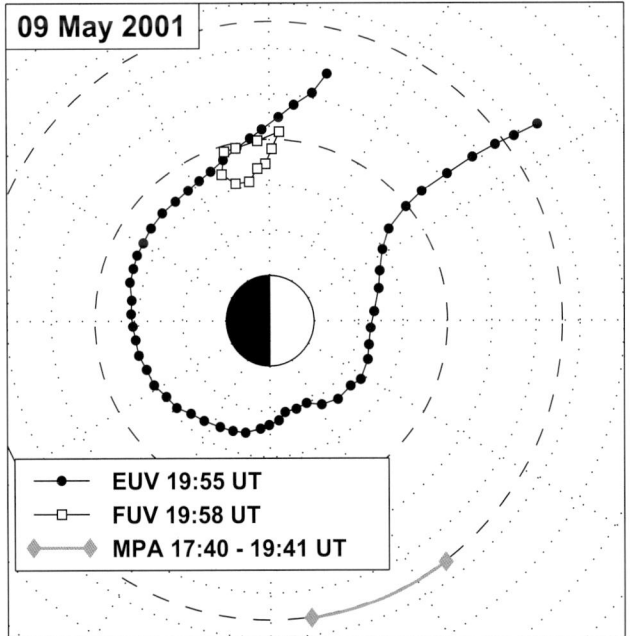

Figure 6. EUV plasmapause locations (black dots) from 09 May 2001 at 19:55 UT are shown along with the mapped proton precipitation region (open squares) from the FUV SI12 image at 19:58 UT. The region over which the MPA hot ion criteria were met is indicated by the gray diamonds.

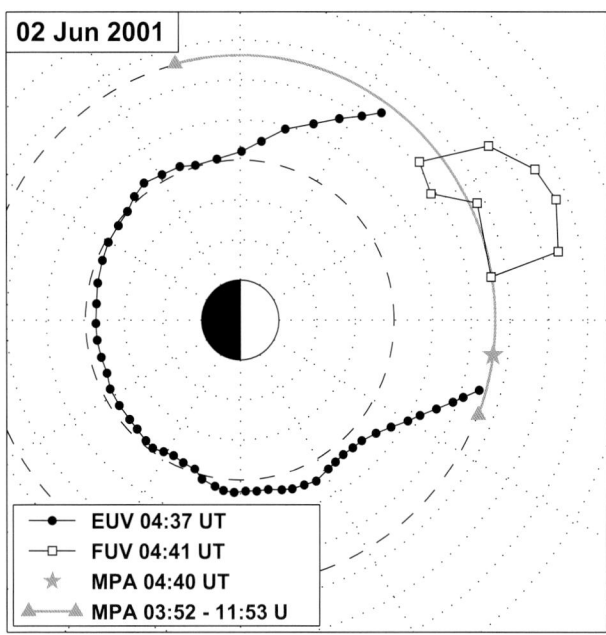

Figure 7. Plasmapause locations (black dots) and the mapped proton precipitation region (open squares) on 02 June 2001. LANL 1994-084 was at the location of the gray star at the time of the EUV and FUV observations, and MPA observed enhanced cold plasma density over the region between the gray triangles as shown in Figure 8.

enhanced hot ion densities and temperature anisotropies is shown in Figure 10a. At this time, the average plasmapause location was at $L \approx 3.6$. A remnant of a small plume can be seen centered at ~ 21 MLT. The EUV observations over the previous several days suggest that the plasmasphere was severely depleted during the large magnetic storm which occurred 19–21 Mar 2001 (Figure 1) and did not recover prior to the subsequent disturbance beginning on 22 Mar 2001. Thus, despite a period of prolonged enhanced convection, a well-defined large scale plume did not form during the second disturbance period. MPA observations on 24 Mar 2001 also confirm the absence of cold plasma at geosynchronous altitude (not shown).

The other MPA hot ion criteria interval for which no subauroral precipitation was observed is 11 July 2001. EUV images from during this time (Figure 10b) indicate that the plasmasphere was inside 4 R_E except for a slight bulge extending to 5 R_E in the afternoon sector. Prior to the EUV data shown in Figure 10b, the IMF had been southward for the past ~ 18 hours, so it is somewhat puzzling that a large scale plume had not formed. One possibility is that convection electric field was effectively shielded from the inner magnetosphere on this day by the inner edge of the plasma sheet and the Region 2 field-aligned current system [e.g.,

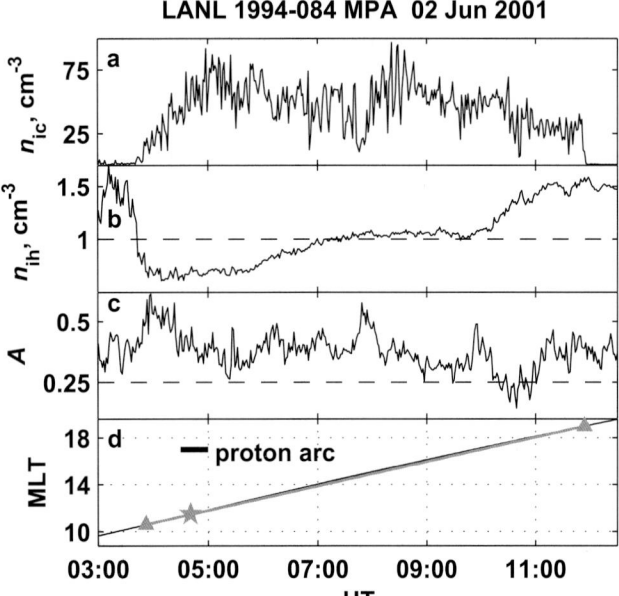

Figure 8. Same as Figure 2, except the gray star in (d) indicates the location of LANL 1994-084 at the time of the EUV and FUV observations in Figure 7.

Jaggi and Wolf, 1973]. This interpretation is supported by observations by the solar wind monitors of a gradual rotation in the IMF from northward to southward over a period of about six hours which could have allowed an effective shielding layer to be established.

Therefore, for the events of 24 Mar 2001 and 11 July 2001, it is possible that the lack of subauroral precipitation may be related to the absence of cold plasma outside $\sim$ 4 to 5 R_E even though the hot ion parameters observed by MPA were similar to those during other subauroral detached proton arc events.

4. ADDITIONAL DETACHED ARC EVENTS

In addition to the subauroral detached arc events discussed thus far, five more arc events were found within the four month study interval at times when the *in situ* hot ion criteria were not met. These events are indicated by the dashed lines in Figure 1 and were identified as a result of a visual survey of the FUV SI12 images for the months of March, May, June and July 2001. In contrast to the arc events discussed in Section 3.1, these arc events on average occurred under quieter geomagnetic conditions and were located at higher magnetic latitudes. For each of these five events, the entire proton precipitation region maps to the equatorial plane outside of geosynchronous orbit, and thus it is not unexpected that the criteria used to identify arc events based on geosynchronous particle observations would be insufficient to identify these events. Also since the arcs map to such large radial distances, it is not possible, using EUV images and MPA observations alone, to unambiguously assess the relationship between the arcs and regions of enhanced cold plasma density for each of these events.

4.1 Proton Arc Event: 22 May 2001

One such example is 22 May 2001 as shown in Figure 11a. The detached proton arc at 20:19 UT maps to a radial distance of between 8 and 10 R_E in the pre-dusk sector. At the time of the arc, the IMAGE satellite was still at relatively low altitude such that the plasmasphere completely filled the EUV field of view. By 22:06 UT, the satellite was closer to apogee and the EUV extracted plasmapause locations indicate a large bulge in the dusk to midnight quadrant which extends to edge of the EUV field of view to at least 7–8 R_E. It is possible this bulge was collocated with the proton arc but in the subsequent two hours corotated eastward. Similarly, for the events of 12 Mar 2001, 03 May 2001, and 23 May 2001, the association with cold plasma using the existing EUV and/or MPA data is difficult to establish precisely.

4.2 Proton Arc Event: 15 May 2001

One detached arc event which is clearly unrelated to any plasmaspheric density structure occurred on 15 May 2001 (Figure 11b). The proton arc at 16:45 UT maps to a radial distance of $\sim$ 10 R_E, further out than any other arc in this

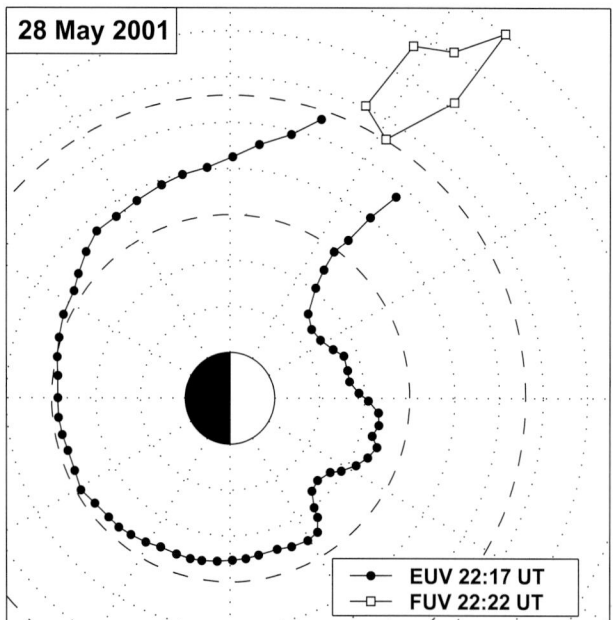

Figure 9. Plasmapause locations (black dots) and the mapped proton precipitation region (open squares) on 28 May 2001.

study. Although high quality EUV data was not available at the time the arc was present, observations at 19:02 UT indicated that the plasmasphere was devoid of any large scale bulges and the plasmapause was located at ~ 4.5 R_E in the dusk sector.

5. RELATIONSHIP BETWEEN DETACHED ARCS AND GEOMAGNETIC ACTIVITY

The spatial distribution of all 16 detached proton arc events is shown in Figure 12. Each dot is the centroid of the the mapped precipitation region determined from one FUV image during the event with the gray dots corresponding to detached arcs identified by the MPA hot ion criteria (solid lines in Figure 1) and the open dots to those that did not meet the specified criteria (dashed lines in Figure 1). As previously mentioned, the open dots map to larger radial distances. The average location of all the proton arcs events is $r = 7.4\ R_E$. There is a clear correlation between the radial distance of the mapped arc centroid and the level of geomagnetic activity as measured by the *Dst* index (Figure 13). The arcs tend to be located at lower latitudes, and thus map closer to the Earth, during geomagnetically disturbed periods. The linear correlation coefficient is 0.69 (high statistical significance). The correlation between arc radial distance and *Kp* was similar ($\rho = 0.70$). Conversely, there was no correlation found between geomagnetic activity and the local time of the arc centroid.

6. Relationship between Detached Arcs and Solar Wind Conditions

Previous studies have reported that subauroral detached proton arcs can be observed after a change from negative to

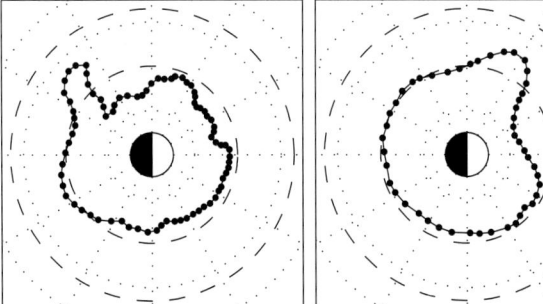

Figure 10. EUV plasmapause locations on a) 24 Mar 2001 and b) 11 July 2001 at times when MPA hot ion parameters exceeded the selected threshold, but FUV did not observe any detached subauroral precipitation. Dotted circles are 2 R_E apart and dashed circles indicate 4 and 6.6 R_E.

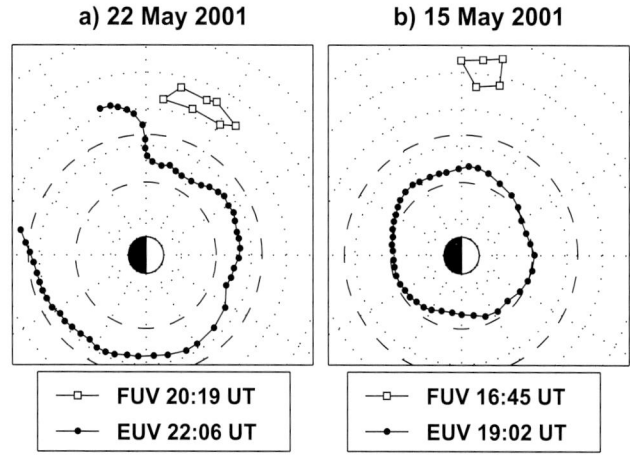

Figure 11. EUV plasmapause locations and FUV mapped precipitation regions for a) 22 May 2001 and b) 15 May 2001. The MPA hot ion criteria were not met during these detached arc events. Dotted circles are 2 R_E apart and dashed circles indicate 4 and 6.6 R_E.

positive of either the B_z or B_y component of the interplanetary magnetic field (IMF) [*Burch et al.*, 2002; *Spasojević et al.*, 2004]. As a result of either IMF transition, the main proton oval in the afternoon sector contracts poleward, while the equatorward part of the oval remains at its original latitude. Thus, a separation of several degrees in latitude is created between the new oval position and the presumably pre-existing proton arc. In addition, the previously reported detached arc events occurred during periods of relatively high solar wind dynamic pressure.

To further explore the relationship between detached proton arcs and solar wind conditions, we performed a superposed epoch analysis of the parameters IMF B_z, B_y, and solar wind dynamic pressure using all 16 arc events (Figure 14). The solar wind data was taken from either the ACE or WIND satellites, and the data for each event interval was appropriately time shifted to account for propagation to the magnetopause. The epoch time of 0 hours (solid vertical line) corresponds to the time the detached arc was first observed. Thus, each panel in Figure 14 represents the average value of each solar wind parameter from eight hours prior to four hours after the detached arc was first observed.

For IMF B_z, there is a clear pattern of southward IMF for about six to eight hours prior to the arc occurrence. This is not unexpected given that most of the arcs occurred during negative excursions of *Dst* and were associated with plasmaspheric plumes which form during periods of enhanced magnetospheric convection. However, at the time the arc is first observed and in the hours that follow, the average value of B_z is close to zero. In examining the 16 individual B_z records, in about half of the events, the arcs appear after a northward

Spatial Distribution of Detached Proton Arc Events

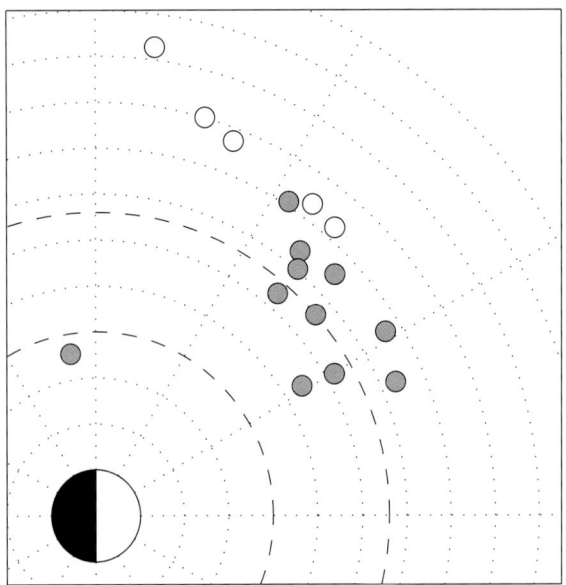

Figure 12. The spatial distribution of all 16 subauroral detached proton arcs identified in the four month study period. Each dot is the centroid of the mapped precipitation region determined from one FUV image during the each event. The gray dots are the events which were identified during intervals which the MPA hot ion criteria were met while the criteria were not met for the open circles. Dotted circles are 1 R_E apart and dashed circles indicate 4 and 6.6 R_E.

IMF turning, and there are particularly abrupt transitions for the events of 05 Mar 2001, 18 June 2001 and 31 July 2001. On the other hand, in the remaining half of the events the IMF remains southward or near zero. Thus, a negative to positive IMF B_z transition is not a necessary condition for the formation of detached proton arcs. However, the fact that some of the arcs are only visible after a northward turning, which causes the main proton oval to contract poleward, suggests that at other times ring current precipitation as a result of interaction with the plasmasphere may contribute to the equatorward portion of the auroral oval even though a distinct and detached arc is not present. There does not appear to be a systematic trend in the IMF B_y in either the superposed epoch analysis, in which the average value before and after the arc observation is close to zero, or in the inspection of the B_y records for individual events.

The detached proton arcs are in general associated with extended periods of enhanced solar wind dynamic pressure consistent with the previously reported case studies. The average dynamic pressure both before and after the arc observation is ~ 3.4 nPa while the average dynamic pressure for the entire four month period is ~ 2.0 nPa. Only one of sixteen arc events did not occur during elevated dynamic pressure, so while high dynamic pressure is perhaps not a necessary condition, arcs are more likely to be seen during the high pressure intervals. The afternoon detached arcs do not appear to be associated with pressure pulses such as has been reported for another class of so-called dayside subauroral proton flashes [*Zhang et al.*, 2002; *Hubert et al.*, 2003].

7. OBSERVATIONS OF ION CYCLOTRON WAVES IN ASSOCIATION WITH DETACHED ARCS

The close association between afternoon subauroral detached proton arcs and regions of cold, dense plasmaspheric material for the majority of arc events in this study supports the previous assertions of *Burch et al.* [2002] and *Spasojević et al.* [2004] that the precipitation may be due to pitch angle scattering of energetic protons by electromagnetic ion cyclotron (EMIC) waves which may preferentially be amplified in regions of enhanced cold plasma density. In addition, *Immel et al.* [this volume] reported ground observations of magnetic pulsations in the Pc1 range in association with four detached arc events. However, ion cyclotron waves are likely generated near the equatorial plane, and may be damped in the off-equatorial regions or affected by ionospheric transmission [*Fraser et al.*, 1996; *Mursula et al.*, 2000].

During March of 2001 Polar crossed the magnetic equatorial plane at radial distances of 7–8 R_E in the post-noon local time sector. We identified conjunction intervals during that month when Polar was near the magnetic equator while the IMAGE spacecraft was positioned at high northern latitudes imaging both the proton aurora and the cold plasma distribution. Fortunately, two of the three detached arc events identified in March 2001 (12 Mar 2001 and 19 Mar 2001) occurred during conjunction times. Although Polar does not pass directly through the mapped precipitation region, the Magnetic Field Experiment (MFE) observed strong EMIC

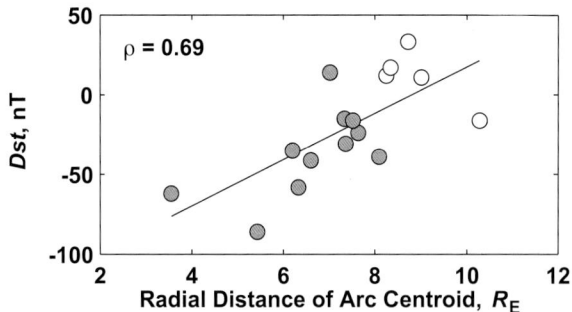

Figure 13. Correlation between the radial distance of the arc centroid and the *Dst* magnetic index at the time of the arc. The circle designations are the same as in Figure 12. The linear correlation coefficient is 0.69 (highly significant).

wave activity during both events. In analyzing the Polar MFE data for the remainder of the month, we found that there are no other periods of strong wave activity during IMAGE conjunctions, nor are there observations of detached arcs during conjunction times in the absence of waves. The third detached arc event of the month, 05 Mar 2001, did not occur during a conjunction with Polar.

7.1 Proton Arc Event with EMIC Waves: 19 Mar 2001

On 19 Mar 2001, FUV observed a detached proton arc during the main phase of a large geomagnetic storm (Figure 1). The detached arc maps to the equatorial plane directly within a plasmaspheric plume observed by EUV (Figure 15). About two hours earlier, the Polar spacecraft crossed the equatorial plane at slightly earlier local time than the mapped precipitation region. In Figure 15, the location of the Polar spacecraft from 20:00 to 22:00 UT was traced to the SM equatorial plane using the T96 magnetic field model. The magnetic field observations over that interval (Figure 16) show the presence of strong ion cyclotron waves that are confined to below the He^+ gyrofrequency (dashed line). We verified that the waves are left-hand polarized and propagate along the magnetic field direction.

7.2 Proton Arc Event with EMIC Waves: 12 Mar 2001

The detached proton arc on 12 Mar 2001 occurred near the onset of a small geomagnetic disturbance and the precipitation region at 09:16 UT maps to the equatorial plane outside of geosynchronous orbit (Figure 17). There were no EUV images available at the time of the detached arc, but about

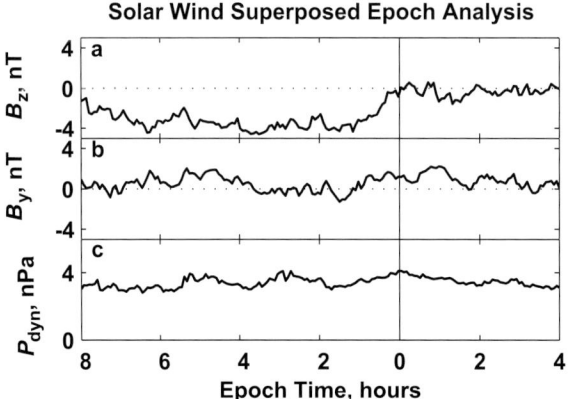

Figure 14. Superposed epoch analysis of a) IMF B_z, b) IMF B_y and c) solar wind dynamic pressure for all 16 subauroral detached proton arcs. The solar wind data was propagated to the magnetopause for each event, and the epoch time of 0 hours refers to the time the detached arc was first observed.

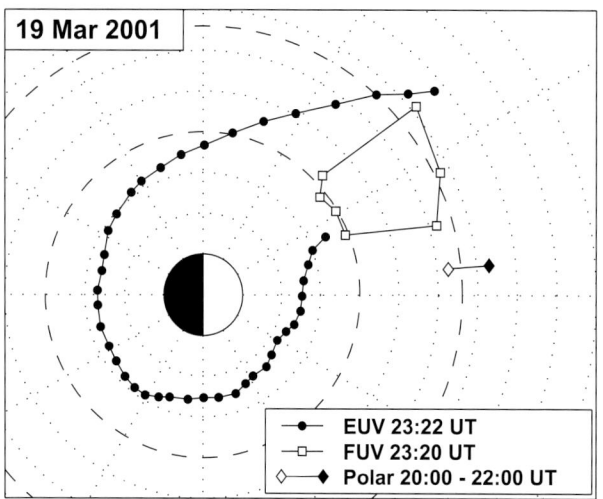

Figure 15. Plasmapause locations (black dots) and the mapped proton precipitation region (open squares) are shown for 19 Mar 2001 along with the location of the Polar spacecraft, mapped to the SM equatorial plane, for a two hour period during which MFE observed strong EMIC waves as shown in Figure 16.

five hours earlier, EUV observed a rather expanded plasmasphere with the plasmapause on the nightside extending to a radial distance of $\sim 6\ R_E$. The plasmapause location on the dayside could not be reliably identified due to sunlight contamination. The MPA instrument on LANL 1994-084 observed cold plasma with densities greater than 10 cm^{-3} in the afternoon sector (region bounded by the gray triangles in Figure 17a,b). At the time the proton arc was observed, MPA measured a cold plasma density of $\sim 35\ cm^{-3}$ just Earthward of the proton arc (location of the gray star).

From 08:15 UT to 10:00 UT, Polar MFE observed ion cyclotron waves at about the same radial distance as the mapped proton precipitation region but at an earlier local time. In this event, ion cyclotron waves were primarily above the He^+ gyrofrequency (dashed line in Figure 18) although near 08:30 UT there is some wave energy below the He^+ gyrofrequency.

In both detached arc events (19 Mar 2001 and 12 Mar 2001), the ion cyclotron waves are observed at about the same radial distance as the mapped precipitation region but Polar was at a slighter earlier magnetic local time. The difference in the wave spectra in the two events can be attributed to differences in the heavy ion content of the plasma [*Kozyra et al.*, 1984]. In the non-storm time event, 12 Mar 2001, the wave energy was above the local He^+ gyrofrequency and thus the energetic ions were likely primarily composed of protons. If a significant amount of cold plasma was present, it was also likely primarily protons. For the storm time event, 19 Mar 2001, waves above the He^+ gyrofrequency were absent and

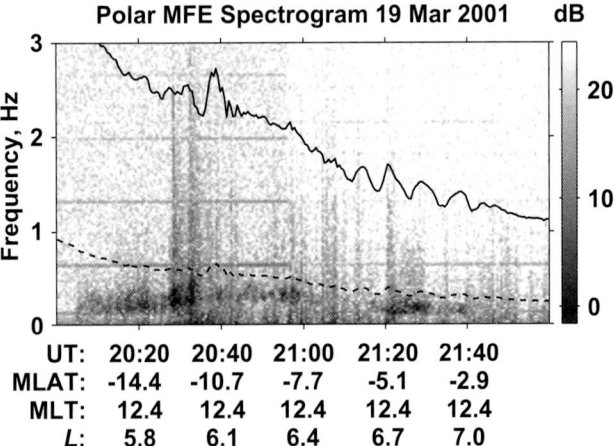

Figure 16. MFE spectrogram for the portion of the Polar orbit shown in Figure 15. The solid and dashed black lines are the H$^+$ and He$^+$ gyrofrequencies respectively. In addition to universal time, magnetic latitude, magnetic local time and L value (based on a dipole magnetic field) are also shown.

high fractions of heavy ions have been shown to suppress wave growth in this range. Clearly, there was a significant amount of He$^+$ in the plume region of the plasmasphere since EUV instrument images the He$^+$ distribution, but there may have also been enhanced heavy ions in the hot plasma distribution as is typical of the storm time ring current.

8. DISCUSSION

We have investigated the occurrence of afternoon subauroral detached proton arcs observed by the IMAGE FUV SI12 instrument and their relationship with regions of enhanced cold plasma density, electromagnetic ion cyclotron waves as well as geomagnetic and solar wind conditions during the months of March, May, June and July 2001.

In situ measurements of energetic ion parameters can be useful in identifying intervals when the detached proton arcs are likely to occur. Over the four month study period, we identified 13 intervals when enhanced hot ion densities and temperature anisotropy were observed at geosynchronous orbit on the dayside. FUV observed subauroral detached proton arcs during 11 of the 13 intervals. Although the exact choice of the hot ion criteria (nih > 1 and A > 0.25) was somewhat subjective, it was based upon measured values of 18 June 2001, a time when wave growth would have been expected based on instability calculations using the *in situ* hot and cold plasma parameters inside the mapped precipitation region [*Spasojević et al.*, 2004]. Only two of the 13 hot ion criteria intervals did not appear to have subauroral precipitation (24 Mar 2001 and 11 July 2001), and EUV and MPA observations for those events indicate that plasmasphere was rather compact and no cold plasma extended to geosynchronous orbit.

Five additional arcs were identified during intervals that did not meet MPA hot ion criteria. In general, these events occurred under quieter geomagnetic conditions and mapped to the equatorial plane well outside of geosynchronous orbit. Therefore, it is not unexpected that the criteria used to identify arc events based on geosynchronous particle observations would be insufficient to identify these events. Statistically, energetic proton distributions become more unstable to wave growth with increasing radial distance [*Anderson et al.*, 1992], but due to a lack of satellite observations outside of geosynchronous orbit, we are unable to verify this effect for these higher latitude arc events.

We found highly significant statistical correlation between the level of geomagnetic activity, using *Dst* and *Kp*, and

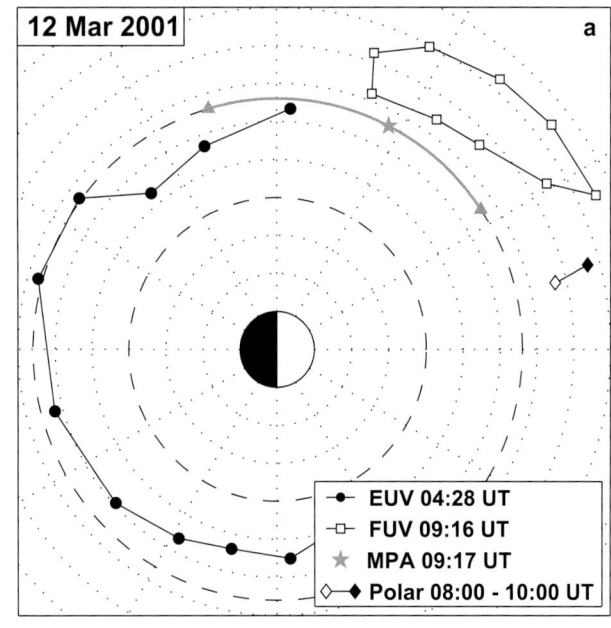
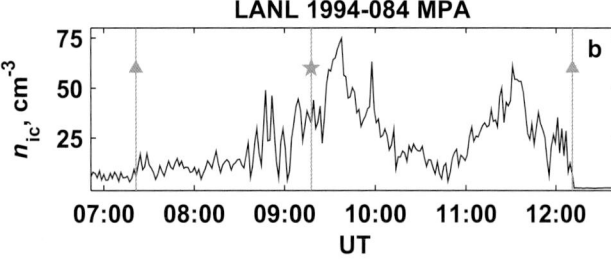

Figure 17. a) Same as Figure 15 for 12 Mar 2001. The MFE data for the segment of the Polar orbit is shown in Figure 18. Also included is the location of the LANL 1994-084 at the time of the FUV observation (gray star). b) LANL 1994-084 MPA measurements of cold ion density. MPA observed cold plasma with density > 10 cm^{-3} in the region between the gray triangles in a) and b).

location of the detached arc. During disturbed conditions, the arcs tend to map in the equatorial plane to smaller radial distances (that is, the arc is located at lower magnetic latitudes) while during quiet conditions, the arcs map to larger radial distances (located at higher latitudes).

Overall, the majority of detached arc events in this study occurred during periods of moderate to strong geomagnetic disturbance. For each of the disturbance-time events, there is a clear association between the mapped proton precipitation region and a global scale plasmaspheric plume. Supporting MPA observations of cold ion densities in the range of 20–50 cm^{-3} within the plume are also available for many of these events.

In contrast, since the events which occurred under quieter conditions map further out, it is not possible to unambiguously confirm the presence of enhanced cold plasma. For the events of 12 Mar 2001, 03 May 2001, 22 May 2001, and 23 May 2001, EUV observes a large, expanded plasmasphere, typical of quiet times, and MPA observes cold plasma at geosynchronous orbit. However, it is still difficult to link these observations of cold plasma to precipitation regions that map out to $> 8 R_E$. In addition, the 15 May 2001 detached arc event appears to be completely unrelated to any plasmaspheric structure. It is possible that the detached arcs which map to larger radial distances are still a result of wave scattering, but the wave amplification proceeds in the absence of cold plasma [*Kozyra et al.*, 1984]. Another possibility is that they may be related to detached blobs of cold plasma trapped in the outer magnetosphere in the aftermath of periods of enhanced convection [*Carpenter et al.*, 1993].

Hot-cold plasma interaction provides an attractive mechanism for the formation of the proton arcs since it could explain why the arcs appear detached from the main proton auroral oval. In absence of cold plasma, the conditions for wave growth become more favorable as the magnetopause is approached due to reduced magnetic field strength and increased temperature anisotropy resulting from drift shell splitting. Enhanced cold plasma in the middle magnetosphere (5–7 R_E) might provide an isolated region closer to the Earth where wave growth and scattering is enhanced. The proton precipitation region in the ionosphere would then appear isolated and detached from the main oval.

We also analyzed the solar wind conditions for each of the sixteen detached arc events. There is a preference for the arcs to occur during periods of enhanced dynamic pressure with the average dynamic pressure before and during the arc events about 1.7 times higher than the overall average dynamic pressure for the entire four month study interval. None of the detached arcs were associated with solar wind pressure pulses. There is also a clear trend of southward IMF for up to 8 hours before the detached arc is first observed. This is not surprising given the fact that most of the events occurred during disturbed periods and were associated with plasmaspheric convection plumes. Some arcs appear only after northward turning of the IMF such as was previously described by *Burch et al.* [2002] and *Spasojević et al.* [2004]. This suggests that the equatorward edge of the proton oval may have contributions from ring current-plasmasphere interactions at other times, but the precipitation region does not appear distinct and detached from the main oval unless a northward turning causes the main oval to retreat to higher latitudes. On the other hand, some arc events are still seen in the absence of northward turnings, such that the value of B_z in the hours after the arc is first observed averages to zero in the superposed epoch analysis of all sixteen arc events.

In order to further link the detached proton arcs with wave scattering, we explored Polar MFE observations of EMIC waves during conjunctions with the IMAGE satellite in the March 2001. Two of the previously identified detached arc events occurred during conjunction intervals and MFE observed strong ion cyclotron waves at about the same radial distance as the mapped precipitation region but at an earlier magnetic local time. For the rest of the month, there were no other periods of strong wave activity as Polar crossed the equatorial plane during conjunctions with IMAGE. Also, no other detached arcs were observed in the absence of waves.

9. CONCLUSIONS

It has long been recognized that energetic protons could be precipitated from the ring current as a result of wave-particle interactions occurring within the duskside plasmasphere. Imaging of the proton aurora and plasmasphere has allowed

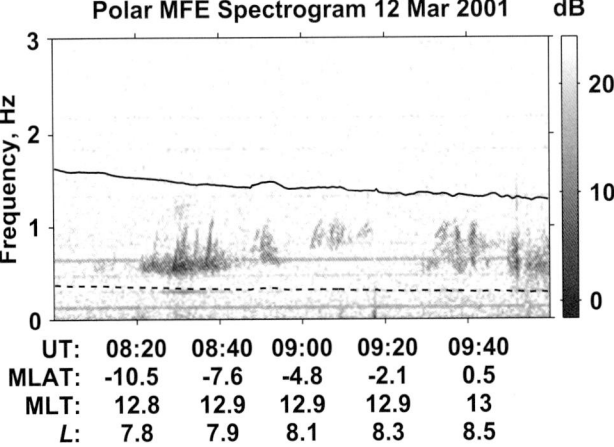

Figure 18. Same as Figure 16 for the segment of the Polar orbit shown in Figure 17.

us to extend previous observational work by exploring this process from a global perspective. We find that arcs of precipitating protons which occur in the afternoon local time sector equatorward of and detached from the main proton auroral oval during geomagnetic disturbances are consistently associated with sunward extending regions of cold plasma or plasmaspheric plumes. We continue to explore the occurrence of electromagnetic ion cyclotron waves in association with the precipitating protons by means of *in situ* observations presented here as well as the ground based measurements reported by *Immel et al.* [this volume]. Finally, the afternoon subauroral proton arcs provide an excellent basis for comparison with predicted proton precipitation patterns from wave scattering included in the increasingly sophisticated global ring current models.

Acknowledgments. We are grateful to Kyoto World Data Center for providing the geomagnetic indices, the ACE MAG and SWE-PAM instrument teams, the ACE Science Center, the CDAWeb, and R. Lepping and K. Oglivie at NASA for providing the solar wind data. The work at Los Alamos was conducted under the auspices of the U.S. Department of Energy. The work at UCLA was supported by NASA contract NAG 5-3171. The work at Univ. of Arizona was supported by a subcontract from Southwest Research Institutte, under NASA contract NAS5-96020.

REFERENCES

Anderson B. J., R. E. Erlandson, and L. J. Zanetti (1992), A statistical study of Pc 1-2 magnetic pulsations in the equatorial magnetosphere. 1. Equatorial occurrence distributions., *J. Geophys. Res., 97*(A3), 3075–3088.

Bame, S. J., et al. (1993), Magnetospheric plasma analyzer for spacecraft with constrained resources, *Rev. Sci. Instrum., 64* 1026–1033.

Burch, J. L. (2000), IMAGE mission overview, *Space Sci. Rev., 91*, 1–14.

Burch, J. L., D. G. Mitchell, B. R. Sandel, P. C. Brandt, and M. Wuest (2001), Global dynamics of the plasmasphere and ring current during magnetic storms, *Geophys. Res. Lett., 28*(6), 1159–1162.

Burch, J. L., et al. (2002), Interplanetary magnetic field control of afternoon- sector detached proton auroral arcs, *J. Geophys. Res., 107*(A9), 1251, doi:10.1029/2001JA007554.

Carpenter D. L. (1970), Whistler evidence of the dynamic behaviour of the duskside bulge in the plasmasphere. *J. Geophys. Res., 75*(19), 3837–3847.

Carpenter D. L., B. L. Giles, C. R. Chappell, P. M. E. Decreau, R. R. Anderson, A. M. Persoon, A. J. Smith, Y. Corcuff, and P. Canu (1993), Plasmasphere dynamics in the duskside bulge region: a new look at an old topic, *J. Geophys. Res., 98*(A11), 19243–19271.

Chappell, C. R. (1974), Detached plasma regions in the magnetosphere. *J. Geophys. Res., 79*(13), 1861–1870.

Chen A. J., and R. A. Wolf (192), Effects on the plasmasphere of a time-varying convection electric field. *Planet. Space Sci., 20*(4), 483–509.

Cornwall, J. M., F. V. Coroniti, and R. M. Thorne (1970), Turbulent loss of ring current protons, *J. Geophys. Res., 75*(25), 4699–4709.

Erlandson, R. E., and A. J. Ukhorskiy (2001), Observations of electromagnetic ion cyclotron waves during geomagnetic storms: Wave occurrence and pitch angle scattering, *J. Geophys. Res., 106*(A3), 3883–3896.

Frank, L. A. (1971), Relationship of the plasma sheet, ring current, trapping boundary, and plasmapause near the magnetic equator and local midnight, *J. Geophys. Res., 76*(10), 2265–2275.

Fraser B. J., H. J. Singer, W. J. Hughes, J. R. Wygant, R. R. Anderson, and Y. D. Hu (1996), CRRES Poynting vector observations of electromagnetic ion cyclotron waves near the plasmapause. *J. Geophys. Res., 101*(A7), 15331–15344.

Fraser B. J., and T. S. Nguyen (2001), Is the plasmapause a preferred source region of electromagnetic ion cyclotron waves in the magnetosphere? *J. Atmos. Sol.-Ter. Phys., 63*(11), 1225–1247.

Fok, M.-C., Y. Ebihara, T. E. Moore, D. M. Ober, and K. A. Keller, (2005), Geospace storm processes coupling the ring current, radiation belt and plasmasphere, this volume.

Gary, S. P., M. F. Thomsen, L. Yin, and D. Winske (1995), Electromagnetic proton cyclotron instability: Interactions with magnetospheric protons, *J. Geophys. Res., 100*(A11), 21961–21972.

Goldstein, J. and B. R. Sandel (2005), The global pattern of evolution of plasmaspheric drainage plumes, this volume.

Goldstein, J., M. Spasojević, P. H. Reiff, B. R. Sandel, W. T. Forrester, D. L. Gallagher, and B. W. Reinisch (2003), Identifying the plasmapause in IMAGE EUV data using IMAGE RPI in situ steep density gradients, *J. Geophys. Res., 108*(A4), 1147, doi:10.1029/2002JA009475.

Goldstein, J., B. R. Sandel, M. F. Thomsen, M. Spasojević, and P. H. Reiff (2004), Simultaneous remote sensing and in situ observations of plasmaspheric drainage plumes, *J. Geophys. Res., 109*, A03202, 10.1029/2003JA010281.

Grebowsky, J. M. (1970), Model study of plasmapause motion, *J. Geophys. Res., 75*(22), 4329–4333.

Hubert, B., J. C. Gérard, S. A. Fuselier, and S. B. Mende (2003), Observation of dayside subauroral proton flashes with the IMAGE-FUV imagers, *Geophys. Res. Lett., 30*(3), 1145, doi:10.1029/2002GL016464.

Immel, T. J., S. B. Mende, H. U. Frey, L. M. Peticolas, C. W. Carlson, J.-C. Gérard, B. Hubert, S. A. Fuselier, and J. L. Burch (2002), Precipitation of auroral protons in detached arc, *Geophys. Res. Lett., 29*(11), 1519, doi:10.1029/2001GL013847.

Immel T. J., S. B. Mende, H. U. Frey, J. Patel, J. W. Bonnell, M. J. Engebretson, and S. A. Fuselier (2005), ULF waves associated with enhanced subauroral proton precipitation, this volume.

Jaggi R. K., and R. A. Wolf (1973), Self-consistent calculation of the motion of a sheet of ions in the magnetosphere. *J. Geophys. Res., 78*(16), 2852–2866.

Jordanova, V. K., C. J. Farrugia, R. M. Thorne, G. V. Khazanov, G. D. Reeves, and M. F. Thomsen (2001), Modeling ring current proton precipitation by electromagnetic ion cyclotron waves during the May 14–16, 1997, storm, *J. Geophys. Res., 106*(A1), 7–22.

Kennel, C. F. and H. E. Petschek (1966), Limit on stably trapped particle fluxes, *J. Geophys. Res., 71*(A1), 1–27.

Khazanov, G. V., K. V. Gamayunov, and V. K. Jordanova (2003), Self-consistent model of magnetospheric ring current and electromagnetic ion cyclotron waves: The 2–7 May 1998 storm, *J. Geophys. Res., 108*(A12), 1419, doi:10.1029/2003JA009856.

Kintner, P. M and D. A. Gurnett (1977), Observations of ion cyclotron waves within the plasmasphere by Hawkeye 1. *J. Geophys. Res., 82*(16), 2314–2318.

Korth H., M. F. Thomsen, J. E. Borovsky, D. J. McComas (1999), Plasma sheet access to geosynchronous orbit, *J. Geophys. Res., 104*(A11), 25047–25061.

Kozyra, J. U., T. E. Cravens, A. F. Nagy, E. G. Fontheim, and R. S. B. Ong (1984), Effects of energetic heavy ions on electromagnetic ion cyclotron wave generation in the plasmapause region, *J. Geophys. Res., 89*(A4), 2217–2233.

Kozyra, J. U., V. K. Jordanova, R. B. Horne, and R. M. Thorne (1997), Modeling of the contribution of electromagnetic ion cyclotron (EMIC) waves to stormtime ring current erosion, in: *Magnetic Storms, Geophys. Monogr. Ser.*, vol 98, edited by B. T. Tsurutani et al., pp. 187–202, AGU, Washington, D. C.

LaBelle, J., R. A. Treumann, W. Baumjohann, G. Haerendel, N. Sckopke, and H. Luhr (1988), *J. Geophys. Res., 93*(A4), 2573–2590.

Lyons, L.R., Thorne R. M. (1972), Parasitic pitch angle diffusion of radiation belt particles by ion cyclotron waves, *J. Geophys. Res., 77*(28), 5608–5616.

Williams D.J., and L. R. Lyons (1974), The proton ring current and its interaction with the plasmapause: Storm recovery phase, *J. Geophys. Res., 79*(28), 4195–4207.

Mauk, B. H. and R. L. McPherron (1980), An experimental test of the electromagnetic ion cyclotron instability within the Earth's magnetosphere, *Phys. Fluids, 23*(10), 211–227.

Mende, S. B., et al. (2000), Far Ultraviolet Imaging from the IMAGE Spacecraft. 3. Spectral Imaging of Lyman-a and OI 135.6 nm, *Space Sci. Rev., 91*, 287–318.

Moldwin M. B., M. F. Thomsen, S. J. Bame, D. McComas, and G. D. Reeves (1995). The fine-scale structure of the outer plasmasphere. *J. Geophys. Res., 100*(A5), 8021–8029.

Mursula K., K. Prikner, F. Z. Feygin, T. Braysy, J. Kangas, R. Kerttula, P. Pollari, T. Pikkarainen T, and O. A. Pokhotelov (2000), Non-stationary Alfven resonator: new results on Pc 1 pearls and IPDP events. *J. Atmos. Sol.-Ter. Phys., 62*(4), 299–309.

Russell C. T., R. C. Snare, J. D. Means, D. Pierce, D. Dearborn, M. Larson, G. Barr, G. Le (1995), The GGS/POLAR fields investigation, *Space Sci. Rev., 71*(1–4), 563–582.

Sandel, B. R., et al. (2000), The Extreme Ultraviolet Imager investigation for the IMAGE mission, *Space Sci. Rev., 91*, 197–242.

Sandel, B. R., R. A. King, W. T. Forrester, D. L. Gallagher, A. L. Broadfoot, and C. C. Curtis (2001), Initial Results from the IMAGE Extreme Ultraviolet Imager, *Geophys. Res. Lett., 28*(8), 1439–1442.

Spasojević, M., J. Goldstein, D. L. Carpenter, U. S. Inan, B. R. Sandel, M. B. Moldwin, and B. W. Reinisch (2003), Global response of the plasmasphere to a geomagnetic disturbance, *J. Geophys. Res., 108*(A9), 1340, doi:10.1029/2003JA009987.

Spasojević, M., H. U. Frey, M. F. Thomsen, S. A. Fuselier, S. P. Gary, B. R. Sandel, and U. S. Inan (2004), The link between a detached subauroral proton arc and a plasmaspheric plume, *Geophys. Res. Lett., 31*, L04803, doi:10.1029/2003GL018389.

Taylor, W. W. L., and L. R. Lyons (1976), Simultaneous equatorial observations of 1- to 30-Hz waves and pitch angle distributions of ring current ions. *J. Geophys. Res., 81*(34), 6177–6183.

Thorne, R. M. and R. B. Horne (1992), The contribution of ion-cyclotron waves to electron heating and SAR-arc excitation near the storm-time plasmapause, *Geophys. Res. Lett., 19*(4), 417–420.

Tsyganenko, N. A., and D. P. Stern (1996), Modeling the global magnetic field of the large-scale Birkeland current systems, *J. Geophys. Res., 101*(A12), 27187–271898.

Yahnina, T. A., A. G. Yahnin, J. Kangas, and J. Manninen (2000), Proton precipitation related to Pc1 pulsations, *Geophys. Res. Lett., 27*(21), 3575–3578.

Young, D. T., S. Perraut, A. Roux, C. De Villedary, R. Gendrin, A. Korth, G. Kremser, and D. Jones (1981), Wave-particle interactions near Ω_{He+} observed on GEOS 1 and 2. I. Propagation of ion cyclotron waves in He^+-rich plasma, *J. Geophys. Res., 86*(A8), 6755–6772.

Zhang, Y., L. J. Paxton, T. J. Immel, H. U. Frey, and S. B. Mende (2002), Sudden solar wind dynamic pressure enhancements and dayside detached auroras: IMAGE and DMSP observations, *J. Geophys. Res., 107*, 8001, doi:10.1029/2002JA009355.

M. Spasojević, Packard Bldg Rm 301, Stanford, CA 94305 USA (maria@nova.stanford.edu)

M. F. Thomsen, Los Alamos National Lab, MS D466, Los Alamos, NM 87544, USA

P. J. Chi, IGPP, UCLA, Box 951567, Los Angeles, CA 90095, USA

B. R. Sandel, Lunar and Planetary Lab, University of Arizona, C. P. Sonett Space Sciences Bldg 63, Tucson, AZ 85721, USA

The Influence of Wave-Particle Interactions on Relativistic Electron Dynamics During Storms

Richard M. Thorne[1], Richard B. Horne[2], Sarah Glauert[2], Nigel P. Meredith[3], Yuri Y. Shprits[1], Danny Summers[4], and Roger R. Anderson[5]

Wave-particle interactions play a fundamental role in the non-adiabatic dynamics of energetic electrons. Plasma waves responsible for such interactions are substantially enhanced during storms, causing rapid pitch-angle scattering (and ultimate loss to the atmosphere) and energy transfer from low to high energies (leading to a hardening of the high-energy tail population). Several wave modes, including electromagnetic ion cyclotron waves and whistler-mode waves, contribute to pitch-angle scattering loss on timescales comparable to a day. Local electron acceleration to relativistic energies is prevalent during the storm recovery, due to interactions with intense whistler-mode chorus emissions outside the plasmapause. Codes have recently been developed to evaluate rates of diffusion in pitch-angle and energy. However, these diffusion rates have yet to be integrated into multi-dimensional diffusion codes, to quantify the role of each wave mode in radiation belt variability during storms. A major obstacle to developing accurate models for radiation belt dynamics is the limited observational data on the power spectral density of each important wave.

1. INTRODUCTION TO RADIATION BELT VARIABILITY DURING STORMS

1.1. Energetic Electron Observations During Storms.

The Earth's energetic (> a few hundred keV) electrons are distributed in two main belts separated by a pronounced quiet-time "slot" between $2 < L < 4$ (Figure 1). The inner belt, which tends to be very stable, is formed by slow inward radial diffusion [*Lyons and Thorne*, 1973] in the presence of loss to the atmosphere due to Coulomb scattering and whistler mode pitch-angle diffusion [*Lyons et al.*, 1972; *Abel and Thorne*, 1998]. The outer belt is extremely variable, especially during geomagnetic storms. During the main phase of a storm (e.g., Figure 1, orbit 186) pronounced flux depletions are observed, which have been attributed to a combination of adiabatic change associated with the formation of a storm-time ring current (the so called Dst effect [e.g., *Kim and Chan*, 1997]) and to rapid pitch-angle scattering losses to the atmosphere [*Albert*, 2003; *Summers and Thorne*, 2003; *O'Brien et al.*, 2004] and drift losses to the magnetopause. *Reeves et al.* [2003] have demonstrated that approximately 50% of the recently monitored magnetic storms leave the outer zone either essentially unaffected or with a net flux depletion at relativistic energies. The remaining 50% of storms cause a net flux enhancement in the outer belt. A small subset of the latter has been attributed to drift resonant acceleration due to penetration into the magnetosphere of a strong interplanetary shock [e.g., *Li et al.*,

[1]Department of Atmospheric and Oceanic Science, University of California, Los Angeles, California.
[2]British Antarctic Survey, Natural Environment Research Council, Cambridge, UK.
[3]Mullard Space Science Laboratory, University College London, Holmbury St Mary, Surrey, UK.
[4]Department of Mathematics and Statistics, Memorial University of Newfoundland, St John's, Newfoundland, Canada.
[5]Department of Physics and Astronomy, The University of Iowa, Iowa City, Iowa.

Inner Magnetosphere Interactions: New Perspectives from Imaging
Geophysical Monograph Series 159
Copyright 2005 by the American Geophysical Union.
10.1029/159GM07

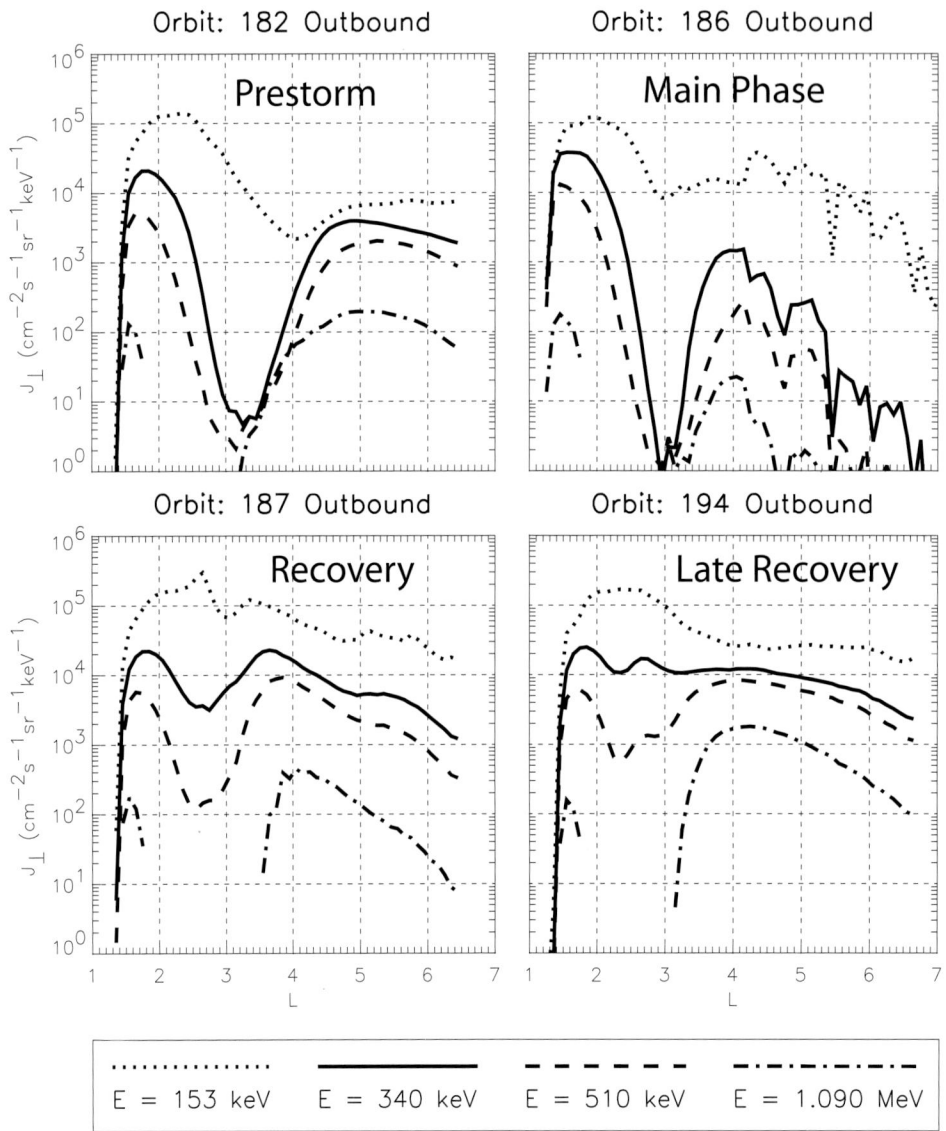

Figure 1. Variation in energetic electron flux observed by the MEA instrument on CRRES during different phases of the October 10–13, 1991 magnetic storm.

1993]. But the majority of storm-associated acceleration follows a temporal evolution similar to that of the October 1990 storm monitored on CRRES (figure 1). Usually, there is a rapid injection of medium energy (few 100 keV) electrons into both the slot region and outer zone following the main phase depletion. However, extremely energetic electrons (> 1 MeV) exhibit a more gradual build up over a period of several days during the storm recovery. Furthermore, during this gradual build up, pronounced peaks in phase space density develop [*Brautigam and Albert*, 2000; *Green and Kivelson*, 2004], indicative of local acceleration.

Here we attempt to quantify the competition between loss and acceleration processes throughout the main and recovery phases of storms and thereby address the distinction between storms that do or do not lead to enhanced outer zone flux [e.g., *Summers et al.*, 2004a]. We will also consider the different dynamical behavior of medium energy and highly relativistic electrons, due to resonant interactions with different magnetospheric plasma waves.

1.2. Storm-time Distribution of Plasma Waves.

In Section 2 we review the basic concepts of resonant scattering of electrons by plasma waves capable of violating the first or second adiabatic invariant. Such wave-particle interactions can lead to pitch-angle scattering and ultimate loss of particles to the atmosphere, or to energy diffusion associated with a net transfer of energy between particles and waves. The spatial distribution of three plasma waves capable of interacting with relativistic electrons during a storm is sketched in Figure 2.

Chorus emissions are intense whistler-mode waves, which are excited in the low-density region outside the plasmapause by the injection of plasmasheet electrons into the inner magnetosphere during enhanced storm-time convection. Chorus emissions are highly non-linear waves that occur in discrete micro-bursts at frequencies between 0.2–0.8 of the equatorial electron gyrofrequency [*Tsurutani and Smith*, 1977; *Santolik et al.*, 2003]. These waves have been associated with intense microburst precipitation (with effective loss times ~ day for 1 MeV electrons) [*Lorentzen et al.*, 2001; *O'Brien et al.*, 2004] and stochastic energy diffusion [*Horne and Thorne*, 1998; *Summers et al.*, 1998; 2002]. A statistical survey has been made of the spatial distribution of chorus emissions seen on CRRES and their dependence on magnetic activity [*Meredith et al.*, 2003b]. Nightside chorus is strongly confined to the equatorial region ($\lambda<15°$), while dayside emissions are stronger at high latitudes ($\lambda>20°$). A significant correlation has been found between the storm-time acceleration of electrons to relativistic energies throughout the entire outer radiation belts and enhanced chorus emissions [*Meredith et al.*, 2002, 2003c] or micro-burst precipitation [*O'Brien et al.*, 2003], suggesting that chorus plays an important role in the acceleration process.

Plasmaspheric hiss is an incoherent whistler-mode wave (in the frequency band between a few hundred Hz and a few kHz), which is generally confined within the plasmapause [*Thorne et al.*, 1973]. Resonant electron interactions with hiss cause pitch-angle scattering and loss of energetic electrons from the slot region [*Lyons and Thorne*, 1973; *Albert*, 1994; *Abel and Thorne*, 1998]. The intensity of plasmaspheric hiss (and corresponding rate of loss) is strongly enhanced during the recovery phase of storms [*Smith et al.*, 1974] and during substorm activity [*Meredith et al.*, 2004]. Scattering by hiss can therefore contribute to the slow decay (over 5–10 days) of enhanced outer zone relativistic electrons flux, as the plasmapause expands outwards to higher L following a storm [*Spjeldvik and Thorne*, 1975].

Electromagnetic ion cyclotron (EMIC) waves are lower frequency (Pc1-2 band) waves (0.1–5 Hz), which are excited in bands below the proton gyrofrequency during the injection of energetic ions into the ring current [*Horne and Thorne*, 1994]. Wave amplification is enhanced by the increase in density along the dusk side plasmapause [*Thorne and Horne*, 1997; *Jordanova et al.*, 1998] and within plasmaspheric drainage plumes that are formed in the afternoon sector during storm conditions [*Spasojevic et al.*, 2003]. EMIC waves can cause rapid ion precipitation [Jordanova et al., 2001; *Spasojevic et al.*, 2004], but can also scatter relativistic electrons [*Thorne and Kennel*, 1971; *Lyons and Thorne*, 1972; *Lorenzen et al.*, 2000; *Summers and Thorne*, 2003; *Meredith et al.*, 2003a].

2. RESONANT WAVE-PARTICLE INTERACTIONS

2.1. Resonant Interactions With Magnetospheric Waves.

As particles undergo their adiabatic gyro, bounce and drift motion in the radiation belts, they can interact with the plasma waves described above. The first invariant of the electron motion can be violated during interactions with plasma waves whose frequency ω is Doppler shifted to a multiple (n = 0, ±1, ±2,...) of the relativistic electron gyrofrequency as expressed below:

$$\omega - k_{\parallel} v_{\parallel} = n\Omega_e / \gamma \qquad (1)$$

where $\gamma = \{1- (v/c)^2\}^{-1/2}$ is the relativistic factor and $k_{\parallel}$ and $v_{\parallel}$ are components of the wave propagation vector and particle velocity along the direction of the ambient magnetic field. During wave-particle interactions, there can be a net exchange of momentum and energy leading to particle scattering in momentum space. An example of momentum space scattering of electrons by a typical band of field-aligned equatorial chorus [*Horne and*

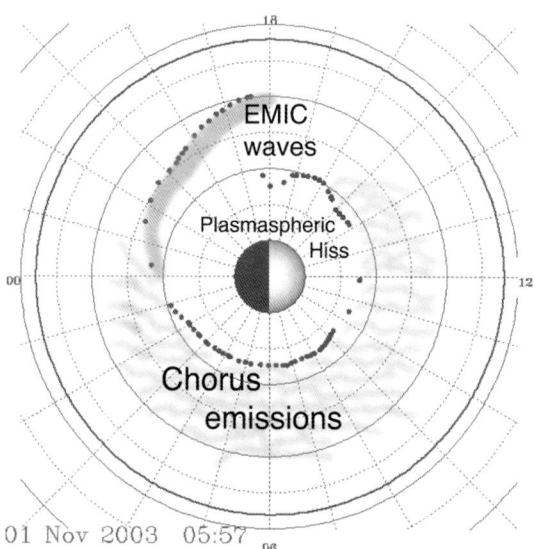

Figure 2. Schematic model for the expected distribution of plasma waves during the recovery phase of the large 2003 Halloween storm. The plasmapause position was obtained from IMAGE observations (courtesy J. Goldstein).

Thorne, 2003] is shown in Figure 3. For a prescribed ratio between the plasma frequency ω_p and electron gyrofrequency Ω_e, the n=1 resonance condition (1), together with the whistler wave dispersion relationship, defines two resonant ellipses in velocity space associated with the minimum $\omega/\Omega_e=0.2$ and maximum $\omega/\Omega_e=0.5$ frequencies in the adopted chorus wave band. During first order resonance, electrons are scattered along prescribed resonant diffusion surfaces [*Summers et al.*, 1998], indicated by the solid bold lines. The preferential direction for scattering is controlled by the gradients in particle phase space density (PSD). As electrons move along these diffusion surfaces, their energy changes. Particle energy diminishes (leading to net wave amplification) during scattering towards the loss cone. Natural gradients (induced by loss to the atmosphere) in medium energy (10–100 keV) electron PSD near the loss cone can therefore provide a source of free energy for chorus excitation. However, the higher-energy electron distribution resonant at larger pitch-angles tends to be essentially isotropic [*Horne et al., 2003b*]. Such electrons preferentially diffuse towards 90° (i.e., towards regions of lower PSD, which occur at higher energy) and thus gain energy (as waves are damped). Such interactions lead to energy transfer from the medium to the high-energy electron population, using chorus waves as intermediaries. Since there are relatively few high-energy particles, the wave attenuation does not substantially affect the net wave growth, but naturally leads to the gradual stochastic acceleration of relativistic electrons in the radiation belts.

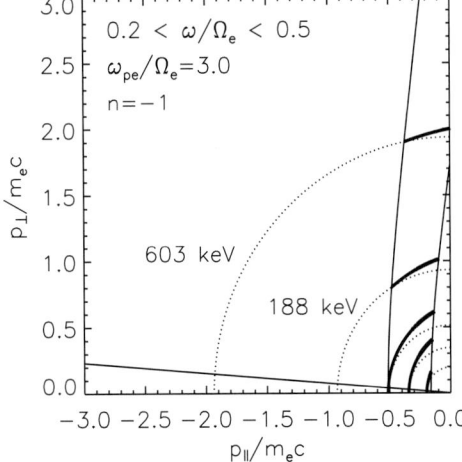

Figure 3. The resonant region in momentum space for first order cyclotron interaction with a typical band of equatorial chorus [*Horne and Thorne*, 2003]. Electrons are constrained to diffuse along the resonant diffusion surface (shown bold).

When the waves propagation vector is oblique, the Landau (n=0) and higher order cyclotron resonances can also occur. This permits resonant scattering over a much broader region of momentum space [*Lyons et al., 1971*] and there is no unique resonant diffusion surface. For small angles of propagation first harmonic scattering will dominate, but higher order scattering (and the Landau resonance) becomes important for highly oblique waves (e.g., plasmaspheric hiss [*Albert*, 1994; *Abel and Thorne*, 1998] or ECH emissions [*Horne and Thorne*, 2000]). During the various permitted resonant interactions, particles will experience a random walk in momentum space, which can be treated by evaluating rates of pitch-angle and energy (or momentum) diffusion.

The efficiency of energy diffusion (compared to pitch-angle scattering) is largely controlled by the ratio between the resonant electron velocity and the wave phase velocity [*Gendrin*, 1981]. Since the wave phase speed is strongly influenced by the ambient plasma density, or more specifically the ratio ω_p/Ω_e, plasmaspheric hiss and EMIC waves mainly cause pitch-angle diffusion and precipitation loss to the atmosphere. Energy diffusion only becomes effective in the low-density region just outside the plasmapause [*Horne et al., 2003a*]. Consequently, when intense chorus emissions ($B_w \sim 100$pT) are sustained for a period of days they are able to provide substantial energy diffusion [*Horne et al., 2005*]. Local acceleration to relativistic energies becomes effective for magnetic storms with prolonged chorus activity in the recovery phase [e.g., *Summers et al.*, 2002] and has even been observed during prolonged substorm activity [*Meredith et al., 2003c; Summers et al., 2004b*].

Radial diffusion, driven by drift resonance with enhanced ULF waves [*Hudson et al.*, 2001; *Elkington et al.*, 2003], also leads to particle acceleration during inward radial transport, in locations where there is a positive radial gradient in particle PSD. The observed temporal variability of the outer zone reflects the competition between the acceleration and loss processes. The Fokker-Planck equation [e.g., *Schulz and Lanzerotti*, 1974] provides a convenient mathematical framework for treating the temporal evolution of particle phase space density. Processes that violate each adiabatic invariant may be described in terms of diffusion coefficients, which scale in proportion to the power spectral density of the relevant resonant waves. Radial diffusion requires ULF fluctuations with periods comparable to the particle azimuthal drift time (~ 10 mins), while pitch-angle and energy diffusion require higher-frequency waves that satisfy (1). In the following section we describe how the quasi-linear formulation [*Kennel and Engelmann*, 1966] can be applied to quantify the bounce-averaged rates of pitch-angle and energy diffusion during resonant wave-particle interactions, which violate the first invariant.

2.2. Numerical Evaluation of Diffusion Coefficients.

For a prescribed band of waves at any given location, resonance (for any harmonic) with a given energy electron will only occur for a limited range of pitch-angles (e.g., Figure 3). Furthermore, as particles move along their bounce orbit from the equator to their mirror point, the condition for resonance (1) varies substantially, due to changes in magnetic field strength, plasma density and the particle parallel velocity. Latitudinal variations in the local pitch-angle diffusion rate $D_{\alpha\alpha} = <(\Delta\alpha)^2>/2\Delta t$ for first harmonic scattering by a field-aligned band of chorus at L=4 are shown in the lower panels of Figure 4, as a function of equatorial pitch-angle. The net diffusion rates when bounce-averaged over the orbit of the electron are shown in the upper panels. For 100 keV electrons, first order cyclotron resonant scattering near the edge of the loss cone (α_o~5.4°) peaks for interactions near 15° latitude, while at 500 keV substantial scattering near the loss cone only occurs above 25° latitude. As a consequence, 100 keV electrons can be scattered into the loss cone by night-side chorus emissions, with a loss time comparable to an hour ($D_{\alpha\alpha}(\alpha_o)$~$3\times10^{-4}$ s^{-1}). In contrast, precipitation loss of relativistic (>500 keV) electrons requires the presence of high latitude chorus emissions. Such waves are only found on the dayside [*Meredith et al.*, 2003b], thus explaining the MLT location of relativistic electron microbursts seen on SAMPEX [*O'Brien et al.*, 2004]. Note also that the scattering loss times for relativistic electrons are comparable to a day ($D_{\alpha\alpha}(\alpha_o)$~$10^{-5}$ s^{-1}), so that microbursts can cause substantial flux depletion during a storm.

The intensity of plasmaspheric hiss [*Meredith et al.*, 2004] and EMIC waves [*Braysy et al.*, 1998; *Erlandson and Ukhorskiy*, 2001] are also substantially enhanced during a storm. Both emissions will therefore contribute to storm-time relativistic electron loss. Hiss is predominantly found on the dayside inside the plasmapause or within drainage plumes. Typical storm-time amplitudes of hiss are 100 pT, and relativistic electrons will be subject to scattering by such waves for about 50% of their drift orbit. To be scattered by left-hand polarized EMIC waves, electrons must overtake the wave (to reverse the effective sense of polarization in the electron frame) with sufficient velocity for the Doppler shift term in (1) to satisfy resonance. Scattering at energies near 1 MeV can only occur in regions where ω_p/Ω_e>10-30 (namely inside the plasmasphere) and also requires the presence of EMIC waves at frequencies just below an ion gyrofrequency

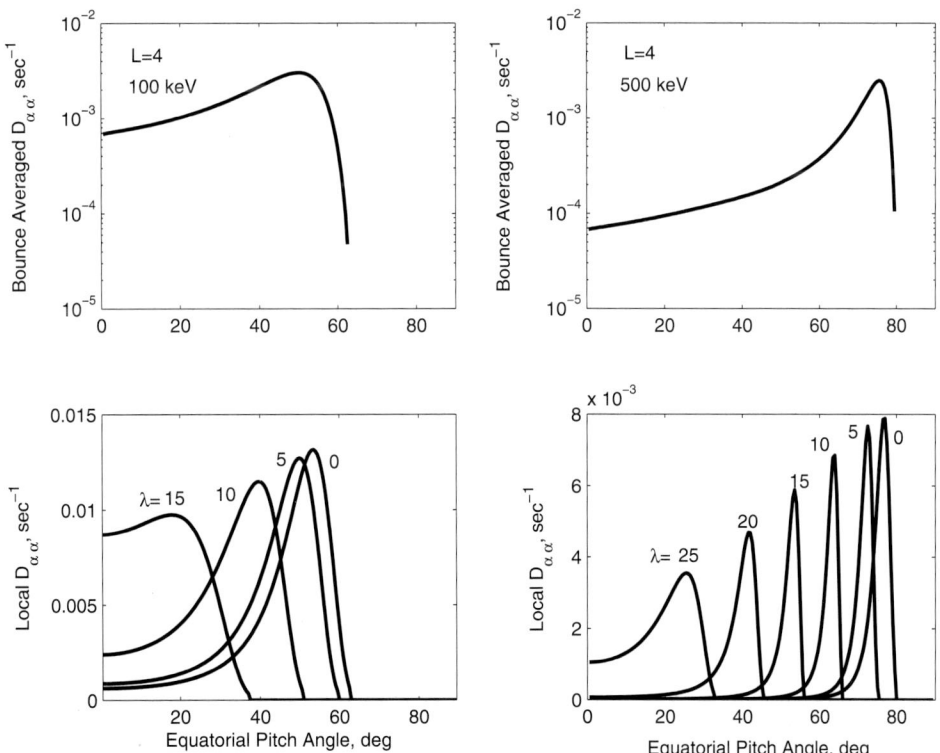

Figure 4. Comparison between the local pitch-angle diffusion rates (lower panels) at specified latitudes and the bounce-averaged values (upper curve) for first order cyclotron resonance between 100 keV and 500 keV electrons and field-aligned chorus.

[*Summers and Thorne*, 2003; *Meredith et al.*, 2003a]. As a consequence of the restricted conditions for resonance, relativistic electron scattering by storm-time EMIC waves probably only occurs for 1% of the particle drift orbit. These properties have been used by *Albert* [2003] to evaluate the net bounce-averaged diffusion rate of MeV electrons by a combination of hiss and EMIC waves during a storm. The results (Figure 5) indicate an electron lifetime ~ 0.8 days compared to 3.5 days for hiss alone. Since EMIC waves during the main phase of a storm can be more intense than the amplitudes (B_w=1nT) adopted by Albert, EMIC scattering could be a dominant mechanism to account for the rapid loss of relativistic electrons during the onset of a storm (Figure 1). EMIC scattering could also cause the rapid electron flux depletions reported by *Onsager et al.* [2002] and *Green et al.* [2004], and the intense hard X-ray events observed on balloons [*Millan et al.*, 2002].

During particularly strong geomagnetic storms, the intensification of the convection electric field causes the plasmapause to be compressed inwards to very low L values. Drainage plumes of high density also develop in the afternoon or dusk sector [*Spasojevic et al.*, 2003]. Such extreme conditions allow chorus emissions to be excited at much lower L (<3) on the dawn side and for EMIC (and hiss) waves to be excited along the dusk side drainage plumes (Figure 2). Recently, a new PADIE code has been developed at the British Antarctic Survey, which is capable of evaluating pitch-angle and energy diffusion rates for multiple-harmonic resonance with any prescribed distribution of waves. Bounce-averaged pitch-angle diffusion rates for a realistic distribution of chorus at L=3 are shown in the left-hand panel of Figure 6. The corresponding bounce-averaged energy diffusion rates $D_{EE} = <(\Delta E)^2>/2E^2\Delta t$ are shown in the left panel of Figure 7. To compute these rates

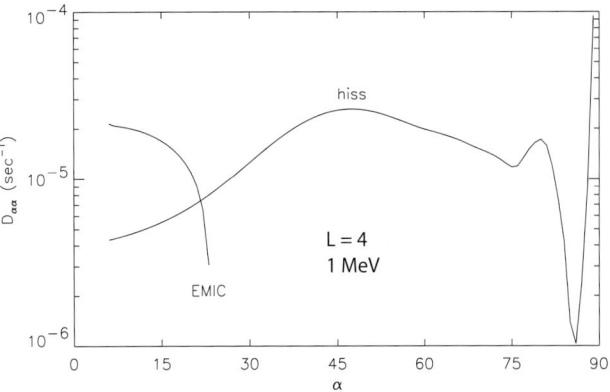

Figure 5. Numerical evaluation of the bounce-averaged pitch-angle diffusion rate of 1 MeV electrons by storm-time plasmaspheric hiss (B_w = 100 pT over 50% of drift path) and EMIC waves (B_w = 1 nT over 1% of drift orbit [*Albert*, 2003].

of diffusion, we adopt a Gaussian distribution of wave frequency peaked at ω/Ω_{eo}=0.35 with width $\delta\omega/\Omega_{eo}$=0.15 based on the equatorial gyrofrequency Ω_{eo}. Following the formalism of Lyons et al. [1972], the wave energy is distributed over a 30° Gaussian distribution of wave normal directions (consistent with observations). We further assume that the chorus wide-band wave intensity is 100 pT within 30° of the equator, and only present on the dawn side (Figure 2). The oblique distribution of waves allows us to include the effect of Landau resonance and the multiple-harmonic cyclotron resonances (for these calculations we include the first five positive and negative harmonics). We assume that the plasma density is 100/cc at L=3 (based on the trough model of *Sheeley et al.*, [2001]), and independent of latitude over the region of interaction.

The bounced-averaged results indicate that both pitch-angle scattering and energy diffusion are extremely dependent on energy and equatorial pitch-angle. Low energy electrons are subject to the most rapid pitch-angle scattering in the vicinity of the loss cone. At energies between 10–30 keV (not shown here) scattering loss times (~ an hour) are shorter than the azimuthal gradient drift time. As a consequence of scattering by chorus and ECH waves [*Horne and Thorne*, 2000], low-energy plasmasheet electrons should develop strong azimuthal gradients as they are injected into the inner magnetosphere during the storm [e.g., *Meredith et al.*, 2004]. Since such particles contribute to the diffuse aurora, the latter should be far more intense at night than on the day side, as typically observed [*Petrinec et al.*, 1999]. Conversely, above 100 keV the computed loss times exceed the electron azimuthal drift times and an azimuthally symmetric distribution should develop. Interestingly, the bounce-averaged diffusion rates near the edge of the loss cone for 100 keV electrons are comparable to those computed from an approximate analytic treatment based on first order cyclotron resonance with field-aligned waves (e.g., Figure 4). This indicates the dominance of first harmonic scattering for the adopted wave characteristics. For energies >1 MeV, first order scattering can only occur at latitudes above the assumed wave cut off at 30°. Consequently, the relativistic electrons require higher harmonic scattering to be precipitated into the atmosphere and the modeled lifetimes from chorus scattering become much longer than a day. The sharp change in the gradient of $D_{\alpha\alpha}$ near α_o~ 30–35° is a model-dependent consequence of our adopted cutoff in the wave power above 30° latitude. Better information on the spatial distribution of chorus intensities will be needed to obtain accurate lifetimes at these relativistic energies.

For E>100 keV, energy diffusion rates tend to maximize over a broad range of equatorial pitch-angles well away from the loss cone, so accelerated particles remain trapped.

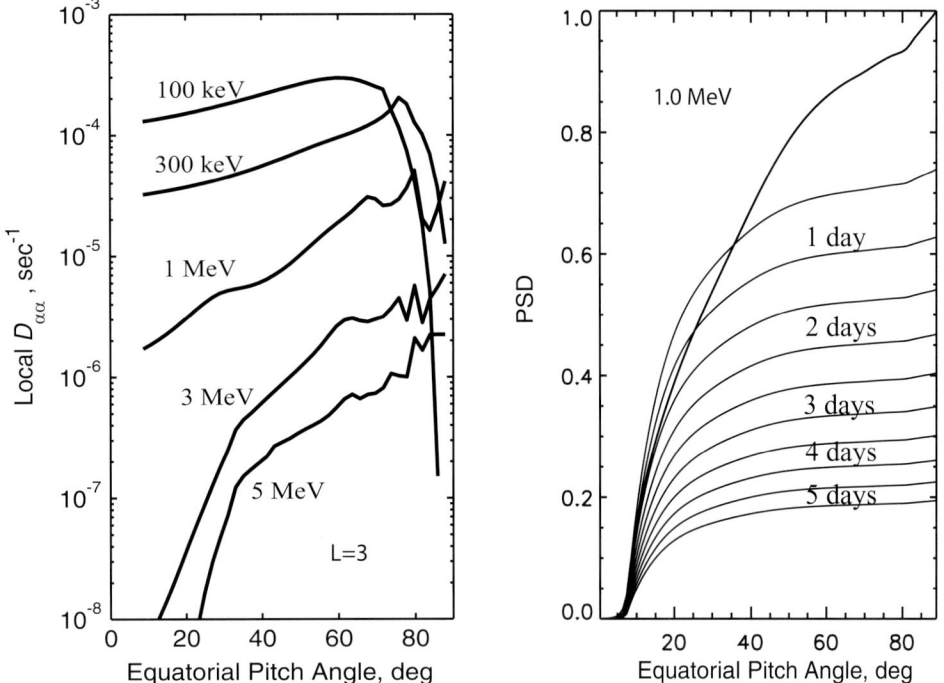

Figure 6. Bounce-averaged pitch-angle diffusion rates at L=3 during resonant interaction with a typical band of chorus (left). Simulation of the temporal evolution (right) of the 1 MeV electron pitch-angle distribution, and decay rates during the November 2003 storm.

Interestingly, relativistic electrons are only subject to the dominant first harmonic scattering at high latitudes, where the ratio of ω_p/Ω_e is reduced. Since energy diffusion becomes far more effective at lower values of ω_p/Ω_e, the values of D_{EE} for highly relativistic electrons tends to increase at lower equatorial pitch-angles, in sharp contrast to their rate of pitch-angle diffusion. The sharp decrease near $\alpha_o \sim$ 30–35° is again a consequence of our adopted cutoff in the wave power above 30° latitude. Nonetheless, although better information on the latitudinal distribution of waves is needed to compute acceleration rates accurately, it is clear that chorus can induce substantial stochastic acceleration over the duration of a storm.

3. TEMPORAL EVOLUTION OF PARTICLE FLUXES DURING A STORM

The temporal evolution of the particle phase space density f(p, α, L, t) can, in principle, be obtained by a numerical integration of the Fokker-Planck equation once all relevant diffusion rates have been specified. Codes such as Salammbo [*Bourdarie et al.*, 1996] and RAM [*Jordanova et al.*, 2001] have been developed to accomplish this, but currently they have not been able to incorporate all relevant physical processes. Because of the greatly different timescales involved in the violation of the first and third adiabatic invariant, one may analyze the consequences of radial diffusion at a rate $D_{LL} = \langle(\Delta L)^2\rangle/2\Delta t$ with a simplified radial diffusion equation in which effects of local energy diffusion are treated as an effective source S, while loss from pitch-angle scattering is represented by a loss time τ_L

$$\frac{\partial f}{\partial t} = L^2 \frac{\partial}{\partial L} \frac{D_{LL}}{L^2} \frac{\partial f}{\partial L} + S - \frac{f}{\tau_L} \qquad (2)$$

Conversely, although radial diffusion can act as a source of PSD at a given L shell, and also modify the pitch-angle distribution, it is convenient to ignore such effects (to first order) in order to assess the effectiveness of processes that violate the first invariant. We will adopt an even simpler approach here: using the diffusion rates obtained from the BAS code to evaluate first the temporal evolution of the resonant particle pitch-angle distribution (and thus obtain lifetimes due to precipitation). We subsequently use this rate of precipitation loss to quantify the post-storm buildup of the high-energy tail population due to energy diffusion. This will allow us to specify the net source rate S(E) for relativistic electron acceleration by magnetospheric chorus emissions, which can subsequently be incorporated into the radial diffusion equation (2). Pitch-angle scattering by other waves can also be included in the loss term.

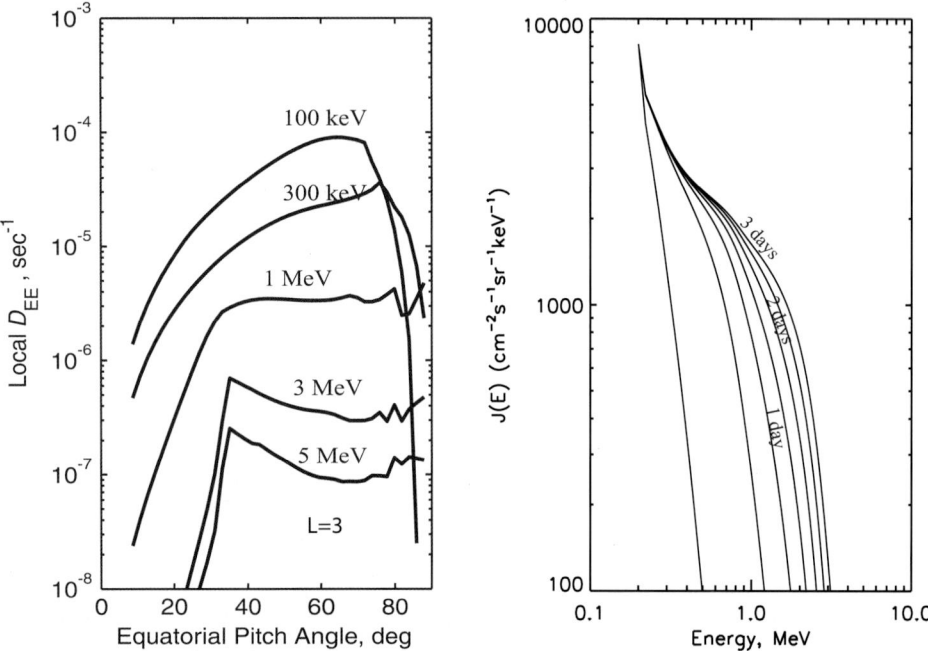

Figure 7. Bounce-averaged energy diffusion rates at L=3 during resonant interaction with chorus (left). Simulation (right) of the buildup of extremely relativistic electrons by local energy diffusion during the November 2003 storm.

3.1. Formation of Relativistic Electron Flat-Topped Pitch-Angle Distributions During Interaction With Chorus.

Under pure pitch-angle diffusion the temporal evolution of the particle PDS $f(\alpha_o, t)$ can be treated by the bounce-averaged pitch-angle diffusion equation:

$$\frac{\partial f}{\partial t} = \frac{1}{s(\alpha_o)y} \frac{\partial}{\partial y} s(\alpha_o) y D_{yy} \frac{\partial f}{\partial y} - \frac{f}{\tau_L} \quad (3)$$

where $D_{yy}(\alpha_o)$ is the bounce-averaged pitch-angle diffusion rate, $y=\cos\alpha_o$, and $s(\alpha_o) \sim 1.3 - 0.56 \sin\alpha_o$ reflects the change in bounce time with equatorial pitch-angle α_o. The results from the PADIE diffusion code (Figure 6) have been used to follow the temporal evolution of $f(\alpha_o, t)$ by numerically integrating (3) assuming that τ_L is infinite for $\alpha_o > \alpha_L$ and equal to the quarter bounce time for $\alpha_o < \alpha_L$. For the initial state we assume that $f(\alpha_o, t=0) \sim \sin\alpha_o$. Solutions for 1 MeV electrons are shown in the right-hand panels of Figure 6 every 1/2 day. After one day the pitch-angle distributions are already beginning to approach their equilibrium flat-topped shape due to resonant interactions with chorus. The approach to this equilibrium shape occurs first at larger pitch-angles, where the rate of diffusion is highest. Subsequently, the fluxes simply decay in time while retaining their flat-topped shape. From the decay rate we obtain a lifetime comparable to 3 days, which is similar to that estimated from SAMPEX micro-bursts [*O'Brien et al.*, 2004]. It is also worth noting that the predicted flat-topped shape is a characteristic feature of relativistic outer zone electrons observed on CRRES during the recovery phase of storms [*Horne et al.*, 2003b]. This agreement substantiates the important role of chorus in scattering relativistic electrons during storms.

3.2. Hardening of High-Energy Tail by Stochastic Acceleration During Interaction With Chorus.

The approach described above can be used to quantify the scattering lifetimes $\tau_L(E)$ due to chorus for all relevant energies. Assuming that the particle phase space density is essentially independent of pitch-angle, the Fokker-Planck equation can be reduced to a one-dimensional form, which can be solved numerically to follow the temporal evolution of the particle phase space density $f(p)$ subject to momentum (or energy) diffusion and precipitation loss:

$$\frac{\partial f}{\partial t} = \frac{1}{p^2} \frac{\partial}{\partial p} p^2 D_{pp} \frac{\partial f}{\partial p} - \frac{f}{\tau_L} \quad (4)$$

where D_{pp} is the bounce-averaged rate of momentum diffusion. Using average values of the energy diffusion rates shown in Figure 7, we have solved (4) numerically for a 3 day period to simulate the recovery phase of the Halloween magnetic storm when the plasmapause remained compressed inside L=3. As an initial condition we adopted a very depleted relativistic population. We also assumed that the flux at 200 keV remained fixed during the storm due to a balance between acceleration and loss processes. The results of the modeling are shown in the right-hand panel of Figure 7. The significant energy diffusion rate, together with the steep initial energy spectrum, causes a rapid enhancement of the high-energy tail population. Fluxes at 1 MeV increase to substantial levels within the first 12 hours of the storm recovery. Hardening of the spectrum continues throughout the recovery and 2–3 MeV electrons become apparent after 1–3 days. This timescale is comparable to the build up of relativistic electron fluxes observed on SAMPEX and HEO during the Halloween storm recovery [*Baker et al.*, 2004].

4. CONCLUDING REMARKS

Substantial progress has been made over the last few years in understanding the non-adiabatic dynamics of outer-zone radiation belt electrons. Observed variability during storms results from a competition between dramatically enhanced source and loss processes. Losses generally dominate during the storm main phase, whereas net acceleration can occur during the extended recovery. Two dominant source processes have been identified: radial diffusion driven by drift resonance with long period ULF waves and local stochastic acceleration resulting from cyclotron resonance with higher frequency waves. Observational evidence indicates that both mechanisms contribute to the enhancement of radiation belt flux [*Mathie and Mann*, 2000; *O'Brien et al.*, 2003]. The rate of radial diffusion increases at higher L, while local acceleration becomes most effective at lower L just outside the storm-time plasmapause. Losses due to pitch-angle scattering into the atmospheric loss cone, and to a lesser extent drift loss into the magnetopause, are also greatly enhanced during storms.

Outer zone radiation belt electrons can interact with several distinct magnetospheric waves, which become enhanced during geomagnetic storms. As a consequence of the interaction, electrons are either scattered in pitch-angle or experience energy diffusion. EMIC waves, excited along the dusk side plasmapause or within storm-time plumes, can induce precipitation loss at MeV energies on timescales less than a day. Such scattering is a potential candidate to account for the rapid depletion of relativistic flux during the storm main phase when EMIC waves are most intense. Storm-time plasmaspheric hiss can also contribute to loss but typical scattering times are longer (several days) and such waves are probably more effective during the extended storm recovery as the plasmapause expands outwards to higher L. High latitude chorus emissions observed outside the plasmapause in the prenoon sector can cause MeV micro-burst precipitation with effective loss times comparable to a day [*Thorne et al.*, 2005]. Such waves also cause energy diffusion, which leads to a net flux increase at relativistic energies even in the presence of micro-burst loss. Substantial acceleration, to energies greater that 1 MeV, can occur over a period of days, during the recovery phase of a magnetic storm.

For particularly strong magnetic storms, when the plasmapause is compressed well inside the normal location of the slot (L<3), local acceleration by chorus emissions can cause the reformation of the relativistic outer belt on L shells normally associated with the quiet-time slot. Following such intense storms, this new belt decays relatively slowly (over several days probably due to scattering by hiss). In the absence of rapid loss, inward radial diffusion should cause subsequent flux enhancements in the inner zone. For more moderate storms, the average intensity of chorus is probably smaller than 100 pT and the plasmapause generally expands outwards past L=3 within 12 hours of the main phase. Consequently, there is insufficient time to allow local acceleration to relativistic energies near L=3. But local acceleration to several hundred keV can occur within 5–10 hours, causing the observed rapid filling of the slot at lower energies (Figure 1). This is probably why the slot is generally only well defined for E> 1 MeV.

Our current theoretical understanding of radiation belt electron dynamics has evolved through the efforts of several research groups who have quantified different aspects of this intriguing puzzle. The important progress achieved to date has required a concentrated effort on each specific physical process, whereas in practice each processes is coupled. For example radial diffusion will modify the pitch-angle distribution and provide a source of particles for the excitation of plasma waves. Losses, due to those waves, will decrease the PSD [*Shprits and Thorne*, 2004] and thus create radial gradients, which enhance the inward radial diffusive flux. Future modeling efforts should be directed towards simultaneous inclusion of each important process. This will require multi-dimensional diffusion codes in which all diffusion coefficients are well specified. It will also involve coupling such kinetic codes that treat the microphysics (including non-linear scattering) with large-scale MHD and transport codes that can specify the injection of the source population. Aside from the numerical stability issues of such complex codes, the major obstacle in achieving such an holistic approach is our current limited knowledge of the spatial distribution and temporal variability of the power spectral intensity of each important wave mode. Future satellite missions, such as the Geosciences LWS radiation belt probes and the proposed ORBITALS mission, should carry instrumentation to address this important issue.

Acknowledgments. This research was supported in part by NASA grants NAG5-11922, NAG5-11826, NNG04GN44G, NSF grant ATM-0402615 and the Natural Sciences and Engineering Research Council of Canada under grant A-0621.

REFERENCES

Abel, B., and R. M. Thorne (1998), Electron scattering loss in Earth's inner magnetosphere, 1, Dominant physical processes, *J. Geophys. Res., 103,* 2385.

Albert, J. M. (1994), Quasi-linear pitch-angle diffusion coefficients: Retaining high harmonics, *J. Geophys. Res., 99,* 23,741.

Albert, J. M. (2003), Evaluation of quasi-linear diffusion coefficients for EMIC waves in a multi-species plasma, *J. Geophys. Res., 108*(A6), 1249, doi:10.1029/2002JA009792.

Baker, D. N., S. G. Kanekal, X. Li, S. P. Monk, J. Goldstein, and J. J. Burch (2004), Extreme Van Allen belt distortion in the 2003 Halloween solar storm, *Nature, 432,* in press.

Bourdarie, S., D. Boscher, T. Beutier, J. A. Sauvand, and M. Blanc (1996), Magnetic storm modeling in the Earth's electron belt by the Salammbô code, *J. Geophys. Res., 101,* 27,171.

Brautigam, D. H., and J. M. Albert (2000), Radial diffusion analysis of outer radiation belt electrons during the October 9, 1990, magnetic storm, *J. Geophys. Res., 105,* 291.

Braysy, T., K. Mursala, and G. Markland (1998), Ion cyclotron waves during a great magnetic storm observed by Freja double-probe electric field instrument, *J. Geophys. Res., 103,* 4145.

Elkington, S. R., M. K. Hudson, and A. A. Chan (2003), Resonant acceleration and diffusion of outer zone electrons in an asymmetric geomagnetic field, *J. Geophys. Res., 108(*A3), 1116, doi:10.1029/2001JA009202.

Erlandson, R. E., and A. J. Ukhorskiy (2001), Observations of electromagnetic ion cyclotron waves during geomagnetic storms: wave occurrence and pitch angle scattering, *J. Geophys. Res., 106,* 3883.

Gendrin, R. (1981), General relationships between wave amplification and particle diffusion in a magnetoplasma, *Rev. Geophys. Sp. Phys., 19,* 171.

Green, J. C., and M. G. Kivelson (2004), Relativistic electrons in the outer radiation belt: Differentiating between acceleration mechanisms, *J. Geophys. Res., 109,* A03213, doi:10.1029/2003JA010153.

Green, J. C., T. G. Onsager, and T. P. O'Brien (2004), Testing loss mechanisms capable of rapidly depleting relativistic electron flux in the Earth's outer radiation belt, *J. Geophys. Res.,* in press.

Horne, R. B., and R. M. Thorne (1994), Convective instabilities of electromagnetic ion cyclotron waves in the outer magnetosphere, *J. Geophys. Res., 99,* 17,259.

Horne, R. B., and R. M. Thorne (1998), Potential wave modes for electron scattering and stochastic acceleration to relativistic energies during magnetic storms, *Geophys. Res. Lett., 25,* 3011.

Horne , R. B., and R. M. Thorne (2000), Electron pitch-angle diffusion by electrostatic electron cyclotron waves: the origin of pancake distributions, *J. Geophys. Res., 105,* 5391.

Horne, R. B., and R. M. Thorne (2003), Relativistic electron acceleration and precipitation during resonant interactions with whistler-mode chorus, *Geophys. Res., Lett., 30*(10), 1527, doi:10.1029/2003GL016973.

Horne, R. B., S. A. Glauert, and R. M. Thorne (2003a), Resonant diffusion of radiation belt electrons by whistler-mode chorus, *Geophys. Res., Lett., 30*(9), 1493, doi:10.1029/2002GL016963.

Horne, R. B., N. P. Meredith, R. M. Thorne, D. Heynderickx, R. H. A. Iles, and R. R. Anderson (2003b), Evolution of energetic electron pitch-angle distributions during storm time electron acceleration to megaelectronvolt energies, *J. Geophys. Res., 108*(A1), 1016, doi:10.1029/2001JA009165.

Horne, R. B., R. M. Thorne, S. A. Glauert, J. M. Albert, N. P. Meredith, and R. R. Anderson (2005), Timescales for radiation belt electron acceleration by whistler mode chorus waves, *J. Geophys. Res., 110,* A03225, doi:10.1029/2004JA10811.

Hudson, M. K., S. R. Elkington, J. G. Lyon, M. J. Wiltberger, and M. Lessard (2001), Radiation belt electron acceleration by ULF wave drift resonance: Simulation of 1997 and 1998 storms, in *Space Weather, Geophys. Monogr. Ser.,* vol 125, edited by. P. Song, H. J. Singer, and G. L. Siscoe, p 289, AGU, Washington, D. C.

Jordanova, V. K., et al. (1998), Effect of wave-particle interactions on ring current evolution for January 10–11, 1997: Initial results, *Geophys. Res. Lett., 25,* 2971.

Jordanova, V. K., C. J. Farrugia, R. M. Thorne, G. V. Khazanov, G. D. Reeves, and M. F. Thomsen (2001), Modeling ring current proton precipitation by EMIC waves during the May 14–16, 1997 storm, *J. Geophys. Res., 106*(A1), 7.

Kennel, C. F., and F. Engelmann (1966), Velocity space diffusion from weak plasma turbulence in a magnetic field, *Phys. Fluids, 9,* 2377.

Kim, H.-J., and A. A. Chan (1997), Fully adiabatic changes in storm time relativistic electron fluxes, *J. Geophys. Res., 102,* 22,107.

Li, X., I. Roth, M. Temerin, J. R. Wygant, M. K. Hudson, and J. M. Blake (1993), Simulation of the prompt energization and transport of radiation belt particles during the March 24, 1991 SSC, *Geophys. Res. Lett., 20,* 2423.

Lorentzen, K. R., M. P. McCarthy, G. K. Parks, J. E. Float, R. M. Millan, D. M. Smith, R. P. Lin, and J. P. Treilhou (2000), Precipitation of relativistic electrons by interaction with electromagnetic ion cyclotron waves, *J. Geophys. Res., 105,* 5381.

Lorentzen, K. R., J. B. Blake, U. S. Inan, and J. Bortnik (2001), Observations of relativistic electron microbursts in association with VLF chorus, *J. Geophys. Res., 106,* 6017.

Lyons, L. R., and R. M. Thorne (1972), Parasitic pitch angle diffusion of radiation belt particles by ion cyclotron waves, *J. Geophys. Res., 77,* 5608.

Lyons, L. R., and R. M. Thorne, Equilibrium structure of radiation belt electrons (1973), *J. Geophys. Res., 78,* 2142.

Lyons, L. R., R. M. Thorne, and C. F. Kennel (1971), Electron pitch-angle diffusion driven by oblique whistler-mode turbulence, *J. Plasma Phys., 6,* 589.

Lyons, L. R., R. M. Thorne, and C. F. Kennel (1972), Pitch angle diffusion of radiation belt electrons within the plasmasphere, *J. Geophys. Res., 77,* 3455.

Mathie, R. A., and I. R. Mann (2000), A correlation between extended intervals of ULF wave power and storm-time geosynchronous relativistic electron flux enhancements, *Geophys. Res. Lett., 27*, 3261.

Meredith, N. P., R. B. Horne, R. H. A. Iles, R. M. Thorne, D. Heynderickx, and R. R. Anderson (2002), Outer zone relativistic electron acceleration associated with substorm enhanced whistler mode chorus, *J. Geophys. Res,, 107*(A7), 1144, doi:10.1029/2001JA900146.

Meredith, N. P., R. M. Thorne, R. B. Horne, D. Summers, B. J. Fraser, and R. R. Anderson (2003a), Statistical analysis of relativistic electron energies for cyclotron resonance with EMIC waves observed on CRRES, *J. Geophys. Res., 108*(A6), 1250, doi:10.1029/2002JA009700.

Meredith, N. P., R. B. Horne, R. M. Thorne, and R. R. Anderson (2000b), Favored regions for chorus-driven electron acceleration to relativistic energies in the Earth's outer radiation belt, *Geophys. Res.Lett., 30*(16), 1871, doi:10.1029/2003GL017698.

Meredith, N. P., M. Cain, R. B. Horne, R. M. Thorne D. Summers, and R. R. Anderson (2003c), Evidence for chorus-driven electron acceleration to relativistic energies from a survey of geomagnetically disturbed periods, *J. Geophys. Res., 108*(A6), 1248, doi:10.1029/2002JA009764.

Meredith, N. P., R. B. Horne, R. M. Thorne, D. Summers, and R. R. Anderson (2004), Substorm dependence of plasmaspheric hiss, *J. Geophys. Res.*, 109, A06209, doi:10.1029/2004JA010387.

Millan, R. M., R. P. Lin, D. M. Smith, K. R. Lorentzen, and M. P. McCarthy (2002), X-ray observations of MeV electron precipitation with a balloon-borne germanium spectrometer, *Geophys. Res. Lett., 29*(24), 2194, doi:10.1029/2002GL015922.

O'Brien, T. P., K. R. Lorentzen, I. R. Mann, N. P. Meredith, J. B. Blake, F. F. Fennell, M. D. Looper, D. K. Milling, and R. R. Anderson (2003), Energization of relativistic electrons in the presence of ULF power and MeV microbursts: Evidence for dual ULF and VLF acceleration, *J. Geophys. Res,, 108*(A8), 1329, doi:10.1029/2002JA009784.

O'Brien T. P., M. D. Looper, and J. B. Blake (2004), Quantification of relativistic electron microbursts losses during the GEM storms, *Geophys. Res.Lett., 31*, L04802, doi:10.1029/2003GL018621.

Onsager, T. G., G. Rostoker, H. J. Kim, T. Obara, H. J. Singer, and C. Smithtro (2002), Radiation belt electron dropouts: Local time, radial, and particle energy dependence, *J. Geophys. Res,, 107*(A7), 1382, doi:10.1029/2001JA000187.

Petrinec, S. M., D. L. Chenette, J. Mobilia, M. A. Rinaldi, and W. L. Imhof (1999), Statistical X-ray auroral emissions--PIXIE observations, *Geophys. Res. Lett., 26*, 1565.

Reeves, G. D. K. L. McAdams, R. H. W. Friedel, and T. P. O'Brien (2003), Acceleration and loss of relativistic electrons during geomagnetic storms, *Geophys. Res. Lett., 30*(10), 1529, doi:10.1029/2002GL016513.

Santolik, O., D. A. Gurnett, J. S. Pickett, M. Parrot, and N. Cornilleau-Wehrlin (2003), Spatio-temporal structure of storm-time chorus, *J. Geophys. Res., 108*(A7), 1278, doi:10.1029/2002JA009791.

Schulz, M., and L. Lanzerotti (1974), *Particle Diffusion in the Radiation Belts*, Spinger, New York.

Sheeley, B. W., M. B. Moldwin, H. K. Rassoul, and R. R. Anderson (2001), An empirical plasmasphere and trough density model: CRRES observations, *J. Geophys. Res., 106*, 25631.

Smith, E. J., A. M. A. Frandsen, B. T. Tsurutani, R. M. Thorne, and K. W. Chan (1974), Plasmaspheric hiss intensity variations during magnetic storms, *J. Geophys. Res., 79*, 2507.

Spasojevic´, M., J. Goldstein, D. L. Carpenter, U. S. Inan, B. R. Sandel. M. B. Moldwin, and B. W. Reinisch (2003), Global response of the plasmasphere to a geomagnetic disturbance, *J. Geophys. Res,, 108*(A9), 1340, doi:10.1029/2003JA009987.

Spasojevic´, M., H. U. Frey, M. F. Thomsen, S. A. Fuselier, S. P. Gary, B. R. Sandel, and U. S. Inan (2004), The link between a detached subauroral proton arc and a plasmaspheric plume, *Geophys. Res,. Lett., 31*, L04803, doi:10.1029/2003GL018389.

Spjeldvik, W. N. and R. M. Thorne (1975), The cause of storm after effects in the middle latitude D-region, *J. Atmos. Terr. Phys., 37*, 777.

Shprits, Y. Y., and R. M. Thorne (2004), Time dependent radial diffusion modeling of relativistic electrons with realistic loss rates, *Geophys. Res., Lett., 31*, L08805, doi:10.1029/2004GL019591.

Summers D, and R. M. Thorne (2003), Relativistic electron pitch-angle scattering by electromagnetic ion cyclotron waves during geomagnetic storms, *J. Geophys. Res., 108*(A4), 1143, doi:10.1029/2002JA009489.

Summers, D., R. M. Thorne, and F. Xiao (1998), Relativistic theory of wave-particle resonant diffusion with application to electron acceleration in the magnetosphere, *J. Geophys. Res., 103*, 20,487.

Summers, D., C. Ma, N. P. Meredith, R. B. Horne, R. M. Thorne, D. Heynderickx, and R. R. Anderson (2002), Model of the energization of outer-zone electrons by whistler-mode chorus during the October 9, 1990 geomagnetic storm, *Geophys. Res,. Lett., 29*(24). 2174, doi:10.1029/2001GL016039.

Summers, D., C. Ma, and T. Mukai (2004a), Competition between acceleration and loss mechanisms of relativistic electrons during geomagnetic storms, *J. Geophys. Res., 109*, A04221, doi:10.1029/2004JA010437.

Summers, D., C. Ma, N. P. Meredith, R. B. Horne, R. M. Thorne, and R. R. Anderson (2004b), Modeling outer-zone relativistic electron response to whistler-mode activity during substorms, *J. Atmos. Sol. Terr. Phys., 66*, 133.

Thorne, R. M., and C. F. Kennel (1971), Relativistic electron precipitation during magnetic storm main phase, *J. Geophys. Res., 76*, 4446.

Thorne, R. M., and R. B. Horne (1997), Modulation of electromagnetic ion cyclotron instability due to interaction with ring current O^+ during geomagnetic storms, *J. Geophys. Res., 102*, 14155.

Thorne, R. M., E. J. Smith, R. K. Burton, and R. E. Holzer (1973), Plasmaspheric hiss, *J. Geophys. Res., 78*, 1581.

Thorne R. M., T. P. O'Brien, Y. Y. Shprits, D. Summers, and R. B. Horne (2005), Timescale for MeV electron microburst loss during geomagnetic storms, *J. Geophys. Res.110*, doi:10.1029/2004JA01882, in press.

Tsurutani, B. T., and E. J. Smith (1977), Two types of magnetospheric ELF chorus and their substorm dependences, *J. Geophys. Res., 82*, 5112.

R. R. Anderson, Department of Physics and Astronomy, University of Iowa, Iowa City, Iowa, IA 52242-1479, USA.

S. Glauert, British Antarctic Survey, Natural Environment Research Council, Madingley Road, Cambridge, CB3 0ET, UK.

R. B. Horne, British Antarctic Survey, Natural Environment Research Council, Madingley Road, Cambridge, CB3 0ET, UK.

N. P. Meredith, Mullard Space Science Laboratory, University College London, Holmbury St Mary, Dorking, Surrey, RH5 6NT, UK.

Y. Y. Shprits, Department of Atmospheric and Oceanic Science, University of California, 7121 MS, Box 951565, Los Angeles, California, CA 90095-1565, USA.

D. Summers, Department of Mathematics and Statistics, Memorial University of Newfoundland, St John's, Newfoundland, A1C 5S7, Canada.

R. M. Thorne, Department of Atmospheric and Oceanic Science, University of California, 7121 MS, Box 951565, Los Angeles, California, CA 90095-1565, USA.

Distribution and Origin of Plasmaspheric Plasma Waves

James L. Green and Shing F. Fung

Goddard Space Flight Center, Greenbelt, Maryland

Scott Boardsen

L3 Communications, Government Services Inc., Largo, Maryland

Hugh J. Christian

Marshall Space Flight Center, Huntsville, Alabama

Recent analysis of electric and magnetic field wave data showing the distribution of plasmaspheric waves is reviewed. These studies find that equatorial electromagnetic (EM) emissions (~30-330 Hz), plasmaspheric hiss (~330 Hz – 3.3 kHz), chorus (~2 kHz – 6 kHz), and VLF transmitters (~10-50 kHz) are the main types of trapped waves within the plasmasphere. Observations of the equatorial EM emissions show that the most intense region is on or near the magnetic equator in the afternoon sector and that during times of negative B_z (interplanetary magnetic field), the maximum intensity moves from L values of 3 to less than 2. These observations are consistent with the origin of this emission being particle-wave interactions in or near the magnetic equator in the outer plasmasphere. Plasmaspheric hiss shows high intensities at high latitudes and low altitudes over L values from 2 to 3 in the early afternoon sector. Plasmaspheric hiss, through particle-wave interactions, maintains the slot region in the radiation belts. The longitudinal distribution of the plasmaspheric hiss intensity is similar to the distribution of lightning: stronger over continents than over the ocean, stronger in the summer than winter, and stronger on the dayside than nightside. A lightning origin for plasmaspheric hiss is also supported by the similarities in the latitudinal distribution of hiss with that of ground transmitters and the quiet-time electron slot region located at slightly higher L (~3) during solar maximum than at solar minimum (L~2.5).

1. INTRODUCTION

The plasmasphere is a high-density region of cold plasma around the earth residing on closed field lines capable of supporting trapped electromagnetic wave modes such as the whistler mode. A whistler mode wave is an electromagnetic wave whose upper frequency cutoff is either the local electron plasma frequency (f_p) or gyrofrequency (f_{ge}); whichever is less [*Stix*, 1992]. Because of the large cold plasma density in the plasmasphere f_p is greater than f_{ge} and supports whistler mode waves up to frequencies of ~ 50 kHz. The index of refraction in the whistler mode is such that these electromagnetic waves have a natural tendency to travel along a magnetic field making repeated journeys from northern to southern hemisphere and vice versa. In the plasmasphere, the main trapped waves include: equatorial electromagnetic (EM) emissions, lightning whistlers, plasmaspheric hiss, chorus, and VLF transmissions. The importance of plasmaspheric whistler mode waves in limiting the flux of energetic radiation belt particles, particularly in the slot region, has been known for sometime, but the origin of these waves has a number of controversial aspects.

The classic theoretical work by *Kennel and Petschek* [1966] held that whistler mode radiation can be amplified by gyroresonance interactions with radiation belt electrons near the magnetic equator, while causing the electrons to be pitch-angle scattered and precipitated (see Plate 1). The continuous interactions between the magnetospherically reflected whistler waves (those that are not purely ducted along the field line) and the trapped electrons are believed to be a contributing mechanism for generating and maintaining the electron slot region between the inner and outer electron belts.

A lightning strike generates a very broad spectrum of electromagnetic waves including the kHz frequency range that, under certain circumstances, couples through the high-density ionosphere to reach the magnetosphere and becomes trapped in the plasmasphere. These signals start out as discrete spherics and disperse in frequency as they propagate nearly along geomagnetic field lines giving the well-known "whistler" spectrum of a tone, decreasing with time [*Storey*, 1953]. The role of lightning as a source of whistler mode waves that cause radiation belt electron precipitation has been studied extensively [see for example; *Goldberg et al.*, 1987; and *Inan et al.*, 1989] but it is not believed that these waves are responsible for maintaining the slot region.

A large number of radio transmitters have been established by a number of countries for the purposes of navigation and communication. The frequencies of these transmitters range from as low as 10 kHz. Like lightning, ground transmitter waves also couple through the ionosphere to the magnetosphere and are trapped in the plasmasphere, contributing to the total plasmaspheric whistler mode spectrum and are also known to precipitate low energy radiation belt electrons with energies less than 50 keV in the slot region between the inner and outer electron radiation belts [see for example: *Imhof et al.*, 1981]. The narrow-band transmitter waves are only able to precipitate electrons at selected energies and as such, could not be responsible for the large decrease in high-energy electron fluxes over a large energy range as observed in the radiation belt slot region.

Lyons et al. [1972], *Albert* [1994], and *Abel and Thorne* [1998a,b] showed that plasmaspheric hiss would be the dominant whistler mode wave responsible for the slot region. For a recent review of the physics of wave-particle interactions between whistler mode waves and the trapped high-energy electrons in the Van Allen Belts see *Thorne et al.* [2005]. Plasmaspheric hiss is a broad diffuse band of electromagnetic radiation in the 100s of Hz to ~3 kHz range that is confined to the plasmasphere [*Taylor and Gurnett*, 1968; *Russell et al.*, 1969; and *Dunckel and Helliwell*, 1969]. The last comprehensive review of plasmaspheric hiss was done by *Hayakawa and Sazhin* [1992]. Two of the most important characteristics of plasmaspheric hiss are its source location and generation mechanism. Even after three decades of space plasma wave research these two characteristics are still controversial as either generated by a gyroresonance process [see for example: *Thorne et al.*, 1973; *Huang et al.*, 1983; *Church and Thorne*, 1983] or by lightning [*Sonwalker and Inan*, 1989; *Draganov et al.*, 1992 and *Bortnik et al.*, 2003a] or both but the relative contribution from these two sources is still unknown even though the literature in this field is extensive. *Thorne et al.* [1979] suggested that plasmaspheric hiss would only grow in intensity from the background thermal noise to its observed intensity from gyroresonance acceleration as the whistler mode wave returned through the equator repeatedly. Based on limited data, *Solomon et al.* [1988] have shown that amplification of background noise to observed hiss intensities is possible. In addition, from ray tracing calculations of magnetospherically reflected whistlers, *Thorne and Horne* [1994] concluded that lightning generated whistlers could not be the source of plasmaspheric hiss because they are subject to significant damping due to Landau resonant interactions with suprathermal electrons with energies greater than about 100 eV.

Observations from the low frequency linear wave receiver on DE by *Sonwalker and Inan* [1989] have shown that lightning-generated whistlers often trigger plasmaspheric hiss. These *in situ* observations were the first to demonstrate that lightning could be the original source of plasmaspheric hiss. Ray tracing calculations by *Draganov et al.* [1992] demonstrated that the refraction of lightning whistlers by the plasmasphere (higher frequencies waves move to lower L shells) produced a natural way to obtain a hiss like spectrum on lower L shells. In addition, *Draganov et al.* [1992] determined that the total wave energy from lightning whistlers could maintain the experimentally observed levels of plasmaspheric hiss. *Green et al.* [2005] has recently shown a strong connection between the distribution of plasmaspheric hiss and that of lightning by matching their geographic distributions. The purpose of this paper is to review the latest results of observations of plasmaspheric whistler mode waves and present new evidence for their origin further supporting the role lightning plays in the generation of plasmaspheric hiss.

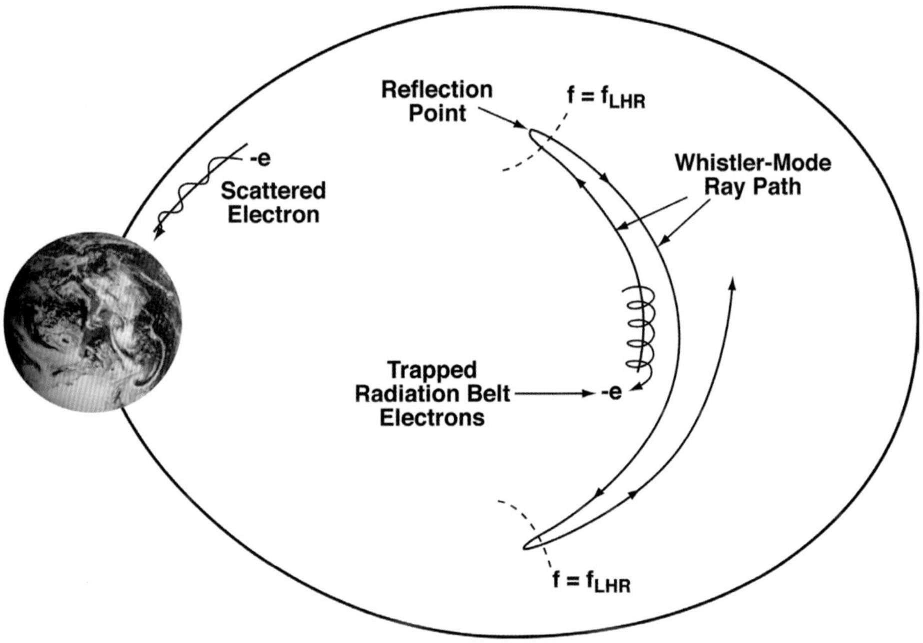

Plate 1. A schematic illustration showing radiation belt electrons interacting with whistler mode waves which scatters electrons into the loss cone.

2. DISTRIBUTION OF TRAPPED EM RADIATION IN THE PLASMASPHERE

Most of the whistler mode wave research in the last 5 years has dealt with various aspects of plasmaspheric hiss and chorus. Chorus is primarily observed near the plasmapause. Resent results of plasmaspheric hiss research will be discussed in this review since chorus emissions and their interaction with the outer-trapped radiation belt is well covered by another paper in this monograph [see *Thorne et al.*, 2005].

The most extensive analysis of trapped EM waves in the plasmasphere has been recently accomplished by *André et al.* [2002] using data from the plasma wave instrument (PWI) on the Dynamics Explorer-1 (hereafter DE) and by *Green et al.* [2005] using DE/PWI and data from the Radio Plasma Imager (RPI) on the Imager for Magnetopause-to-Aurora Global Exploration (IMAGE) spacecraft.

André et al. [2002] presented the average plasmaspheric magnetic field spectral density showing frequency and radial distance distribution of whistler mode waves in the plasmasphere for disturbed and quiet conditions. However, that study did not take into account local time variations in producing their radial distributions (see their Plate 4 and 5). Using the same data set as *André et al.* [2002], significant local time and latitudinal variations were found by *Green et al.* [2005] over the entire wave spectrum. Latitudinal and local time variations in the distribution of trapped EM waves are an important clue as to their origin.

The DE/PWI and IMAGE/RPI data used in the *Green et al.* [2005] study were separated into bins of 5° in geomagnetic latitude and 12° in geomagnetic longitude for all radial distances < 3.5 R_E, saving the total of the log of the spectral density and total numbers of measurements in each bin for each frequency. No normalization of spectral power densities as a function of radius or distance along a flux tube was attempted. The data values used for each bin in the wave maps are a weighted average over a bin and its 8 nearest neighbors (total spectral density summed over nine bins divided by the total number of individual measurements summed over the nine bins). The DE/PWI data are from the period September 16, 1981 to June 23, 1984. The IMAGE data coverage is from January 1, 2001 to August 6, 2003. For additional details of the wave map technique used see *Green et al.* [2005].

Plate 2 has been adapted from *Green et al.* [2005] and shows significant intensity variations in the average DE/PWI magnetic field spectral densities at three frequencies (176.1 Hz, 1.2 kHz, and 11.8 kHz) in both latitude and local time. White pixels indicate no DE measurements. The wave measurements for the latitudinal distribution (left side) in Plate 2 are sorted by using the sign of the z component of the interplanetary magnetic field (IMF B_z). It is well known that the IMF B_z parameter is an important factor in the generation of geomagnetic substorms. The average magnetic field spectral density is binned according to the absolute value of the magnetic latitude, times the sign of the IMF B_z. Therefore, the +z axis in Plate 2 left hand panels are the subset of observations for which IMF B_z is positive, while the -z axis are the subset of observations for which the IMF B_z is negative. This method has a number of advantages and is applicable to whistler mode waves due to their expected generation and propagation symmetry about the magnetic equator. *Green et al.* [2005] identified the waves in Plate 2 as Equatorial EM emissions (A, B), plasmaspheric hiss (C, D), and VLF transmitters (E, F). All the characteristic features of the magnetic field spectral density distributions shown in Plate 2 are also found over the entire frequency range from ~30 to ~330 Hz for the equatorial EM emissions, from ~330 Hz to ~3.3 kHz for plasmaspheric hiss, and from ~10 to 50 kHz for VLF transmitters.

Previous studies of the equatorial EM waves have been limited after first being pointed out by *Russell et al.* [1970] and have left the impression that the EM equatorial emissions have "no marked dependence on local time" [*Gurnett*, 1976]. The EM equatorial waves are believed to play an important role in transferring energy from energetic protons convecting earthward from the plasmasheet to the thermal plasmaspheric ions flowing along the geomagnetic field lines [*Gurnett*, 1976; *Boardsen et al.*, 1992]. The 176.1 Hz waves shown in Plate 2A are for measurements in the 1230–0030 MLT meridian plane, and Plate 2B is a plot of measurements in the equatorial plane for the same frequency band in solar magnetospheric (SM) coordinates. The distribution of wave spectral densities shown in Plate 2A confirms previous studies of the near equatorial nature of these waves in the outer plasmasphere. In addition, Plate 2A strongly suggests that particle-wave interactions could occur within a few tens of degrees of the magnetic equator over a large range in radial distance (1.5–4 R_E geocentric, typically outside the slot region) and that specific L shells for both positive and negative B_z values (non-storm and storm conditions) have the highest intensities of the emission. Plate 2B shows that the most intense portion of the equatorial EM emissions is located in the mid to late afternoon local time region and that the emission is rarely measured in the pre-midnight sector.

Although found essentially everywhere in the plasmasphere at some intensity, plasmaspheric hiss is most intense throughout the local afternoon sector (Plate 2C and 2D) and on L shells which contain the slot region in the electron radiation belts. In the frequency range from ~330 Hz to ~3.3 kHz both magnetic meridian and equatorial plane plasma wave magnetic field spectral density maps from PWI data show

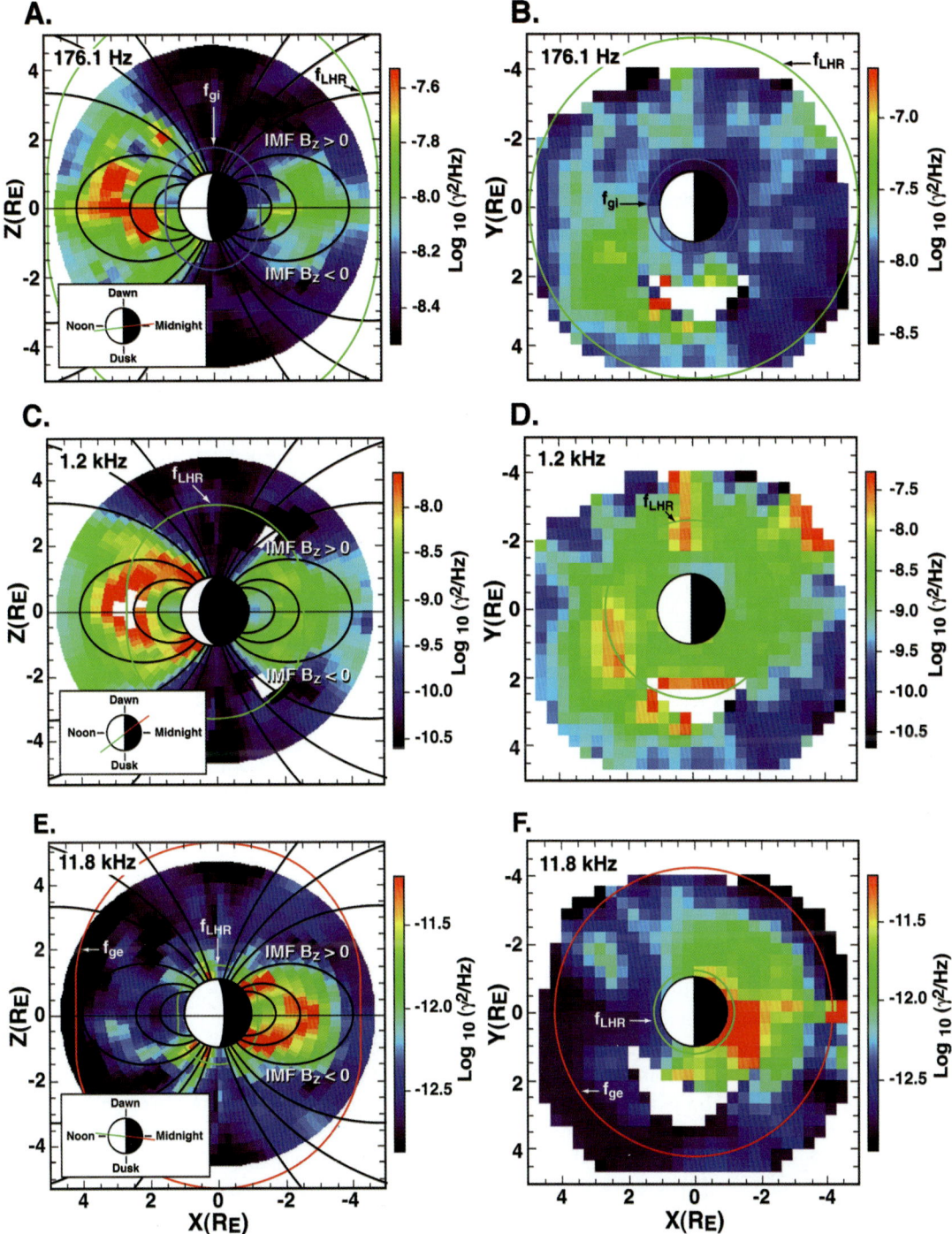

Plate 2. The latitudinal (left side) and local time (right side) distributions of the average magnetic field spectral density of low frequency EM waves in the plasmasphere at 176.1 Hz, 1.2 kHz, and 11.8 kHz [adapted from *Green et al.,* 2005]. Closed L shells of 1.5, 3, and 4 are shown in Panels A, C, E along with field lines at 70°, 75°, and 80°. The local time plane for each of the latitudinal distributions is shown in the inset Plate and corresponds to the most intense region as determined by the local time distribution. These waves have been identified as EM equatorial emissions, plasmaspheric hiss, and radiation from ground transmitters.

essentially the same distributions as in Plate 2 for 1.2 kHz. Over the frequency range from 2.7 to 3.3 kHz the intensity and spatial extent of the emission decrease significantly. The transition from the EM equatorial emission to plasmaspheric hiss occurs over the frequency range from ~230 Hz to ~380 Hz. The wave activity in Plate 2D, occurring in the early morning and near dawn (~0200-0300 hours MLT) sector is just the beginning of the chorus emissions that is discussed in *Thorne et al.* [2005].

VLF ground transmitters over the frequency range from ~10 to 50 kHz operate with a narrow bandwidth (usually < 1 kHz) and generate wave energy that typically couples through the ionosphere and into the plasmasphere in the whistler mode. Plate 2E and 2F show the latitudinal and local time distribution in DE magnetic field spectral densities at 11.8 kHz with f_{ge} and f_{LHR} indicated by red and green lines, respectively. The distributions shown in this Plate are similar to those at frequencies of all other ground transmitters from 10-50 kHz. Panel E shows the latitudinal distribution of the magnetic field spectral density of the waves that follow mid-latitude field lines. Panel F shows that the maximum magnetic field spectral density of waves from ground transmitters occurs on the nightside. It is important to note the similarity in the latitudinal distribution of plasmaspheric hiss (Plate 2C) and that of transmitters (Plate 2E) even though the most intense portions of these waves have different local time distributions. The similarity in the latitudinal distributions of these EM waves supports the *Green et al.* [2005] claim for a lightning origin for plasmaspheric hiss.

3. ON THE ORIGIN OF PLASMASPHERIC HISS

In order to investigate what contribution lightning may play in providing whistler mode radiation into the plasmasphere, *Green et al.* [2005] remapped the plasmaspheric hiss observations of DE/PWI and IMAGE/RPI into geographic coordinates and compared them with the average distribution of lightning. Plasmaspheric hiss that grows in intensity owing to particle-wave interactions from the background thermal noise, as suggested by *Thorne et al.* [1979], would have a distribution completely independent of any geographic mapping. In order for lightning to be even considered to be an element of plasmaspheric hiss, a geographic relationship would have to be established between lightning and the observed distribution of plasmaspheric hiss.

The average distribution of lightning strikes worldwide has been measured from the Optical Transient Detector on board the MicroLab-1 satellite by *Christian et al.* [2003] and is shown in Plate 3 along with the main slot region L shells [after *Rodger et al.*, 2003]. Plate 3A [adapted from *Christian et al.*, 2003] shows the average lightning distribution for 3 months in the summer (June through August) and Plate 3B is for 3 months in the winter (December–February). Both lightning distributions in Plate 3 clearly show that the lightning distributions are largely confined to the continents. Owing to the tilt of the magnetic pole, the continents of North America, Europe, and Russia, provide the vast majority of lightning whistlers into the slot region during the summer with eastern Australia and the southern tip of South American contributing during the winter months [*Christian et al.*, 2003]. In addition, these seasonal distributions of lightning activity show significantly more lightning in the summer hemisphere then in the corresponding winter hemisphere. From an analysis of their annual lightning distributions, *Christian et al.* [2003] determined that lightning occurs mainly over land areas, with an average land/ocean ratio of 10:1. Statistically, these authors found that there are an average of 44 (+/-5) lightning flashes occurring around the globe every second. In another recent study of lightning, *Mazany et al.* [2002] measured the day-night effect at mid-latitudes and found that as much as 10 times more lightning events occur in the post noon than in the post midnight sector. Note that in Plate 2D hiss is most intense in the post noon sector.

Plate 4 from *Green et al.* [2005] shows the average wave electric field spectral density from all DE/PWI measurements of plasmaspheric hiss at 3 kHz mapped to magnetic coordinates and also shows the continents, The left panels are dayside and the right panels, nightside. The top two panels are summer data and the bottom two panels present the winter data. The equinox months of March and September were excluded in order to facilitate the comparison with strictly the summer and winter lightning distributions of Plate 3 while separating out the contributions of whistler mode lightning into the plasmasphere from the northern and southern hemispheres. The electric field wave measurements within 10° of the magnetic equator or in locations in which the gyrofrequency was less than the wave frequency were excluded.

The resulting wave maps of Plate 4 shows the correspondence between the enhanced plasmaspheric hiss intensities and continents. It is important to note that the plasmaspheric hiss distributions look very similar over the ~500 Hz to about 3 kHz frequency range. The geographic distribution of plasmaspheric hiss, like the geographic distribution of lightning, qualitatively follows the continents and is stronger on the dayside than the nightside and is stronger in summer than in winter. *Green et al.* [2005] also found the same distributions were observed with the IMAGE/RPI data.

During the summer (Plate 5A and 5B) peak intensities of plasmaspheric hiss are shown mainly over northern hemisphere land with minimums over the ocean. During the winter (Plate 5C), the peak in the plasmaspheric hiss corresponds

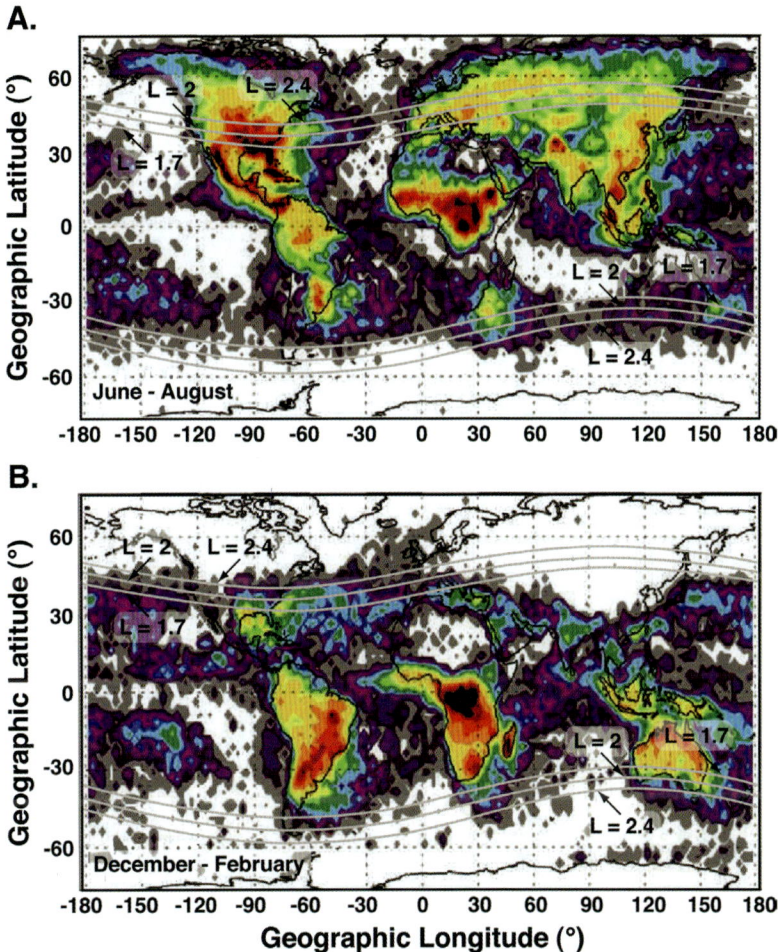

Plate 3. Panels A and B show the average distribution of lightning strikes world-wide measured from the Optical Transient Detector on board the MicroLab-1 satellite during summer (June–August) and winter (December–February) months, respectively [adapted from *Christian et al.,* 2003]. The main slot region L shells are also drawn at L = 2.4, 2, and 1.7.

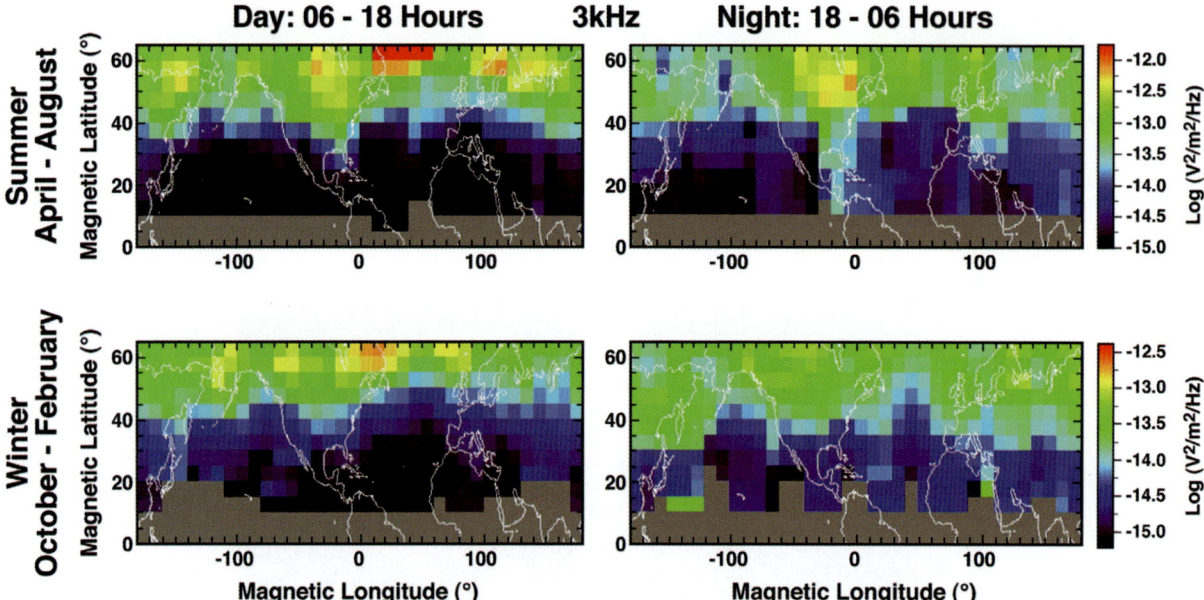

Plate 4. The average electric wave spectral density from all DE/PWI data of plasmaspheric hiss mapped to geographic coordinates at 3 kHz. The left panels are dayside and the right panels, nightside observations. The top panels are summer and bottom panels are winter with the exclusion of the equinox months March and September. Like the world distribution of lightning, the world distribution of plasmaspheric hiss follows the continents and is stronger on the dayside than the nightside and is stronger in summer than in winter [from *Green et al., 2005*].

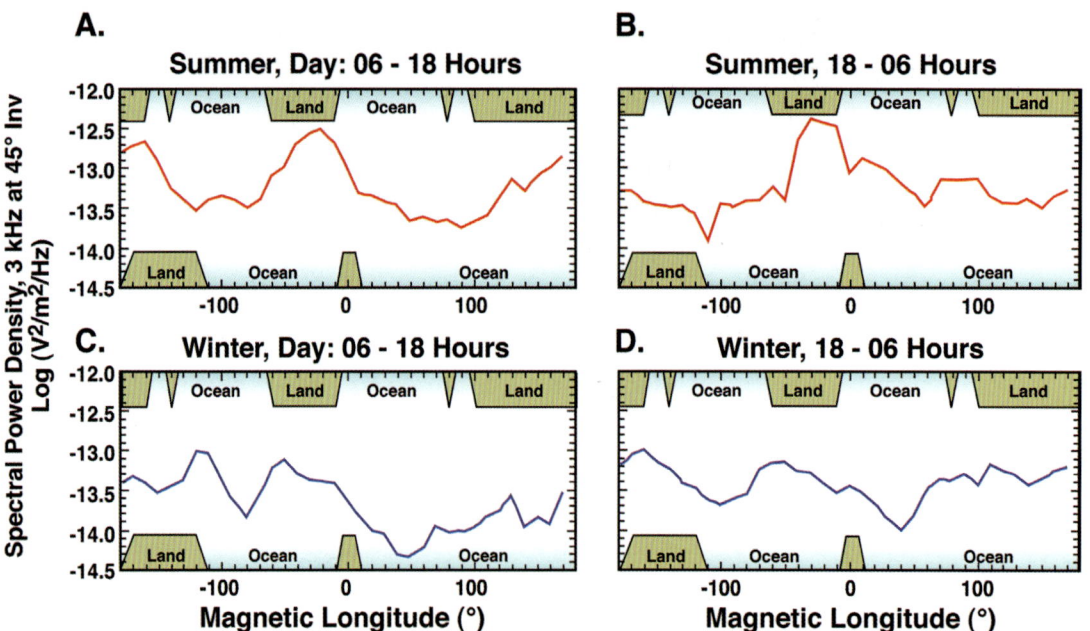

Plate 5. Electric field spectral density measurements and landmasses at 45° invariant latitude of data sorted in Plate 4. The summer data is shown panels A (dayside) and B (nightside) with winter data in panels C (dayside) and D (dayside). Peak intensities of plasmaspheric hiss are over land with minimums over the ocean with summer intensities being stronger than winter [adapted from *Green et al., 2005*]. Contributions from the southern hemisphere occur primarily during the winter (panels C and D).

2. The mainly geographic mapping of plasmasphere hiss is strikingly similar to lightning distributions for both day/night and summer/winter variability and asymmetries (Plate 3, 4, and 5). The lightning and hiss intensities are stronger over continents than over the ocean, stronger in the summer than winter, and stronger on the dayside than nightside.
3. Solar-cycle movement of the slot region (Plate 7) is consistent with a corresponding shift in the plasmaspheric hiss distribution caused by the raised atmospheric densities during solar maximum intervals when attenuation of lightning whistlers is increased along the lower L range of the slot region (below ~L = 3), effectively raising the slot to higher L.

Acknowledgments. The authors gratefully acknowledge discussions with W. W. L. Taylor. The National Aeronautics and Space Administration supported this research under SECGIP03-0085 and the IMAGE project. The DE and IMAGE data used in this paper are from the National Space Science Data Center archive.

REFERENCES

Abel, B., and R. M. Thorne, Electron scattering loss in Earth's inner magnetosphere: 1. Dominant physical processes, *J. Geophys. Res., 103*, 2385–2396, 1998a.

Abel, B., and R. M. Thorne, Electron scattering loss in Earth's inner magnetosphere: 2. Sensitivity to model parameters, *J. Geophys. Res., 103*, 2397–2407, 1998b.

Albert, J. M., Quasi-linear pitch angle diffusion coefficients: retaining high harmonics, *J. Geophys. Res., 99*, 23741–23746, 1994.

André, R., F. Lefeuvre, F. Simonet, and U. S. Inan, A first approach to model the low-frequency wave activity in the plasmasphere, *Annales Geophys., 20*, 981–996, 2002.

Boardsen, S. A., D. L. Gallagher, D. A. Gurnett, W. K. Peterson, and J. L. Green, Funnel-shaped, low frequency equatorial waves, *J. Geophys. Res., 97*, 14967–14976, 1992.

Bortnik, J., U. S. Inan, and T. F. Bell, Frequency-time spectra of magnetospherically reflecting whistlers in the plasmasphere, *J. Geophys. Res., 108*(A1), 1030, doi:10.1029/2002JA009387, 2003a.

Bortnik, J., U. S. Inan, and T. F. Bell, Energy distribution and lifetime of magnetospherically reflecting whistlers in the plasmasphere, *J. Geophys. Res., 108*(A5), 1199, doi:10.1029/2002JA009316, 2003b.

Christian, H. J., R. J. Blakeslee, D. J. Boccippio, W. L. Boeck, D. E. Buechler, K. T. Driscoll, S. J. Goodman, J. Hall, W. Koshak, D. Mach, and M. Stewart, Global frequency and distribution of lightning as observed from space by the Optical Transient Detector, *J. Geophys. Res., 108,* doi:10.1029/2002JD002347, 2003.

Church, S. R., and Thorne, R. M., On the origin of plasmaspheric hiss: Ray path integrated amplification, *J. Geophys. Res., 88*, 7941–7957, 1983.

Draganov, A. B., U. S. Inan, V. S. Sonwalkar, and T. F. Bell, Magnetospherically reflected whistlers as a source of plasmaspheric hiss, *Geophys. Res. Lett., 19*, 233–236, 1992.

Draganov, A. B., U. S. Inan, V. S. Sonwalkar, and T. F. Bell, Whistlers and plasmaspheric hiss: Wave directions and three-dimensional propagation, *J. Geophys. Res., 98*, 11401–11410, 1993.

Dunckel, N., and R. A. Helliwell, Whistler-mode emissions on the Ogo 1 satellite, *J. Geophys. Res., 74*, 6371–6385, 1969.

Fung, S. F., E. V. Bell, L. C. Tan, R. M. Candey, M. J. Golightly, S. L. Huston, J. H. King, and R. E. McGuire, Development of A Magnetospheric State-Based Trapped Radiation Data Base, *Adv. Space Research*, in press, 2005

Goldberg, R. A., S. A. Curtis, and J. R. Barcus, Detailed spectral structure of magnetospheric electron bursts precipitated by lightning, , *J. Geophys. Res., 92*, 2505–2513, 1987.

Green, J. L., S. A. Boardsen, L. Garcia, W. W. L. Taylor, S. F. Fung, B. W. Reinisch, On the Origin of Whistler Mode Radiation in the Plasmasphere, *J. Geophys. Res., 110*, doi: 10.1029/2004JA010495, 2005.

Gurnett, D. A., Plasma wave interactions with energetic ions near the magnetic equator, *J. Geophys. Res., 81*, 2765–2770, 1976.

Hayakawa, M., and S. S. Sazhin, Mid-latitude and plasmaspheric hiss: a review, *Planet Space Sci., 40*, 1325, 1992.

Helliwell, R. A., "Whistlers and Related Ionospheric Phenomena," Stanford University Press, pp. 61–72, 1965.

Huang, C. Y., C. K. Goertz, and R. R. Anderson, A theoretical study of plasmaspheric hiss generation, J. Geophys. Res., 88, 7927–7940, 1983.

Imhof, W., R. R. Anderson, J. Reagan, and E. Gaines, The significance of VLF transmitters in the precipitation of inner belt electrons, *J. Geophys. Res., 86*, 11225–11234, 1981.

Inan, U. S., M. Walt, H. D. Voss, and W. L. Imhof, Energy spectra and pitch angle distributions of lightning-induced electron precipitation: Analysis of an event observed on the S81-1 (SEEP) satellite, *J. Geophys. Res., 94*, 1379–1401, 1989.

Kennel, C. F., and H. E. Petschek, Limit on stably trapped particle fluxes, *J. Geophys. Res., 71*, 1–28, 1966.

Lyons, L. R., R. M. Thorne, and C. F. Kennel, Pitch-angle diffusion of radiation belt electrons within the plasmasphere, *J. Geophys. Res., 77*, 3455–3474, 1972.

Mazany, R., S. Businger, S.I. Gutman, and W. Roeder, A lightning prediction index that utilizes GPS integrated precipitable water vapor. *Weather and Forecasting, 17*, 1034–1047, 2002.

Pierce, E. T., *Chapter 10: Atmospherics and Radio Noise*, in "Lightning", edited by R. H. Golde, Academic Press, 1977.

Rodger, C. J., M. A. Cilverd, and R. J. McCormick, Significance of lightning-generated whistlers to inner radiation belt electron lifetimes, *J. Geophys. Res., 108(A12)*, 1462, doi:10.1029/2003JA009906, 2003.

Russell, C. T., R. E. Holzer, and E. J. Smith, OGO 3 observations of ELF noise in the magnetosphere: 1. Spatial extent and frequency of occurrence, *J. Geophys. Res., 74*, 755–777, 1969.

Russell, C. T., R. E. Holzer, and E. J. Smith, OGO 3 observations of ELF noise in the magnetosphere, 2. The nature of the equatorial noise, *J. Geophys. Res., 75*, 755–768, 1970.

Solomon, J., N. Cornilleau-Wehrlin, A. Korth, and G. Kremser, An Experimental study of ELF/VLF hiss generation in the Earth's magnetosphere, *J. Geophys. Res.*, *93*, 1839–1847, 1988.

Sonwalker, V. S., and U. S. Inan, Lightning as an embryonic source of VLF hiss, *J. Geophys. Res.*, *94*, 6986–6994, 1989.

Stix, T. H., *Waves in Plasmas*, AIP Press, New York, p. 27–28, 1992.

Storey, L. R. O., An investigation of whistling atmospherics, *Phil. Trans. Roy. Soc.*, *246*, 113–141, 1953.

Taylor, W. W. L., and D. A. Gurnett, The morphology of VLF emissions observed with the Injun 3 satellite, *J. Geophys. Res.*, *73*, 5615–5626, 1968.

Thorne, R. M., E. J. Smith, R. K. Burton, and R. E. Holzer, Plasmaspheric hiss, *J. Geophys. Res.*, *78*, 1581–1595, 1973.

Thorne, R. M., and J. N. Barfield, Further observational evidence regarding the origin of plasmaspheric hiss, *Geophys. Res. Letts.*, *3*, 29–32, 1976.

Thorne, R. M., S. R. Church, and D. J. Gorney, On the origin of plasmaspheric hiss: the importance of wave propagation and the plasmapause, *J. Geophys. Res.*, *84*, 5241–5247, 1979.

Thorne, R. M., and R. B. Horne, Landau damping of magnetospherically reflected whistlers, *J. Geophys. Res.*, *99*, 17249–17258, 1994.

Thorne, R. M, R. B. Horne, S. Glauert, N. P. Meredith, Y. Y. Shprits, D. Summers, and R. R. Anderson, The Influence of Wave-Particle Interactions on Relativistic Electron Dynamics During Storms, This volume, 2005.

Voss, H. D., W. L. Imhof, J. Mobilia, E. E. Gaines, M. Walt, U. S. Inan, R. A. Helliwell, D. L. Carpenter, J. P. Katsufrakis, and H. C. Chang, Lightning-induced electron precipitation, *Nature*, *312*, 740, 1984.

Voss, H. D., M. Walt, W. L. Imhof, J. Mobilia, and U. S. Inan, Satellite observations of lightning-induced electron precipitation, *J. Geophys. Res.*, *103*, 11725–11744, 1998.

S. A. Boardsen, L3 Communications, Government Services Inc., 1801 McCormick Drive, Largo, Maryland, 20774, USA

H. J. Christian, NASA Marshall Space Flight Center, Huntsville, Alabama, 35805, USA

S. F. Fung, NASA Goddard Space Flight Center, Code 612.4, Greenbelt, Maryland, 20771, USA

J. L. Green, NASA Goddard Space Flight Center, Code 605, Greenbelt, Maryland, 20771, USA (James.Green@NASA.gov)

Direct Effects of the IMF on the Inner Magnetosphere

R. A. Wolf[1], S. Sazykin[1], X. Xing[1], R. W. Spiro[1], F. R. Toffoletto[1],
D. L. De Zeeuw[2], T. I. Gombosi[2], and J. Goldstein[3]

Several direct and well-established inner-magnetospheric effects following changes in Interplanetary Magnetic Field direction find natural explanations in terms of the Rice Convection Model. A southward turning of the IMF normally causes an increase in both cross-polar-cap potential drop and in polar-cap size. In RCM simulations, these two factors combine to produce a condition termed "undershielding," characterized by increased penetration of the dawn-dusk convection electric field into the inner magnetosphere, erosion of the nightside plasmapause, and drainage of plasmaspheric plasma via plumes that stretch to the dayside magnetopause. A northward turning of the IMF causes a condition known as "overshielding," characterized by dusk-to-dawn directed electric fields across the inner magnetosphere. A recent run with the coupled BATSRUS/RCM computer code suggests that the overshielding electric field peaks 12–25 minutes after the IMF direction at the dayside magnetopause turns northward, and that this time delay is about the same on both the day- and night-sides of the Earth. Changes in solar wind and IMF conditions may also influence the inner magnetosphere in more subtle ways, through their influence on the plasma-sheet plasma distribution represented by the specific entropy. Results of combining a Tsyganenko magnetic field model and Tsyganenko-Mukai representation of plasma-sheet plasma suggest that a northward turning of the IMF may reduce the specific entropy in the plasma sheet and cause interchange instability in the plasma sheet and auroral ionosphere.

1. INTRODUCTION

Penetration to the low- and mid-latitude ionosphere and inner magnetosphere of electric fields associated with the interaction of the flowing solar wind plasma with Earth's magnetosphere is known to have important practical effects on the near-Earth plasma environment and on manmade systems. This paper attempts a brief summary of the effects that changes in the solar wind and Interplanetary Magnetic Field (IMF) have on the inner magnetosphere and underlying ionosphere. Of particular interest is the response of the inner magnetosphere to changes in the direction and strength of the IMF. In the inner magnetosphere, electric field (E×B) and gradient/curvature drifts determine the dynamics of the inner edge of the plasma sheet and the buildup and evolution of the ring current. At subauroral latitudes magnetospherically generated electric fields control the dynamics of the plasmapause and the formation and dynamics of the main ionospheric trough. Under steady solar wind and IMF conditions, the inner magnetosphere becomes shielded from the effects of the cross-tail magnetospheric convection electric field by magnetic field-aligned currents (the so-called Region 2 current system) that couple the inner plasma sheet to the underlying ionosphere. Disruption of this current system in response to changes in solar-wind/magne-

[1] Physics and Astronomy Dept., Rice University, Houston, Texas
[2] Center for Space Environment Modeling, University of Michigan, Ann Arbor, Michigan
[3] Space Science and Engineering Division, Southwest Research Institute, San Antonio, Texas

Inner Magnetosphere Interactions: New Perspectives from Imaging
Geophysical Monograph Series 159
Copyright 2005 by the American Geophysical Union.
10.1029/159GM09

tosphere coupling causes inward penetration of convection electric fields that can have profound ionospheric implications, changing ionospheric layer heights and leading to the generation and evolution of a number of different plasma instability processes.

Our discussion proceeds mainly from the viewpoint of the Rice Convection Model (RCM), a computer model specifically formulated to model the physics of the inner magnetosphere and its coupling to the ionosphere. In Section 2 we describe how empirical models of solar-wind/magnetosphere coupling are used to estimate the RCM's principal electromagnetic inputs, the cross-polar-cap electric potential drop and the structure of the magnetospheric magnetic field. In Sections 3 and 4 we summarize well-established ways in which the inner magnetosphere reacts to southward and northward turnings of the IMF. In addition, Section 4 also includes new results from a coupled MHD/RCM code that address the time delay of the inner magnetospheric response to a northward turning of the IMF. Section 5 discusses possible effects of IMF-associated changes in the specific-entropy function $PV^{5/3}$ at the RCM's tailward boundary, and the inner-magnetospheric implications of these changes.

2. RCM–ASSUMPTIONS AND INPUT PARAMETERS

The RCM was specifically designed for accurate treatment of the closed-field-line, slow-flow part of the magnetosphere. In the model, inertial currents are assumed negligible, precluding the inclusion of MHD waves and limiting the model to regions where the flows are highly subsonic. Magnetic flux tubes are assumed to contain plasma in bounce equilibrium. Cross-field motions of charged particles are assumed to consist of E×B, gradient, and curvature drift. The model assumes elastic pitch-angle scattering, resulting in isotropic pitch-angle distributions. For a detailed discussion of the formulation of the RCM, see *Toffoletto et al.* [2003] and references therein.

The ionospheric conductance distribution due to sunlight and other non-auroral processes is calculated by field-line-integrating an IRI-90 ionospheric model [*Bilitza et al.*, 1993] combined with the MSIS-90 empirical neutral atmosphere [*Hedin*, 1991]. Conductance enhancements associated with auroral electron precipitation are computed assuming that electrons scatter in pitch angle at a fixed fraction of the strong-pitch-angle-scattering rate and using the empirical formulas of *Robinson et al.* [1987]. The effects of neutral winds are currently neglected.

In the RCM runs discussed here, the magnetosphere is assumed to be initially empty, and the upward flow of ionospheric ions directly into the RCM-modeled inner magnetosphere is neglected. Magnetospheric plasma is assumed to enter the RCM modeling region through its tailward boundary, where the distribution is assumed to have the form of either a Maxwellian or kappa distribution. For standard RCM runs, the values of $PV^{5/3}$ and nV at the tailward boundary are estimated empirically, a subtle issue that will be discussed further in Section 5.

The cross-polar-cap potential drop, which is a crucial input for the RCM because it measures the total strength of convection, is estimated from solar-wind data using an empirical formula developed by *Boyle et al.* [1997], but set to linearly saturate at 200 kV, in accordance with the observations of *Hairston et al.* [2003]. That formula implies that the polar-cap potential drop depends strongly on the southward component of the IMF. Magnetospheric magnetic fields within the RCM modeling region have been estimated in various ways. The semi-empirical *Hilmer-Voigt* [1995] magnetic field model was used for the older RCM runs discussed in Sections 3 and 4, while a T96 model [*Tsyganenko and Stern*, 1996] was used to obtain the results shown in Section 5. For the coupled MHD-RCM run discussed in Section 4, the MHD code [*De Zeeuw et al.*, 2004] supplies theoretically computed magnetic fields.

3. ELECTROMAGNETIC EFFECTS OF A SOUTHWARD TURNING OF THE IMF

As predicted by *Schield et al.* [1969], *Vasyliunas* [1970], and *Wolf* [1974], magnetic-field aligned currents (the so-called "region-2 Birkeland currents") flow between the inner plasma sheet and ionosphere and act to shield the inner magnetosphere from the main effects of the dawn-dusk convection electric field. However, the shielding is clearly ineffective in times when convection is changing with time. For example, suppose that solar wind and IMF conditions are steady for a long period of time, resulting in a long period of steady convection such that the inner edge of the plasma sheet is able to adjust itself to shield the inner magnetosphere effectively. Then the IMF turns southward, causing a sudden increase in convection and leaving the region-2 currents inadequate to shield the inner magnetosphere under the changed conditions. Much of the dawn-dusk convection field will now penetrate into the near-Earth inner magnetosphere. The westward electric field across the night side will cause the plasma-sheet inner edge to move sunward, resulting in a gradual increase of the region-2 currents and more effective inner-magnetosphere shielding, but that takes time. The temporary penetration of the dawn-dusk field into the inner magnetosphere after an increase in convection is termed "undershielding." Figure 1 shows two electric equipotential patterns, as computed by the RCM for a time of good shielding just before a sudden increase in cross-polar-cap

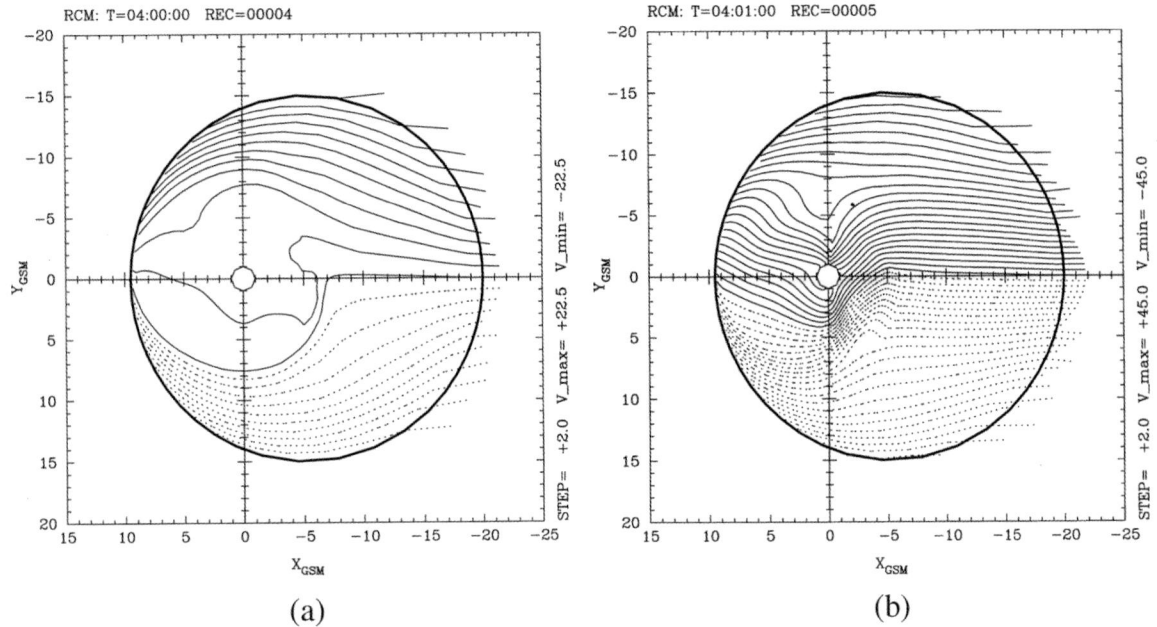

Figure 1. Equipotential diagrams just before and just after a sudden increase in polar cap potential from 45 keV to 90 kV. The view is of the equatorial plane, with the Sun to the left. The corotation electric field is not included in the display, though it was included in the simulation. From *Sazykin* [2000].

potential drop, and just after the potential drop increase. The potential patterns are shown in the magnetospheric equatorial plane. The electric field that penetrates into the inner magnetosphere after the increase departs significantly from a simple dawn-dusk orientation, mainly because of the effects of the sharp conductance jumps associated with the dawn and dusk terminators; note the computed westward electric field near Earth in the midnight-to-dawn sector and eastward electric field in the dawn-to-noon sector. Figure 2 shows very good agreement between RCM-computed downward E×B drift near the ionospheric equator and corresponding average velocities derived by *Fejer and Scherliess* [1997] from Jicamarca radar data, using statistical analysis of many convection-increase events. Detailed theoretical discussions of shielding can be found in *Wolf* [1983] and *Spiro et al.* [1988].

Recent years have produced one major change in the 1970s picture of the physics of undershielding. It is now clear that the undershielding that occurs after a southward IMF turning is not due entirely to the potential drop increase, but also to a change in the magnetic configuration. *Wolf et al.* [1982] presented theoretical arguments showing that effective shielding requires that dayside magnetic flux tube volume values exceed nightside values, while *Fejer et al.* [1990] applied these arguments to the effect of magnetic configuration changes on low-latitude perturbation electric fields. Figure 3 displays magnetic field lines in the noon-midnight plane computed for

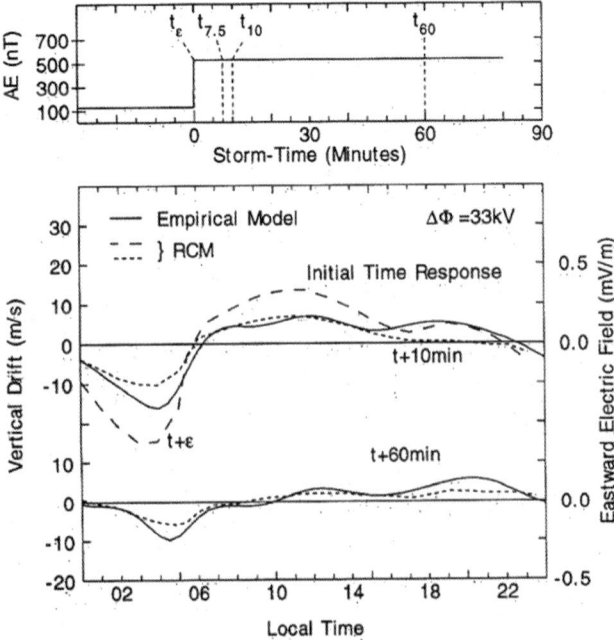

Figure 2. Comparison of RCM-calculated prompt-penetration-induced downward drifts with those derived statistically from observations with the Jicamarca radar. Observational results are shown for times just after an increase in AE and 60 minutes later (bottom plot). RCM results are shown for times just after a 33 kV increase in potential and 10 minutes later (middle plot), then 60 minutes later (bottom plot). Adapted from *Fejer and Scherliess* [1997].

130 IMF EFFECTS ON INNER MAGNETOSPHERE

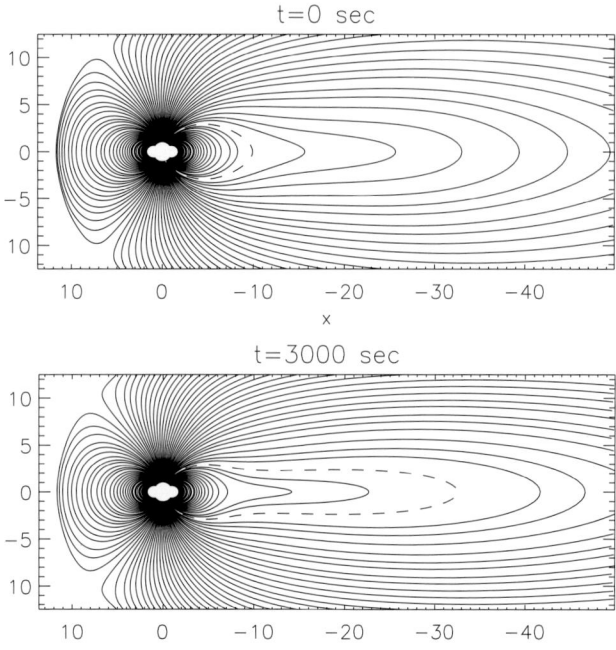

Figure 3. Magnetic field lines in the noon-midnight meridian. The top diagram shows a Tsyganenko model, relaxed to equilibrium. The lower diagram shows how 50 minutes of strong adiabatic convection affects the configuration. The dashed curve in each diagram maps to 68° latitude. From *Toffoletto et al.* [2000].

steady-state conditions using the RCM coupled to an equilibrium magnetic field solver that maintains approximate force balance between the RCM-computed plasma distribution and the inner-magnetosphere magnetic field [*Toffoletto et al.*, 2000]. The effect of a southward IMF turning is two-fold: (1) an increase in the dawn to dusk convection electric field; and (2) an increase in the tail-lobe magnetic field and stretching of inner-plasma-sheet flux tubes. Because of this stretching, the equatorial mapping point of a given point arising from the nightside auroral ionosphere moves tailward: this stretching corresponds to E×B drift in an eastward induction electric field that exists in the tail but does not map to the ionosphere. This eastward induction electric field tries to move the inner edge of the plasma sheet tailward, increasing the flux tube volume and reducing or eliminating the shielding and thus contributing to the undershielding. The eastward induction field and westward potential field (undershielding) oppose each other near the equatorial plane; however, an ionospheric observer sees only the westward potential field, because the induction field does not map to the ionosphere. Thus, both the increase in polar-cap potential and the tail field-line stretching and associated increase in flux tube volume contribute substantially to undershielding following a southward turning of the IMF.

Plate 1 shows the evolution of the plasmapause in an RCM simulation of the event of 10 July 2000, for which IMAGE EUV data have been analyzed by *Goldstein et al.* [2003a]. The plasmapause evolution was modeled by representing the assumed initial plasmapause in terms of a string of test particles, then following the string as the particles E×B drift in the model electric field; test particles are added by interpolation when adjacent points get too far apart. The following features should be noted:

(i) There is always a drainage tail. Theoretical models of plasmapause shape based on the assumption of time-dependent convection have predicted such drainage tails for many years (e.g., *Grebowsky* [1970]), and early satellite observations showed regions of high density outside the main plasmasphere (*Chappell* [1972] and references therein). However, there was a long-running controversy about whether those outer high-density regions were detached from the main plasmapause or connected to it, as suggested by the models. EUV observations from IMAGE seem to have settled that controversy in favor of drainage plumes that extend out from the main plasmapause [*Sandel et al.*, 2001].

(ii) In this simulation, a plasmapause protuberance that was evident near local noon at 0000 UT rotated east over the next four hours and grew somewhat. Then the strong convection caught it and made it into a second drainage tail. Though the protuberance at 0000 UT depended on the way the code was initialized and is not necessarily physical, some instances of double drainage tails have been observed [*Goldstein et al.*, 2004a].

(iii) The long period of strong convection on 10 July 2000 caused the simulated drainage tail to become very thick, filling a large part of the dayside magnetosphere, a phenomenon that was first noted many years ago [e.g., *Grebowsky*, 1970; *Chappell*, 1972].

(iv) The plasma that filled the drainage tail and escaped to the magnetopause boundary layer came from the main plasmasphere, which has eroded substantially on the night side. The simulated plasmapause radius at local midnight decreased almost a factor of two between 0420 and 0800. EUV observations [*Goldstein et al.*, 2003a] documented the severe erosion of the real plasmasphere in this event. The observed motion of the plasmapause to within about 3 R_E of Earth's center is well reproduced by the model.

A southward turning of the IMF that leads to sustained and substantial southward IMF causes severe convection effects in the inner magnetosphere. Fresh plasma is injected into the ring current from the plasma sheet, a process that has been extensively studied with several different ring-current models [e.g., *Fok and Moore*, 1997; *Kozyra et al.*, 2002; *Jordanova et al.*, 2003; *Chen et al.*, 2003], and also by a set of models which calculate electric fields that are self-con-

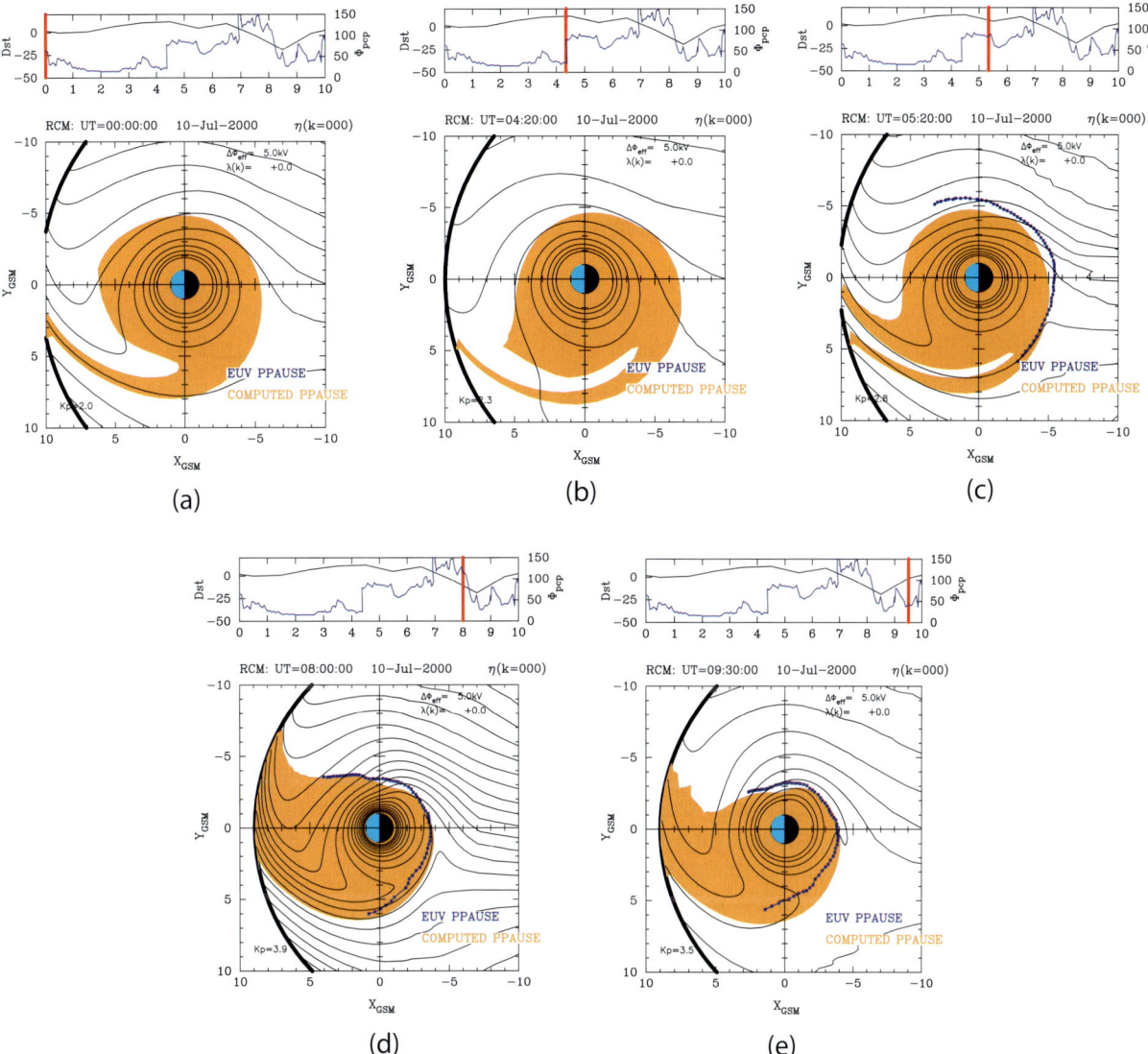

Plate 1. RCM simulation of plasmapause evolution for an event that occurred 10 July 2000. In each panel, the upper plot shows model inputs (polar cap potential (blue) and *Dst* (black)) as a function of time, with the red bar indicating the time for the plasmapause plot. In the lower part of each panel, orange indicates the computed plasmasphere, and black contours show instantaneous streamlines for cold plasma. The view is of the equatorial plane, with the Sun to the left. The connected points shown in blue are IMAGE EUV measurements of the plasmapause location. The five panels represent times (a) UT=0, more than 4 hours before the beginning of the strong convection; (b) UT=0420, just before the strong convection; (c) UT=0520, after about an hour of strong convection; (d) UT=0800, the end of strong convection; (e) UT=0930, in a low-convection period.

sistent with the ring-current plasma—the Rice Convection Model [*Garner et al.*, 2004], the Comprehensive Ring Current Model [*Fok et al.*, 2001; *Ebihara et al.*, 2004], and the RAM model [*Liemohn and Ridley*, 2002]. Most of the self-consistent calculations exhibit two features that are in striking agreement with recently discovered observational features. The storm simulations of *Garner et al.* [2004], for example, show strong poleward/outward electric fields in the dusk-midnight sector, which were observed near the equatorial plane [*Rowland and Wygant*, 1998; *Burke et al.*, 1998] and also in the subauroral ionosphere [*Foster and Vo*, 2002], where they are termed Subauroral Polarization Stream (SAPS) events. Self-consistent simulations [e.g., *Fok et al.*, 2003] as well as calculations based on a high-resolution version of the AMIE empirical model supplemented by a penetration electric field [*Jordanova et al., 2003*] have shown a tendency for the main-phase ring-current-ion pressure to peak near local midnight (and sometimes east of midnight), in agreement with energetic-neutral-atom results from IMAGE [*Brandt et al.*, 2002].

4. ELECTROMAGNETIC EFFECTS OF A NORTHWARD TURNING OF THE IMF

From an inner magnetosphere (RCM) point of view, the effect of a northward turning of the IMF is basically the opposite of the effect of a southward turning. The cross-polar-cap potential drop decreases, and the magnetic field configuration changes in such a way that the polar cap shrinks and the inner-plasma-sheet magnetic field lines become less stretched. Both of these changes lead to a condition of "overshielding." In this condition, the region-2 Birkeland currents are stronger than necessary to shield the inner magnetosphere from the dawn-dusk convection electric field, resulting in a dusk-to-dawn-directed electric field in the inner magnetosphere. Figure 4 shows a typical RCM-computed overshielding pattern. The overshielding phenomenon was first identified in Jicamarca radar data by *Kelley et al.* [1979].

Although RCM calculations predict typical lifetimes of undershielding electric fields that are in good agreement with observations, RCM-computed overshielding electric fields have noticeably shorter lifetimes [*Fejer et al.*, 1990], even when the time-dependent magnetic field effect is included [*Sazykin*, 2000]. In other words, when the polar-cap potential drop changes and magnetic field changes are symmetric for undershielding and overshielding cases, the response of the ionospheric penetration electric field is not. A possible reason for this discrepancy is our lack of knowledge of how the magnetic field responds to IMF northward turnings. The problem can only be addressed with coupled RCM-MHD

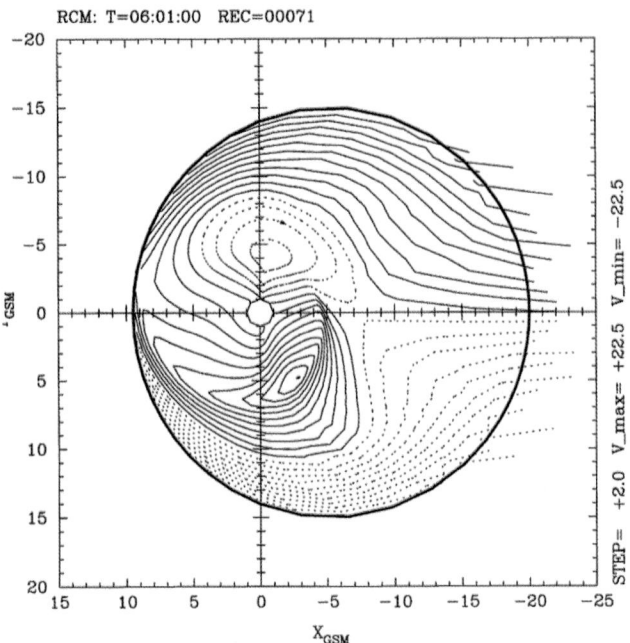

Figure 4. RCM-computed equatorial equipotentials immediately after a factor-of-two reduction in polar-cap potential. From *Sazykin* [2000].

codes that compute the magnetic field in a self-consistent manner (see below).

One of the first discoveries from the IMAGE EUV instrument was the occasional development of a shoulder on the plasmapause [*Goldstein et al.*, 2002], which was quickly interpreted as the result of overshielding. The near-Earth electric potential pattern of Figure 4 shows eastward electric field before dawn and westward electric field after dawn. *Goldstein et al.* [2002, 2003b] used the Magnetospheric Specification Model, which does not compute the electric field but uses a prescribed electric field pattern based on typical RCM results, to show that the pre-dawn eastward electric field moves the plasmapause out in that local time sector, while the westward post-dawn field moves it earthward; the result is development of a shoulder in the plasmapause during the period of overshielding. If northward IMF continues for some hours after the period of overshielding, convection near the plasmapause is weak, and the shoulder approximately corotates around into the day side. *Goldstein et al.* [2002] showed the effectiveness of overshielding in a case where the IMF switched quickly from southward to northward, while *Goldstein et al.* [2003b] showed a larger shoulder that resulted when the IMF underwent a longer and more gradual transition from southward to northward IMF.

Although the RCM has been used for many years to study prompt-penetration electric fields, it has never been possible using the RCM alone to make detailed predictions about the timing of these fields in response to solar-wind changes, because we lacked detailed information on the time-dependence of the RCM boundary conditions. For example, we typically assumed that the potential at the RCM poleward boundary switched suddenly in response to a sudden change in IMF, but the truth must be more complicated: the IMF change must be felt first near local noon and then gradually spread toward the night side. We have similarly lacked a quantitative representation of how the magnetic field configuration responds to an IMF change, as a function of time.

Embedding Rice Convection Model machinery in the BATS-R-US global MHD code [*De Zeeuw et al.*, 2004] promises to allow more detailed theoretical analysis of how the inner-magnetospheric electric field reacts to a specified change in IMF, since the new coupled BATSRUS/RCM code treats solar-wind/magnetosphere coupling and the inner and outer magnetospheres self-consistently. The MHD code computes the time evolution of both the high-latitude potential distribution and the magnetic reconfiguration as a function of time and provides those inputs to the RCM. The RCM uses its many-species representation to keep track of the inner-magnetospheric particle distribution and passes its pressure distribution back to the MHD code, which nudges its pressures to maintain approximate agreement with the RCM.

Plate 2 shows the time-dependent ionospheric-potential distribution after a northward turning of the IMF. The results displayed are similar to those reported by *De Zeeuw et al.* [2004], except that, in the present run, the RCM uses its own computational machinery to compute Birkeland currents and electric potential in its modeling region. For the earlier run, the RCM used MHD-computed Birkeland currents and potentials. The two procedures should give the same results in principle, but the RCM uses a finer grid in the ionosphere, for better numerical accuracy. In order to study the penetration-electric-field problem, the present run computes ionospheric potentials down to the equator, whereas the low-latitude boundary for the earlier run was set at about 51 degrees latitude.

For the present simulation, the solar wind was assumed steady for eight hours, with $n=5$ cm^{-3} (protons), $V=400$ km/s, $B_x=B_y=0$, $B_z=-5$ nT. At 0800 UT, B_z was switched to +5 nT at the sunward boundary of the MHD modeling region ($X=32\ R_E$). In the simulation, the northward IMF hit the dayside magnetopause at about 0809 UT. The simplest way to get a theoretical estimate for when the northward IMF should arrive at the dayside magnetopause would be to divide the geocentric distance of the sunward edge of the modeling region by the solar wind speed, which gives a time delay of 8.5 minutes. There are two obvious sources of error in this simple estimate: (i) the magnetopause stands upstream from the Earth; (ii) the speed slows drastically as the flow passes through the bow shock and approaches the magnetopause. Since the simple estimate agrees well with the simulation, apparently effects (i) and (ii) approximately cancel in the present case.

It is clear from Plate 2 that the effect of the northward turning on the polar-cap electric field distribution begins at local noon and gradually propagates across the polar cap to the night side. Remarkably, the same is not true of the penetration field: it intensifies and then de-intensifies but maintains basically the same spatial pattern throughout the event. This may be due, in part, to the fact that the overshielding electric potential is basically a 2D dipole pattern. The higher multipole moments that describe the concentration of the effect in the local-noon region shortly after 0809 UT die off more rapidly with increasing colatitude than the dipole term does and consequently do not strongly affect the field that penetrates to the low-latitude ionosphere.

Figure 5 shows a quantitative index of electric-field penetration, in the form of the total potential difference along the low-latitude ionospheric boundary of the RCM at 9.84 degrees latitude. The penetration potential increases gradually over the first ten minutes after the northward turning at the magnetopause, peaks 12–25 minutes after the northward turning, then decays on a time scale ~30 minutes.

Considerable work has been done with the EUV instrument on the IMAGE spacecraft, comparing the time-dependence of electric fields at the nightside plasmapause with IMF B_z or solar wind E_y (e.g., *Goldstein et al.*, 2003a, 2003c, 2004c, *Goldstein and Sandel*, 2004). That work concentrates mostly on southward turnings of the IMF and concludes that plasmasphere erosion is delayed about 30 minutes from the arrival of the IMF turning at the magnetopause. Examination of sharp northward turnings in the plots from *Goldstein et al.* [2004c] suggests that a time delay ~30 minutes applies also to northward turnings. It is not clear how much significance to attach to the difference between the IMAGE-observed time delay (~30 minutes) and our preliminary theoretical estimate (12–25 minutes). About 2–3 minutes of the difference can be explained by the fact that Goldstein and collaborators assume that plasma travels at full solar-wind speed between the upstream spacecraft and the magnetopause, whereas we use the MHD model to calculate the magnetopause arrival time. More realistic modeling with realistic ionospheric conductances, plus more detailed study of measured time profiles of the plasmapause electric field for specific northward-turning events, will be needed to determine whether there is a meaningful discrepancy between theory and observations with regard to the magnitude of time delay.

Dayside and nightside plasmapause responses often seem to occur with comparable delays (J. Goldstein, private communication), in agreement with the simulation. In some

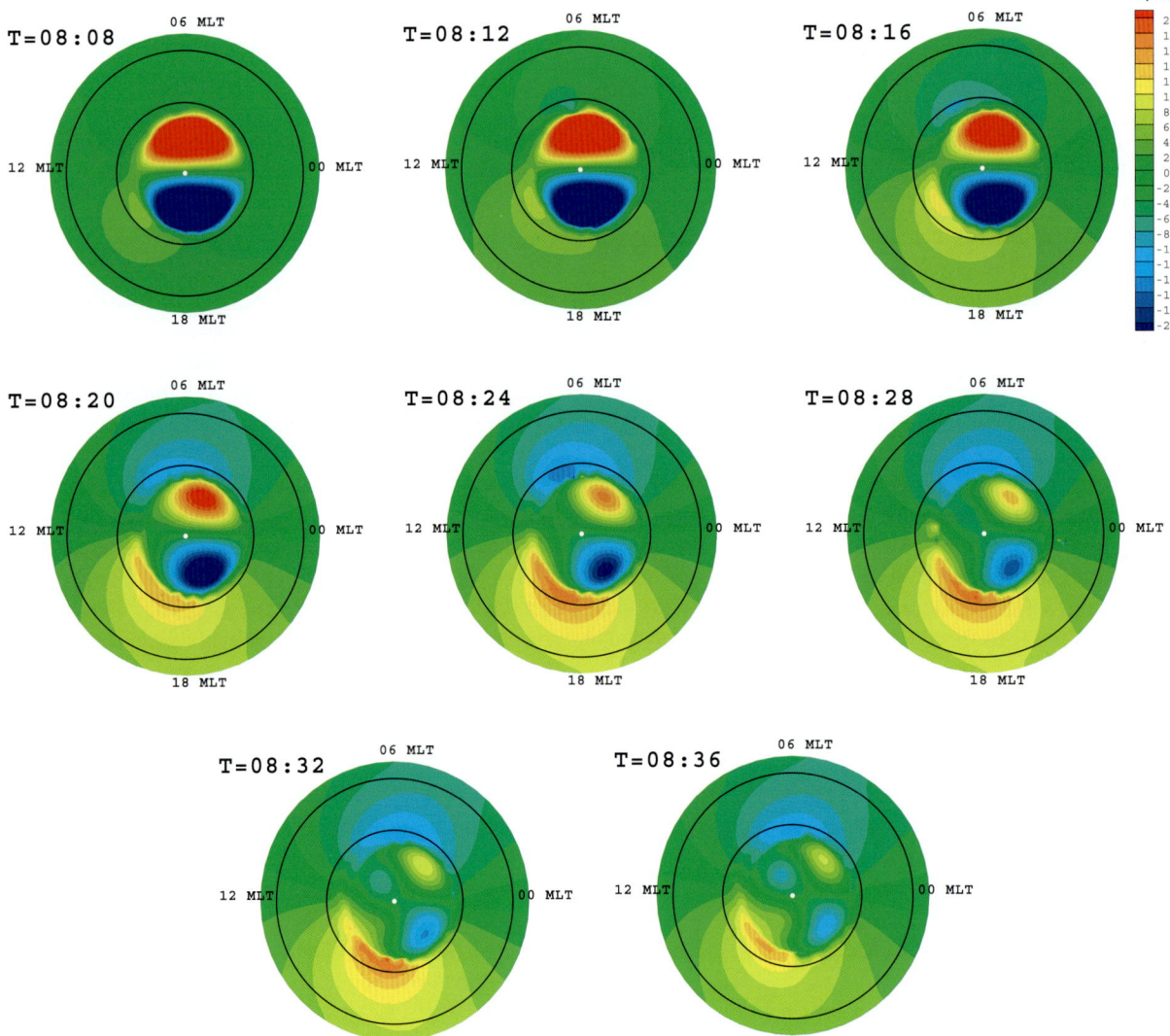

Plate 2. Ionospheric potentials computed by the coupled BATSRUS/RCM code after a northward turning of the IMF. The views are from high above the north pole, with the Sun to the left. The black circles represent 30 and 60 degrees latitude. The northward turning was imposed at the sunward boundary of the simulation box ($X=32\ R_E$) beginning at 0800 UT. The color bar shows electric potential values ranging from –20 kV to +20 kV. The orange-yellow colors on the dusk side at low- and mid-latitudes represent overshielding, as do the blue colors on the dawn side.

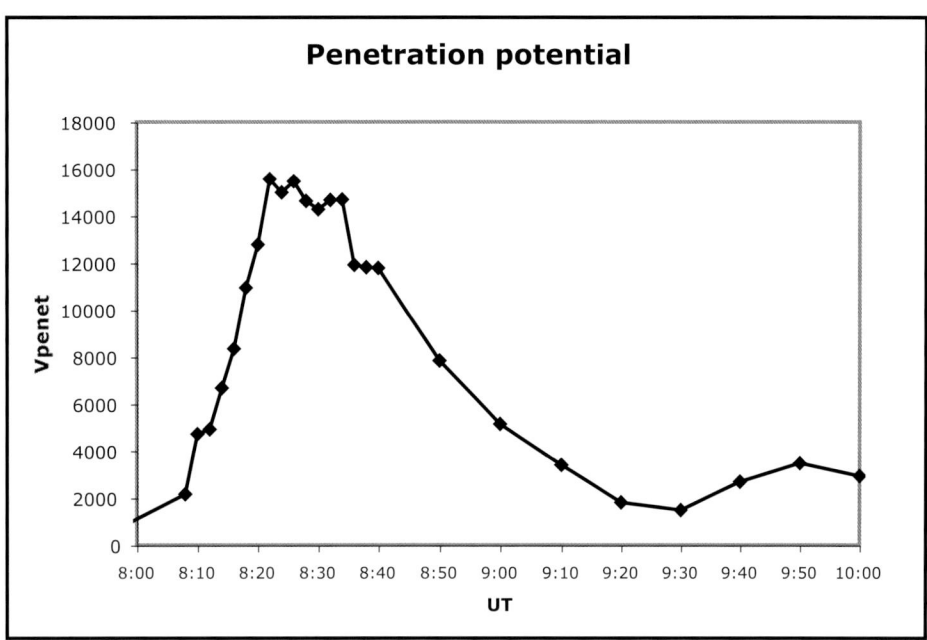

Figure 5. Total magnetospherically generated potential drop along the equator as a function of UT, after the northward turning of the IMF.

substorm events, the effect seems to propagate from night side to day side [*Goldstein et al.*, 2004b, c, *Goldstein and Sandel*, 2004]. Of course, our present simulation does not include substorm effects.

It should be noted that this simulation, the first in which the coupled BATSRUS/RCM calculated electric fields in the low-latitude ionosphere, assumed uniform ionospheric conductance (4 S per hemisphere Pedersen conductance, zero Hall). Thus the results shown in Figures 6–7 must be regarded as preliminary, since experience has shown the importance of conductance variations on the local-time distribution of the electric penetration potential. Nonetheless, these results represent an important first step in fully self-consistent modeling of prompt-penetration electric fields.

5. EFFECT OF PLASMA-SHEET $PV^{5/3}$

As discussed in Section 2, considerable effort has been expended in development of empirical algorithms for estimating polar-cap potential drop and magnetospheric magnetic fields as functions of solar-wind parameters. Much less effort has been devoted to algorithms that would allow empirical estimation of nV and $PV^{5/3}$ at the RCM's tailward boundary. *Garner et al.* [2003] produced a first set of estimates by combining a *Tsyganenko* [1989] magnetic field model with several statistical studies of plasma-sheet parameters. That study has recently been redone using a *Tsyganenko and Stern* [1996] magnetic field and the new *Tsyganenko-Mukai* [2003] empirical plasma model. Sample results for the most crucial parameter $PV^{5/3}$ are shown in Figure 6 for two different solar-wind conditions, one with southward IMF and one with northward.

Figure 6 illustrates two difficulties involved in using presently available empirical models to estimate $PV^{5/3}$ at any given time:

(i) For southward IMF, $PV^{5/3}$ shows a huge maximum near $X=-28$, $Y=0$ in the case of southward IMF, which is due to very weak equatorial magnetic fields in that region in the T96 model; that makes the flux-tube volume large. It is not clear whether such a peak is ever typical of the real plasma sheet for southward IMF. Its occurrence in the *Tsyganenko and Stern* [1996] magnetic field model might result from inclusion, in the averaging process, of periods when B_z is southward at $X\sim -28$ due to occurrence of a neutral line earthward of the observing spacecraft. A configuration where $PV^{5/3}$ decreases tailward, as in a region tailward of $X=-28$ in the bottom panel of Figure 6, would be strongly interchange unstable, so it is not clear that such a gradient in $PV^{5/3}$ could actually exist for extended periods.

(ii) $PV^{5/3}$ systematically increases tailward through the inner plasma sheet, which means that the value supplied to the RCM depends substantially on where the boundary is placed. This is a symptom of the pressure-balance inconsistency [*Erickson and Wolf*, 1980].

Nevertheless, it is clear from Figure 6 that $PV^{5/3}$ values in the inner plasma sheet ($-20 < X < -10$) are generally larger

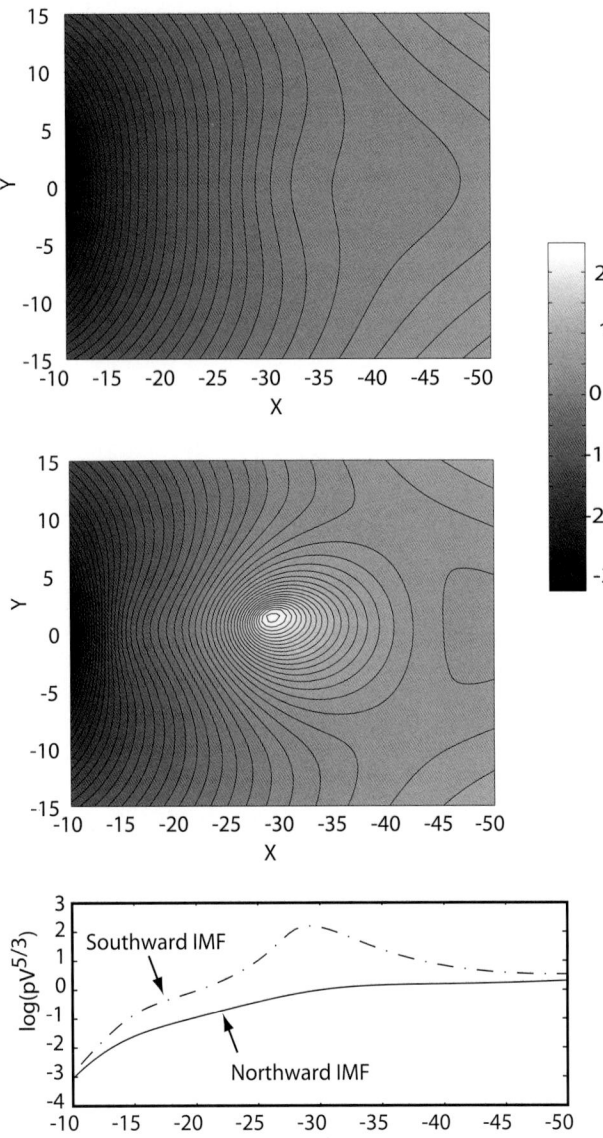

Figure 6. The top and middle panels show equatorial contour plots of $\log_{10}(PV^{5/3})$ for northward and southward IMF, in units of $nPa(R_E/nT)^{5/3}$. The assumed solar wind parameters are $n=5$ cm^{-3}, $V=400$ km/s, $B_x=B_y=5$ nT, for both panels, but $B_z=+5$ nT in the top panel and -5 nT for the middle panel. The bottom panel shows $\log_{10}(PV^{5/3})$ along the $-X$ axis.

for southward IMF than for northward. This suggests that a northward turning of the IMF would cause a reduction in $PV^{5/3}$ at the RCM tailward boundary, if that boundary is held fixed in space. This implies that, shortly after a northward turning of the IMF, low-$PV^{5/3}$ flux tubes may exist tailward of higher-$PV^{5/3}$ tubes. Such a configuration is interchange unstable.

Plate 3 shows results from an RCM simulation of the March 31, 2001 magnetic storm, in which $PV^{5/3}$ at the tailward boundary was varied in time, based on estimates from the *Tsyganenko* [2003] storm-time magnetic field model and the *Tsyganenko-Mukai* [2003] plasma model. Strong interchange instability occurs several times in the simulation, and Plate 3 shows snapshots from one such period. Low-$PV^{5/3}$ flux tubes (green in the figure) are supplied through the tailward boundary, and interchange fingers develop, with the low-$PV^{5/3}$ flux tubes swept rapidly into the near-Earth part of the plasma sheet. Differential gradient/curvature drift, which becomes increasingly important near the Earth, tends to blur the fingers when they get to the inner magnetosphere. It should be remarked that the north-south component of the IMF is not the only solar-wind parameter that affects $PV^{5/3}$, according to the Tsyganenko/Tsyganenko-Mukai combined models. Solar-wind density and velocity also have effects, which were included in the simulation shown in Plate 3.

We should also note that if the reduction in plasma-sheet $PV^{5/3}$ occurs near the end of the main phase of a magnetic storm, when $PV^{5/3}$ is high in the ring-current region, then the interchange instability extends deeper into the inner magnetosphere and is able to transport the higher density main-phase ring current plasma outward [*Sazykin et al.*, 2002], speeding storm recovery.

From these results we infer that a northward turning of the IMF may generate interchange instability in the plasma sheet, a possibility that was pointed out earlier by *Golovchanskaya et al.* [2002] in the context of explaining auroral-arc structures. However, our theoretical argument rests on the highly undertain assumption that the Tsyganenko/Tsyganenko-Mukai combined models can provide a valid estimate of time variations in plasma-sheet $PV^{5/3}$ in the period immediately after a northward turning of the IMF.

6. SUMMARY

Some effects of IMF B_z changes on the inner magnetosphere are well-established and can be understood, at least to zero order, in terms of changes in the electromagnetic inputs to the RCM, particularly the polar-cap potential and the magnetic-field configuration. These effects include overshielding and undershielding, SAPS events, ring-current injection, as well as plasmaspheric erosion, drainage tails, and shoulders.

One detail that is not yet well established, but is ripe for both theoretical and observational investigation, is the exact time dependence of the prompt-penetration electric field in response to a step function change in IMF B_z. We have made a theoretical prediction concerning this time dependence, based on a simulation with the new BATSRUS/RCM coupled code.

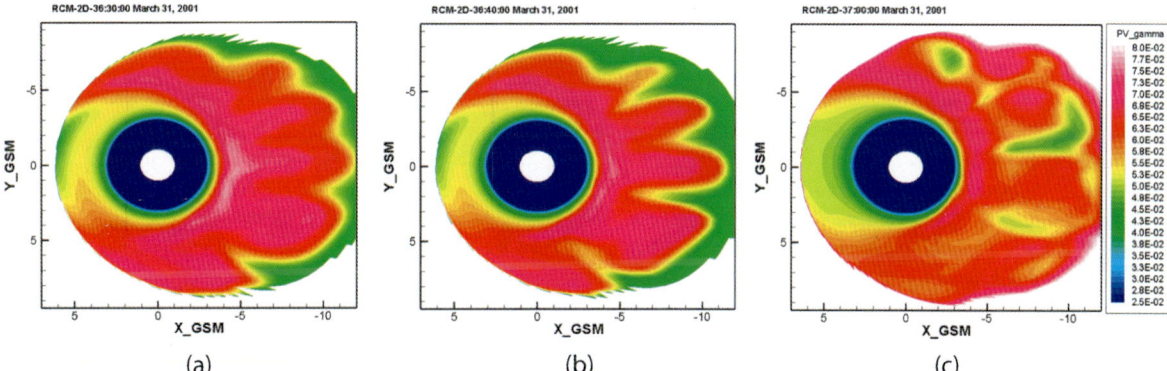

Plate 3. Equatorial contour plots of RCM-computed $PV^{5/3}$ for 12:30, 12:40, and 13:00 UT in a simulation of the March 31, 2001 magnetic storm.

Another area of current study is the effect of IMF changes on plasma parameters in the plasma sheet. Empirical models suggest that a northward turning of the IMF may reduce inner-plasma-sheet $PV^{5/3}$ and trigger interchange instability in the plasma sheet.

Acknowledgements. Research with the Rice Convection Model is supported by the NASA Sun-Earth-Connection Theory Program under Grant NAG5-11881 and by NSF grant ATM-20101349. Work with the coupled BATSRUS/RCM has been supported by NSF ITR grant ATM-0325255 at Rice and by Air Force MURI grant F49620-01-1-0359 and NASA ESS CT cooperative agreement NCC5-614 to the University of Michigan.

REFERENCES

Bilitza, D., K. Rawer, L. Bossy, and T. Gulyaeva, International Reference Ionosphere—Past, Present, and Future, 1, Electron-Density, *Advances in Space Research*, *13* (3), 3–13, 1993.

Boyle, C. B., P. H. Reiff, and M. R. Hairston, Empirical polar cap potentials, *J. Geophys. Res.*, *102* (A1), 111–125, 1997.

Brandt, P. C., S. Ohtani, D. G. Mitchell, M.-C. Fok, E. C. Roelof, and R. Demajistre, Global ENA observations of the storm main-phase ring current: Implications for skewed electric fields in the inner magnetosphere, *Geophys. Res. Lett.*, *29* (20), 1962, DOI 10.1029/2002GL015160, 2002.

Burke, W. J., N. C. Maynard, M. P. Hagan, R. A. Wolf, G. R. Wilson, L. C. Gentile, M. S. Gussenhoven, C. Y. Huang, T. W. Garner, and F. J. Rich, Electrodynamics of the inner magnetosphere observed in the dusk sector by CRRES and DMSP during the magnetic storm of June 4–6, 1991, *J. Geophys. Res.*, *103* (12/98), 29399, 1998.

Chappell, C. R., Recent satellite measurements of the morphology and dynamics of the plasmasphere, *Rev. Geophys. Space Phys. Space Phys.*, *100*, 951–972, 1972.

Chen, M. W., M. Schulz, G. Lu, and L. R. Lyons, Quasi-steady drift paths in a model magnetosphere with AMIE electric field: Implications for ring current formation, *J. Geophys. Res.*, *108* (A5), 1180, 10.1029/2002JA009584, 2003.

De Zeeuw, D. L., S. Sazykin, R. A. Wolf, T. I. Gombosi, A. J. Ridley, and G. Toth, Coupling of a global MHD code and an inner magnetosphere model: Initial results, *J. Geophys. Res. (in press)*, 2004.

Ebihara, Y., M.-C. Fok, R. A. Wolf, T. Immel, and T. E. Moore, Ionospheric conductivity control of ring-current development, *J. Geophys. Res. (in press)*, 2004.

Erickson, G. M., and R. A. Wolf, Is steady convection possible in the Earth's magnetotail?, *Geophys. Res. Lett.*, *7*, 897–900, 1980.

Fejer, B. G., and L. Scherliess, Empirical models of storm time equatorial zonal electric fields, *J. Geophys. Res.*, *102* (A11), 24,047–24,056, 1997.

Fok, M.-C., and T. E. Moore, Ring current modeling in a realistic magnetic field configuration, *Geophys. Res. Lett.*, *24* (7/15/97), 1775–1778, 1997.

Fok, M.-C., R. A. Wolf, R. W. Spiro, and T. E. Moore, Comprehensive computational model of Earth's ring current, *J. Geophys. Res.*, *106* (A5), 8417–8424, 2001.

Fok, M.-C., T. E. Moore, G. R. Wilson, J. D. Perez, X. X. Zhang, P. C:son Brandt, D. G. Mitchell, E. C. Roelof, J.-M. Jahn, C. J. Pollock, and R. A. Wolf, Global ENA IMAGE simulations, *Space Sci. Rev.*, *109*, 77–103, 2003.

Foster, J. C., and H. B. Vo, Average characteristics and activity dependence of the subauroral polarization stream, *J. Geophs. Res.*, *107* (A12), 1475, DOI 10.1029/2002JA009409, 2002.

Garner, T. W., R. A. Wolf, R. W. Spiro, M. F. Thomsen, and H. Korth, Pressure balance inconsistency exhibited in a statistical model of magnetospheric plasma, *J. Geophys. Res.*, *108* (A8), 1331, doi:10.1029JA009877, 2003.

Garner, T. W., R. A. Wolf, R. W. Spiro, W. J. Burke, B. G. Fejer, S. Sazykin, J. L. Roeder, and M. R. Hairston, Magnetospheric electric fields and plasma sheet injection to low L-shells during the 4–5 June 1991 magnetic storm: Comparison between the Rice Convection Model and observations, *J. Geophys. Res.*, *109* (A2), A02214, doi: 10.1029/2003JA010208, 2004.

Goldstein, J., R. W. Spiro, P. H. Reiff, R. A. Wolf, B. R. Sandel, J. W. Freeman, and R. L. Lambour, IMF-driven overshielding electric field and the origin of the plasmaspheric shoulder of May 24, 2000, *Geophys. Res. Lett.*, *29* (16), 1819, DOI 10.1029/2001GL014534, 2002.

Goldstein, J., B. R. Sandel, W. T. Forrester, and P. H. Reiff, IMF-driven plasmasphere erosion of 10 July 2000, *Geophys Res. Lett.*, *30* (3), 1146, doi10.1029/2002GL0 16478, 2003a.

Goldstein, J., R. W. Spiro, B. R. Sandel, R. A. Wolf, S.-Y. Su, and P. H. Reiff, Overshielding event of 28–29 July 2000, *Geophys. Res. Lett.*, *30*(8), 1421, doi:10.1029/2002GL0 16644, 2003b.

Goldstein, J., B. R. Sandel, M. R. Hairston, and P. H. Reiff, Control of plasmaspheric dynamics by both convection and sub-auroral polarization stream, *Geophys. Res. Lett.*, *30* (23), 2243, doi: 10.1029/2003GL018390, 2003c.

Goldstein, J., B. R. Sandel, M. F. Thomsen, M. Spasojevic, and P. H. Reiff, Simultaneous remote sensing and in situ observations of plasmaspheric drainage plumes, *J. Geophys. Res.*, *109* (A3), A03202, doi:10.1029/2003JA 010281, 2004a.

Goldstein, J., B. R. Sandel, M. R. Hairston, and S. B. Mende, Plasmapause undulation of 17 April 2002, *Geophys. Res. Lett.* (in press), 2004b.

Goldstein, J., J. L. Burch, B. R. Sandel, S. B. Mende, P. C. Brandt, and M. R. Hairston, Coupled response of the inner magnetosphere and ionosphere on 17 April 2002, to be submitted to *J. Geophys. Res.*, 2004c.

Goldstein, J., and B. R. Sandel, The global pattern of evolution of plasmaspheric drainage plumes, submitted to AGU Monograph: Proceedings of the Yosemite Workshop, 2004.

Golovchanskaya, I. V., Y. P. Maltsev, and A. A. Ostapenko, High-latitude irregularities of the magnetospheric electric field and their relation to solar wind and geomagnetic conditions, *J. Geophys. Res.*, *107* (A1), doi:10.1029/2001 JA900097, 2002.

Grebowsky, J. M., Model study of plasmapause motion, *J. Geophys. Res.*, *75*, 4329–4333, 1970.

Hairston, M. R., T. W. Hill, and R. A. Heelis, Observed saturation of the ionosphere polar cap potential during the 31 March 2001 storm, *Geophys. Res. Lett. 30,* doi:10.1029/ 2002GL015894, 2003.

Hedin, A. E., Extension of the MSIS thermosphere model into the middle and lower atmosphere, *J. Geophys. Res., 96,* 1159–1172, 1991.

Hilmer, R. V., and G.-H. Voigt, A magnetospheric magnetic field model with flexible current systems driven by independent physical parameters, *J. Geophys. Res., 100* (A4), 5613–5626, 1995.

Jordanova, V. K., A. Boonsiriseth, R. M. Thorne, and Y. Dotan, Ring current asymmetry from global simulations using a high_ resolution electric field model, *J. Geophys. Res., 108* (A12), 1443, doi: 10.1029/2003JA009993, 2003.

Kelley, M. C., B. G. Fejer, and C. A. Gonzales, An explanation for anomalous ionospheric electric fields associated with a northward turning of the interplanetary magnetic field, *Geophys. Res. Lett., 6,* 301–304, 1979.

Kozyra, J. U., M. W. Liemohn, C. R. Clauer, A. J. Ridley, M. F. Thomsen, J. E. Borovsky, J. L. Roeder, V. K. Jordanova, and W. D. Gonzalez, Multistep Dst development and ring current composition changes during the 4–6 June 1991 magnetic storm, *J. Geophys. Res., 107* (A8), doi:10.1029/ 2001JA000023, 2002.

Liemohn, M. W., and A. J. Ridley, A model-derived storm time asymmetric ring current driven electric field description, *J. Geophys. Res., 107* (A8), DOI 10.1029/2001JA000051, 2002.

Robinson, R. M., R. R. Vondrak, K. Miller, T. Dabbs, and D. Hardy, On calculating ionospheric conductances from the flux and energy of precipitating electrons, *J. Geophys. Res., 92,* 2565–2569, 1987.

Rowland, D. E., and J. R. Wygant, Dependence of the large-scale, inner magnetospheric electric field on geomagnetic activity, *J. Geophys. Res., 103* (A7), 14959–14964, 1998.

Sandel, B. R., R. A. King, W. T. Forrester, D. L. Gallagher, A. L. Broadfoot, and C. C. Curtis, Initial results from the IMAGE Extreme Ultraviolet Imager, *Geophys. Res. Lett., 28* (4/15/01), 1439–1442, 2001.

Sazykin, S. Y., Theoretical Studies of Penetration of Magnetospheric Electric Fields to the Ionosphere, Ph. D. thesis, Utah State University, Logan, Utah, 2000.

Sazykin, S., R.A . Wolf, R. W. Spiro, T. I. Gombosi, D. L. De Zeeuw, and M. F. Thomsen, Interchange instability in the inner magnetosphere associated with geosynchronous particle flux decreases, *Geophys. Res. Lett., 29* (10), doi:10.1029/2001GL014416. Corrections *Geophys. Res. Lett., 29* (16), doi:10.1029/2002GL015846, 2002 and *31* (16), L05803, doi:10.1029/2003GL019191, 2004.

Schield, M. A., J. W. Freeman, and A. J. Dessler, A source for field-aligned currents at auroral latitudes, *J. Geophys. Res., 74,* 247–256, 1969.

Spiro, R. W., R. A. Wolf, and B. G. Fejer, Penetration of high-latitude-electric-field effects to low latitudes during SUNDIAL 1984, *Ann. Geophys., 6,* 39–50, 1988.

Toffoletto, F. R., J. Birn, M. Hesse, R. W. Spiro, and R. A. Wolf, Modeling inner magnetospheric electrodynamics, in *Proceedings of the Chapman Conference on Space Weather*, edited by P. Song, G. L. Siscoe, and H. .J. Singer, Am. Geophys. Un., Washington, D. C., 2000.

Toffoletto, F., S. Sazykin, R. Spiro, and R. Wolf, Inner magnetospheric modeling with the Rice Convection Model, *Space Sci. Rev., 107,* 175–196, 2003.

Tsyganenko, N. A., A magnetospheric magnetic field model with a warped tail current sheet, *Planet. Space Sci., 37,* 5–20, 1989.

Tsyganenko, N. A., and D. P. Stern, Modeling the global magnetic field of the large-scale Birkeland current systems, *J. Geophys. Res., 101* (A12), 27,187–27,198, 1996.

Tsyganenko, N. A., H. J. Singer, and J. C. Kasper, Storm-time distortion of the inner magnetosphere: How severe can it get?, *J. Geophys. Res., 108* (A5), 1209, doi:10.1029/2002 JA009808, 2003.

Tsyganenko, N. A., and T. Mukai, Tail plasma sheet models derived from Geotail particle data, *J. Geophys. Res., 108* (A3), doi:10.1029/2002JA009707, 2003. Correction: *J. Geophys. Res., 108* (A10), 1374, doi:10.1029/2003 JA010127, 2003.

Vasyliunas, V. M., Mathematical models of magnetospheric convection and its coupling to the ionosphere, in *Particles and Fields in the Magnetosphere*, edited by B. M. McCormac, pp. 60–71, D. Reidel, Hingham, M. A, Hingham, MA, 1970.

Wolf, R. A., Calculations of magnetospheric electric fields, in *Magnetospheric Physics*, edited by B. M. McCormac, pp. 167–177, D. Reidel, Dordrecht, Netherlands, 1974.

Wolf, R. A., The quasi-static (slow-flow) region of the magnetosphere, in *Solar Terrestrial Physics*, edited by R. L. Carovillano, and J. M. Forbes, pp. 303–368, D. Reidel, Hingham, MA, 1983.

D. L. De Zeeuw and T. I. Gombosi, Center for Space Environment Modeling, University of Michigan, 2455 Hayward Street, Ann Arbor, MI 48109.

J. Goldstein, Space Science Division, Southwest Research Institute, San Antonio, TX 78228.

S. Sazykin, R. W. Spiro. F. R. Toffoletto, R. A. Wolf and X. Xing, Physics and Astronomy Department, Rice University MS108, P.O. Box 1892, Houston, TX 77251-1892.

Non-Potential Electric Field Model of Magnetosphere-Ionosphere Coupling

Igor V. Sokolov, Tamas I. Gombosi, and Aaron J. Ridley

Center for Space Environment Modeling, University of Michigan, Ann Arbor, Michigan, USA

A new model is proposed to describe the electrodynamic coupling between the magnetosphere and ionosphere. In contrast with existing models, the ionospheric electric field is not assumed to be a potential field. The equation coupling the electric currents flowing into the ionosphere to the ionospheric electric fields is integrated analytically. This approach results in a simple, local and physically reasonable boundary condition, coupling the local tangential plasma velocity values in the magnetosphere to the tangential magnetic field through a properly integrated ionospheric conductivity. For simplified test cases the simulation results are in good agreement with those obtained with a traditional ionospheric electric potential model. The proposed approach improves computational efficiency and also allows prediction of the electromotive forces acting on closed electric current loops at the surface of the Earth. Such electric current loops can be induced by non-potential electric fields, generated by rapid changes in the ionosphere and magnetosphere.

INTRODUCTION

To a large extent the state of the magnetosphere is controlled by conditions in the solar wind and in the ionosphere. In a first approximation the distant solar wind is unaffected by the presence of the magnetosphere: therefore a "one-way" coupling adequately describes the interaction with the magnetosphere-ionosphere system. The magnetosphere-ionosphere (M-I) coupling, on the other hand, is a highly non-linear two-way interaction, which strongly affects the large-scale behavior of both domains.

Self-consistent global magnetosphere models include some kind of dynamic ionosphere model which interacts with the magnetosphere and provides ionospheric boundary conditions that actively respond to changing magnetospheric conditions [*Ogino and Walker,* 1984; *Lyon et al,* 1986; *Tanaka,* 1995; *Raeder et al,* 1995; *Janhunen,* 1996; *White et al,* 1998; *Powell et al,* 1999].

While mass exchange between the ionosphere and the magnetosphere is undoubtedly of major importance, the dominant component of M-I coupling is a system of field-aligned currents (Birkeland currents) connecting the magnetosphere and the high-latitude ionosphere. These Birkeland currents carry momentum and energy along stretched magnetic field lines connecting the ionosphere and the magnetosphere. Self-consistent global magnetosphere models need to describe the generation and closure of these Birkeland currents through appropriate boundary conditions and embedded non-MHD models.

The most important current systems coupling the ionosphere and the magnetosphere are the so called Region 1 and Region 2 Birkeland currents. Region 1 currents, flowing near the open-closed magnetic field boundary, connect the magnetopause current to the ionosphere where they are closed through ionospheric Pedersen currents. Region 2 currents flow along closed magnetic field lines and connect to the ionosphere at lower magnetic latitudes than the Region 1 current. Region 2 currents are generated in the inner part of the plasma sheet and in the ring current region.

Most global MHD magnetosphere models use the so called electrostatic ionosphere approximation. The MHD code has an inner boundary at a radius of R_B (most codes use values of $R_B = 2.5–4.5\ R_E$). At this inner boundary, the MHD model is coupled to the ionosphere model with the help of appropriate boundary conditions. In practice, either plasma velocities or corresponding electric fields are imposed at the boundary that are calculated in the ionosphere in a three-step process:

1. Field aligned currents are calculated in the magnetosphere from the curl of the magnetic field near R_B, and these Birkeland currents are mapped down to the ionosphere along unperturbed (intrinsic) magnetic field lines.

2. A height-integrated ionospheric conductance pattern is generated and the ionospheric potential is calculated from the equation:

$$j_R(R_1) = [\nabla_t \cdot (\Sigma \cdot \nabla_t \psi)_t]_{R=R_1} \qquad (1)$$

This describes the relationship between the height integrated conductance tensor, Σ, the ionospheric electric potential, ψ, and the radial component of the current, j_R (here R_1 is the radius of the ionosphere and the subscript t denotes the two tangential components of a 3D vector along the spherical surface).

3. The electric potential is mapped out along unperturbed field lines to the inner boundary at R_B where electric fields and velocities are generated. The corotation velocity field is added to the ionosphere generated velocity field.

The details of this method were summarized by *Goodman* [1995] with some corrections by *Amm* [1996].

The electrostatic ionosphere approximation captures some of the fundamental features of the M-I coupling process, but it suffers from several shortcomings, including: (1) inconsistency between the Birkeland currents and the dynamic (non-intrinsic) component of the magnetic field, and (2) neglect of skin-effect currents in the ionosphere (which cannot be described by potential electric fields).

There are additional limitations to the electric potential description of the ionosphere: (1) near strong electric fields, such as auroral arcs, the electric field is not a potential field, so the potential description is incorrect. (2) For a space weather prediction model, the ground induced currents (GICs) are an important consideration. These GICs are driven by electric fields by strongly varying magnetic fields, which are related to varying magnetic fields by Faraday's law:

$$\frac{\partial \mathbf{B}}{\partial t} = -[\nabla \times \mathbf{E}], \qquad (2)$$

where $\mathbf{E}$ is the electric field and $\mathbf{B}$ is the magnetic field. If the electric field is described as a potential field, it has no curl. Therefore it cannot have a $\partial \mathbf{B}/\partial t$ associated with it. (3) The calulation of the potential solution is costly in terms of computational time, and typically do not spread out well over large numbers of processors. On massively parallel machines, this can slow the main MHD solver down significantly. Therefore, the coupling is done only every few seconds, instead of every MHD iteration, which is the way the coupling should be done.

Another methodology is to relate the electric field (and therefore ion velocity) directly to the magnetic field structure at the boundary. This method allows the coupling to take very little computation time, so it can be completed every iteration. In addition, it has the physical properties which can clearly describe the dB/dt term, so it can be used for space weather applications. Furthermore, it will better physically model strong electric field sources, such as those that occur near auroral arcs.

In this paper we describe an electrodynamic ionosphere model that addresses these limitations.

MODEL ASSUMPTIONS AND EQUATIONS

Here we describe the main features of the proposed M-I coupling technique and compare it step by step with presently used methods.

M-I Boundary Is at $R_B = R_I$

In this model the interface between the ionosphere and the magnetosphere is placed at the top of the ionosphere. Therefore, there is no need for any mapping of either the field aligned currents or of the electric field potential between the ionosphere and the magnetosphere. The MHD equations are solved above the ionosphere $(R > R_B = R_I)$.

It should be emphasized that in most global MHD models relatively large values were chosen for R_B in order to exclude regions with very high Alfven speeds (and consequently with very small explicit time-steps). However, moving the inner boundary away from the actual M-I interface necessitates mapping physical quantities between the ionosphere and the inner boundary of the MHD simulation region. This mapping process includes additional simplifying assumptions that are not well justified. For example, it is usually assumed that the *total* electric current density in the mapping region ($R_I < R < R_B$) is directed along the unperturbed magnetic field line, $\mathbf{B}_0$. The conservation of the *total* current density yields a mapping relation that couples the electric current density $j_R(RI)$ flowing into the ionosphere (see Eq.(1)) to the Birkeland current at the magneto spheric inner boundary $R = R_B$:

$$j_R = \mathbf{n}_R \cdot \mathbf{j}_\parallel(R_1) \qquad (3)$$

where $\mathbf{n}_R$ is the unit vector in the radial direction and the field aligned current at the inner boundary of the simulation is given by:

$$j_\parallel(R_I) = \frac{\mathbf{B}_0(R_I)}{\mu_0}\left[\frac{\mathbf{B}_0(R_B)\cdot[\nabla\times\mathbf{B}(R_B)]}{B_0^2(R_B)}\right] \quad (4)$$

Here a point on the sphere with radius of R_I is connected by the magnetic field line $\mathbf{B}_0$ to a corresponding point on the sphere with radius R_B. On the other hand, if the *total* electric current is not field aligned, then there is no conservation law for the field aligned component and Eq. (4) is no longer valid.

The condition for the total electric current to be field aligned ("force-free" magnetic configuration) appears to require that there is no motion and absolutely no pressure gradient in the region $R_I < R < R_B$. Obviously, this condition is not satisfied in the real M-I system. For this reason it is advantageous not to impose any assumption on the orientation of the electric current and to derive it directly from the MHD numerical solution. We also think that the problem of very small explicit time steps when the inner boundary is in the region of high magnetic field $(R_B = R_I)$ can and should be solved using available numerical technology, such as local or implicit timestepping [*Hirsch*, 1990] or a physics-based convergence acceleration [*Gombosi et al.*, 2002], rather than by using assumptions that are not well justified.

Arbitrarily Directed Coupling Current

The boundary condition for coupling the solution of the MHD equations in the magnetosphere to the electric current density distribution in the ionosphere (see Eq. 1) is obtained from the radial component of the simplified Ampere's law by neglecting the displacement current:

$$\mu_0 \mathbf{j} = \nabla \times \mathbf{B} \quad (5)$$

The radial component of this equation is:

$$j_R(R_I) = \frac{\mathbf{n}_R\cdot[\nabla\times\mathbf{B}]}{\mu_0} = \frac{\nabla_t\cdot[\mathbf{B}_t\times\mathbf{n}_R]_t}{\mu_0} \quad (6)$$

The easiest way to prove the transformation from the radial component of $\nabla \times \mathbf{B}$ to the two-dimensional divergence of a two-dimensional vector is to compare the appropriate expressions for the components of an arbitrary vector A (see Appendix in *Landau et al.* [1985]):

$$\mathbf{n}_R \cdot [\nabla \times \mathbf{A}] = \frac{1}{R\sin\theta}\left[\frac{\partial}{\partial\theta}(\sin\theta A_\phi) - \frac{\partial A_\theta}{\partial\phi}\right] \quad (7)$$

$$\nabla_t \cdot \mathbf{A} = \frac{1}{R\sin\theta}\left[\frac{\partial}{\partial\theta}(\sin\theta A_\theta) + \frac{\partial A_\phi}{\partial\phi}\right] \quad (8)$$

and note that $\mathbf{A}\times\mathbf{n}_R = (0, A_\phi, -A_\theta)$ for $\mathbf{A} = (A_R, A_\phi, A_\theta)$.

We note two important points. First, Eq. (6) is not equivalent to Eq. (4), even in the limit when $R_B \to R_I$. In this limiting case Eq. (4) becomes identical to Eq. (6) only if the electric current near R_I is field aligned. Second, equation Eq. (6) automatically ensures that the total radial current vanishes when integrated over the entire outer boundary of the ionosphere: $R j_R dR_t$ = R $r_t \cdot [B_t \times n_R]_t dRt$ = 0 as a complete integral of the divergence (here $d\mathbf{R}_t$ is a spherical surface element). This is significant because the corresponding integral for the field aligned currents in Eq. (4) does not have to vanish.

It should also be mentioned that the magnetic field in Eq. (6) is the magnetic field perturbation (the deviation from $\mathbf{B}_0$). Although it is typically small compared to the unperturbed $\mathbf{B}_0$ field, it controls the electric current density ($\mathbf{B}_0$ is usually a potential field).

Non-Potential Electric Fields in the Magnetosphere and Ionosphere

In a good approximation the motional electric field near the boundary surface is

$$\mathbf{E}(R_I) = [\mathbf{B}_0 \times \mathbf{u}]_{R=R_I} \quad (9)$$

where u is the plasma bulk velocity. Here the contribution from the magnetic field perturbation B is neglected. The electric field is perpendicular to the magnetic field line ($\mathbf{E}\cdot\mathbf{B}_0 = 0$), so that its radial component can be expressed in terms of the tangential components:

$$E_R(R_I) = -\frac{\mathbf{E}_t(R_I)\cdot\mathbf{B}_{0t}}{B_{0R}} \quad (10)$$

A consequence of Faraday's law is that the tangential components of the electric fields are continuous at any surface. This means that the same tangential electric field $\mathbf{E}_t(R_I)$ should be used in Eq. (1) (the general electric field now replaces $-\nabla_t \psi$):

$$j_R(R_I) = -[\nabla_t \cdot (\Sigma \cdot \mathbf{E}_t)_t]_{R=R_I} \quad (11)$$

In Eq. (11) it is not assumed that the electric field has a scalar potential, and consequently, the electric field is not necessarily curl-free, $\nabla \times \mathbf{E} = -\partial\mathbf{B}/\partial t \neq 0$. We will discuss the magnetospheric and the ionospheric consequences of this point separately.

The present generation of global MHD models assume that magnetic field lines are equipotentials between the ionosphere and the inner boundary of the magnetospheric simulation domain ($(\mathbf{B}_0 \cdot \nabla)\psi = 0$). We know that only curl-free electric fields can be characterized by scalar potentials, therefore in the mapping region the $\nabla \times \mathbf{E} = 0$ condition must hold. Using Faraday's law (Eq. (2)) one can readily see that the assumption of a potential electric field is equivalent to the assumption of a time independent (steady-state) magnetic field in the region between the ionosphere and the magnetospheric inner boundary. Since we place the inner boundary at $R_B = R_u$ we do not need to assume that the magnetic field lines in the inner magnetosphere are time stationary equipotentials.

In the ionosphere the assumption of a potential electric field is more restrictive resulting in mathematical and physical inconsistencies. Non-potential ionospheric electric fields also lead to important physical and technological effects. Metallic tubes, power transmission lines, huge transformers, land, sea-water and oceans form monumental closed conducting contours, that are not connected to the ionosphere by *pairs of wires*. However, they all are *inductively* coupled to the non-potential electric fields in the ionosphere. The non-potential part of the electric field, $\mathbf{E}_t(R_1)$, induces the time variation of the radial magnetic field, B_R, which can be obtained as long as $\mathbf{E}_t(R_1)$ is known (see Eq. 2):

$$\frac{\partial \mathbf{B}_R}{\partial t} = -\nabla_t \times \mathbf{E}_t \qquad (12)$$

The voltage induced in a closed contour can be obtained by integrating Eq. (12) over the area enclosed by the contour or alternatively, by integrating the ionospheric electric field along the contour:

$$U = -\int \frac{\partial \mathbf{B}_R}{\partial t} dS = \oint \mathbf{E}_t(R_1) \cdot d\mathbf{l} \qquad (13)$$

This interesting effect is completely ignored and cannot be included in a potential field model for the ionospheric electric field: it follows from the assumption of $\mathbf{E}_t(R_1) = -\nabla \psi$, that the value of Eq. (13) is exactly zero.

The use of the Eq. (1) with a potential electric field as a boundary condition for the MHD equations also results in a mathematical problem. Substituting a potential electric field for $\mathbf{E}_t$ in Eq. (12) one can find that this equation requires that $\partial B_R(R_1)/\partial t = 0$. Therefore, the potential boundary condition is applicable only if $B_R(R_1)$ is known and it is constant in time (steady state).

Without assuming a potential electric field, the ionospheric field $\mathbf{E}_t$ can be written as the sum of a potential and non-potential component:

$$\begin{aligned} \mathbf{E}_t &= -\nabla_t \psi + [\mathbf{n}_R \times \Pi_E] \\ \nabla_t \cdot \mathbf{E}_t &= -\nabla_t^2 \psi \\ \nabla_t \times \mathbf{E}_t &= \mathbf{n}_R \nabla_t^2 \Pi_E \end{aligned} \qquad (14)$$

Here Π_E is a 2D scalar potential. Eqs. (6) and (11) can be combined to yield:

$$\nabla_t \cdot \left(\frac{[\mathbf{n}_R \times \mathbf{B}_t]}{\mu_0} - \Sigma \cdot \mathbf{E}_t \right) = 0 \qquad (15)$$

Eqs. (14) and (15) are the key to the new M-I coupling method.

Height Integrated Ionosphere With No Radial Electric Currents

We treat the ionosphere as a thin conducting layer occupying a very narrow altitude region, $h_L \leq h \leq h_U$ (here h_L and h_U represent the lower and upper boundaries of this layer). In addition, we ignore any altitude dependence. The physical and chemical processes responsible for the ionospheric conductivity are parametrized in terms of the height integrated components of the 2×2 conductance tensor Σ. The well known way to construct this tensor is described in Appendix A.

In addition, Eq. (15) implies that all the radial electric current flowing into or out of the ionosphere is only due to the M-I coupling and ignores all other radial electric currents, just as in the TIEGCM model [*Richmond et al, 1992*]. Until these additional currents are properly incorporated into the model, it cannot provide a quantitative description of some important geophysical phenomena.

Under these assumptions we integrate Eq. (15) over altitude for the ionosphere:

$$\frac{[\mathbf{n}_R \times \mathbf{B}_t]}{\mu_0} - \Sigma \cdot \mathbf{E}_t = \frac{[\mathbf{n}_R \times \mathbf{B}'_t]}{\mu_0} \qquad (16)$$

where $\mathbf{B}'_t$ is the tangential magnetic field below the ionosphere. From Eq. (15) it follows that $\mathbf{B}'_t$ satisfies the condition

$$\nabla_t \cdot [\mathbf{n}_R \times \mathbf{B}'_t] = -\mathbf{n}_R \cdot [\nabla_t \times \mathbf{B}'_t] = 0 \qquad (17)$$

i.e., the tangential magnetic field below the ionosphere is a *ID* potential field:

$$\mathbf{B}'_t = [\nabla_t \Pi_M]_{R = R_E + h_L} \qquad (18)$$

where Π_M is a 2D scalar.

The boundary condition relates the ionospheric electric field to that of below the ionosphere. From the continuity of the radial electric field and from Eqs. (12 and 14) we see that the non-potential part of the electric field should be continuous through the lower boundary of the ionosphere:

$$\mathbf{B}'_R = \mathbf{B}_R; \quad [\nabla \times \mathbf{E}'_t] = \nabla_t^2 \Pi_E \qquad (19)$$

The electrostatic potential part of the ionospheric electric field is strongly distorted by charge separation at the lower ionospheric boundary and it is mostly radial: $\mathbf{E}'_R = -\psi/h_L$.

We note that the potential electric field does not induce currents and it only produces the charge separation.

THE ELECTRODYNAMIC PROCESSES BELOW THE IONOSPHERE ARE DECOUPLED

So far the assumptions relating to the M-I coupling model have been formulated in terms of variables, parameters and equations associated with the magnetosphere and/or ionosphere. However, eq.(16) goes beyond this and it couples the ionospheric electric field to the tangential magnetic field below the ionosphere. This magnetic field, in turn, depends on physical processes below the ionosphere.

Generally, there is a two-way coupling in this region because the magnetic field below the ionosphere is a function of $[\nabla_t \times \mathbf{E}_t] = \nabla_t^2 \Pi_E$ and can be thought of as a linear response of some very complicated electrodynamic system (the Earth plus everything conducting on it) to the rotational part of the electric field. Using Eqs. (18) and (19), the most general linear response function can be written in the following form:

$$\Pi_M = \iint F(\Pi_E)\, dt\, d\mathbf{R}_t \quad (20)$$

Investigating such current systems is very interesting but complicated and we intend to explore this subject in subsequent publications. Here we discuss some simple limiting cases, in which the M-I coupling problem can be more or less readily decoupled from the electrodynamics below the ionosphere. All the discussed models allow some evaluations for the magnitude of the ground induced currents (GIC).

Concentrated Ground Impedance

Assume that the magnetic field variations and the electric fields are shielded by Earth's conductivity in a thin skin layer of $h_S \sim (t_V/\sigma_E \mu_0)^{1/2}$, where t_V is a time scale for magnetic variations and $\acute{o}_E$ is the ground conductivity. Hence, at $h \approx -h_s$ the magnetic field vanishes, $B \to 0$.

For this case the physical interpretation of the $[\mathbf{n}_R \times \mathbf{B}'_t]$ term in Eq.(16) is as follows. Our formulation takes into account all the currents and electric fields in the ionosphere and allows for non-potential electric fields that can induce closed electric current loops in the ground, ocean and so on (see Eq. 13). These currents in turn can induce some additional magnetic fields.

An explicit and unambiguous expression for $\mathbf{B}'_t$ can be directly obtained from Ampere's law (Eq. 5). The height integrated tangential components of Eq. (5) can be written as follows:

$$[\mathbf{n}_R \times \mathbf{B}'_t] - [\mathbf{n}_R \times \mathbf{B}_t]_{h=-h_S}$$
$$+ \left[\nabla_t \times \int_{-h_S}^{h_L} \mathbf{n}_R B_R\, dh\right] = \mu_0 \int_{-h_S}^{0} \mathbf{j}_t\, dh \quad (21)$$

The induced magnetic field at $h = -h_s$ is assumed to tend to zero, hence the second term vanishes in the left hand side of Eq. (21). Comparing Eqs. (18 and 21), one can represent Eq. (21) in the form of Eq. (16) with the following unambiguous expression for $\mathbf{B}'_t = [\mathbf{n}_R \times \nabla_t \Pi_M]$:

$$\left[\mathbf{n}_R \times \nabla_t \left(\Pi_M - \int_{-h_S}^{h_L} B_R\, dh\right)\right] = \mu_0 \int_{-h_S}^{0} \mathbf{j}_t\, dh \quad (22)$$

Our simple model of a concentrated ground impedance assumes that the integral of B_R in the left hand side can be evaluated as $(h_L + h_s)B'_R$, while the height integrated ground induced current density $\mathbf{I}_{gr} = \int_{-hs}^{0} \mathbf{j}_t dh$ that is a function of $\partial \mathbf{B}_R/\partial t = -\nabla \times \mathbf{E}$ [cf. *Viljanen et al.*, 1999], is assumed to be linearly proportional to the rotational part of the electric field:

$$\mathbf{I}_{gr} = \Sigma_E [\mathbf{n}_R \times \nabla_t \Pi_E] \quad (23)$$

where the integrated surface conductivity is $\Sigma_E \sim \sigma_E h_S$. With these simplifications Eq.(22) becomes:

$$\Pi_M = \mu_0 \Sigma_E \Pi_E + (h_S + h_L) B_R, \quad \frac{\partial B_R}{\partial t} = -\nabla_t^2 \Pi_E \quad (24)$$

Together with these relationships, the rotational part of Eq.(16)

$$\nabla^2 \Pi_E + \nabla \times \Sigma^{-1} [\mathbf{n}_R \times \nabla_t \Pi_M]/\mu_0$$
$$= \nabla \times \Sigma^{-1} [\mathbf{n}_R \times \mathbf{B}_t]/\mu_0 \quad (25)$$

forms a closed system of equations that allows us to find Π_E, Π_M and then the ionospheric electric field $\mathbf{E}_t$ for any given $\mathbf{B}_t$ above the ionosphere.

The total jump in the magnetic field, $[\mathbf{n}_R \times \mathbf{B}_t]/\mu_0$, throughout the Earth skin layer, atmosphere and ionosphere can be related to the total current using Eqs. (15, 16, 17 and 23) in the following form ([cf. *Untiedt and Baumjohann*, 1993]):

$$\frac{\mathbf{n}_R \times \mathbf{B}_t}{\mu_0} = \mathbf{I}_{iono} + \mathbf{I}_{gr} + \mathbf{I}_{ind}$$
$$\mathbf{I}_{iono} = \Sigma \cdot \mathbf{E}_t$$
$$\mathbf{I}_{gr} = \Sigma_E [\mathbf{n}_R \times \nabla_t \Pi_E] \quad (26)$$
$$\mathbf{I}_{ind} = \left[\mathbf{n}_R \times \nabla_t \frac{(h_L + h_S)B_R}{\mu_0}\right]$$
$$\nabla_t \cdot \left[\frac{\mathbf{n}_R \times \mathbf{B}_t}{\mu_0} - \mathbf{I}_{iono}\right] = \nabla_t \cdot (\mathbf{I}_{gr} + \mathbf{I}_{ind}) = 0$$

Without neglecting the displacement current, the magnetic field $\mathbf{B}'$, would be equal to the magnetic field in the Earth wave-guide (see [*Yoshikawa and Itonaga, 2000*]). The magnetic field above the ionosphere is separated from the effects of the ionospheric current (that is proportional to the total ionospheric electric field), from the GIC (proportional to the rotational part of the ionospheric electric field) and from the small reactive impedance (proportional to the radial magnetic field). Usually the latter term is small as compared to the input from GIC, Appendix B discusses the opposite limiting case for the interested reader.

Potential Ionosphere

In two limiting cases the M-I coupling can be decoupled from the processes taking place below the ionosphere and it becomes independent of the poorly defined Σ_E.

First let us assume an infinite Earth conductivity: $\Sigma_E \gg |\Sigma|$. According to Eq.(25), such a conductivity completely eliminates the rotational part of the ionosphere electric field ($\Pi_E = 0$). In this case Eq. (1) describes the total electric field that becomes a purely potential field. From Eqs. (1, 16, 24 and 25) with $\Pi_E = 0$ one can find an expression describing the GIC for this model:

$$\mathbf{I}_{gr} = -\frac{[\mathbf{n}_R \times \mathbf{B}_t]}{\mu_0} \Sigma \cdot \nabla_t \psi, \quad \Pi_E = 0, \quad B_R = 0. \quad (27)$$

We note that according to Eqs. (15, 16 and 27) the resulting GIC are divergence-free.

In this approximation we obtained the traditional potential ionosphere and the GIC as the limiting case of the concentrated impedance model. Mathematically, the potential ionosphere model is not anyway easier than the more realistic concentrated impedance model, because Eqs. (24 and 25) are of the same type and of the same complexity as Eq. (1). At the same time the physical model is still oversimplified and inconsistent because it includes GIC, but does not include the non-potential electric field. To improve the model, more realistic non-potential electric fields and related GIC should be incorporated.

Non-Potential Ionosphere With No Ground Induced Currents

The processes below the ionosphere also can be decoupled in the opposite limiting case when $\Sigma_E \ll |\Sigma|$. This assumption means that we neglect the GIC, or more precisely, we neglect the feedback of GIC on the ionospheric electric field through the magnetic field below the ionosphere. The $Đ_M = 0$ condition makes it possible to obtain an explicit expression for the ionospheric electric field from Eq. (25):

$$\mathbf{E}_t = \frac{\Sigma^{-1}}{\mu_0} [\mathbf{n}_R \times (\mathbf{B}_t - (h_S + h_L)\nabla_t B_R)],$$
$$\frac{\partial B_R}{\partial t} = -[\nabla_t \times \mathbf{E}_t] \quad (28)$$

For brevity we denote $\mathbf{B}_t - (h_S + h_L)\nabla_t B_R$ as $\tilde{\mathbf{B}}_t$. The non-potential part of the electric field can be recovered, if desired, from the $\nabla_t^2 \Pi_E = [\nabla_t \times \mathbf{E}_t]$ equation. One can also obtain the small GIC by using $\Pi_M = \Sigma_E \Pi_E$.

This model is again oversimplified and one needs the incorporation of a more realistic GIC description and related non-potential electric fields. However, this GIC-free non-potential model is very simple and computationally beneficial because it does not require the solution of an elliptic equation like Eq. (1) or Eq. (25) We perfomed a large number of numerical tests (see below) and found that the potential ionosphere model and the non-potential GIC-free model give similar results for the overall M-I coupling, so we can take advantage of the mathematical and numerical simplicity and physical reasonability of the GIC-free model. We choose this model to be used in our M-I coupling model. We note that improving these models requires more mathematical and computational sophistication.

We thus chose a simple boundary condition for the mag-netosphere relating the tangential components of the magnetic field perturbation to those of the electric field. It describes the M-I coupling in a very simple, local manner. Comparing Eq. (28) to Eqs. (87.1–6) of *Landau et al.* [1985] we see that our boundary condition is basically identical to the Leontovich boundary condition that was introduced for describing the interaction of electromagnetic waves with a thin conducting layer (ionosphere or, originally, the skin layer of metals).

RESISTIVE SLIP VELOCITY

No Mass Exchange Solution

Eqs. (9, 10 and 28) can be combined to express the plasma velocity at the M-I boundary. However, the field aligned component of the plasma velocity (the component along the intrinsic field, B_0) is undefined since the cross product has no contribution from the parallel component. This uncertainty had been noted by *Goodman* [1995] and means that additional assumptions need to be made about the field aligned plasma velocity component at the M-I interface.

In this paper we focus on the electrodynamic coupling between the ionosphere and magnetosphere. While we recognize the importance of ionospheric outflow to magnetospheric composition and dynamics, we focus on the simple case when the mass exchange between the ionosphere and magnetosphere is neglected. We will generalize our approach in a subsequent publication.

For the sake of simplicity we choose the radial plasma velocity to be zero at the M-I boundary. By combining Eqs. (9, 10 and 28), we find the boundary condition coupling the velocity and the magnetic field perturbation:

$$B_{0R}[\mathbf{n}_R \times \mathbf{u}_t] = \frac{\Sigma^{-1}}{\mu_0} \cdot [\mathbf{n}_R \times \mathbf{B}_t] \quad (29)$$

In addition, we have the $u_R = 0$ condition at the boundary, describing no mass exchange between the ionosphere and the magnetosphere.

Boundary Conditions at the M-I Interface

Combining Eqs. (29 and B8) we finally find the boundary condition at the magnetosphere - ionosphere interface in a simple vector form:

$$u_R = 0 \quad (30)$$

$$\mathbf{u}_t = \mathbf{v}_{rot} - \frac{\mathbf{n}_R \times \mathbf{B}_t}{\mu_0 e N_e} + \frac{\mathbf{B}_{t\perp}}{B_{0R}\mu_0\Sigma_\parallel} + \frac{\mathbf{B}_{t\parallel}}{B_{0R}\mu_0\Sigma_\perp} \quad (31)$$

The condition for the tangential velocity is surprisingly simple and has a transparent physical interpretation: the tangential velocity component of the nearly perfectly conducting plasma near the ionosphere is equal to the sum of corotation velocity plus the current velocity $-\mathbf{J}_t/(eN_e)$ (in the general case this term can differ from both the electron and ion current velocities but it is of the same physical nature), plus the "resistive slip" velocity:

$$\mathbf{u}_{slip} = \frac{\mathbf{B}_{t\perp}}{B_{0R}\mu_0\Sigma_\parallel} + \frac{\mathbf{B}_{t\parallel}}{B_{0R}\mu_0\Sigma_\perp}$$

The "slip velocity" is due to finite ionospheric plasma resistance because the magnetic field lines can slip through the plasma as long as the magnetic field is not completely frozen in.

NUMERICAL MODEL

We use the BATSRUS code of the University of Michigan [*Powell et al.,* 1999] to simulate the magnetosphere. The ideal MHD equations with full energy equation are solved using a conservative finite volume scheme with second order of accuracy. An adaptive block grid is used, the control volumes ("cells") being rectangular boxes.

The values of the MHD variables are interpolated to the faces of the control volume using the van Leer /3-limiter [*Hirsch,* 1990] with $\beta = 1.2$, applied to the increments of the primitive variables (density, velocity, magnetic field B and pressure), as in the MUSCL scheme [cf. *Toro,* 1999].

The monotone numerical fluxes through the faces of the control volume are upwinded using the Artificial Wind scheme [*Sokolov et al.,* 1999, 2002]. Second order time update is used as in the second order Runge-Kutta scheme.

The boundary conditions given by Eqs. (30 and 31) are used to construct the first order monotone boundary condition (via a Riemann solver) at the inner boundary for the MHD computational domain at $R = R_I$, that is interpreted as the interface between a moving perfectly conducting fluid (magnetosphere) and a rotating thin spherical shell of finite conductivity (ionosphere).

For comparison we also used the potential ionosphere model. The equation for the potential Eq. (1) with the same conductivity tensor as in the Eq. (B6) can be written in the following form:

$$\nabla_t \cdot \left((\Sigma'_\parallel - \Sigma'_P) \frac{(\nabla_t \psi \cdot \mathbf{B}_{0t})\mathbf{B}_{0t}}{B_{0t}^2} + \Sigma'_P \nabla_t \psi \right) \quad (32)$$
$$- [\nabla_t \psi \times \mathbf{n}_R] \cdot \nabla_t (\Sigma'_H \cos\alpha) = j_R(R_I)$$

Eq. (32) is numerically solved using GMRES algorithm.

SIMULATION RESULTS

The main purpose of the present simulation is to compare the newly introduced non-potential model for the M-I coupling with the existing potential model. That is why we choose a test problem for which the widely used potential model is applicable and should give a physically reasonable answer. That is we simulate the steady-state plasma flow around the Earth, the direction of the magnetic field dipole axis being perpendicular to the velocity of solar wind. For our non-potential model chosing a steady-state problem is not important, however, the potential model is strictly applicable only for a steady-state problem. In the simualtions the solar wind parameters are: $n_e = 5$ cm^{-3}, velocity=500 km/s, temperature $T = 1.5 \cdot 10^5$ K and southward IMF=-5 nT.

In the first test we consider the case of a single constant conductivity, $\Sigma_\perp = \Sigma_\parallel = 4$ Ohm^{-1}, $\Sigma'_H = 0$. Again, for the non-potential model any conductivity gradients are unimportant, while for the potential model the results obtained with non-constant conductivity depend on a particular choice of the scheme for the conductivity gradients.

Plate 1 shows the field aligned current distribution calculated at the $R = 2R_E$ surface. This seems to be a representative characteristics for M-I coupling because, although calculated in the magnetosphere, it is known to be sensitive to any change of the ionosphere parameters [cf. *Ridley et al.,* 2003].

In the top panel of Plate 1 we present the distribution of the ionosphere current obtained using non-potential M-I

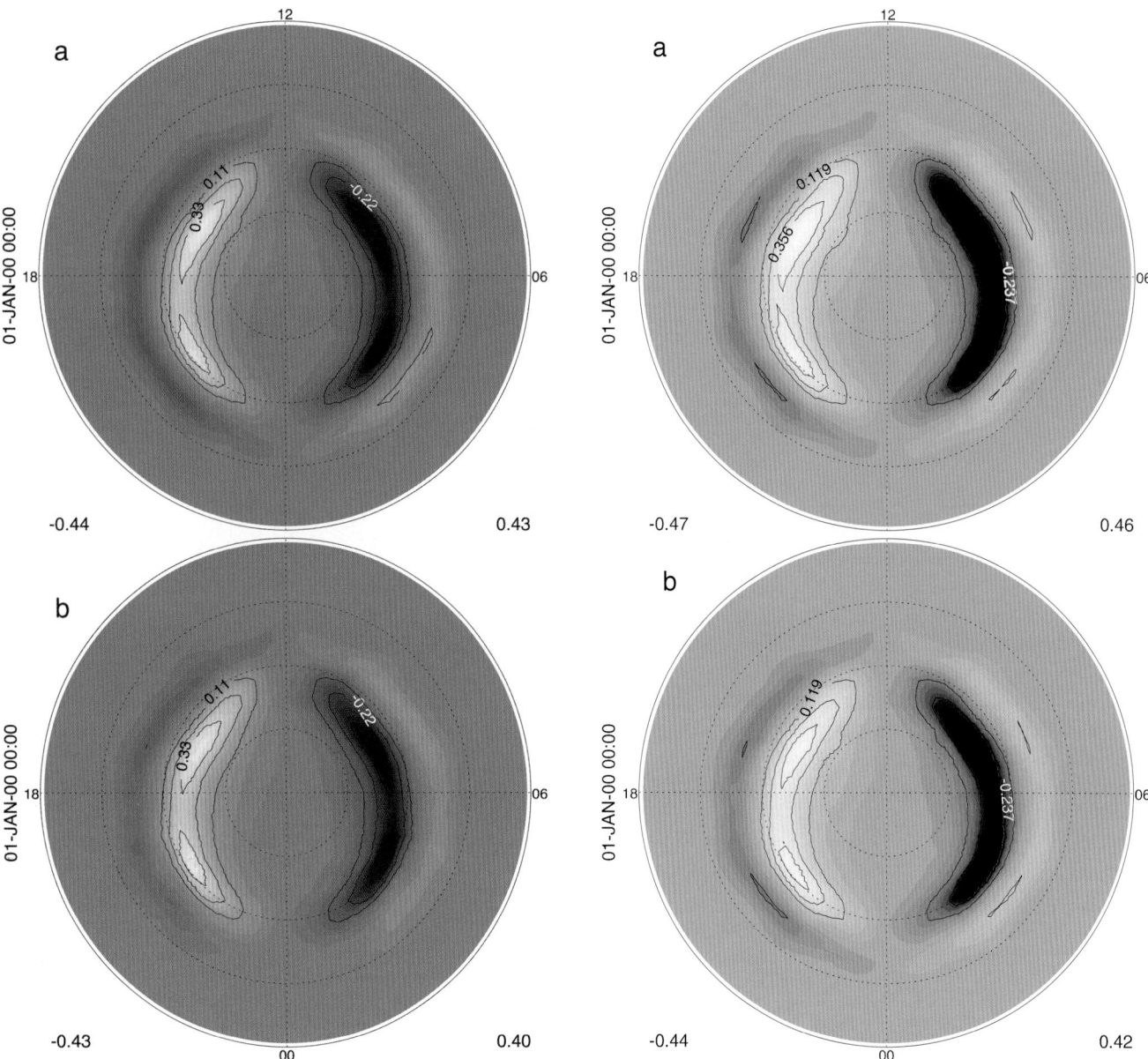

Plate 1. Ionospheric radial current distribution obtained with the non-potential technique for M-I coupling (upper panel) and the same current density obtained with a potential model (lower panel).

Plate 2. Ionospheric radial current distribution obtained with the non-potential technique including Hall conductivity (top panel) and the same, obtained with a potential model (lower panel).

coupling model with the boundary condition in the form of Eq. (31). The result seems to be reasonable. In the bottom panel of Fig. 1 the same distribution is presented obtained from the potential model in solving the Eq. (32). The agreement between the results is very good.

In Plate 2 we present the results of the same simulations but with the Hall conductivity taken into account. In the simulations we used $\int \sigma_\parallel dh = 1000 \, \Omega^{-1}$ and $\int \sigma_P dh = \int \sigma_H dh = 4 \, \Omega^{-1}$ with the proper latitude dependence according to Eqs. (B2-B5). The difference in the results is again reasonably small, mainly due to the significant differences in the numerical algorithms we used in the potential and non-potential models.

DISCUSSION AND CONCLUSION

We demonstrated that the results obtained using the new non-potential model for M-I coupling are very similar to compared to the results obtained with the potential model.

The non-potential model seems to be physically more relevant when describing fast and abrupt processes in the M-I system, for which the essentially non-potential perturbations of the electric field are not applicable.

The non-potential model is of higher computational efficiency, since the electric potential is not needed for describing the M-I coupling, and consequently there is no need to solve Poisson's equation Eq. (32).

The non-potential model allows us to find the non-potential electric field, calculate and predict the electromotive forces, which are induced in the natural and technological large-scale closed electric circuits at the Earth during fast geomagnetic processes.

APPENDIX A: SOURCE SURFACE METHOD AND ACCELERATED DAMPING RATE FOR NON-POTENTIAL ELECTRIC FIELD

Consider the limiting case when Earth is a perfect insulator (with zero conductivity) and there are no GICs. In this case, the magnetic field below the ionosphere is purely potential, hence, it can be unambiguously expressed in terms of its radial component (B_R) at a sphere, using the so-called source surface method. This method is widely used in solar physics, allowing to reconstruct 3D spatial distribution of the coronal magnetic field from measurements of the radial component of the field at the solar surface [*Altschuler et al*, 1977]. Recently our group applied this method to simulate the 3D solar wind driven by solar magnetogram observations [*Roussev et al., 2003*].

In the case of the terrestrial ionosphere with no ground conductivity, the normal component of the magnetic field (that is continuous across the ionosphere in the thin ionosphere approximation) can be expressed as a series of spherical functions, Y_{nm}:

$$B_{R|R=R_E+h_L} = \sum_{n,m} a_{nm} Y_{nm}(\theta, \phi) \quad \text{(A1)}$$

Once the expansion coefficients, a_{nm}, are specified, the full magnetic field vector at the sphere can be expressed in the following form:

$$\mathbf{B}' = \nabla_t \psi, \; \psi = (R_E + h_L) \sum_{n,m} \frac{a_{nm}}{n} Y_{nm}(\theta, \phi) \quad \text{(A2)}$$

Next we can substitute Eq. (A2) into Eq. (16) and express the tangential electric field in the ionosphere in terms of all the three components of the magnetic field:

$$\Sigma \cdot \mathbf{E}_t = \frac{1}{\mu_0}[\mathbf{n}_R \times \mathbf{B}_t] \\ - \frac{1}{\mu_0}(R_E + h_L)[\mathbf{n}_R \times \nabla_t] \sum_{n,m} \frac{a_{nm}}{n} Y_{nm}(\theta, \phi) \quad \text{(A3)}$$

A physically instructive simplification of Eq. (A3) can be obtained by neglecting the factor $1/n$ in the series expansion, even though this approach overestimates the influence of the magnetic field B^0 below the ionosphere. Now one can recognize that the series expansion is the radial magnetic field component at the sphere and we can combine Eqs. (A1) and (A3) to obtain the following:

$$\mu_0 \Sigma \cdot \mathbf{E}_t = [\mathbf{n}_R \times \mathbf{B}_t] - (R_E + h_L)[\mathbf{n}_R \times \nabla_t] B_R \quad \text{(A4)}$$

The physical interpretation of this equation becomes clear if we consider the case of a uniform and scalar conductivity Σ_0, and take the time derivative of Eq. (A4):

$$\frac{\partial \mathbf{E}_t}{\partial t} = \frac{1}{\mu_0 \Sigma_0}\left[\mathbf{n}_R \times \frac{\partial \mathbf{B}_t}{\partial t}\right] \\ - \frac{R_E + h_L}{\mu_0 \Sigma_0}[\mathbf{n}_R \times \nabla_t]\frac{\partial B_R}{\partial t} \quad \text{(A5)}$$

This equation can be further simplified using Eq. (12):

$$\frac{\partial \mathbf{E}_t}{\partial t} = \frac{1}{\mu_0 \Sigma_0}\left[\mathbf{n}_R \times \frac{\partial \mathbf{B}_t}{\partial t}\right] \\ + \frac{R_E + h_L}{\mu_0 \Sigma_0}[\mathbf{n}_R \times \nabla_t](\nabla_t \times \mathbf{E}_t) \quad \text{(A6)}$$

Finally, we apply the $\nabla_t \times$ operator to Eq. (A6):

$$\frac{\partial}{\partial t}(\nabla_t \times \mathbf{E}_t) - \frac{R_E + h_L}{\mu_0 \Sigma_0} \nabla_t^2 (\nabla_t \times \mathbf{E}_t) = \frac{\partial}{\partial t} \frac{\nabla_t \cdot \mathbf{B}_t}{\mu_0 \Sigma_0} \quad \text{(A7)}$$

The solutions of Eq.(A7) with vanishing right hand side would decrease in time, not slower than $\sim \exp(-\nu t)$, where the damping rate ν is:

$$\nu = \frac{1}{2(R_E + h_L)\mu_0 \Sigma_0} \approx \frac{1}{4\Sigma_0 [mhos]} \; [s^{-1}] \quad \text{(A8)}$$

This means that the perturbation of the non-potential electric field caused by a variation of the magnetic field above the ionosphere (which comes from the magnetosphere) spreads over the ionosphere and decreases in time. Approximation (A4) overestimates the damping rate for spherical harmonics with larger n values. However, Eq. (A8) is correct in any case, because it is derived for $n = 1$ spherical harmonics and the omission of $1/n$ factor is justified for it. The maximum damping rate approximation given by Eq. (A4) is useful both from the theoretical (it allows to treat non-potential time-varying electric fields, while it rapidly converges to the steady-state solution for purely potential electric field) and from the practical points of view (it results in a simple and fast converging numerical method).

On the other hand, the physical reasoning under this model is rather poor. The magnetic field is assumed to freely penetrate to the ground down to the Earth center due to infinitely small conductivity, while with realistic finite Earth conductivity σ_E, which is especially high at larger depths, the time $\mu_0 R_E^2 \sigma_E$ needed for such the penetration is enormously long and is a way greater than time scale t_V for any geomagnetic variation: $t_V \ggg \mu_0 R_E^2 \sigma_E$.

APPENDIX B: CONDUCTANCE TENSOR

In this section we discuss some properties of the height integrated electric conductance tensor, Σ. Our observations are simple and we make them to avoid misunderstandings about the physical nature of the various terms. In the literature both the conductance and resistance tensors are used, and, in addition, in the M-I coupling problem the height integrated 2D conductance tensor is used, rather than the 3D conductivity tensor.

The generalized Ohm's law in the ionosphere plasma is:

$$\mathbf{j} = (\sigma_\| - \sigma_P) \frac{(\mathbf{E}' \cdot \mathbf{B}_0)\mathbf{B}_0}{B_0^2} + \sigma_P \mathbf{E}' - \sigma_H \frac{[\mathbf{E}' \times \mathbf{B}_0]}{B_0} \quad \text{(B1)}$$

where σ_P and σ_H the Pedersen and Hall conductivity, $\sigma_\|$ is the conductivity along the magnetic field, $\mathbf{E}' = \mathbf{E} + [v_{rot} \times \mathbf{B}_0]$ is the electric field in the rotating frame of reference, v_{rot} is the velocity of rotation (the ionosphere is assumed to rotate with the Earth). Eq. (B1) is discussed in textbooks [cf. *Gombosi*, 1998] and is well-known.

Next we consider the ionosphere as a *thin* conducting layer. This assumption allows us to neglect the radial component of the current *density* j_R in Eq. (B1) that is small compared to the tangential components (see *Amm* [1996]). By finding the radial electric field E'_R from the condition $j_R = 0$, one can reduce Eq. (B1) to an equation for the tangential current density

$$\mathbf{j}_t = (\sigma'_\| - \sigma'_P) \frac{(\mathbf{E}'_t \cdot \mathbf{B}_{0t})\mathbf{B}_{0t}}{B_{0t}^2} + \sigma'_P \mathbf{E}'_t - \sigma'_H \cos\alpha [\mathbf{E}'_t \times \mathbf{n}_R] \quad \text{(B2)}$$

where

$$\sigma'_\| = \frac{\sigma_\| \sigma_P}{\sigma_\| \cos^2\alpha + \sigma_P \sin^2\alpha}, \quad \text{(B3)}$$

$$\sigma'_P = \frac{\sigma_\| \sigma_P \cos^2\alpha + (\sigma_P^2 + \sigma_H^2)\sin^2\alpha}{\sigma_\| \cos^2\alpha + \sigma_P \sin^2\alpha}, \quad \text{(B4)}$$

$$\sigma'_H = \frac{\sigma_\| \sigma_H}{\sigma_\| \cos^2\alpha + \sigma_P \sin^2\alpha} \quad \text{(B5)}$$

and

$$\cos\alpha = \frac{B_{0R}}{B_0}$$

We note that the conductivity $\sigma'_\|$, which is usually denoted as $\sigma_{\theta\theta}$, is 'parallel' only with respect to the (small) projection of the full magnetic field on the ionosphere layer $\mathbf{B}_t$, but the physical processes under it are mostly the same as for the Pedersen conductivity, because the B_{0R} component is dominant. Eqs. (B2–B5) are identical to the results obtained by *Amm* [1996]. Eq. (B2) can be height integrated assuming that the tangential electric field in the thin conducting layer does not depend on altitude:

$$\mathbf{J}_t = (\Sigma'_\| - \Sigma'_P) \frac{(\mathbf{E}'_t \cdot \mathbf{B}_{0t})\mathbf{B}_{0t}}{B_{0t}^2} + \Sigma'_P \mathbf{E}'_t - \Sigma'_H \cos\alpha [\mathbf{E}'_t \times \mathbf{n}_R] \quad \text{(B6)}$$

$$\Sigma'_\| = \int \sigma'_\| dh, \; \Sigma'_P = \int \sigma'_P dh, \; \Sigma'_H = \int \sigma'_H dh \quad \text{(B7)}$$

Note, that in the 2D Ohm's law (Eq. B6) the difference between the "parallel" and "Pedersen" conductances is small, especially at high latitudes (at $\sin^2\alpha \to 0$): $\Sigma'_\| \approx \Sigma'_P \approx$

$\int \sigma_p dh$, although for actual 3D conductivity coefficients usually $\sigma_\| \gg \sigma_p$ in the ionosphere. This point is physically evident, because, due to the large radial magnetic field, the tangential current is not field aligned at all and is not much sensitive to the comparatively small tangential component of the magnetic field. Particularly, in our first numerical test (see above) we use $\Sigma'_p = \Sigma'_\| = 4\ Ohm^{-1}$.

Finally, we should consider the right hand side of Eq. (B6) as the product of the 2 x 2 matrix, Σ, multiplied by the vector Σ'_t and invert this matrix (see Eq. (29)) by expressing the tangential electric field in terms of the tangential current:

$$\mathbf{E}'_t = \Sigma^{-1} \mathbf{J}_t = \frac{\mathbf{J}_{t\|}}{\Sigma_\|} + \frac{\mathbf{J}_{t\perp}}{\Sigma_\perp} + \frac{[\mathbf{J}_t \times \mathbf{n}_R]B_{0R}}{eN_e} \quad (B8)$$

where

$$\Sigma_\| = \frac{\Sigma'_\| \Sigma'_P + (\Sigma'_H \cos\alpha)^2}{\Sigma'_P}, \quad (B9)$$

$$\Sigma_\perp = \frac{\Sigma'_\| \Sigma'_P + (\Sigma'_H \cos\alpha)^2}{\Sigma'_\|}, \quad (B10)$$

$$N_e = \frac{B_0 \left(\Sigma'_\| \Sigma'_P + (\Sigma'_H \cos\alpha)^2\right)}{e\Sigma'_H} \quad (B11)$$

here the parallel and perpendicular components of the tangential currents are defined with respect to the tangential magnetic field: $\mathbf{J}_{t\|} = \mathbf{B}_{0t} (\mathbf{J}_t \cdot \mathbf{B}_{0t})/B_{0t}^2$, $\mathbf{J}_{t\perp} = \mathbf{J}_t - \mathbf{J}_{t\|}$.

Here $\Sigma_\|$ and $\Sigma_\perp$ are proper non-linear combinations of the height integrated conductances. Nevertheless, if only the electrons are responsible for the conductivity and their conductivity coefficients can be considered as height independent variables in the integration, then $\Sigma_\| = \Sigma_\perp = \int \sigma_\| dh$ and they become the height integrated Lorentz conductivity and do not depend on the magnetic field. Analogously N_e is also some formally introduced non-linear combination of the height integrated conductances, which has nothing to do with the electron density in a general case, but in the same particular case it becomes $N_e = \int n_e dh$, n_e being the electron density. This simplification is not used in our numerical solutions, it is invalid for the Earth ionosphere, because the ions are mainly responsible for the Pedersen conductivity, but it is useful in analyzing the physical interpretation of the boundary conditions described in the body of this paper.

Acknowledgments. The authors appreciate the helpful comments of Prof. V. Vasyliunas. This work was supported by NSF ITR grant ATM-0325332, by DoDMURI grant F49620-01-1-0359, and by NASA ESS CT cooperative agreement NCC5-614. TIG was partially supported by a BSF (Binational Science Foundation) grant.

REFERENCES

Altschuler, M. D., R. H. Levine, M. Stix, and J. Harvey, High-resolution mapping of the magnetic field of the solar corona, *Sol. Phys., 51,* 345–375, 1977.

Amm, O., Comment on "A three-dimensional, iterative, mapping procedure for the implemenation of an ionosphere-magnetosphere anisotropic Ohm's law boundary condition in global magnetohydrodynamic simulations" by Michael L. Goodman, *Ann. Geophys., 14,* 773–774, 1996.

Gombosi, T. I., *Physics of the Space Environment,* Cambridge University Press, Cambridge, UK, 1998.

Gombosi, T. I., G. Tóth, D. L. De Zeeuw, K. C. Hansen, K. Kabin, and K. G. Powell, Semi-relativistic magneto-hydrodynamics and physics-based convergence acceleration, *J. Comput. Phys., 177,* 176–205, 2002.

Goodman, M. L., A three-dimensional, iterative, mapping procedure for the implementation of an ionosphere-magnetosphere anisotropic Ohm's law boundary condition in global magnetohydrodynamic simulations, *Ann. Geophys., 13,* 843–853, 1995.

Hirsch, C, *Numerical Computation of Internal and External Flows, Volume 2, Computational Methods for Inviscid and Viscous Flows,* John Wiley & Sons, Toronto, 1990.

Janhunen, P., GUMICS-3: A global ionosphere-magnetosphere coupling simulation with high ionospheric resolution, in *Proceedings of the ESA 1996 Symposium on Environment Modelling for Space-Based Applications,* pp. 233–239, ESA SP-392, 1996.

Landau, L. D., E. M. Lifshitz, and L. P. Pitaevskii, *Electrodynamics of Continuous Media,* Butterworth-Heinemann, Oxford, UK, 2nd edn., 1985.

Lyon, J. G., J. Fedder, and J. Huba, The effect of different resistivity models on magnetotail dynamics, *J. Geophys. Res., 91,* 8057–8064, 1986.

Ogino, T, and R. J. Walker, A magnetohydrodynamic simulation of the bifurcation of tail lobes during intervals with a northward interplanetary magnetic field, *Geophys. Res. Lett, 11,* 1018–1021, 1984.

Powell, K. G., P. L. Roe, T. J. Linde, T. I. Gombosi, and D. L. D. Zeeuw, A solution-adaptive upwind scheme for ideal magnetohydrodynamics, *J. Comput. Phys., 154*(2), 284–309, 1999.

Raeder, J., R. J. Walker, and M. Ashour-Abdalla, The structure of the distant geomagnetic tail during long periods of northward IMF, *Geophys. Res. Lett., 22,* 349–352, 1995.

Richmond, A. D., E. C. Ridley, and R. G. Roble, A thermosphere/ionosphere general circulation model with coupled electrodynamics, *Geophys. Res. Lett., 19,* 601–604, 1992.

Ridley, A. J., T. I. Gombosi, and D. L. De Zeeuw, Ionospheric control of the magnetosphere: Conductance, *J. Geophys. Res., 00,* 00–00, 2003, submitted.

Roussev, I., T. Gombosi, I. Sokolov, M. Velli, W Manchester, D. DeZeeuw, P. Liewer, G. Toth, and J. Luhmann, A three-dimensional model of solar wind incorporating solar magnetogram observations, *Astrophys. J., 595,* L57–L61, 2003.

Sokolov, I., E. V. Timofeev, J. ichi Sakai, and K. Takayama, Artificial wind—a new framework to construct simple and efficient upwind shock-capturing schemes, *J. Comput. Phys., 181,* 354–393, 2002.

Sokolov, I. V, E. V. Timofeev, J. I. Sakai, and K. Takayama, On shock capturing schemes using artificial wind, *Shock Waves, 9,* 423–426, 1999.

Tanaka, T, Generation mechanisms for magnetosphere-ionosphere current systems deduced from a three-dimensional MHD simulation of the solar wind-magnetosphere-ionosphere coupling process, *J. Geophys. Res., 100*(A7), 12,057–12,074, 1995.

Toro, E. F, *Riemann solvers and numerical models for fluid dynamics,* Springer-Verlag, New York, 1999.

Untiedt, J., and W Baumjohann, Studies of polar current system using the IMS Skandinavian magnetometer array, *Space Sci. Rev, 63,* 245–390, 1993.

Viljanen, A., O. Amm, and R. Pirjola, Modeling geomagnetically induced currents during different ionospheric situations, J. *Geophys. Res., 104,* 28,059–28,071, 1999.

White, W W, G. L. Siscoe, G. M. Erickson, Z. Kaymaz, N. C. Maynard, K. D. Siebert, B. U. 6. Sonnerup, and D. R. Weimer, The magnetospheric sash and the cross-tail S, *Geophys. Res. Lett, 25*(10), 1605–1608, 1998.

Yoshikawa, A., and M. Itonaga, The nature of reflection and mode conversion of MHD waves in the inductive ionosphere: Multistep mode conversion between divergent and rotational electric fields, *J. Geophys. Res., 105,* 10,565–10,584, 2000.

Tamas I. Gombosi, Aaron J. Ridley, and Igor V. Sokolov, Center for Space Environment Modeling, Department of Atmospheric, Oceanic, and Space Sciences, University of Michigan, Ann Arbor, MI 48109. (e-mail: igorsok@umich.edu, tamas@umich.edu, ridley@umich.edu)

Pressure-Driven Currents Derived from Global ENA Images by IMAGE/HENA

Edmond C. Roelof

Johns Hopkins University/Applied Physics Laboratory, Laurel, Maryland

The global ion distributions extracted from ENA images obtained by IMAGE/HENA can be used to calculate the distribution of partial plasma pressure throughout the ring current region. Knowing the plasma pressure and the magnetic field, one can calculate the three-dimensional current system that is driven by the pressure gradient, assuming force equilibrium in the absence of convection with $\nabla \cdot \mathbf{J} = 0$. For an isotropic pressure tensor $\mathbf{P} = \mathrm{P}\mathbf{I}$, the complete current intensity $\mathbf{J}$ can be calculated from Euler potentials, one of which is the pressure (P) itself, while the other (Q) is the partial volume of the magnetic flux tube measured from the minimum-B equator. For a dipole field, the latter is a simple analytic function, but in principle Q can be calculated from any magnetic field model. If $\mathbf{P}$ is not isotropic, a solution still exists, but it is not expressible directly in terms of Euler potentials. Examples of effects in estimating the ion pressure from IMAGE/HENA images will be presented in order to explain the underestimate of the actual Region 2 field-aligned currents by the magnetometers on the constellation of Iridium communication satellites.

INTRODUCTION

Ion intensity distributions have been extracted from IMAGE/HENA ENA images ~10–200 keV using a conditioned linear inversion algorithm [*Demajistre et al., 2004*]. These ENA image inversions have been validated by comparison with *in situ* ion measurements from Cluster/CIS during 6 geomagnetic storms [*Vallat et al., 2004*].

Another type of "ground truth" comparison can be made with ion intensity distributions obtained from these ENA image inversions. The corresponding partial pressure distributions calculated from the inverted ion intensities can be used to estimate the global pressure-driven electric current system [*Roelof, 1989*]. Field-aligned (Region 2) currents that should be driven into the ionosphere by the pressure distribution have been computed and compared with simultaneous FAC (field-aligned currents) measured by magnetometers on the Iridium communication satellites [*Roelof et al., 2004*].

The purpose of this paper is to summarize the present method for obtaining pressure-driven currents from ENA images and to discuss the effects that lead to underestimates of the measured Iridium Region 2 FAC values.

COMPUTING PRESSURE-DRIVEN ELECTRIC CURRENTS FROM HENA IMAGES

For an isotropic pressure tensor $\mathbf{P} = \mathrm{P}\mathbf{I}$, the complete current intensity $\mathbf{J}$ satisfying $\nabla \cdot \mathbf{J} = 0$ can be calculated from Euler potentials, one of which is the pressure (P) itself, and the other (Q) is the partial volume of the magnetic flux tube measured from the minimum-B equator [*Roelof, 1989*]. If $\mathbf{P}$ is not isotropic, a solution still exists, but it is not expressed directly in Euler potentials.

With the assumption of isotropic pressure balance in the absence of convection

$$\mathbf{J} \times \mathbf{B} = \nabla P$$

there is an easy method for obtaining the complete pressure-driven current system. That method [*Roelof, 1989*] introduces Euler potentials for the *electrical current*, thus automatically satisfying the steady state condition $\nabla \cdot \mathbf{J} = 0$. We chose two arbitrary Euler potentials Q and R so that

$$\mathbf{J} = \nabla Q \times \nabla R$$

Not to be coy about it, R will indeed turn out to be the pressure P, but that is not assumed to be so at the outset. It is determined when we require that $\mathbf{J} \times \mathbf{B} = \nabla P$. Note that this force balance requires that $\mathbf{B} \cdot \nabla P = 0$, or that the pressure is constant along a field line.

$$\mathbf{J} \times \mathbf{B} = (\mathbf{B} \cdot \nabla Q)\nabla R - (\mathbf{B} \cdot \nabla R)\nabla Q$$

We now observe that if we set R = P that $\mathbf{B} \cdot \nabla R = \mathbf{B} \cdot \nabla P = 0$, so the second term vanishes. Then we are left with only the first term, so that

$$(\mathbf{B} \cdot \nabla Q)\nabla P = \nabla P$$

which will be satisfied if

$$\mathbf{B} \cdot \nabla Q = 1 \quad \text{or} \quad \partial Q/\partial s = 1/B$$

The formal solution can be written

$$Q = {}_0\!\int^s ds'/B(s')$$

where it is understood that the path integration is along a field line. Therefore, the correct condition for the existence of a pressure-driven current is that the surfaces Q = constant (not B = constant) and P = constant must intersect.

BASIC PROPERTIES OF EULER POTENTIALS

The lines of current intensity **J** lie in the intersections of the surfaces Q = constant and P = constant. This is because the gradients ∇Q and ∇P are normal to their two surfaces, respectively. However, the vector cross-product of the normals must lie in both surfaces, because it must be locally perpendicular to either normal. Therefore it must lie in the common line formed by the intersection of the two surfaces.

The net current ΔI integrated across an area of surface S normal to the current itself is $\Delta Q \Delta P$, where ΔI, ΔQ, and ΔP may be finite increments. To prove this, we use the equivalency of Euler and vector potentials

$$_S\!\int d\mathbf{S} \cdot \mathbf{J} = {}_S\!\int d\mathbf{S} \cdot \nabla Q \times \nabla P = -{}_S\!\int d\mathbf{S} \cdot \nabla \times (P\nabla Q)$$

and then apply Stokes' theorem around the curve C bounding S.

$$_S\!\int d\mathbf{S} \cdot \nabla \times (P\nabla Q) = {}_C\!\int d\mathbf{c} \cdot P\nabla Q = {}_C\!\int dQP$$

Define the area S normal to **J** and as being bounded by the four surfaces (Q, Q+ΔQ, P, and P+ΔP) = constant. Along the first two, dQ = 0, and along the second two, P has a constant value.

$$_C\!\int dQP = P\Delta Q - (P+\Delta P)\Delta Q = -\Delta P \Delta Q$$

$$\Delta I = {}_S\!\int d\mathbf{S} \cdot \mathbf{J} = \Delta Q \Delta P$$

EULER POTENTIALS FOR CURRENT IN A DIPOLE MAGENTIC FIELD

For a dipole field (again assuming negligible perturbations by the electric currents), the integral of ds/B can be done analytically and Q turns out to be a separable function of L and θ. The simple formula follows from

$$ds = \mathbf{b} \cdot d\mathbf{r} = -(2\cos\theta dr + \sin\theta r d\theta)/\gamma$$

$$= -d\theta\,(2\cos\theta dr/d\theta + r\sin\theta)/\gamma$$

where $\gamma = (3\cos^2\theta + 1)^{1/2}$. For a dipole field $B = B_0(a/r)^3\gamma$ and the field line is given by $r = aL\sin^2\theta$ (holding L constant) so that $dr/d\theta = 2aL\sin\theta\cos\theta$. Now we can integrate the equation $dQ = ds/B$, holding L constant:

$$Q(L,\mu) = (a/B_0)L^4 \,{}_0\!\int^\mu d\mu\,(1-\mu^2)^3$$

$$= (a/B_0)L^4\,\mu p(\mu)$$

where $p(\mu) = 1 - \mu^2 + (3/5)\mu^4 - (1/7)\mu^6$, an even polynomial. The separability of $Q(L,\mu)$ with regard to the (L,μ) dependence greatly facilitates the analytical calculation of ∇Q in spherical polar coordinates.

The vector cross-product $\mathbf{J} = \nabla Q \times \nabla P$ can either be formed directly from the (r,θ,ϕ) components, or it can be expressed in terms of the coefficients of ∇L, $\nabla \mu$, and $\nabla \phi$. Both calculations give the final result for the components of the current density:

$$J_r = -(L^2/aB_0)[8\mu^2 p(\mu)/\sin^6\theta + \sin^2\theta]\partial P/\partial \phi$$

$$J_\theta = -(4L^2/aB_0)[\mu p(\mu)/\sin^5\theta]\partial P/\partial\phi$$

$$J_\phi = +(L^3\sin^3\theta/aB_0)\partial P/\partial L$$

A straightforward but tedious calculation shows that these components satisfy the condition $\nabla\cdot\mathbf{J} = 0$ for any pressure distribution of the form $P(L,\phi)$. Note that this form satisfies the ancillary condition $\mathbf{B}\cdot\nabla P = 0$ because $B_\phi = 0$ and $\mathbf{B}\cdot\nabla L = 0$ for a dipole magnetic field.

The current density in the equatorial plane ($\mu = 0$) is simply

$$J_r = -(L^2/aB_0)\,\partial P/\partial\phi$$

$$J_\theta = 0$$

$$J_\phi = +(L^3/aB_0)\,\partial P/\partial L$$

These equations may be rewritten (again, for $\mu = 0$ so that $r = aL$ and $\mathbf{e}_\theta = -\mathbf{e}_z$)

$$\mathbf{J}_{eq} = (r^3/B_0 a^3)\,\mathbf{e}_\theta\times\nabla_{eq}P = [-1/B_{eq}(r)]\,\mathbf{e}_z\times\nabla_{eq}P$$

This form shows that the current lines in the equator follow the contours of constant pressure.

DIRECT CURRENT CALCULATION USING CONSTANT-PRESSURE SURFACES

We can generate the partial pressure surfaces $P(L,\phi) =$ constant from the ion intensities obtained from the HENA image inversions, and then directly calculate their intersections with the analytic surfaces $Q(L,\mu) =$ constant. If we generate surfaces at equal intervals in Q and P, then their intersections will give current "lines", all carrying the same current ΔI (in amperes) $= \Delta Q\Delta P$. This gives an easy way to visualize the global pressure-driven current system.

However, in the ionosphere where $r\cong a$ so that $L = \sin^{-2}\theta$, Q depends purely on the cosine of the co-latitude ($\mu = \cos\theta$)

$$Q_i = (a/B_0)\,\mu p(\mu)/(1-\mu^2)^4$$

Take the contour (used above in the proof of Euler's theorem) to be a box bounded by $(\theta,\theta+\Delta\theta)$ and $(\phi,\phi+\Delta\phi)$ on the surface $r = a$ in the ionosphere. On the lines of constant co-latitude, $dQ = 0$, so the line integral

$$\Delta I = -\oint_c d\mathbf{c}\cdot P\nabla Q = -\oint_c dQ\,P$$

becomes

$$\Delta I = \int_\theta^{\theta+\Delta\theta}d\mu\,(dQ_i/d\mu)\,[P(\sin^{-2}\theta,\phi+\Delta\phi) -$$

$$P(\sin^{-2}\theta,\phi)]$$

and the normal area into which this current flows is

$$A_n \cong (a^2\sin\theta\cos\chi)\Delta\theta\Delta\phi \qquad \tan\chi = \tan\theta/2$$

so the current intensity is $J_i(\text{amp/m}^2) = \Delta I/A_n$

The integral for ΔI can, in principal, be evaluated numerically, but $dQ_i/d\mu$ contains a term that varies as $\sin^{-10}\theta$. The current into the ionosphere is $J_i(\theta,\phi) = -J_r$ evaluated at $r = a$ so that $L = 1/\sin^2\theta$. The analytic expression for the current into the ionosphere is

$$J_i = (1/aB_0)[8\mu^2 p(\mu)\sin^{-10}\theta+\sin^{-2}\theta]\partial P/\partial\phi$$

The very strong inverse dependence on $\sin\theta$ shows that the ionospheric current density will be extremely sensitive to the azimuthal pressure gradient in the near-Earth plasma sheet. However, we should remember that the magnetic field in the same region becomes significantly deformed from that of a simple dipole, and our calculation is based on the simple geometry of a dipole field. At best, this divergence as $\sin\theta\to 0$ (if $\partial P/\partial\phi$ remains finite for large values of L) is an indication that azimuthal pressure gradients in the near-Earth plasma sheet play an important role in driving the Region 2 current system.

ERROR FACTORS IN ESTIMATING FIELD-ALIGNED CURRENTS FROM IMAGE/HENA PRESSURE DISTRIBUTIONS

My colleagues and I have made preliminary quantitative comparisons [*Roelof et al., 2004*] between the pressure-driven currents deduced from IMAGE/HENA image inversions (following the method described in this paper) with the FAC intensities measured by the magnetometers on the constellation of Iridium communication satellites (data courtesy of Dr. Brian Anderson of JHU/APL). We find that our estimates of the low-altitude (Region 2) FAC intensities yield the same general morphology in the sub-auroral regions, but the estimated intensities are a factor of 5–50 smaller than those estimated from the Iridium measurements. We now identify four effects that we believe are contributing to the present discrepancies.

Partial pressure. The energy range of the HENA TOF system should capture most (but not all) of the pressure in the ring current region (10–200 keV/nuc). The partial pressure for a given ion species $P(E_1,E_2)$ between two energies E_1 and E_2 is given by

$$P(E_1,E_2) = {}_{E1}\!\int^{E2} dE p j(E)$$

where p is the particle momentum and j is the ion uni-directional differential intensity. If P is the total pressure for that species, then

$$P(E_1,E_2) = f_P\, P$$

where $f_P = P(E_1,E_2)/P$ is the error factor due to the partial pressure. Clearly we have $f_P<1$ for each species.

Gradient smoothing. The algorithm we use [Demajistre et al., 2004] for obtaining the ion distribution $j_{ion}(L,\phi)$ imposes a smoothness condition. Consequently, the gradients in the (partial) pressure will always be underestimated to some degree. It is primarily the azimuthal gradient $(\partial P/\partial \phi)$ that drives the low-altitude field-aligned currents measured by Iridium. Consequently

$$|\partial P(E_1,E_2)/\partial\phi|_{true} = f_G\, |\partial P(E_1,E_2)/\partial\phi|_{measured}$$

where we will always have $f_G<1$.

Anisotropic pitch-angle distribution. If the ion pitch-angle distribution is not isotropic, it is most likely to be a trapping distribution $\boldsymbol{P} = \boldsymbol{I}P_\perp + \boldsymbol{BB}(P_\parallel - P_\perp)/B^2$ with $P_\parallel < P_\perp$. ENA images obtained at moderate to high latitudes would respond more to $P_\parallel$ than $P_\perp$. Therefore when we assume that P is isotropic for the inversion, we will obtain a pressure that is too low

$$P_{inversion} = f_A\, P$$

where $f_A<1$. We outline in the Appendix a computational approach that can accommodate anisotropic ion pitch-angle distributions. It involves the same function Q that is an Euler potential (along with the pressure P in the isotropic case). Unfortunately, however, the more general solution for the current **J** is not expressible purely in terms of Euler potentials and thus does not possess their considerable computational advantages.

Stretched field lines. The assumption of dipolar field lines is certainly not valid in the outer ring current region. At an equatorial distance r, the dipole field $B_d(r)$ will be smaller than the actual stretched field $B(r)$. Suppose we characterize the stretching by a dipolar increment $\Delta L>0$ such that $B(r) = B_d(r-\Delta L)$.

For a dipole field, we have shown that the low-altitude FAC $J_\parallel$ is proportional to $L^5 \partial P/\partial \phi$. Assuming this relation holds for the stretched field lines, we should correct the L-value in the formula to $r-\Delta L$ so as to make the dipole field as strong as the stretched field at $r = L$.

$$J_\parallel^{measured} \cong (1-\Delta L/r)^5\, J_\parallel^{true}$$

So we have $J_\parallel^{measured} = f_S J_\parallel^{true}$ and $f_S \cong (1-\Delta L/r)^5 < 1$.

Cumulative effects. In our initial attempt [Roelof et al., 2004] to compare the calculated low-altitude FACs with those observed on the Iridium satellites (data courtesy of Dr. Brian Anderson, JHU/APL), we found that the FACs measured by Iridium showed a similar MLT dependence in the sub-auroral (Region 2) latitudes, but they were a factor of 5–50 larger than our values calculated from the ion partial-pressure distributions obtained by inverting the IMAGE/HENA images. We now believe that this is a reasonable result in the light of the four error effects analyzed above, each of which is characterized by a factor <1. Since the effects are essentially independent of each other, we can write

$$J_\parallel^{HENA} \sim (f_P f_G f_A f_S)\, J_\parallel^{Iridium} \sim J_\parallel^{Iridium}/(5\text{–}50)$$

Now suppose the factors all have comparable values (f). They will explain the discrepancy if

$$f \sim 1/(5\text{–}50)^{1/4} = 0.67\text{–}0.38$$

Since we consider that each effect is significant, a typical value of about 1/2 for any of the four factors seems reasonable in explaining the observed discrepancy between $J_\parallel^{HENA}$ and $J_\parallel^{Iridium}$ (particularly since we included only the H^+ pressure in our estimates and completely neglected that of O^+)

SUMMARY AND CONCLUSIONS

The Euler potential representation of pressure-driven currents offers a global visualization of this magnetospheric electric current system and its interaction with the ionosphere through field-aligned (Region 2) currents. At present, we have used it only in the approximation of a dipole magnetic field. There is no reason why the same technique cannot be used for a more realistic empirically-based magnetic field model. All that is required in using, for example, a Tysganenko-type model is the additional computational burden for the Euler potential (Q) because one can no longer make use of the simple analytic expressions in the dipole approximation for the Euler potential

(Q). The step up to an anisotropic pressure tensor is also possible in principle (see the Appendix), but it requires extracting the pitch-angle dependence of the ion distribution from the ENA images.

Our first attempts [*Roelof et al., 2004*] to compare our pressure-driven current system with the low-latitude (Region 2) FAC intensities measured by the Iridium magnetometers revealed a significant discrepancy in magnitude that we believe we now understand in terms of the four effects discussed in the previous section.

In the near future, we hope to make a new type of comparison between our estimates of the pressure-driven currents with the curlometer measurements of electric currents made *in situ* within the heart of the ring current region by the four Cluster spacecraft [*Vallat et al., 2005*].

Acknowledgments. This work was supported under NASA contract NAS5-96020, NASA grant NAG5-12722, and NSF grant ATM-0302529. Dr. Brian Anderson of JHU/APL has graciously been providing the Iridium FAC measurements prior to publication for comparison with the estimates from IMAGE/HENA.

APPENDIX: COMPUTATION OF ELECTRIC CURRENTS DRIVEN BY ANISOTROPIC PRESSURE

The pressure balance equation for an anisotropic pressure tensor $\boldsymbol{P}$ is

$$\mathbf{J} \times \mathbf{B} = \nabla \cdot \boldsymbol{P}$$

A solution to this equation is suggested by the existence of the Euler potentials (Q,P) for an isotropic pressure tensor $\text{Tr}(\boldsymbol{P}) = 2P_\perp + P_\parallel \rightarrow 3P$. The Euler potential Q then satisfied the relation

$$\mathbf{B} \cdot \nabla Q = 1$$

Thus the Euler potential was found by integrating $dQ = ds/B$ along differential intervals (ds) along a given field line. For a dipole field, this could be done analytically.

A possible solution for the general case of an anisotropic pressure tensor ($\boldsymbol{P}$) is

$$\mathbf{J} = \nabla Q \times \nabla \cdot \boldsymbol{P} + W\mathbf{B}$$

where W is a (so far) arbitrary scalar function of position. Substitution in the pressure balance equation gives

$$\mathbf{J} \times \mathbf{B} = (\mathbf{B} \cdot \nabla Q)\nabla \cdot \boldsymbol{P} - (\mathbf{B} \cdot \nabla \cdot \boldsymbol{P}) \cdot \nabla Q$$

However (also from the pressure balance equation), we must have $\mathbf{B} \cdot \nabla \cdot \boldsymbol{P} = 0$, so the first term vanishes. Thus if $\mathbf{B} \cdot \nabla Q = 1$, the second term satisfies $\mathbf{J} \times \mathbf{B} = \nabla \cdot \boldsymbol{P}$. This means that Q is the *same* Euler potential we derived for the case of an isotropic pressure, defined completely by the spatial dependence of the magnetic field $\mathbf{B}(\mathbf{r})$.

The reason that the (parallel) term W$\mathbf{B}$ was added to $\mathbf{J}$ is in order to satisfy $\nabla \cdot \mathbf{J} = 0$.

$$\nabla \cdot \mathbf{J} = (\nabla \times \nabla Q) \cdot \nabla \cdot \boldsymbol{P} - (\nabla \times \nabla \cdot \boldsymbol{P}) \cdot \nabla Q + \mathbf{B} \cdot \nabla W = 0$$

Thus W must satisfy the equation

$$\mathbf{B} \cdot \nabla W = (\nabla \times \nabla \cdot \boldsymbol{P}) \cdot \nabla Q$$

which has the same form as that for Q. In other words, the equation

$$dW = (ds/B)(\nabla \times \nabla \cdot \boldsymbol{P}) \cdot \nabla Q$$

can also be integrated along any field line, because all the functions on the RHS are presumed known.

REFERENCES

Demajistre, R., E. C. Roelof, P. C:son Brandt, and D. G. Mitchell, Retrieval of global magnetospheric ion distributions from high-energy neutral atom measurements made by the IMAGE/HENA instrument, *J. Geophys. Res.*, **109**, doi:10.1029/2003JA010322, A04214, 2004.

Roelof, E. C., Remote sensing of the ring current using energetic neutral atoms, *Adv. Space Res.*, Vol. 9, (12)195–(12)203, 1989.

Roelof, E. C., P. C:son Brandt, and D. G. Mitchell, Derivation of currents and diamagnetic effects from global plasma pressure distributions obtained by IMAGE/HENA, *JASP*, **33**(2004), 747–751.

Vallat, C., I. Dandouras, P. C:son Brandt, R. Demajistre, D. G. Mitchell, E. C. Roelof, H. Reme, J.-A. Sauvaud, L. Kistler, C. Mouikis, M. Dunlop, and A. Balogh, First comparisons of local ion measurements in the inner magnetosphere with energetic neutral atom magnetospheric image inversions: Cluster-CIS and IMAGE-HENA observations, *J. Geophys. Res.*, **109**, doi:1029/2003JA010224, A04213, 2004.

Vallat, C., I. Dandouras, M. Dunlop, A. Balogh, E. Lucek, G. K. Parks, M. Wilber, E. C. Roelof, G. Chanteur, and H. Reme, First use of the curlometer technique in the ring current using multi-spacecraft CLUSTER-FGM data: 3D mapping of the inner magnetosphere currents, *Ann. Geophys., submitted*, 2005.

Edmond C. Roelof, Johns Hopkins University/Applied Physics Laboratory, 1100 Johns Hopkins Road, Laurel, MD 20723-6099 (edmond.roelof@jhuapl.edu)

On the Relation Between Electric Fields in the Inner Magnetosphere, Ring Current, Auroral Conductance, and Plasmapause Motion

P. C. Brandt[1], J. Goldstein[2], B. J. Anderson[1], H. Korth[1], T. J. Immel[3],
E. C. Roelof[1], R. DeMajistre[1], D. G. Mitchell[1], B. Sandel[4]

At around 19:30 UT on 17 April 2002, an undulative motion of the plasma-pause was observed by the Extreme Ultraviolet (EUV) camera on board the IMAGE satellite. The motion expanded westward into the duskside and at the same time a ring current pressure increase was observed by the High Energy Neutral Atom (HENA) imager on board IMAGE. The pressure increase was due to a substorm injection with a localized auroral onset at 19:05 UT. It has been shown that the motion of the plasmapause was caused by an electric field set up in the sub-auroral, duskside ionosphere, by the closure of the Region-2 currents, poleward to Region-1 as originally proposed by Anderson et al. (1993). In this paper we discuss the cause of the westward propagation of the motion of the plasmapause, and suggest that it may be caused by the westward expansion of the aurora.

INTRODUCTION

The electric field of the inner magnetosphere determines to a significant extent the dynamics of ions around and below ring current energies. Observations of the global ring current during storms by the High Energy Neutral Atom (HENA) (*Mitchell et al.*, 2000) imager onboard IMAGE (*Burch*, 2000) have revealed an eastward rotation of the peak of the ring current for many storms (*Brandt et al.*, 2002b). It is clear from these observations that the electric field of the inner magnetosphere is not just a superposition of the dawn-to-dusk electric field and the corotational electric field, which has come to be the classical picture. Already *Wolf* (1970) found that the electric field of the inner magnetosphere is a product of the self-consistent closure of the current systems through the ionosphere. During enhanced convection, the electric potential pattern is rotated eastward and there are regions of enhanced and decreased electric fields, as compared to the classical picture. All these deviations from the classical picture are an effect of the coupling between the magnetosphere and ionosphere. In order to understand the cause and effect of the coupling processes, we need to study the global response of the magnetosphere-ionosphere system.

In this paper we study a region of enhanced electric field in the evening side magnetosphere/ionosphere that occurs during enhanced convection. The phenomenon has been known to the ionospheric community for quite some time (*Foster*, 1993; *Foster and Rich*, 1998) and has been termed the Sub-Auroral Polarization Stream (SAPS) (*Foster and Burke*, 2002). It manifests itself as fast (>1000 m/s), sunward, ionospheric drifts at sub-auroral latitudes. This implies that there is a strong (~60 mV/m at ionospheric heights) poleward electric field in the same region. *Anderson et al.* (1993) suggested a production mechanism for the SAPS, which is in agreement with both models and observations. Figure 1 illustrates this mechanism. Enhanced pressure gradients in the ring current region leads to an enhanced downward Region-2 field-aligned current (FAC) in the duskside, sub-auroral ionosphere. Here, the Pedersen conductance dominates over

---

[1] The Johns Hopkins University Applied Physics Laboratory, Laurel, MD
[2] Southwest Research Institute, San Antonio, TX
[3] Space Science Laboratory, Berkeley, CA
[4] Lunar and Planetary Laboratory, Tucscon, AZ

Inner Magnetosphere Interactions: New Perspectives from Imaging
Geophysical Monograph Series 159
Copyright 2005 by the American Geophysical Union.
10.1029/159GM12

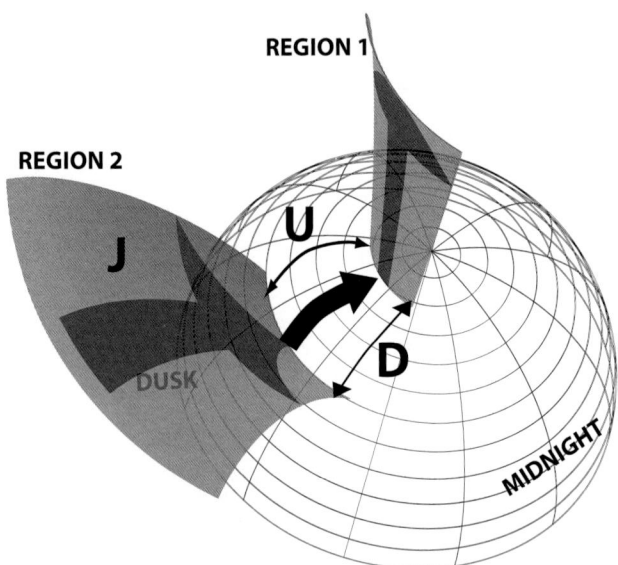

Figure 1. Illustration of the production mechanism of the Sub-Auroral Polarization electric field as proposed by Anderson et al. (1993). The duskside Region-2 FAC is driven by pressure gradients in the ring current. The Region-2 FAC closes through Pedersen currents in the ionosphere through the sub-auroral ionosphere, poleward to the Region-1 current system. Since the poleward currents flow through a region with low conductance, the induced poleward electric field can be significant.

the Hall conductance, so the Region-2 current closes poleward to the Region-1 current system. The conductance in the sub-auroral region is much lower than in the auroral region, which means that even a modest ionospheric closure current flowing across this low-conductance gap, will produce a large potential drop, or, strong poleward electric field. This will translate to a radially outward electric field in the equatorial magnetosphere around the inner edge of the ring current (outer edge of the plasmasphere).

It was not until the Extreme Ultraviolet (EUV) (*Sandel et al.*, 2000) camera onboard IMAGE returned global images of the plasmasphere that the effects of the electric field on the inner magnetosphere could be appreciated on a global scale. Since the dynamics of the plasmasphere are solely governed by electric fields, tracking the motion of the plasmasphere constitutes a powerful tool to retrieve information about the electric field (*Goldstein et al.*, 2004b). Global enhancements of the convection electric field as well as local enhancements of the SAPS electric field modify the dynamics of the plasmapause. *Goldstein et al.* (2003) showed that the motion of the duskside plasmapause required a SAPS-like electric field in addition to the convection electric field to be consistent with observations.

At 19:00–20:00 UT on 17 April 2002, a peculiar motion of the plasmapause was caught by IMAGE/EUV. The motion was consistent with a localized onset of a SAPS-like electric field that expanded in local time as shown in more detail by *Goldstein et al.* (2004a). All instruments on the IMAGE satellite were operating during this event, which allows us to study what controls the evolution of the SAPS-electric field.

First we present the observations of the plasmasphere for this event. Second, we describe the ring current oservations by HENA and how electrical currents can be derived. Third, we discuss how the electric field relates to the pressure-driven FAC and ionospheric conductance, inferred from auroral Far-Ultraviolet (FUV) observations.

PLASMASPHERE OBSERVATIONS—INDICATIONS OF PROPAGATING ELECTRIC FIELD

Figure 2 shows the observations from the EUV imager. A more detailed description of this event has been given by *Goldstein et al.* (2004a). The images show the EUV images projected onto the equatorial plane with the outer dashed line at geosynchronous. The EUV camera consists of three heads, and during this observation the middle head suffered from sunlight contamination. The plasmapause has been traced manually and is marked with a solid line in each image. The dashed lines are the plasmapause locations for each previous image. The image sequence shown tries to capture the undulative motion of the plasmapause.

The first indentation of the pre-midnight plasma-pause was detected at around 19:05 UT. At 19:26–19:36 UT the plasmapause showed a mild outward bulge in the 19–21 MLT sector. Such a motion is indicative of an eastward electric field, which may be related to the sudden increase of ring current pressure and inflation of the magnetic field. This peculiarity will not be discussed here. At 19:46 UT EUV observed an indentation of the eastern edge of the outward bulge, which was formed previously. Then, until 20:30 UT the indentation propagated westward, stripping away the outward bulge. Since it is only an electric field that can modify the shape of the plasmapause on these timescales, this suggests that a localized, radially outward electric field was initiated in the pre-midnight sector, and then gradually propagated or expanded westward.

SOLAR WIND CONDITIONS

The interplanetary magnetic field (IMF) had been fluctuating most of the day of 17 April, but became more southward around 16:00 UT. For our period of interest, the GSE z-component of the IMF increased from southward -20 nT, to a couple of nT's (positive) at around 19:00 UT. This means that a modest ring current pressure was in place before our event. At or shortly after 19:00 UT the IMF B_z decreased to about -14 nT. At the same time (~19:05 UT) a substorm onset could easily be seen in the FUV images onboard IMAGE. The IMF Bz increased gradually after this, reaching about -2 nT around 22:00 UT. This means that we can expect a strong increase in ring current pressure after 19:00 UT.

RING CURRENT OBSERVATIONS

Plate 1 shows the global ring current observations by HENA at the same times as for the EUV images in Figure 2. The upper panel shows the hydrogen ENA images in the 10–60 keV range. These images are projected in an equidistant azimuthal projection extending 100°×100°. This projection is almost identical to how a human eye would see it. Two set of dipole field lines, L=4 and 8, are drawn at noon, dawn, midnight and dusk, marked with the local time. The limb of the Earth is represented by the circle in the center of each image and the terminator line is drawn across it. Note that there is a sharp, and artificial, cut-off in ENA flux a couple of ten degrees in on the dayside. This is caused by the shutter on the HENA imager that protects the detectors from direct sunlight. It blocks out the ENA fluxes coming from the dayside in this particular orbit geometry, but leaves the nightside complete. We can note that later in the orbit the coverage becomes gradually more complete of the dayside.

The lower panel Plate 1 shows the proton distribution in the equatorial plane, obtained by applying a constrained linear inversion technique to the above images. The proton pressure has been derived for 10–81 keV energy range, using images with a 10-minute integration time. The inversion technique has been described in detail by *DeMajistre et al.* (2004) and the results have been validated against Cluster in-situ in data in the 27–39 keV range (*Vallat et al.*, 2004).

At 19:04 UT there was already a modest (partial) ring current in place, and at 19:37 UT we note a significant increase in ENA flux and corresponding increase in the nightside proton pressure. From careful inspection of the ENA images and inversion results the increase started around 19:20 UT and was most likely the result of the substorm injection that had its auroral onset at around 19:05 UT. *Brandt et al.* (2002a) and *Mende et al.* (2002) have shown that the ring current fluxes increase about 10–20 min after the auroral onset.

At 19:57 UT the fluxes (and pressure) remained unchanged at large, which is probably due to the continuous convection still maintained by the significantly southward component of the IMF. At 20:58 UT we can note that the injected proton population has drifted slightly westward, which is expected keeping in mind that the IMF B_Z is gradually increasing during this time, and so we can expect a gradually decreasing convection electric field, which is expected to make the ring current morphology more symmetric.

PRESSURE-DRIVEN CURRENTS

The ring current connects to the ionosphere via the Region-2 current system. This system can be viewed as the three dimensional current system driven by the ring current and plasma sheet pressure distribution. For a typical ring current configuration during enhanced convection as we observe here, the currents follow the isopressure contours in the equatorial plane and there are no FAC in the equatorial plane. Further off from the equatorial plane, the field-aligned component of the currents increases and horse-shoe like current lines connect the magnetospheric current to the ionospheric currents. The pressure-driven FACs flow up from the dawn side ionosphere, connect at lower latitudes with the outer edge of the ring current pressure, and then flow back into the ionosphere on the duskside.

Ionospheric FACs

In order to quantitatively relate the ring current pressure distribution to ionospheric phenomena, it is essential to be able to compute at least a part of the 3D current system from the pressure distributions. With the global ring current observations at hand we have an unprecedented opportunity to perform this calculation. Although this calculation is still in its early development, it will be a good measure of the temporal development of the current intensity with a time resolution of about ten minutes.

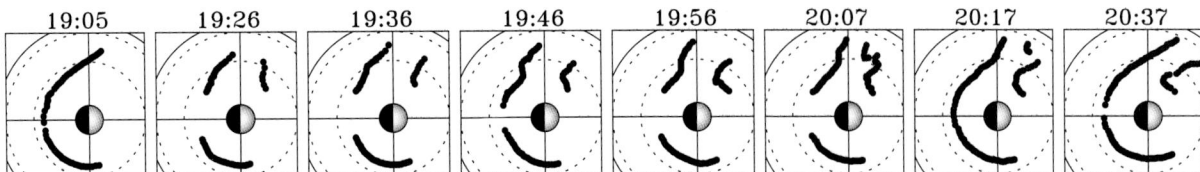

Figure 2. Plasmapause locations in the equatorial plane extracted from observations of the plasmasphere by the Extreme Ultraviolet (EUV) camera onboard IMAGE on 17 April 2002. Note the duskside undulation of the plasmapause 19:26–20:37 UT.

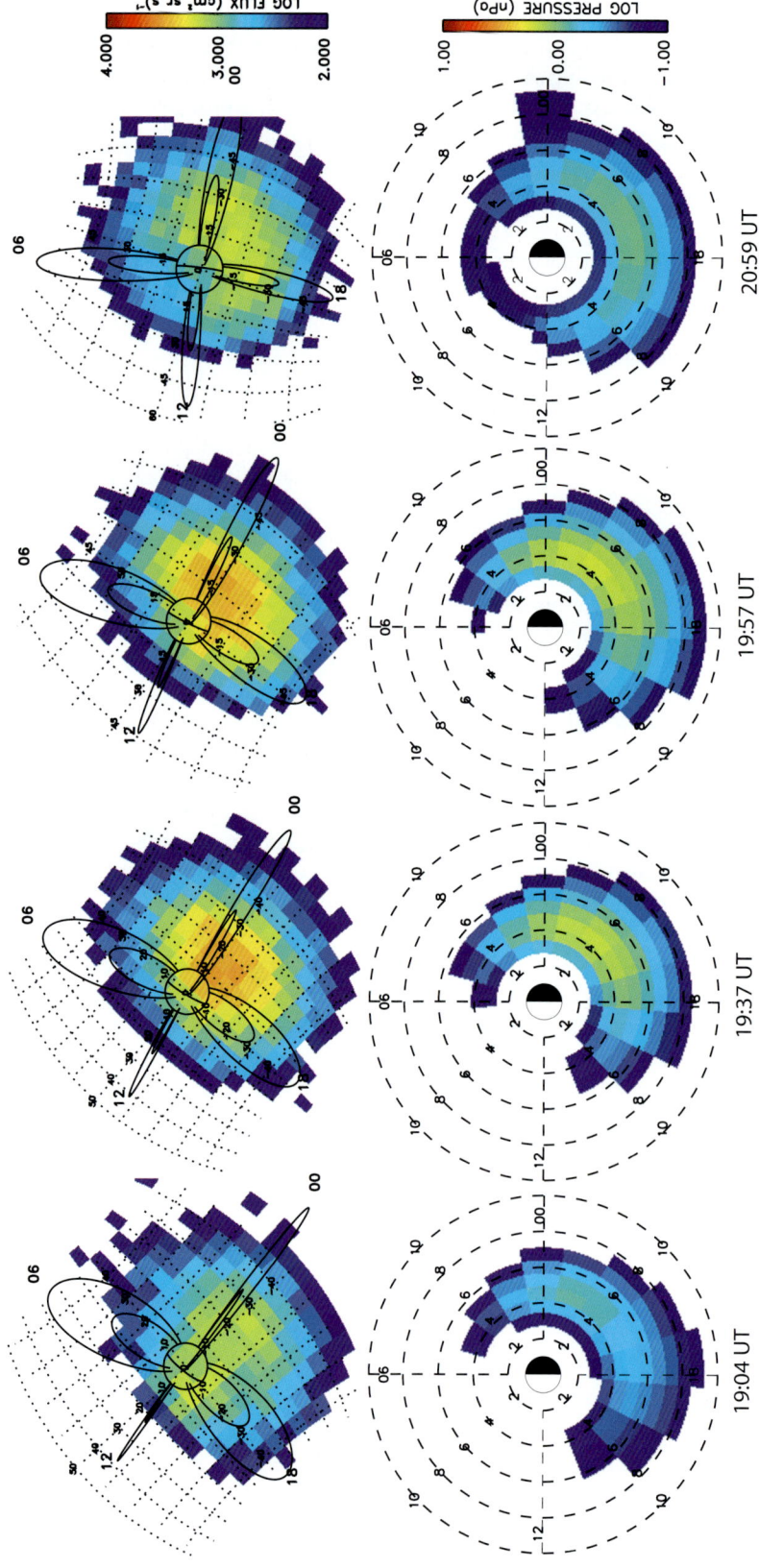

Plate 1. The upper panel shows hydrogen ENA images in the 10–60 keV range. The lower panel shows the result of applying a constrained linear inversion algorithm (*DeMajistre et al.*, 2004) to the ENA images above. The plots show the proton pressure distribution in the 10–81 keV range in the equatorial plane.

The following method was developed by *Roelof* (1989) and applied to HENA data by *Roelof et al.* (2004) and *Brandt et al.* (2004). Below follows a brief description. Once a pressure distribution is obtained, one may compute the 3D current system by solving the force balance equation below.

$$\mathbf{J} \times \mathbf{B} = \nabla \cdot \mathbf{P} \qquad (1)$$

Here **J** is the current density vector, **B** is the vector magnetic flux density, which we assume to be dipolar in this case. Preliminary results using a more realistic magnetic field model will soon be available. The general pressure tensor is denoted **P**, but in this case we will assume it to be isotropic and therefore a scalar P. *Roelof* (1989) realized that if one assumes that the current can be expressed in terms of the following Euler potentials,

$$\mathbf{J} = \nabla Q \times \nabla P \qquad (2)$$

then it turns out that the Euler potential P will simply be the scalar pressure and the potential Q will express the flux-tube volume for a unit magnetic flux, or,

$$Q = \int_s \frac{ds}{B} \qquad (3)$$

where the integral is taken along the magnetic field line.

Plate 2a shows the FAC pattern obtained from the Plate 2 magnetometer data on board the Iridium satellites (*Anderson et al.*, 2000). Red is upgoing (positive) FAC density, and blue is downgoing (negative) FAC density. Only FAC densities above the two-sigma level are displayed. The magnetic latitude is marked and noon is to the left. The pattern was obtained during a 60 min accumulation during 19:30–20:30 UT. The Region-2 FACs are below approximately 60.latitude and are upgoing (red) on the dawnside and downgoing (blue) on the duskside. Plate 2b shows the result the above calculation using the derived HENA pressures in the 10–81 keV range at 19:37 UT. For this we assumed a dipole field and isotropic pressure. A dipole L-shell of 6 maps to about 65.invariant latitude and we see that the HENA-derived FACs are at too high latitudes compared to the Iridium FACs. This is an effect of the dipole approximation used and an implementation of a more realistic magnetic field is being tested during the writing of this paper. The HENA-derived FACs also underestimate the FAC density by about a factor of 5–10, which is an effect of the dipole approximation as well as the lack of complete energy and mass coverage of the ion population that contributes to the plasma pressure. Nevertheless, the HENA-derived FAC should provide a proxy for the true FAC, with a much higher time resolution than Iridium currently can provide.

RELATION BETWEEN PLASMAPAUSE MOTION, IONOSPHERIC ELECTRIC FIELD, CURRENT AND CONDUCTANCE

From Figure 2 we know that the westward motion of the indentation in the plasmapause starts around 19:26 UT. In order to compare to the temporal evolution of the Region-2 FAC we have integrated the HENA-derived FACs over the duskside, northern, ionosphere below 65° magnetic latitude. The result is shown in Figure 3. 65° latitude corresponds to about dipole L-shell Figure 3 of 6, beyond which the uncertainties in the proton pressure retrieved from the HENA images becomes larger. The times of the snapshots of the plasmapause in Figure 2 are indicated by dashed lines in this figure. We see that there is a slight elevation in FAC intensity already at 19:05. At 19:18 UT the FAC intensity display a sharp rise and about 8 min later the indentation of the plasmapause started propopagating westward. The FAC intensity stayed elevated until about 20:30 UT. This temporal sequence is consistent with the mechanism proposed by *Anderson et al.* (1993).

At this point, the question is what governs the undulative motion as seen in Figure 2. Since it is most likely an electric field enhancement that propagates in local time at the plasmapause, we can assume that the same must be true for the electric field pattern at ionospheric heights. There are two factors that govern the electric field in the ionosphere: ionospheric conductance and the FAC entering and exiting the ionosphere. This is expressed by the following equation

$$\nabla \cdot (\Sigma \cdot \nabla \phi) = J_{\parallel}, \qquad (4)$$

where Σ is the height integrated conductance tensor, ϕ is the electric potential, and $J_{\parallel}$ is the scalar FAC perpendicular to the ionosphere.

The morphology and dynamics of the ionospheric electric field therefore depend on these two factors. A propagating electric field can therefore be explained in terms of a propagating conductance pattern and/or FAC pattern. In the simplest example, consider a uniform, downward Region-2 FAC and an isolated region of high conductance at higher latitudes. Since any current closes in such a way to minimize the resistance along its path, the resulting ionospheric current in this case will flow from the footprint of the Region-2 FAC towards the isolated region of conductance. Now, if the region of conductance expands or propagates in local time, the ionospheric current flowing poleward to the conductance region will also expand/propagate in the same direction.

An approximate estimate of the conductance can be obtained through the Robinson formula (*Robinson et al.*, 1988) using the mean energy and energy flux of the aurora.

164 ELECTRIC FIELD IN THE INNER MAGNETOSPHERE

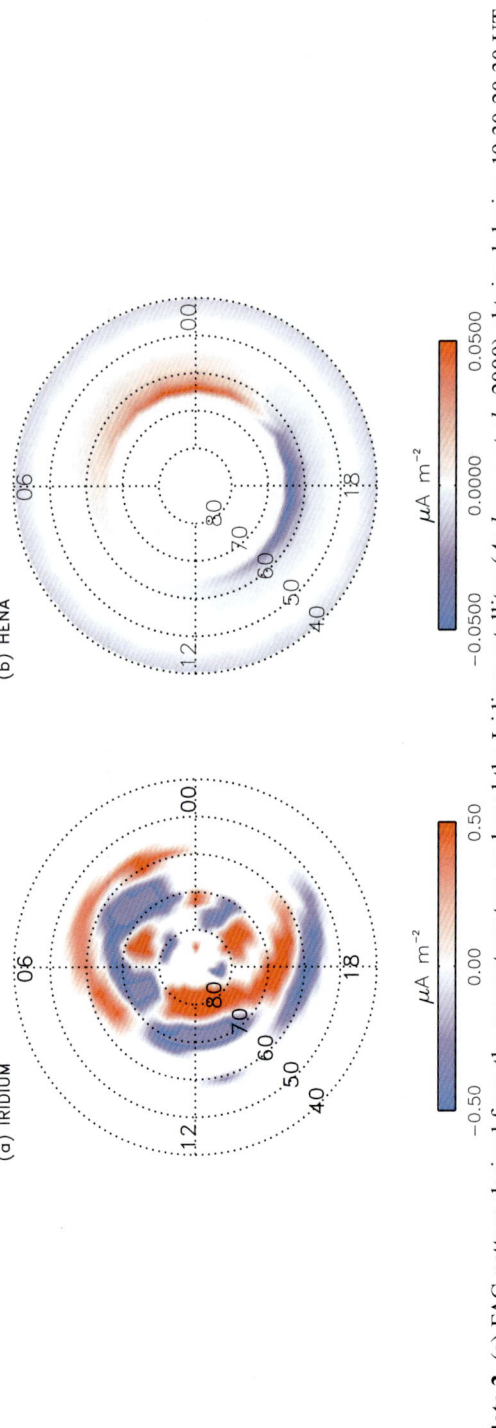

Plate 2. (a) FAC pattern derived from the magnetometers on board the Iridium satellites (*Anderson et al.*, 2000) obtained during 19:30-20:30 UT on 17 April 2002. (b) FAC pattern derived from the HENA-derived global proton pressure (10–81 keV) at 19:37 UT, also shown in the lower panel of Plate 1.

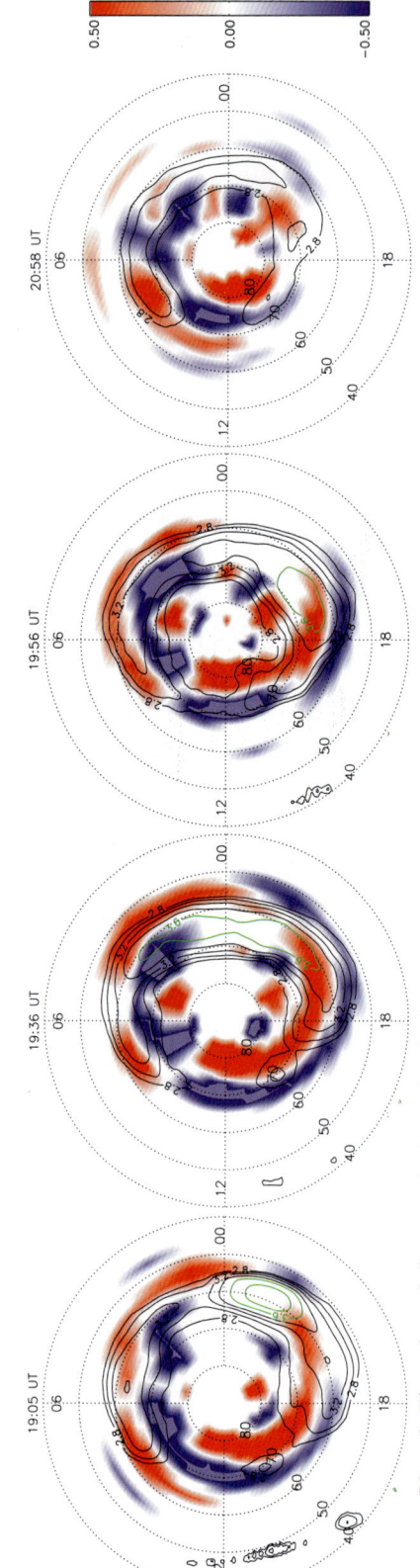

Plate 3. The colorcoded pattern is the FACs obtained by Iridium with a 60 min resolution in a similar format to Figure 2 starting at 18:30, 19:00, 19:30, and 20:30 UT. The overplotted contours show the logarithmic countrates of the FUV/WIC and is a representation of the ionospheric conductance. Note how the contour of maximum countrate (green contour) expands westward into the duskside. This is consistent with the motion of the plasmapause.

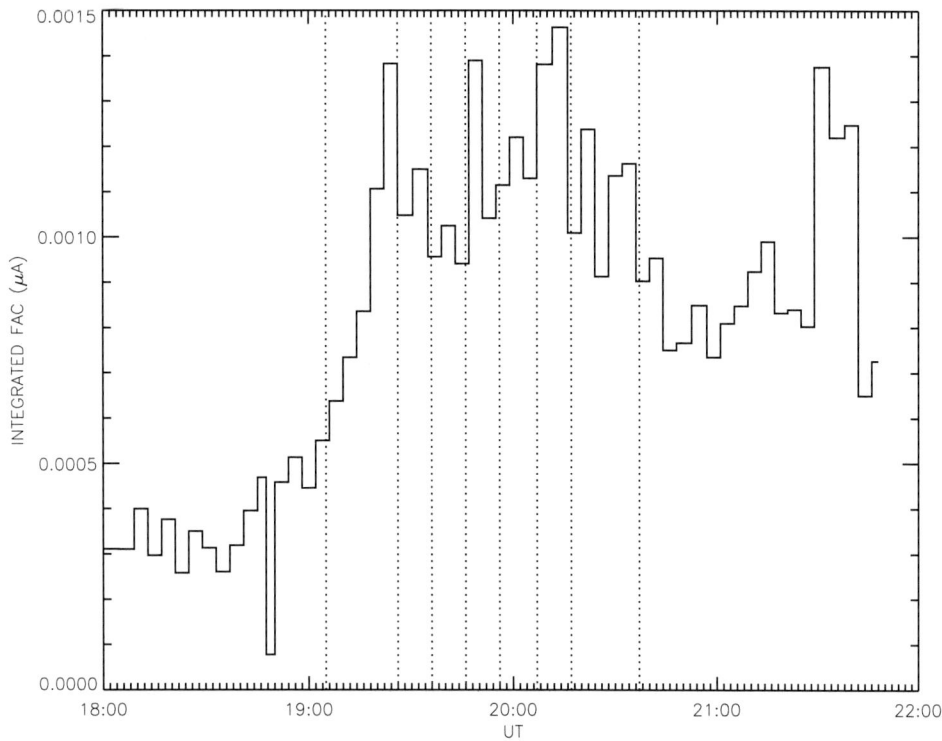

Figure 3. The FAC intensity derived from HENA measurements, integrated over the duskside ionosphere below 65° magnetic latitude. The dashed lines mark the times of the plasmapause snapshots shown in Figure 2.

In this paper we will not estimate the conductance but assume that the conductance is proportional to the intensity of the aurora as measured by the Wideband Imaging Camera (WIC) in the IMAGE/FUV experiment (*Mende et al.*, 2000).

Plate 3 shows for similar plots of the northern ionosphere including the FAC pattern with the FUV/WIC countrate contours in logarithmic scale are overplotted. The FAC pattern was obtained with a 60 min accumulation starting at 18:30, 19:00, 19:30, and 20:30 UT for each image and is displayed in the same format as for Figure 2. Only FAC densities above the two-sigma level are displayed. The FUV/WIC contours were obtained with a 2 min integration time, and a dayglow subtraction has been applied. The contours of the $\geq 10^{3.6}\,\text{s}^{-1}$ are marked in green. We see that the auroral onset starts out around midnight in a localized region and then spreads westward towards the duskside.

The duskside, downward FAC is below the aurora in all the FAC patterns. For the first image (18:30–19:30 UT) we see that there is a very weak downward current. In the second image (19:00-20:00 UT), a significant increase of the downward current can be seen extending from midnight to noon. This is consistent with increase of ring current pressure observed by IMAGE/HENA at the same time. In the third image (19:30–20:30 UT) there is still a downward current, but the most intense region has moved slightly towards the dayside. This is consistent with the ring current pressure drifting slightly in the westward direction as pointed out in the two last images of Figure 1. In the last image (20:30–21:30 UT) the downward current has decreased significantly, which is consistent with the decrease in the pressure gradients as can be seen in the last image of Figure 1.

SUMMARY

We discussed the westward motion of an indentation in the plasmapause around 19:30 UT on 17 April 2002, as was reported by *Goldstein et al.* (2004a), who also showed that the indentation of the plasmapause was caused by the SAPS electric field driven by the Region2 current (*Foster and Burke*, 2002). We showed simultaneous observations of the ring current, the plasmasphere, the aurora, and the ionospheric FACs. Ring current observations were obtained from IMAGE/HENA; plasmasphere observations from IMAGE/EUV, and auroral observations from IMAGE/FUV/WIC. The FACs were obtained from Iridium magnetometer data (*Anderson et al.*, 2000). The ring current proton pressure in the 10-81 keV range was derived from the HENA data using a constrained linear inversion technique (*DeMajistre et al.*, 2004). The pressure was assumed to be isotropic and the FACs in and out of the ionosphere was computed in a dipole field approximation using the force balance equation with an Euler potential solution (*Roelof*, 1989). The HENA derived FACs were integrated over the duskside and their temporal behavior in relation to the plasmapause motion was found to be consistent with the scenario proposed by *Anderson et al.* (1993).

In order to invetigate the cause of the westward motion, we overplotted auroral intensity contours on the FAC pattern obtained by Iridium. During this event, the duskside, sub-auroral, downward Region-2 FAC is enhanced and the auroral onset displays a localized intensification that expands westward. Since the SAPS electric field is the produced by the poleward closure of the Region-2 FAC to the enhanced auroral conductance, these observations indicate that it could be the local-time expansion of the enhanced auroral conductance that leads to the expansion of the electric field, and, hence, the westward expansion of the indentation in the plasmapause.

Acknowledgments. This paper has been funded by NASA and NSF grants NAG5-12722 and ATM-0302529.

REFERENCES

Anderson, B. J., K. Takahashi, and B. A. Toth, Sensing global Birkeland currents with Iridium engineering magnetometer data, *Geophys. Res. Lett., 27* (24), 4045–4048, 2000.

Anderson, P. C., W. B. Hanson, R. A. Heelis, J. D. Craven, D. N. Baker, and L. A. Frank, A proposed production model of the rapid subauroral ion drifts and their relationships to substorm evolution, *J. Geophys. Res., 98* (A4), 6069–6078, 1993.

Brandt, P. C., R. DeMajistre, E. C. Roelof, D. G. Mitchell, and S. Mende, IMAGE/HENA: Global ENA imaging of the plasmasheet and ring current during substorms, *J. Geophys. Res., 107* (A12), 1454, doi: 10.1029/2002JA009307, 2002a.

Brandt, P. C., S. Ohtani, D. G. Mitchell, M. C. Fok, E. C. Roelof, and R. DeMajistre, Global ENA observations of the storm main-phase ring current: Implications for skewed electric fields in the inner magnetosphere, *Geophys. Res. Lett., 29* (20), 1954, doi: 10.1029/2002GL015160, 2002b.

Brandt, P. C., E. C. Roelof, S. Ohtani, D. G. Mitchell, and B. Anderson, IMAGE/HENA: Pressure and current distributions during the 1 October 2002 storm, *Adv. Space Res., 33/5*, 719–722, doi:10.1016/S0273-1177(03)00633-1, 2004.

Burch, J. L. (Ed.), *The IMAGE mission*, Kluwer Academic, reprinted from *Space Sci. Rev.*, vol. 91, Nos. 1–2, 2000.

DeMajistre, R., E. C. Roelof, P. C. Brandt, and D. G. Mitchell, Retrieval of global magnetospheric ion distributions from high energy neutral atom (ENA) measurements by the IMAGE/HENA instrument, *J. Geophys. Res., 109* (A04214), doi:10.1029/2003JA010322, 2004.

Foster, J. C., Storm time plasma transport at middle and high latitudes, *J. Geophys. Res., 98* (A2), 1675–1689, 1993.

Foster, J. C., and W. J. Burke, SAPS: A new categorization for sub-auroral electric fields, *EOS Transactions, 83* (36), 393–394, 2002.

Foster, J. C., and F. J. Rich, Prompt mid-latitude electric field effects during severe geomagnetic storms, *J. Geophys. Res., 103* (A11), 26,367–26,372, 1998.

Goldstein, J., B. R. Sandel, and P. H. Reiff, Control of plasmaspheric dynamics by both convection and sub-auroral polarization stream, *Geophys. Res. Lett., 30*(24), 2243, doi:10.1029/2003GL018,390, 2003.

Goldstein, J., B. R. Sandel, M. R. Hairston, and S. B. Mende, Plasmapause undulation of 17 April 2002, *Geophys. Res. Lett., 31* (L15801), doi:10.1029/2004GL019959, 2004a.

Goldstein, J., R. A. Wolf, B. R. Sandel, and P. H. Reiff, Electric fields deduced from plasmapause motion in IMAGE EUV images, *Geophys. Res. Lett., 31* (1), L01,801, doi:10.1029/2003GL018386, 2004b.

Mende, S. B., H. U. Frey, T. J. Immel, D. G. Mitchell, P. C. Brandt, and J. C. Gerard, Global comparison of magnetospheric ion fluxes and auroral precipitation during a substorm, *Geophys. Res. Lett., 29* (12), 2002.

Mende, S. B., et al., Far ultraviolet imaging from the IMAGE spacecraft. 1. System design, *Space Sci. Rev., 91*, 243–270, 2000.

Mitchell, D. G., et al., High energy neutral atom (HENA) imager for the IMAGE mission, *Space Sci. Rev., 91*, 67– 112, 2000.

Robinson, R. M., R. Vondrak, K. Miller, T. Dabbs, and D. Hardy, On calculating ionospheric conductances from the fkux and energy of precipitating electrons, *J. Geophys. Res., 92*, 2265, 1988.

Roelof, E. C., Remote sensing of the ring current using energetic neutral atoms, *Adv. Space Res., 9* (12), 12,195– 12,203, 1989.

Roelof, E. C., P. C. Brandt, and D. G. Mitchell, Derivation of currents and diamagnetic effects from global pressure distributions obtained by IMAGE/HENA, *Adv. Space Res., 33/5*, 747–751, doi:10.1016/S0273-1177(03)00633-1, 2004.

Sandel, B. R., et al., The extreme ultraviolet imager investigation for the IMAGE mission, *Space Sci. Rev., 91, (1/2)*, 197–242, 2000.

Vallat, C., I. Dandouras, P. C. Brandt, and D. G. Mitchell, First comparison between ring current measurements by Cluster/CIS and IMAGE/HENA, *J. Geophys. Res., 109* (A04213), doi:10.1029/2003JA010224, 2004.

Wolf, R. A., Effects of ionospheric conductivity on convective flow of plasma in the magnetosphere, *J. Geophys. Res., 75* (25), 4677–4698, 1970.

P. C. Brandt, B. J. Anderson, H. Korth, E. C. Roelof, R. DeMajistre, D. G. Mitchell, The Johns Hopkins University Applied Physics Laboratory, 11100 Johns Hopkins Road, Laurel, MD 20723, pontus.brandt@jhuapl.edu

J. Goldstein, Southwest Research Institute, 6220 Culebra Rd, San Antonio, TX 78238, jgoldstein@swri.edu

T. J. Immel, Space Science Laboratory, Centennial Dr @Grizzly Peak, Berkeley, CA 94720-7450, immel@ssl.berkeley.edu

B. R. Sandel, Lunar and Planetary Laboratory, 1541 E University Blvd, Tucson, AZ 85721-0077, sandel@arizona.edu

Small-Scale Structure in the Stormtime Ring Current

Michael W. Liemohn

Atmospheric, Oceanic, and Space Sciences Department, University of Michigan, Ann Arbor

Pontus C. Brandt

Johns Hopkins University Applied Physics Laboratory, Laurel, Maryland

The partial ring current tries to limit itself through the electric field associated with the ionospheric closure currents. The relationship between the hot ion pressure peak and the electric potential structure is examined, and the negative feedback of the stormtime ring current on itself is discussed. To investigate this issue, simulation results for the magnetic storm of April 17, 2002 were analyzed. It was found that the small-scale well-and-peak potential pairs, formed when magnetotail plasma is injected or convected in to the inner magnetosphere, significantly change the near-Earth plasma distribution. Though the potential structures are not always visible in the potential distribution plots or energetic neutral atom images, their effect is noticeable because the main pressure peak is broken into many smaller peaks and the flow pattern of the hot ions is altered. The consequences of these partial ring current-induced potential structures were also discussed. In particular, subauroral polarization streams and injection flow channels were seen in the potential distributions at various times during the storm, formed when the small-scale electric field is aligned with the large-scale electric field in some local region.

1. INTRODUCTION

When hot plasma moves from the magnetotail into the inner magnetosphere, it adiabatically accelerates and thus undergoes additional gradient-curvature drift [*Alfvén and Fälthammar*, 1963]. During times of strong convection, this "partial ring current" easily dominates the plasma pressure in near-Earth space [e.g., *Akasofu and Chapman*, 1964; *Frank et al.*, 1970]. It is a partial ring because the current does not go all the way around the Earth. Instead, the plasma is convected away from Earth on the dayside, is adiabatically de-energized, and then flows to the dayside magnetopause [*Takahashi et al.*, 1990]. The resulting plasma morphology is a pressure crescent that extends across the nightside/duskside of the Earth [e.g., *Ejiri*, 1978; *Liemohn et al.*, 2001a].

The perpendicular current flows along the pressure isobars, but it is inversely proportional to the local magnetic field strength [*Parker*, 1957, 2000]. When the current flow (that is, a pressure isobar) is not parallel to the magnetic field isocontours, currents must flow along the magnetic field to balance this inequality [*Vasyliunas*, 1970]. These field-aligned currents (FACs) flow in to and out of the ionosphere. There, the electric conductivity is high enough that the current can cross the field lines and close the circuit loop [e.g., *Farley*, 1959; *Banks and Kokarts*, 1973]. The latitude and longitude distribution of the conductivity, as well as the distribution of the sources and sinks (the FACs), determine the ionospheric current circulation [e.g., *Heppner*, 1972].

These currents are directly related to the ionospheric electric fields through Ohm's law. Because the inner magnetospheric field lines can be treated as perfect conductors (no electric potential drop along them), the ionospheric electric fields map out to the inner magnetosphere and exert a force on the charged particles there. So, moving hot plasma into the inner magnetosphere by some electric field changes the electric field.

Several studies have examined the self-consistent feedback loop between the subauroral ionosphere and the inner magnetosphere. *Jaggi and Wolf* [1973] predicted severe distortions of the potential pattern near the Earth around local midnight. *Southwood and Wolf* [1978] showed that this feedback takes a few hours to readjust the plasma and the electric field into a stable balance. This final, quasi-equilibrium state is known as the shielding effect [*Spiro and Wolf*, 1984]. During this adjustment interval, undershielding occurs where the high-latitude potential pattern penetrates to low latitudes [e.g., *Spiro et al.*, 1988; *Fejer et al.*, 1990]. Not only does the high-latitude pattern shift to lower latitudes, the subauroral potential pattern is distorted and often amplified, resulting in a different form of penetration electric field [e.g., *Ridley and Liemohn*, 2002]. The main feature of second kind of penetration field is a potential well near midnight. This potential well can be large and causes a severe alteration in the plasma drifts in this region [*Fok et al.*, 2001, 2003; *Khazanov et al.*, 2003; *Jordanova et al.*, 2003; *Garner et al.*, 2004; *Liemohn et al.*, 2004].

Indeed, with every injection of a blob of plasma into the inner magnetosphere, a potential well develops at the eastern end of the pressure crescent and a potential peak develops at the western end (according to the physical process described above). Figure 1 shows a schematic illustration of this phenomenon. The electric fields point in toward the well and out away from the peak. Because the magnetic field in the magnetospheric equatorial plane points northward, the resulting plasma drifts are counter-clockwise around the well and clockwise around the peak. So, with every injection, a vortex pair is formed that alters the local flow of plasma. *Sazykin et al.* [2002] investigated similar vortices as a cause of structure in the hot plasma morphology. In that case, however, the vortices were from the interchange instability. The flow around the western end of the plasma pressure peak is said to be the cause of the skewed ion distributions (shifting the peak to the post-midnight sector) seen in the energetic neutral atom images [*Brandt et al.*, 2002].

As seen in Figure 1, the flows from the twin vortices merge in the middle, creating an outward flow co-located with the peak of the pressure crescent. Therefore, the mere existence of the pressure peak creates a flow pattern

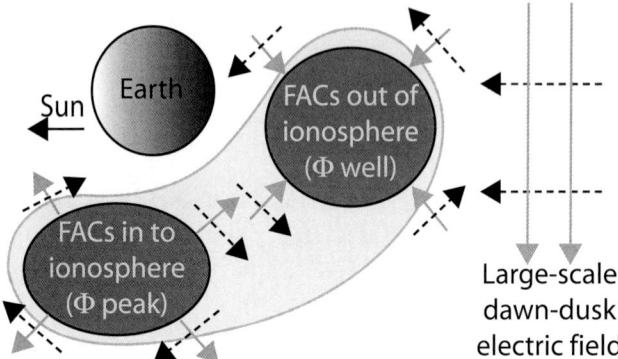

Figure 1. Schematic of the relationship between the partial ring current, the field-aligned closure currents, and the strong electric fields in the inner magnetosphere. The large, lightly-shaded region is an example partial ring current (out to some arbitrary isobar). The two darker shaded regions are the regions of strong field-aligned current (again, drawn out to some arbitrary FAC level). The lightly-shaded arrows show the electric fields associated with the large-scale potential difference (big arrows on the right), and the potential peak and well associated with the FAC regions. The dark, dashed arrows show the ExB drift direction for each electric field arrow.

that acts to expel the peak from the inner magnetosphere. However, only in the middle of the pressure crescent, where the two vortices meet, do strong outward flows exist. The rest of the pressure crescent does not experience expelling forces, and the ends of the crescent actually are pushed in towards the Earth.

The story above describes the basic scenario of what happens when hot plasma is moved into the inner magnetosphere. The details of this process, however, are not well understood. For instance, are any of these electric fields significant relative to the large-scale dawn-dusk convection electric field? Remember that at this time, shielding is broken down and the large-scale field can penetrate deep into the inner magnetosphere (that is, to low latitudes). If so, when are the vortex electric fields significant and how big are they?

The study presented here addresses these questions with the use of a kinetic transport model. A particular case study, the April 17, 2002 magnetic storm, will be examined in detail in the next section. The following section gives a general discussion of these results, and relates it to previous work on the inner magnetospheric electric field structure. It is found that the ring current is self-limiting; the fields are locally quite strong and the net result of the negative feedback loop is that the initially-large pressure peaks are broken up into small-scale peaks and valleys, eventually forming a rough but symmetric ring current in the late recovery phase of the storm.

2. EXAMPLE CASE STUDY: THE APRIL 17, 2002 STORM

The April 17, 2002 magnetic storm was caused by the initial interplanetary shock and sheath passage of the multi-storm sequence that month. At least 4 interplanetary coronal mass ejections (ICMEs) struck the magnetosphere from the 17th to the 24th, with varying degrees of geomagnetic activity associated with each hit. The first shock arrived at ~11:55 UT on April 17, and the subsequent sheath passage lasted over 12 hours (followed by a magnetic cloud, which caused the second storm in the sequence). The solar wind velocity jumped from ~350 km/s to ~500 km/s, and the solar wind density exceeded 20 cm^{-3} for over 6 hours, with spikes up to 60 cm^{-3}. The interplanetary magnetic field (IMF) oscillated wildly during the sheath, especially during the first 8 hours, with southward excursions of up to −25 nT. This solar wind disturbance caused a magnetic storm with a Dst minimum value of −105 nT (at 17:00 UT), recovering to roughly -50 nT early on April 18 (when the cloud arrived and the second storm began). *Liemohn et al.* [2004] provides more details of the solar wind and geophysical conditions during this event.

This storm was modeled with the version of the ring current-atmosphere interaction model (RAM) described by *Liemohn et al.* [2004]. This version of RAM, based on earlier versions by *Fok et al.* [1993], *Jordanova et al.* [1996], and *Liemohn et al.* [1999], solves the time-dependent, gyration- and bounce-averaged kinetic equation for the phase-space density $f(t, R, \varphi, E, \mu_0)$ of one or more ring current species. The five independent variables are time, geocentric radial distance in the equatorial plane, magnetic local time, kinetic energy, and cosine of the equatorial pitch angle. The code includes collisionless drifts, energy loss and pitch angle scattering due to Coulomb collisions with the thermal plasma (densities from the *Ober et al.* [1997] model), charge exchange loss with the hydrogen geocorona (densities from the *Rairden et al.* [1986] model), and precipitative loss to the upper atmosphere. Solution of the kinetic equation is accomplished by replacing the derivatives with second-order accurate, finite volume, numerical operators. Note that this is not a particle-tracking code but actually a "fluid" calculation, with the "fluids" being the several million grid cells in phase space for each plasma species.

The source term for the phase space density calculated by RAM is the outer simulation boundary, where observed particle fluxes from the magnetospheric plasma analyzer (MPA) [*McComas et al.*, 1993] and synchronously orbiting plasma analyzer (SOPA) [*Belian et al.*, 1992] instruments on the LANL geosynchronous-orbit satellites are applied as input functions. Variations in the observed plasma sheet density at a single satellite are assumed to represent temporal variations of a spatially uniform nightside plasma sheet. Data gaps are filled in using data from earlier and later local times if the ion data is not significantly degraded by losses. The composition of the inner plasma sheet is assumed to vary with solar and magnetic activity according to the statistical relationship derived by *Young et al.* [1982]. Additional details of the present state of RAM are presented by *Liemohn et al.* [1999, 2001a, b, 2004].

The electric field description is an important component of RAM. Simulations with two different field descriptions are discussed in this study. The first is a self-consistent electric field inside the simulation domain. That is, field-aligned currents calculated from the hot ion results are used as a source term in a Poisson equation solution for the ionospheric potential. A high-latitude boundary condition from the Weimer potential model [*Weimer*, 1996] is used at ~72°. A time-varying conductance pattern is also prescribed. Specifically, a static but spatially nonuniform dayside and nightside conductance pattern is defined, and then a dynamic "smooth auroral oval" ring of conductance is also applied. These rings (north and south hemispheres) vary with the location and strength of the field-aligned currents calculated from RAM. Please see *Ridley et al.* [2004] and *Liemohn et al.* [2004] for additional details of the conductance pattern. The second electric field description used in this study is the *Weimer* [1996] pattern, applied everywhere instead of just at the high-latitude boundary. Therefore, the two potential descriptions have the same large-scale, dawn-dusk electric field, but different inner magnetospheric patterns.

2.1. Hot Ion Pressures

Plates 1 and 2 show the calculated hot ion (H$^+$ plus O$^+$) pressure distributions at 12 times throughout the April 17 magnetic storm. The maximum total energy content (that is, the integral of this pressure over the simulation domain) occurs at 17:00 UT for the self-consistent electric field run (Plate 1) and at 17:30 UT for the Weimer-96 run (Plate 2). The main similarity between Plates 1 and 2 is that the pressure peaks in the evening sector throughout the main phase and first few hours of the recovery phase, but then this peak weakens and moves around the Earth during the late recovery phase. A primary difference between these simulation results is that Plate 1 shows far more small-scale structure than seen in Plate 2. The pressure peak in Plate 2 is always a well-defined, very smooth crescent. It is highly asymmetric during the main phase of the storm (before the peak), and relaxes to a nearly symmetric local time pressure distribution by 21:00 UT.

In contrast, Plate 1 reveals a somewhat larger symmetric ring current component during the main phase but a stronger

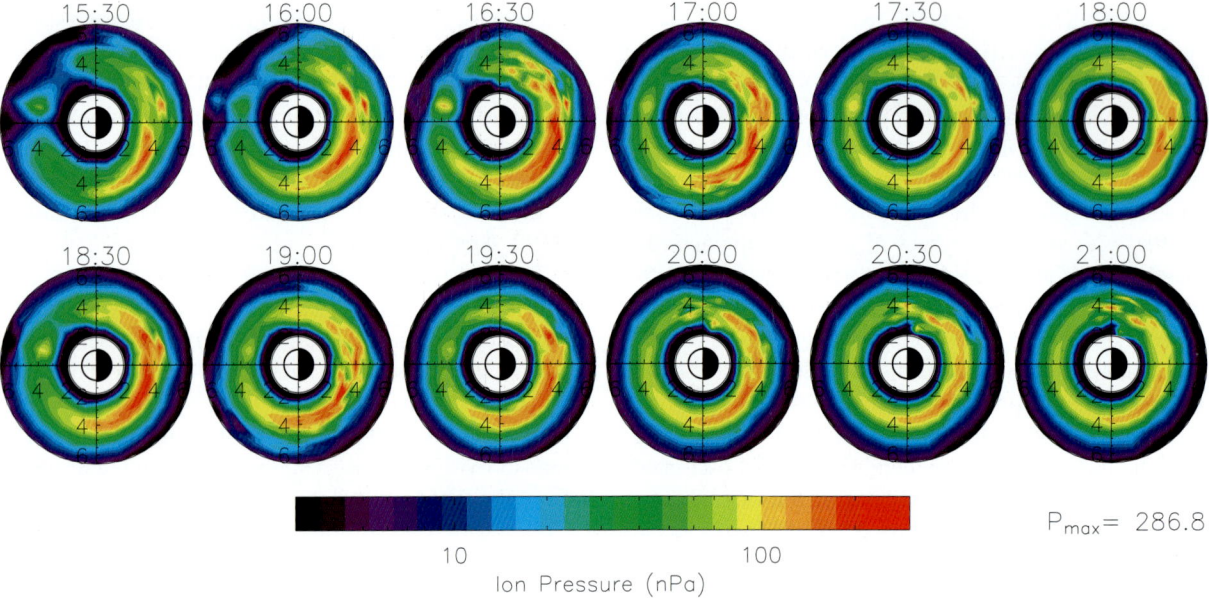

Plate 1. Equatorial plane plasma pressures in the inner magnetosphere for the simulation with the self-consistent electric field description throughout the simulation domain. The view in each subplot is over the north pole, with the sun to the left and dusk down, with distances in Earth radii. The maximum hot ion pressure in these 12 dial plots is listed in the lower right.

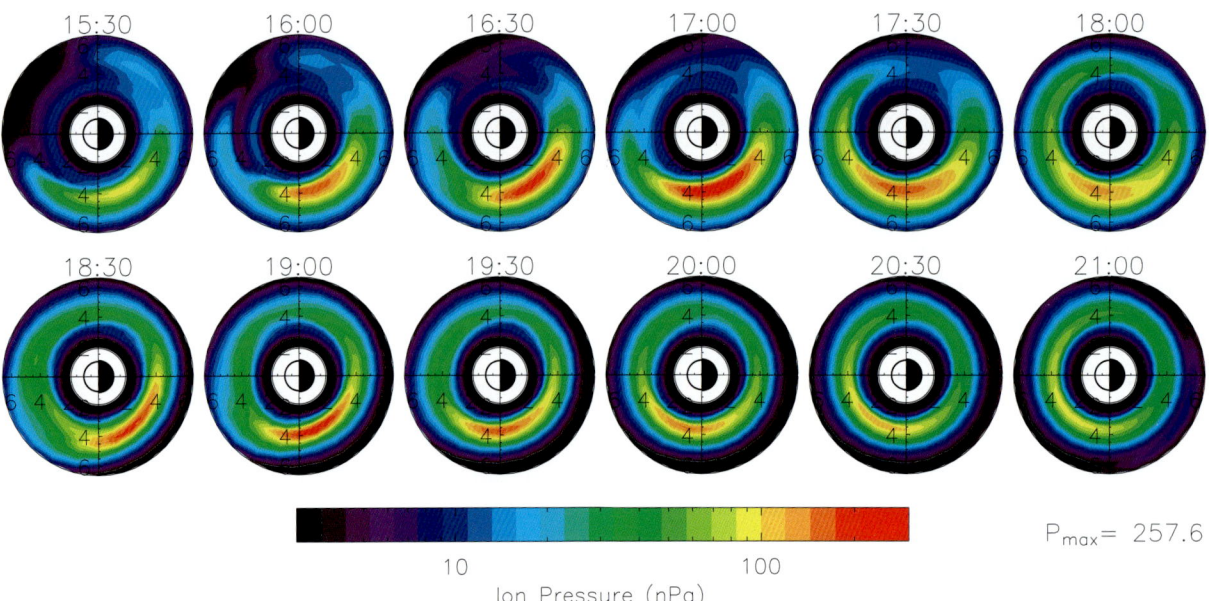

Plate 2. Like Plate 1, plasma pressures in the inner magnetosphere for the simulation with the Weimer-96 electric field description throughout the simulation domain.

partial ring current component in the recovery phase (relative to that in Plate 2). There are many localized peaks within the main pressure peak. In particular, it appears that the pressure peaks are azimuthally long and radially narrow on the duskside inner magnetosphere but more spot-like in structure on the dawnside. Another primary difference between Plates 1 and 2 is that the pressure peak in the self-consistent run (Plate 1) extends much farther eastward (toward dawn) than that created by the Weimer-96 electric fields.

In general, the hot ion pressure values from the two results are quite similar in magnitude. The peak value for all 12 times shown (listed in the lower right corner of each plate) is ~10% bigger for the self-consistent electric field result than for the Weimer-96 electric field result. The reason for this is that the vortices trap some of the plasma on the nightside and don't let it convect as easily through the inner magnetosphere. There is less convective trapping with the Weimer E-field. The rotational flow of the vortices also means that particles are injected more deeply in localized regions of the nightside, which leads to extra adiabatic acceleration and a (slightly) higher peak pressure value.

2.2. Electric Potentials

Figures 2 and 3 show the electric potential distribution in the simulation domain for the self-consistent and Weimer-96 simulations, respectively. Figure 2 contains far more small-scale potential structure than does Figure 3. The main feature in the plots of Figure 3 that resembles the potential well-and-peak pairs from plasma injection is the extension of the duskside potential well over to the midnight meridian near 4 R_E. Otherwise, Figure 3 shows the standard two-cell convection pattern (a dawn-to-dusk electric field) that is very strong in the main phase of the storm and tapers off to much lower values in the recovery phase.

Figure 2 also has this basic pattern, with a strong large-scale convection field in the main phase that weakens in the recovery phase. However, it also contains several (sometimes many) local maxima and minima. At 17:00 UT, the brief interval of weakened large-scale convection reveals a rather strong overshielding potential pattern throughout the inner magnetosphere (causing reverse convection) and significant small-scale potential structure near dusk. When the large-scale convection field is strong, however, the only consistent feature is that seen in the Weimer-96 potential patterns: the extension of the duskside potential well towards (or past) midnight near the Earth. A substantial difference between Figure 2 and 3 is that the complexity of the self-consistent potential distribution persists throughout the recovery phase of the storm. In fact, because of the weakening large-scale convection field, the potential wells and peaks from the localized pressure peaks are much more noticeable in the lower row of plots in Plate 4 than in the upper row.

2.3. Observational Perspective

It is useful to consider what data might be able to tell us about the existence of these small-scale features in the hot ion

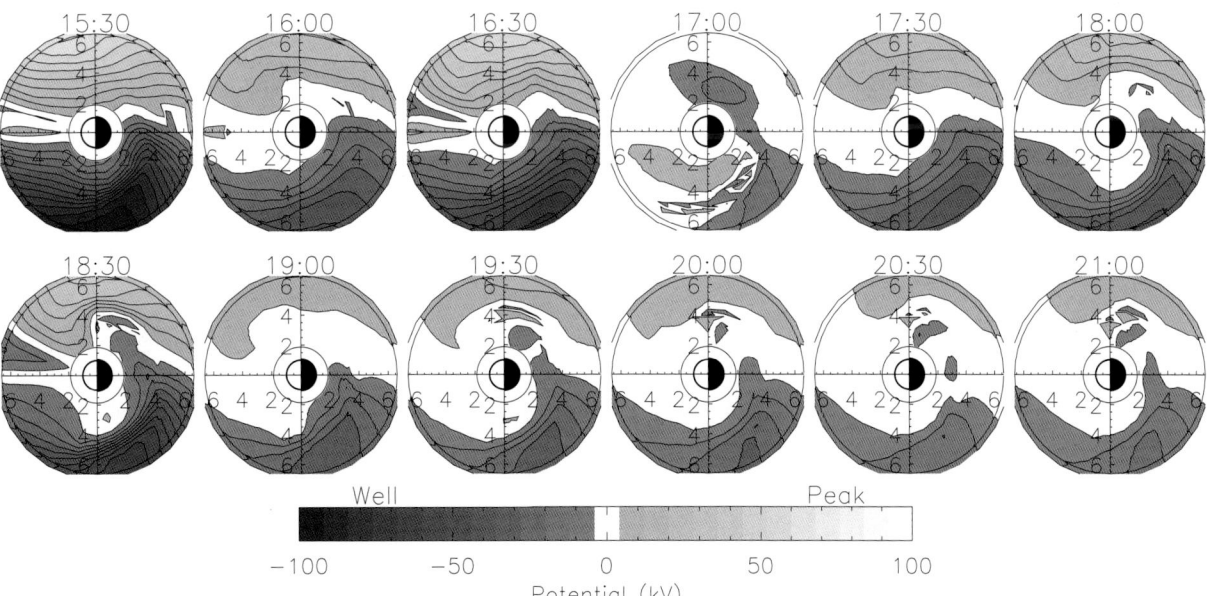

Figure 2. Electric potentials in the inner magnetosphere for the simulation with the self-consistent electric fields. The format is like Plate 1. Contours are drawn every 8 kV.

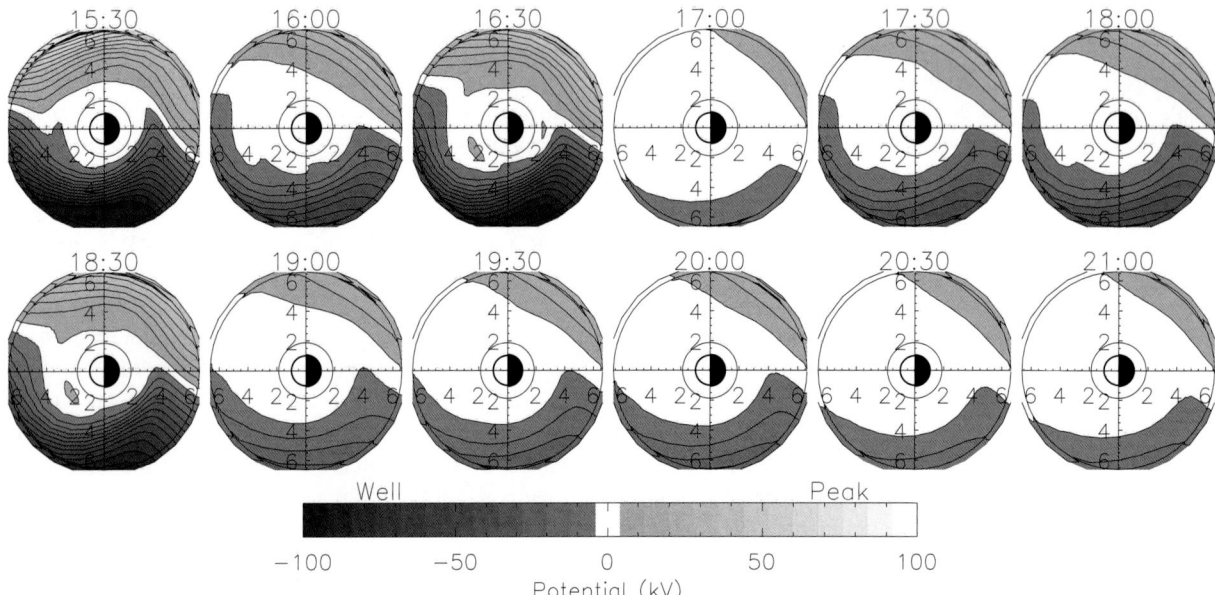

Figure 3. Like Figure 2, electric potentials in the inner magnetosphere for the simulation with the Weimer-96 electric fields.

pressure distribution of the stormtime ring current. A convenient data set is that from the HENA instrument [*Mitchell et al.*, 2000] onboard the IMAGE spacecraft [*Burch*, 2000]. This instrument records energetic neutral atoms, a bi-product of the charge exchange decay mechanism of the ring current, and is able to view the entire inner magnetosphere at a relatively high time cadence. For this study, a single example data-model comparison will be shown in order to highlight a few critical points.

The selected data for this comparison is a 10-minute integral image of hydrogen ENAs from 20:00 UT on April 17 in the 39–60 keV energy range. Plate 3 presents the data-model comparisons. Plates 3a and 3c show the equatorial plane hot ion fluxes from the Weimer-96 and self-consistent electric field simulations, respectively (averaged over pitch angle and energy in the 39–60 keV range). Plate 3b and 3d show the simulated ENA fluxes that HENA would have seen from the ion distributions in Plates 3a and 3c, respectively, by passing the ion fluxes through a forward-modeling routine (assuming pitch-angle isotropy and a dipole magnetic field). Plate 3e shows the HENA observations for this time and energy range. Note that the 3 ENA images in Plate 3 only show pixels in which the line of sight passes through the magnetic equatorial plane between 2 and 5.5 R_E geocentric distance. This truncation of the images focuses the presentation on the spatial region where the data-model comparison is most valid. For more details on the forward-modeling routine, please see *DeMajistre et al.* [2004]. For more details on the data-model comparison technique, please see *Liemohn et al.* [2005].

There are several features of Plate 3 that should be highlighted. One significant result is that the small scale features in the ion flux distribution of Plate 3c do not appear in the corresponding forward-modeled ENA flux distribution (Plate 3d). The lines of sight for each pixel pass through many flux tubes, and small-scale features are lost. The HENA image also shows no small-scale structure and it is likely that HENA cannot resolve it, so it cannot be used to prove or disprove the existence of these small-scale features.

A second point is that the ion flux values are very similar between the 2 simulations (compare Plates 3a and 3c). The self-consistent results have a flux peak in the evening sector and show significant structure while the Weimer results have a flux peak in the afternoon sector without much small-scale structure. The peak magnitudes, however, are almost identical.

A third point is that the simulated ENA fluxes from the self-consistent result (Plate 3d) are much closer in magnitude and morphology to the observed ENA fluxes (Plate 3e) than are those from the Weimer electric field simulation (Plate 3b). This isn't always the case (that is, for other times and energy channels, the results using the Weimer-96 field are sometimes a better match), but the example result shown in Plate 3 provides some amount of validation for the self-consistent result.

3. DISCUSSION

The purpose of this study was to assess the relative magnitude and the timing of the potential well-and-peak pairs

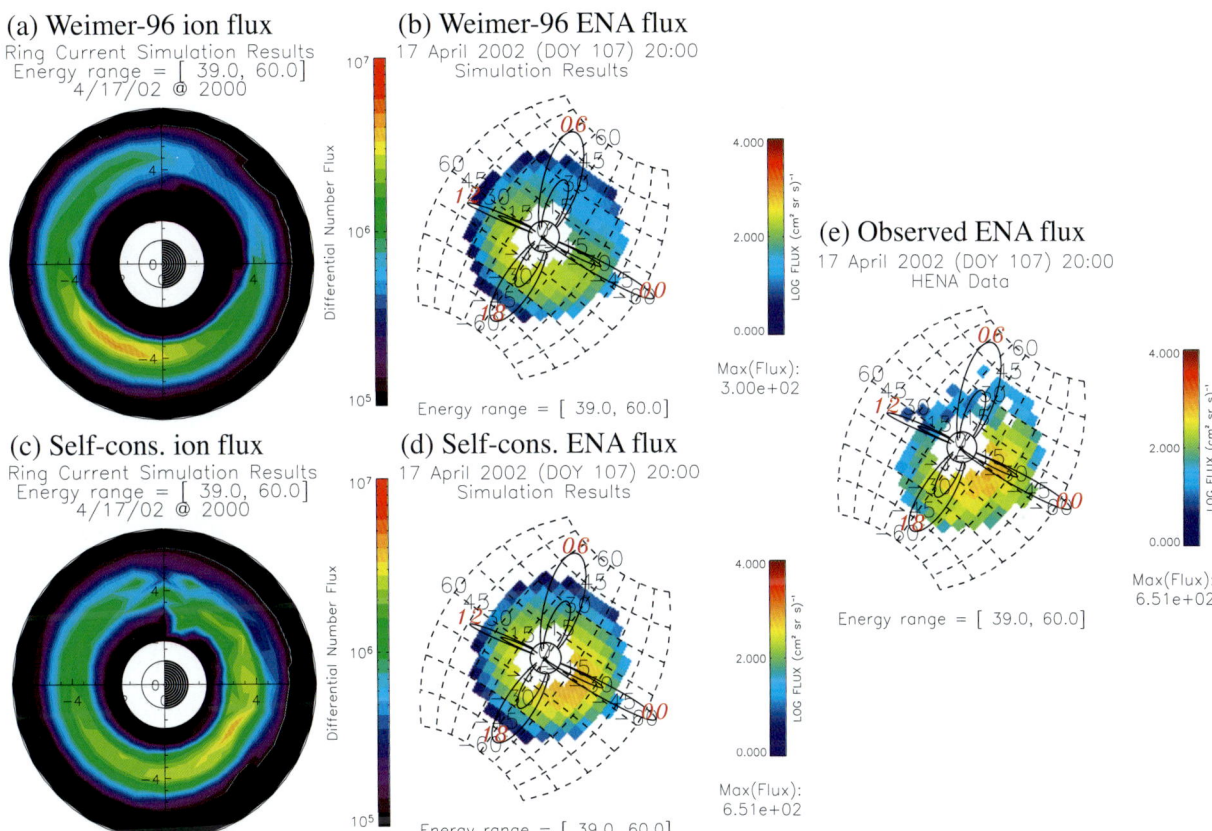

Plate 3. Data-model comparisons between RAM results and HENA observations for 20:00 UT on April 17. (a) Differential ion fluxes in the 39-60 keV energy range for the simulation with the Weimer-96 electric field description, and (b) the corresponding forward-modeled ENA fluxes from the IMAGE-HENA perspective. Similar plots for the simulation with the self-consistent electric field are shown in (c) and (d), respectively. Shown in (e) are the observed ENA fluxes from HENA for this energy channel. The ENA images only show pixels with lines of sight that cross the magnetic equatorial plane between 2 and 5.5 R_E geocentric distance. The view is from over the North Pole and slightly anti-sunward, with the Sun direction to the left and slightly upward, as indicated.

generated around localized plasma pressure peaks in the inner magnetosphere. The answer is that they appear to always be important, as evidenced by the structure in the plasma pressure distributions plotted in Figure 1. However, they are not particularly visible in the potential pattern (at least on the chosen colorscale, with contours every 8 kV) when the large-scale convection field is strong. Furthermore, the resulting small-scale features in the ion flux and pressure distributions are not visible in the corresponding simulated ENA images.

The net effect of the self-consistent potential structures in the inner magnetosphere is that the injected plasma cannot simply convect around the duskside of the planet as a smooth and continuous pressure crescent. Some regions stagnate, others are pushed outward or inward, and still others are westwardly convected faster than expected. While the self-consistent electric fields are not strong enough to completely stop the large-scale Earthward flow of plasma sheet material into the inner magnetosphere, they are strong enough to destroy the coherence of the pressure peak. Potential vortices can exist right in the heart of the pressure peak and, assuming a certain conductance distribution in the inner magnetosphere, these potential well-and-peak pairs can significantly change the flow of plasma in the inner magnetosphere.

These differences in the hot ion pressure distribution between these two simulations are entirely consistent with the narrative in the previous section. The Weimer-96 potential model was compiled from observations but is sorted by solar wind and IMF conditions. Therefore, the transient, small-scale potential vortices associated with plasma injections are not resolved in the model. In fact, they largely cancel themselves out, because the injections can be localized or broad, and the plasma (and also the potential vortices) sweep through the inner magnetosphere.

Note that there can be very strong "inward" gradients in the potential in the self-consistent simulation (that is, westward flows). For instance, at 15:30, the electric fields in the evening sector near 4 R_E reach 3 mV/m, and they reach 2 mV/m at 16:30 and 18:30 UT. These strong dawn-dusk electric fields have been observed both in the topside ionosphere [*Yeh et al.*, 1991; *Anderson et al.*, 1991, 2001; *Foster and Vo*, 2002] and near the equatorial plane [*Rowland and Wygant*, 1998; *Wygant et al.*, 1998; *Burke et al.*, 1998]. They are so strong because at these places and at these times, a strong pressure-peak-induced flow is adding to a strong large-scale convection field. These regions of strong westward flow (that is, outward electric field in the magnetosphere) are known as sub-auroral polarization streams (SAPS) [*Foster and Burke*, 2002]. In these simulations, they are formed as a direct consequence of the additive superposition of the large-scale field and one of the small-scale fields in the inner magnetosphere during the storm. The SAPS can wax and wane and move around over the course of the storm, but the place where two fields add together to form the strongest westward drifts is near dusk.

Similarly, strong westward electric fields are occasionally generated in localized regions (particularly the evening sector) near the outer simulation boundary. This are exactly analogous to the "flow channels" described by *Chen et al.* [2003] and *Khazanov et al.* [2003, 2004] in both self-consistent electric field calculations and in AMIE potential distributions. Note that, in these simulation results, they are not induced electric fields formed by substorm dipolarizations, but rather potential electric fields formed by the additive effect of the large-scale field and a localized pressure peak field. In Figure 2, these flow channels are mainly seen in the evening sector and sometimes in the pre-dawn sector, just where they were seen in the previous studies.

It is interesting that the self-consistent potential pattern shows more fine-scale structure in the recovery phase than in the main phase (Figure 2). This is because some of the injected plasma is essentially trapped on the nightside, unable to convect sunward because of the pressure peak induced electric fields. There is a timescale for the nightside peak to eventually dissolve into a symmetric ring. In the meantime, though, the pressure peak is like a hydra: as the vortex flows break up one pressure peak, the result is two smaller peaks, each with their own vortex pairs. Westward drift is, of course, occurring throughout the storm (both main phase and recovery phase). The small-scale vortices simply alter these flows and hinder some of the plasma from completing its circuit.

Let us return to the schematic diagram of the process being discussed (Figure 1). The largest shaded region denotes the pressure peak (out to some arbitrary isobar). The partial ring current then flows westward in the outer half of this region and eastward in the inner half, turning around at each end. The inward-directed magnetic field gradient, however, means that the westward partial ring current is stronger than its eastward counterpart. The two smaller (and darker) shaded regions give the approximate locations of the FACs in to and out of the ionosphere that close the unbalanced portion of the partial ring current. There is a potential well coincident with the FACs flowing out of the ionosphere, and there is a potential peak coincident with the FACs flowing in to the ionosphere. Electric fields then radiate out from the potential peak and in towards the potential well. The resulting ExB drift is then a pair of vortex flows, counterclockwise around the well and clockwise around the peak. This is exactly like the high-latitude two-cell convection pattern formed by the region 1 current system, except these vortex pairs are much smaller. Right over the central region of the pressure peak is an outward flow where the two vortices overlap. The strength of the vortices, of course, depends on many factors, most notably the plasma intensity, energy and pitch angle distribution, and the ionospheric conductivity distribution. For the nominal conditions chosen for the simulation presented in this study, the vortices were strong enough to create significant and noticeable structure in the plasma pressure distribution in the inner magnetosphere throughout the storm event.

4. CONCLUSIONS

This study addressed the question of the timing and magnitude of the potential structures formed by the closure of the partial ring current. To investigate this issue, simulation results for the magnetic storm of April 17, 2002 were analyzed. It was found that the small-scale well-and-peak potential pairs, formed when magnetotail plasma is injected/convected in to the inner magnetosphere, significantly change the near-Earth pressure distribution. In particular, the main pressure peak is broken into many smaller peaks. The main phase asymmetric pressure peak also extends farther eastward towards dawn. In addition, relative to non-self-consistent results, the symmetric component of the ring current is larger in the main phase (although it is still smaller than the partial ring current) and the partial ring current is larger in the recovery phase (although it is still smaller than the symmetric ring current at this time). While the small-scale potential structures were not always visible in the plots, especially when the large-scale convection electric field was strong, the fine structure in the plasma pressure distribution was always visible. However, these features are not present in the simulated ENA images, and the HENA observations are also devoid of such features. Therefore, it is unresolved whether these small-scale structures in the hot ion distribution of the stormtime ring current are real. The magnitude and morphology of the simulated ENA fluxes are quite close to the observed values, though. This lends support to the concept of ring-current generated potential vortices inhibiting the rapid flow of particles through the inner magnetosphere.

The consequences of these partial ring current-induced potential structures were also discussed. In particular, SAPS and flow channels were seen in the potential distributions at various times during the storm. They form when the small-scale electric field is aligned with the large-scale electric field in some local region. The primary regions of formation (dusk and evening, respectively) are consistent with previous studies of these phenomena. While both the SAPS and flow channels wax and wane and move around throughout the event, they are a persistent feature of the stormtime potential pattern.

Acknowledgments. The authors would like to thank the sources of funding for this study: NASA grants NAG5-10297, NAG-10850, and NAG-12772 and NSF grants ATM-0090165 and ATM-0302529. Liemohn would also like to thank the organizers of the Yosemite 2004 Workshop for inviting him to present these results at the meeting. In addition, Liemohn thanks Drs. J. U. Kozyra, A. J. Ridley, and G. V. Khazanov for very useful discussions on this topic. Finally, we would like to thank all of the data providers who made the ring current simulations possible, especially M. F. Thomsen and G. D. Reeves at the Los Alamos National Laboratory, the Kyoto World Data Center for the Kp and Dst index, and CDAWeb for allowing access to the Level 2 plasma and magnetic field data of the ACE spacecraft (and to D. J. McComas and N. Ness for providing their data to CDAWeb).

REFERENCES

Akasofu, S.-I., and S. Chapman, On the asymmetric development of magnetic storm fields in low and middle latitudes, *Planet. Space Sci., 12,* 607, 1964.

Alfvén, H., and C.-G. Fälthammar, *Cosmical Electrodynamics,* Oxford University Press, London, 1963.

Anderson, P. C., W. B. Hanson, and R. A. Heelis, The ionospheric signatures of rapid subauroral ion drifts, *J. Geophys. Res., 96,* 5785, 1991.

Anderson, P. C., D. L. Carpenter, K. Tsuruda, T. Mukai, and F. J. Rich, Multisatellite observations of rapid subauroral ion drifts (SAID), *J. Geophys. Res., 106,* 29,599, 2001.

Banks, P. M., and G. Kokarts, *Aeronomy,* Parts A and B, Academic Press, New York, 1973.

Belian, R. D., G. R. Gisler, T. Cayton, and R. Christensen, High-Z energetic particles at geosynchronous orbit during the great solar proton event series of October 1989, *J. Geophys. Res., 97,* 16,897, 1992.

Brandt, P. C., S. Ohtani, D. G. Mitchell, M.-C. Fok, E. C. Roelof, and R. Demajistre, Global ENA observations of the storm main-phase ring current: Implications for skewed electric fields in the inner magnetosphere, *Geophys. Res. Lett., 29*(20), 1954, doi: 10.1029/2002GL015160, 2002.

Burch, J. L., IMAGE mission overview, *Space Sci. Rev., 91,* 1, 2000.

Burke, W. J., N. C. Maynard, M. P. Hagan, R. A. Wolf, G. R. Wilson, L. C. Gentile, M. S. Gussenhoven, C. Y. Huang, T. W. Garner, and F. J. Rich, Electrodynamics of the inner magnetosphere observed in the dusk sector by CRRES and DMSP during the magnetic storm of June 4-6, 1991, *J. Geophys. Res., 103,* 29,399, 1998.

Chen, M. W., M. Schulz, G. Lu, and L. R. Lyons, Quasi-steady drift paths in a model magnetosphere with AMIE electric field: Implications for ring current formation, *J. Geophys. Res., 108*(A5), 1180, doi: 10.1029/2002JA009584, 2003.

DeMajistre, R., E. C. Roelof, P. C:son Brandt, and D. G. Mitchell, Retrieval of global magnetospheric ion distributions from high-energy neutral atom measurements made by the IMAGE/HENA instrument, *J. Geophys. Res., 109,* A04214, doi: 10.1029/2003JA010322, 2004.

Ejiri, M., Trajectory traces of charged particles in the magnetosphere, *J. Geophys. Res., 83*, 4798, 1978.

Farley, D. T., A theory of electrostatic fields in a horizontally stratified ionosphere subject to a vertical magnetic field, *J. Geophys. Res., 64*, 1225, 1959.

Fejer, B. G., R. W. Spiro, R. A. Wolf, and J. C. Foster, Latitudinal variation of perturbation electric fields during magnetically disturbed periods: 1986 SUNDIAL observations and model results, *Ann. Geophys., 8*, 441, 1990.

Fok, M.-C., J. U. Kozyra, A. F. Nagy, C. E. Rasmussen, and G. V. Khazanov, Decay of equatorial ring current ions and associated aeronomical consequences, *J. Geophys. Res., 98*, 19,381, 1993.

Fok, M.-C., R. A. Wolf, R. W. Spiro, and T. E. Moore, Comprehensive computational model of Earth's ring current, *J. Geophys. Res., 106*, 8417, 2001.

Fok, M.-C., et al., Global ENA image simulations, *Space Sci. Rev., 109*, 77, 2003.

Foster, J. C., and H. B. Vo, Average characteristics and activity dependence of the subauroral polarization stream, *J. Geophys. Res., 107*(A12), 1475, doi: 10.1029/2002JA009409, 2002.

Foster, J. C., and W. J. Burke, SAPS: A new categorization for subauroral electric fields, *EOS Trans. AGU, 83*, 393, 2002.

Frank, L. A., Direct detection of asymmetric increases of extraterrestrial 'ring current' proton intensities in the outer radiation zone, *J. Geophys. Res., 75*, 1263, 1970.

Garner, T. W., R. A. Wolf, R. W. Spiro, W. J. Burke, B. G. Fejer, S. Sazykin, J. L. Roeder, and M. R. Hairston, Magnetospheric electric fields and plasma sheet injection to low L-shells during the 4–5 June 1991 magnetic storm: Comparison between the Rice Convection Model and observations, *J. Geophys. Res., 109*, A02214, doi:10.1029/2003JA010208, 2004.

Heppner, J. P., Polar cap electric field distributions related to the interplanetary magnetic field direction, *J. Geophys. Res., 77*, 4877, 1972.

Jaggi, R. K., and R. A. Wolf, Self-consistent calculation of the motion of a sheet of ions in the magnetosphere, *J. Geophys. Res., 78*, 2842, 1973.

Jordanova, V. K., L. M. Kistler, J. U. Kozyra, G. V. Khazanov, and A. F. Nagy, Collisional losses of ring current ions, *J. Geophys. Res., 101*, 111, 1996.

Jordanova, V. K., A. Boonsiriseth, R. M. Thorne, and Y. Dotan, Ring current asymmetry from global simulations using a high-resolution electric field model, *J. Geophys. Res., 108*(A12), 1443, doi: 10.1029/ 2003JA009993, 2003.

Khazanov, G. V., M. W. Liemohn, T. S. Newman, M.-C. Fok, and R. W. Spiro, Self-consistent magnetosphere-ionosphere coupling: Theoretical studies, *J. Geophys. Res., 108*(A3), 1122, doi: 10.1029/2002JA009624, 2003.

Khazanov, G. V., M. W. Liemohn, T. S. Newman, M.-C. Fok, and A. J. Ridley, Magnetospheric convection electric field dynamics and stormtime particle energization: Case study of the magnetic storm of 4 May 1998, *Ann. Geophys., 22*, 497, 2004.

Liemohn, M. W., J. U. Kozyra, V. K. Jordanova, G. V. Khazanov, M. F. Thomsen, and T. E. Cayton, Analysis of early phase ring current recovery mechanisms during geomagnetic storms, *Geophys. Res. Lett., 25*, 2845, 1999.

Liemohn, M. W., J. U. Kozyra, M. F. Thomsen, J. L. Roeder, G. Lu, J. E. Borovsky, and T. E. Cayton, Dominant role of the asymmetric ring current in producing the stormtime Dst*, *J. Geophys. Res., 106*, 10,883, 2001a.

Liemohn, M. W., J. U. Kozyra, C. R. Clauer, and A. J. Ridley, Computational analysis of the near-Earth magnetospheric current system, *J. Geophys. Res., 106*, 29,531, 2001b.

Liemohn, M. W., A. J. Ridley, D. L. Gallagher, D. M. Ober, and J. U. Kozyra, Dependence of plasmaspheric morphology on the electric field description during the recovery phase of the April 17, 2002 magnetic storm, *J. Geophys. Res., 109*(A3), A03209, doi: 10.1029/2003JA010304, 2004.

Liemohn, M. W., A. J. Ridley, P. C. Brandt, D. L. Gallagher, J. U. Kozyra, D. M. Ober, D. G. Mitchell, E. C. Roelof, and R. DeMajistre, Parametric analysis of nightside conductance effects on inner magnetospheric dynamics for the 17 April 2002 storm, *J. Geophys. Res.*, to be submitted, 2005.

McComas, D. J., S. J. Bame, B. L. Barraclough, J. R. Donart, R. C. Elphic, J. T. Gosling, M. B. Moldwin, K. R. Moore, and M. F. Thomsen, Magnetospheric plasma analyzer: initial three-spacecraft observations from geosynchronous orbit, *J. Geophys. Res., 98*, 13,453, 1993.

Mitchell, D. G., et al., High energy neutral atom (HENA) imager for the IMAGE mission, *Space Sci. Rev., 91*, 67, 2000.

Ober, D. M., J. L. Horwitz, and D. L. Gallagher, Formation of density troughs embedded in the outer plasmasphere by subauroral ion drift events, *J. Geophys. Res. 102*, 14,595, 1997.

Parker, E. N., Newtonian development of the dynamical properties of ionized gases of low density, *Phys. Rev., 107*, 924, 1957.

Parker, E. N., Newton, Maxwell, and magnetospheric physics, in *Magnetospheric Current Systems, Geophys. Monogr. Ser.*, vol. 118, ed. by S.-I. Ohtani, R. Fujii, M. Hesse, and R. L. Lysak, Am. Geophys. Un., Washington, D. C., p. 1, 2000.

Rairden, R. L., L. A. Frank, and J. D. Craven, Geocoronal imaging with Dynamics Explorer, *J. Geophys. Res., 91*, 13,613, 1986.

Ridley, A. J., T. I. Gombosi, and D. L. De Zeeuw, Ionospheric control of the magnetosphere: Conductance, *Ann. Geophys., 22*, 567, 2004.

Rowland, D., and J. R. Wygant, The dependence of the large scale electric field in the inner magnetosphere on magnetic activity, *J. Geophys. Res., 103*, 14,959, 1998.

Sazykin, S., R. A. Wolf, R. W. Spiro, T. I. Gombosi, D. L. De Zeeuw, and M. F. Thomsen, Interchange instability in the inner magnetosphere associated with geosynchronous particle flux decreases, *Geophys. Res. Lett., 29*(10), doi 10.1029/2001GL014416, 2002.

Southwood, D. J., and R. A. Wolf, An assessment of the role of precipitation in magnetospheric convection, *J. Geophys. Res., 83*, 5227, 1978.

Spiro, R. W., and R. A. Wolf, Electrodynamics of convection in the inner magnetosphere, in *Magnetospheric Currents*, Potemra, T. A., ed., p. 247. Amer. Geophysical Union, Washington, DC, 1984.

Spiro, R. W., R. A. Wolf, and B. G. Fejer, Penetration of high-latitude-electric-field effects to low latitudes during SUNDIAL 1984, *Ann. Geophys., 6*, 39, 1988.

Takahashi, S., T. Iyemori, and M. Takeda, A simulation of the storm-time ring current, *Planet. Space Sci., 38*, 1133, 1990.

Vasyliunas, V. M., Mathematical models of magnetospheric convection and its coupling to the ionosphere, in *Particles and Fields in the Magnetosphere*, edited by B. M. McCormac, p. 60, D. Riedel, Norwell, Mass., 1970.

Weimer, D. R., A flexible, IMF dependent model of high-latitude electric potentials having "space weather" applications, *Geophys. Res. Lett., 23*, 2549, 1996.

Wygant, J., D. Rowland, H. J. Singer, M. Temerin, F. Mozer, and M. K. Hudson, Experimental evidence on the role of the large spatial scale electric field in creating the ring current, *J. Geophys. Res., 103*, 29,527, 1998.

Yeh, H.-C., J. C. Foster, F. J. Rich, and W. Swider, Storm time electric field penetration observed at midlatitude, *J. Geophys. Res., 96*, 5707, 1991.

Young, D. T., H. Balsiger, and J. Geiss, Correlations of magnetospheric ion composition with geomagnetic and solar activity, *J. Geophys. Res., 87*, 9077, 1982.

P. C. Brandt, Johns Hopkins University, Applied Physics Laboratory, 11100 Johns Hopkins Road, Laurel, MD 20723 (pontus.brandt@jhuapl.edu).

M. W. Liemohn, Atmospheric, Oceanic, and Space Sciences Department, University of Michigan, 2455 Hayward St., Ann Arbor, MI 48109-2143 (liemohn@umich.edu).

the O^+ in the energetic ring current has been observed to be much more strongly modulated by substorm dynamics than the proton component [*Mitchell et al.*, 2003].

The ionosphere has been known to supply cold light ion plasma to the magnetosphere since the discovery of the plasmasphere [*Freeman* et al., 1977]. On openly convecting field lines, polar wind flows continuously as convection opens the field lines and empties their accumulations into the polar lobes and downstream solar wind, so that they never reach equilibrium pressures. Dense plasmaspheric plasmas result from polar wind-like light ion outflows into the nearly co-rotating inner magnetosphere, where they are trapped and accumulate to equilibrium pressures, and are therefore called "refilling flows." Recently, the outer plasmasphere has been shown to flow sunward during magnetospheric disturbances [*Elphic* et al., 1997; *Sandel* et al., 2001; *Goldstein* et al., 2003]. The roots of plasmaspheric plume flux tubes create a corresponding plume of enhanced plasma density in the ionosphere proper [*Foster* et al., 2002]. Moreover, these cold plasmas have been discovered to be present in the subsolar low latitude magnetopause region under a wide variety of conditions [*Su* et al., 2001; *Chandler and Moore*, 2003; *Chen and Moore*, 2004]. When strong convection drains away part of the plasmasphere, the supply of plasma is enhanced in a transient way by the rapid release of accumulated plasma. Under steady conditions, however, the plasmasphere remains trapped and the magnetosphere is supplied only from the higher latitude polar wind outflow regions.

The contribution of ionospheric light ions to magnetospheric hot plasmas is less well established and is complicated by the difficulty of discriminating protons of solar or geogenic origin. Christon et al. [1994] used energy spectral features, in comparison with He^{++} assumed to be of solar origin, to estimate the relative contributions. They found that both solar wind and polar wind contributed comparable densities to the hot magnetospheric plasmas, with a somewhat greater ionospheric contribution for high solar activity levels.

In this paper we simulate the entry of solar wind plasmas into the inner magnetosphere under growth phase conditions of southward interplanetary field. We also consider the light ion polar wind outflows that are pervasive, continuous, and at most weakly responsive to solar wind intensity or magnetospheric activity. Finally, we consider the outflow of heavy ions from the active auroral zones, associated with electromagnetic and kinetic energy dissipation within the ionosphere proper, referring to the latter as "auroral wind". We defer consideration of more typical "Parker spiral" IMF with By dominant. Light ion auroral outflows have fluxes similar to polar wind, while the heavy ion outflows have fluxes that range from much less than to much greater than polar wind outflows depending on the free energy available [*Moore* et al., 1999b]. Using these simulations, we address the question of how solar and polar wind protons, and auroral wind oxygen ions, are distributed in the magnetosphere during conditions that lead toward storm time ring current development.

2. OBSERVATIONS

We begin by presenting, in Plate 1, velocity distribution observations from the Polar spacecraft, the orbital apogee of which reached to the equatorial plasma sheet at 9-10 Re and swung through the midnight region during the Fall of 2001, and 2002. The plasma sheet is highly dynamic on time scales that cannot be sampled continuously using the 18 hour orbit of Polar. In the present paper, we focus instead on the persistent structure of this region and the ion velocity distributions that define that structure in relatively quiet times, but we also point toward variations that would be expected during more active periods.

A pervasive feature of the Polar observations in this region is the existence of cold high Mach number field-aligned flows away from the Earth in the lobe regions. We term these "lobal wind", defining them as a mixture of polar wind and dayside auroral winds from the cleft ion fountain region. A second pervasive feature is the hot and therefore low Mach number flow of ions, embedded within the colder lobal wind outflows. These are identified as nightside auroral wind outflows of either the beam or conic varieties, depending on the ratio of perpendicular to parallel thermal speed. These appear with varying prominence, presumably dependent upon the conjugate auroral activity at the point of observation. They have higher parallel and perpendicular energy, with a broader angular pattern, than the relatively cold lobal wind flows. While the range of energies is continuous, such auroral outflows are usually distinguished from the lobal wind flows within which they are embedded.

Plate 1 summarizes these observations schematically, showing the various velocity distribution types in their typical arrangement along a Polar orbit, relative to the current sheet, plasma sheet, and lobes. The plasma sheet appears as a layered structure of velocity distribution features, as described above. There is considerable variability in the extent and prominence of the various features from pass to pass, presumably associated with convection and auroral activity. A substantial repeatability of this pattern is observed over many passes through the region, and it can be considered as an underlying structure, upon which variations are superposed. In the following sections, we investigate the degree to which this structure can be understood in terms of the particle populations that enter and travel through the plasma sheet region.

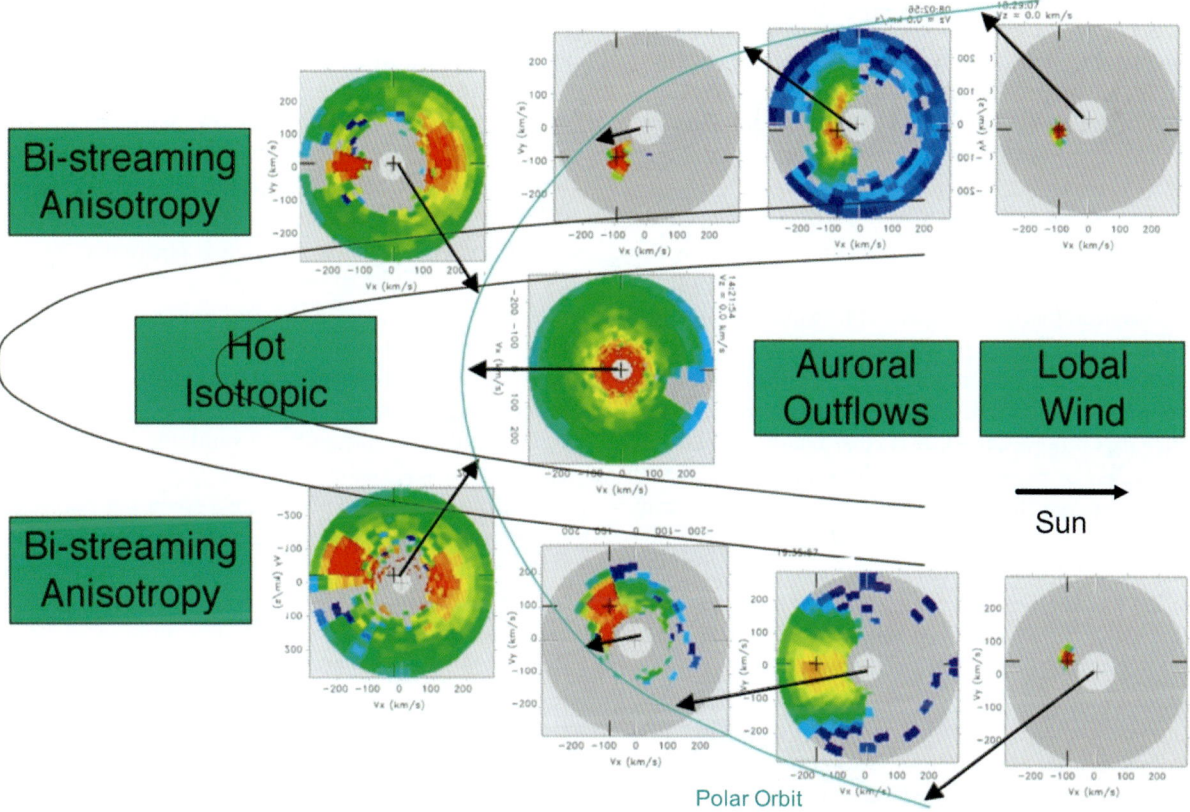

Plate 1. A schematic collage of the velocity distribution types and their association with observing position relative to the current sheet, along a typical Polar Orbit during Fall 2001-2002.

Motivation for considering the contributions of polar and auroral winds to the ring current has been found in recent observations from the IMAGE mission. In Plate 2, we reproduce an illustration of the compositional variability of the ring current, from Mitchell et al. [2003]. The main point of this study is that ring current oxygen ions are strongly modulated by substorm activity, while the lighter protons are less influenced by individual substorms, appearing to respond primarily on longer time scales to the storm growth phase. A possible interpretation of this observation is that O^+ injection to the ring current is relatively direct and driven by individual substorms while proton injections are related to larger scale magnetospheric phenomena such as global convection. Another related hypothesis is that substorm energization mechanisms are more effective on higher mass species. Here, we focus on global convection and transport simulations, rather than substorm dynamics. However, our results will turn out to have implications for this observation.

Intimately related to the injection of new plasmas from the nightside plasma sheet, into the ring current region, is the stripping away of the outer plasmaspheric plasmas to form a cold plasma plume. This in turn is transported sunward toward the magnetopause [*Goldstein* et al., 2003]. In Plate 3, we exhibit recent observations that clearly reveal the arrival of those plasmas in the subsolar magnetopause region [*Chandler and Moore*, 2003]. It is clear that these plasmas flow into subsolar reconnection regions and are entrained into the downstream flows of the low latitude boundary layer and high latitude mantle [*Chen and Moore*, 2004]. They thus provide a source of relatively slow ions that may later be recycled into the flows returning through the inner magnetosphere, serving as a delayed source of the ring current, to the degree that they are energized by the process. In this process the dominant light ions may be expected to mimic the behavior of entering solar wind protons.

Our objective in what follows is to contribute toward understanding of the ionospheric plasma circulation cycle, driven by its contact with the solar wind flows. We seek to describe the nature of solar wind proton flows around and through the magnetospheric system, to accomplish the same for polar wind and auroral wind flows, and then to develop a basis for comment on observations of detailed velocity distributions and of substorm variations within storms. These will serve as a basis to plan further studies involving dynamic fields during the growth phase of storms.

3. MODELING

For this study, we implemented a 3D full particle motion calculation in fields that are specified on a regular spatial grid of points to accept field prescriptions from dynamic models. This calculation is based on the full particle simulation of Delcourt et al. [1993]. The particles are propagated in self-consistently computed fields from the magnetohydrodynamic simulation of Lyon, Fedder, and Mobarry [*Fedder et al.*, 1995; *Mobarry et al.*, 1996]. The advantage of using MHD fields is that we can consider the entry of solar wind through realistic boundary layer fields with reconnection operative in at least a qualitatively realistic way. Ionospheric outflows were not included in this calculation, but the F region ionosphere is modeled so that there is drag on the system at the roots of the flux tubes, represented by field-aligned current systems and ionospheric conductivity. For computational efficiency, we resampled the LFM fields onto a spherical grid with polar axis aligned to the GSM-X axis. The spacing in polar angle is a uniform $2°$ and the resolution in azimuthal angle on the GSM YZ plane is ~5.5 deg. The grid spacing in radius varies with polar and azimuth angles with higher spatial resolution on the dayside and lower on the nightside. This approach allows very rapid identification of the current cell in which a particle is positioned, as it moves, conserving computing resources.

Interpolation between the grid points of the MHD simulation enables calculation of field values at any point, as a particle moves about within the simulation space. This is accomplished by simple linear interpolation in place of more sophisticated techniques that would allow continuous field gradients at the grid points. The inevitable field gradient discontinuities at grid points are a source of numerical diffusion, decreasing with the spacing of the grid used.

We step the particles in time many cycles per gyro period, typically 72 times or every 5 deg. of gyrophase, using Delcourt's double precision implementation of a 4th order Runge-Kutta algorithm. We have previously shown [*Moore et al.*, 2001], that the trajectories are precisely reversible over flight paths of many 10s of R_E, and many hours (when used with analytically continuous fields).

Our results are sensitive to grid and temporal spacing, indicating numerical errors. To minimize such effects, we used the finest practical grid spacing for our MHD simulation fields, and determined that the numerical effects were substantially reduced. The main effect of such errors is a diffusive effect on the particle trajectories that mimics real field variability, but is not currently guided by observed field fluctuations. We plan to introduce observed or simulated field fluctuations and reduce numerical effects to a comparatively negligible level, as soon as practicable.

For the solar wind, initial positions were randomly selected from a uniform distribution over the GSM YZ plane at Xgsm = 15 R_E, upstream of the simulated bow shock. For the polar wind, we started protons at 4 R_E altitude with invariant latitudes randomly distributed above $55°$ and over all local

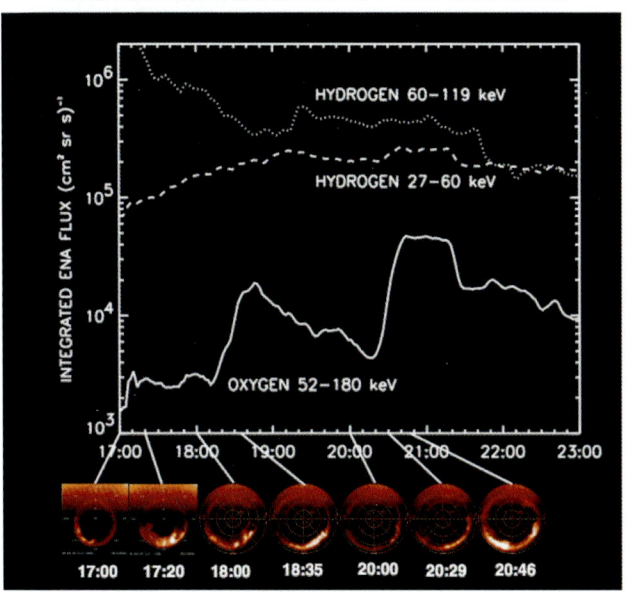

Plate 2. Variation of H+ and O+ ring current fluxes in association with substorms as documented by auroral imagery [after Mitchell et al., 2003].

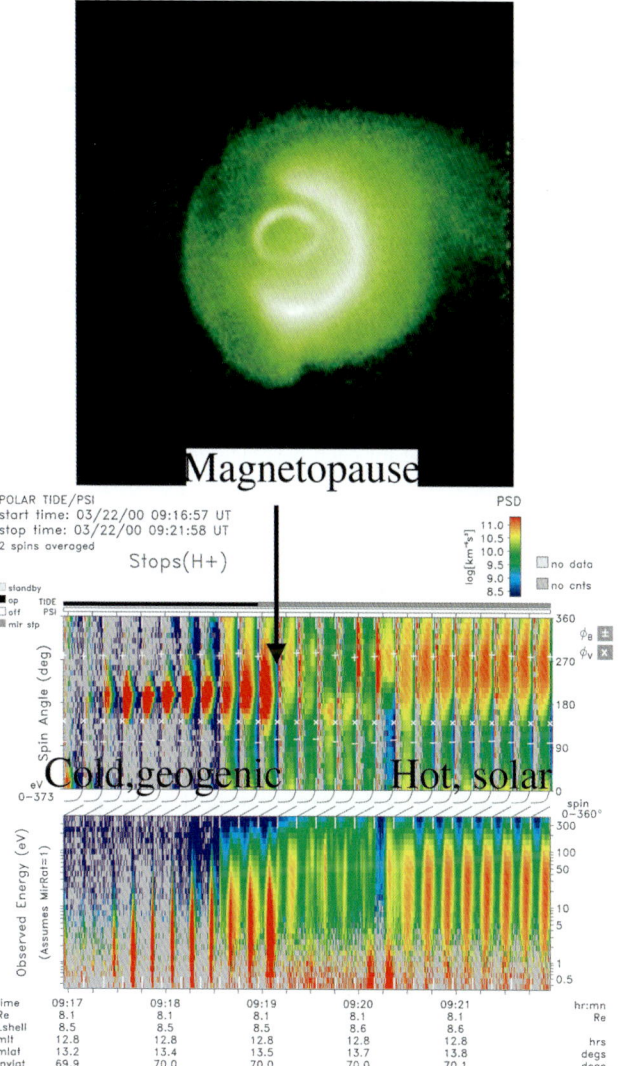

Plate 3. Observations of plasmaspheric plume formation (upper panel after Sandel et al. [2001], and of the presence of plasmaspheric plasma at the dayside magnetopause (lower panel, after Chandler and Moore [2003].

times. Auroral acceleration processes have not been applied to polar wind protons originating from that region. Though the total escaping flux of protons is relatively unaffected by such processes [Moore et al., 1999a], the circulation of protons accelerated by auroral processes will be more like that of O⁺ ions treated as accelerated nightside auroral wind. For the auroral wind, we started O⁺ at 1000 km altitude within a nominal auroral oval that was divided into dayside and nightside parts along the dawn-dusk meridian, as described in Table 1. Initial velocities were selected randomly from a uniform distribution of boxcar width equal to the specified bulk and thermal speeds, as shown in Table 1. Particles were run until they precipitated into the atmosphere, escaped from the simulation volume, or exceeded a time limit of 10–24 hours to assure that each particle makes at least one circumnavigation of the Earth, regardless of energy. A given particle contributes to a given cell only upon its first entry into that cell. Subsequent returns to the same cell are ignored.

A large number of particle trajectories was run and accumulated into a database of bins with resolution of 1 R_E^3, for all particles. The record for each particle consists of one record describing the particle initial conditions, and many records describing the particle state as it crosses each boundary in physical space. Entry and exit times are recorded as particles transition between cells, providing a time of flight through each cell. Particles were run with randomly selected initial conditions (within specified ranges) until the most populated bins contained >1000 particles. Some bins tend to remain empty, particularly for solar particles, because most of them pass through the system without entering the magnetosphere. To counter this, additional solar particles were run, focusing on the upstream regions with the highest probability of entry, until the most populated inner magnetosphere bins contained >100 solar particles. Requiring 100 particles in each bin provides reasonable statistical errors (<10%) for assessing the dominant transport paths, the shape of the particle velocity distribution, and for bulk properties of the plasma.

Bulk properties were estimated following an extension of the method described by Chappell et al. [1987] and Delcourt et al., [1989]. Examples of both are exhibited in subsequent figures. For each particle in a given spatial bin, the particle velocity and transit time for that bin are calculated. For a particle (i) passing through a particular bin (j), the contribution of density in this bin by this particle is:

$$n_{ij} = F_i \cdot T_{ij} / V_j \qquad (1)$$

where F_i is the ion source flux in ion/s for particles of the specified velocity, T_{ij} is the residence time of particle i in bin j, and V_j is the volume of bin j, that is 1 R_E^3 in our case.

F_i is specified directly by means of the density and flow of the source plasma across the source boundary.

$$F_i = n_s * v_s * dA \ ; \ dA = A/N_T \qquad (2)$$

Here dA is the area of the source surface allocated to each particle, which is the total area of the source divided by the number of particles emitted, assuming a uniform distribution of particle emission on the source surface, which is must be assured when randomizing the initial locations. The source number density and flow velocity may be specified, or the product of those two is just as useful, if better known.

Substituting (2) into (1), we have

$$n_{ij} = n_s \cdot v_s \cdot A \cdot T_{ij} / (V_j \cdot N_T) \qquad (3)$$

The density at bin j is just the summation of n_{ij} over all particles that are passing through bin j:

$$n_j = \sum_i n_{ij} \qquad (4)$$

These relations can be applied to any source flowing across a boundary surface. The density from the ionospheric outflow can be calculated in a similar way. In that case, V_{sw} should by replaced by V_{pw} or V_{aw} and other parameters are replaced with values appropriate to the polar wind or auroral wind.

Once densities are calculated, the total (dynamic plus kinetic; we don't separate the two here) pressure at bin j is given by,

$$P_j = \sum_i P_{ij} \qquad (5)$$

$$P_{ij} = n_{ij} * E_{ij} \qquad (6)$$

Table 1. Source region particle initial conditions

Parameter	Value	Comment
Solar Wind Density	6.5 cm-3	typical
Thermal speed (Temp)	31 km/s (5 eV)	"
Velocity	400 km/s	"
Flux	3 x 10⁸ cm⁻²s⁻¹	"
Polar Wind Density	0.5 cm-3	Su et al., 1998
Thermal speed (Temp)	17 km/s (1.5 eV)	"
Velocity	100 km/s	"
Flux	3 x 10⁸ cm⁻²s⁻¹	"
Auroral Wind O+ (day)		75° to 65° ILat range
Density	1000 cm-3	Pollock et al., 1990
Thermal speed (Temp)	10 km/s (10 eV)	"
Velocity	10 km/s	"
Flux	1 x 10⁹ cm⁻²s⁻¹	"
Auroral Wind O+ (night)		75° to 60° ILat range
Density	10 cm-3	Yau et al., 1988
Thermal speed (Temp)	100 km/s (1000 eV)	"
Velocity	100 km/s	"
Flux	1 x 108 cm⁻²s⁻¹	"

where E_{ij} is the average energy of particle i in bin j.

For the solar wind case:
n_s = 6.5 cm^{-3}; v_s = 400 km/s or 4×10^7 cm/s; A = 3600 R_E^2; N_T = 800000

For the polar wind case:
$n_s \cdot v_s = 3 \times 10^8$ $(1.15/4)^3$ cm^{-2}s^{-1} = 7.1×10^6 cm^{-2}s^{-1}; A = area of sphere 4 Re radius for invariant latitudes above 55 deg; NT = 20000

For the auroral wind dayside case:
$n_s \cdot v_s = 1 \times 10^9$ cm^{-2}s^{-1}; A = area of sphere 1.15 R_E within specified auroral oval; NT = 20000

For the auroral wind nightside case:
$n_s \cdot v_s = 1 \times 10^8$ cm^{-2}s^{-1}; A = area of sphere 1.15 R_E within specified auroral oval; N_T = 20000

This method distributes the source weighting properly with the number of particles emitted from any particular part of the boundary so that the results are unaffected by the number of trajectories initiated. This allows us, for example, to concentrate solar wind particles in the region with maximum probability of entry (subsolar), while using a smaller number of particles to define the magnetosheath flow.

The method also allows for the diffusive filling of velocity space from source populations that become highly structured in velocity space at some spatial locations [*Moore et al.*, 2000]. If the relatively low resolution fields we use were realistic on all spatiotemporal scales, and there were no diffusive processes present, observed velocity distributions would be very finely structured with narrow features. In practice, magnetospheric fields include fluctuations over a wide range of frequencies, which are evidently diffusive, since extremely fine features are not observed within hot plasmas, though of course certain anisotropies are observed, as discussed above. Our bulk properties calculation attributes to each particle both a mean phase space density and a velocity space volume over which each particle is representative of the source. This allows for the particle weighting to be spread over the full region of velocity space at each location in the simulation space, rather than characterizing only a single point in velocity space.

The MHD fields for these simulations come from a time sequence that involves a few hours of northward Bz, an abrupt southward turning for two hours, and finally, a return to northward IMF [*Slinker et al.*, 1995]. We selected a time about 45 minutes after the southward turning, well after the formation of a distant reconnection X line, but before the appearance of a near-Earth X line and ejection of a plasmoid. That is, we chose a state representative of the substorm growth phase, when the magnetotail closely resembles a Level-2 T89 field (corresponding to Kp ~ 1-2) earthward of the distant neutral line at ~40 R_E.

In Plate 4, the field snapshot is visualized in the XY and XZ planes. For the SBz case, subsolar reconnection and distant plasma sheet reconnection are both present, but are not in balance, and the plasma sheet is growing. Subsolar reconnection drives a high latitude flow that reinforces and becomes part of the double cell circulation flow in the equatorial plane. The magnetotail pressure distribution drives an earthward flow up to about 150 km/s in the inner plasma sheet and especially along the flanks. The action of the distant neutral line helps to inflate the plasma sheet during this period, but it is being convected tailward and has relatively little influence on driving sunward convection at this point in the growth phase.

4. RESULTS

In Plate 5 we display the plasma pressure in the GSM-XZ or noon-midnight meridian plane for the three sets of particles: solar wind, polar wind, and auroral wind. The bow shock and magnetosheath are prominent features, as are the cusps and the cross section of the inner ring current-like region. Solar wind protons tend to avoid the lobes and plasma sheet Earthward of -40 R_E, and it is unclear, from this view, how they are reaching the plasma sheet and the inner magnetosphere to form the ring current-like structure, where pressure of these protons reaches about 1 nPa. A clear plasma mantle has formed, from the lowest energy solar wind particles, which meet at the plasma sheet beyond 30-50 Re.

The polar wind outflow fills the lobes, some of it reaching the plasma sheet Earthward of about -40 Re and then convecting back toward the Earth to form a region of drifting ring current-like protons that has a pressure of < 0.1 nPa. Polar wind protons are also seen to convect to the magnetopause where they are jetted up over the poles, participating in the magnetosheath flow as they escape downstream with a low partial pressure.

The auroral wind flows through the polar lobes (dayside part) or directly into the plasma sheet (nightside part), and reaches the plasma sheet earthward of about 25 Re. Most of this auroral wind flows sunward through the inner magnetosphere, forming a ring current population. It continues sunward to the magnetopause where it is entrained into the mantle and polar cap flows again, but this time at higher energy than when initially leaving the ionosphere. Consequently, it travels substantially farther down tail, in part beyond the limit at 70 Re in the simulation space, and in part returning Earthward through the plasma sheet a

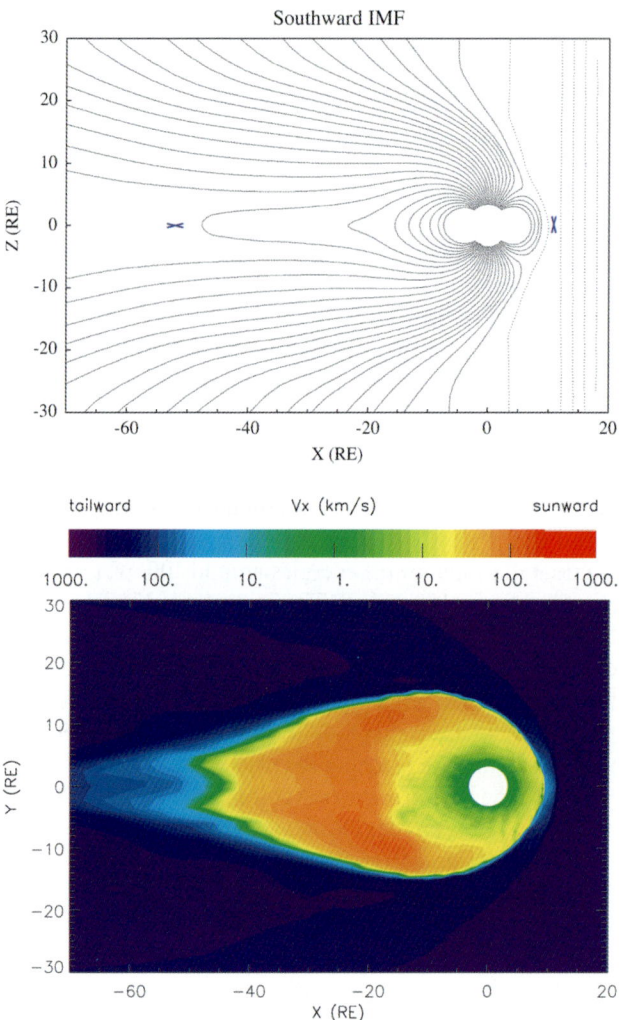

Plate 4. The computed magnetohydrodynamic fields for the SBz case. (top) The magnetic (B) lines are indicated in the noon-midnight meridian. (bottom) The electric field is indicated as color contoured values of the Vx in the GSM-XY plane, coded to discriminate sunward (reddish) from tailward (bluish) flows, with green indicating regions of relatively low velocity flows.

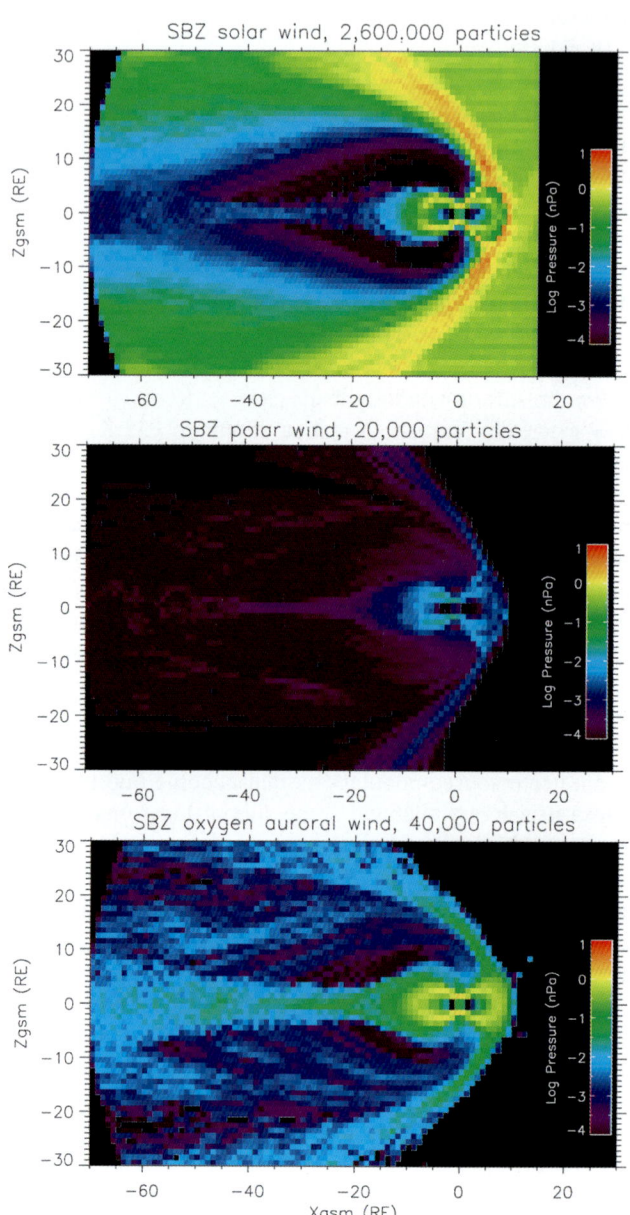

Plate 5. Plasma pressure is color contoured in the GSM-XZ or noon-midnight meridian plane. Solar wind particle pressures are shown in the upper panel; polar wind particle pressures are shown in the middle panel; and auroral wind pressures are shown in the lower panel.

second time. The ring current region contains contributions from both first and second pass O⁺ ions from the auroral wind, the latter being substantially more energetic, but less numerous.

In Plate 6, we display the plasma pressure in the XY or ecliptic-equatorial plane, comparing solar wind, polar wind, and auroral wind contributions in the three panels. Features of interest for the solar wind include the bow shock, the magnetosheath, and the low latitude boundary layer flows, and formation of a cavity in the wake region, within which few trajectories penetrate, while those that do have a relatively low associated pressure compared with the solar wind proper. This probably reflects a requirement for entering solar wind ions to be of relatively low energy to convect to the neutral sheet before their parallel motion causes them to escape the system. The figure (with overlaid path/arrow) clearly reveals the principal path along which solar wind protons enter the inner magnetosphere, through the magnetopause along the dawn flank, then looping back to form a drifting ring current-like population, with a pressure reaching ~1 nPa. This feature is not as close to Earth as a full storm-time ring current, reflecting relatively weak inner magnetospheric convection. This owes in part to the moderate SBz conditions for this simulation.

Polar wind ions populate the plasma sheet, escaping downstream where they land beyond the convection reversal of Plate 5, and returning Earthward where they land within the Earthward flow. The latter illuminate a clear plasma sheet structure that connects directly with the inner magnetospheric closed drift region. Ionospheric plasma that is convected to the sunward magnetopause region is then entrained into the inner magnetosheath along flanks of the magnetosphere and downstream as part of the low latitude boundary layer.

The auroral wind behaves similarly to the polar wind in this view, but with substantially greater pressure contribution, especially in the inner magnetospheric ring current region, where pressure reaches ~2 nPa for the outflow fluxes assumed here. In addition, there is a larger dawn dusk asymmetry of both the plasma sheet and ring current pressure distributions for the auroral wind. Substantial downstream plumes are formed in the low latitude boundary layers here as in the polar wind case.

In Plate 7, we display the plasma pressure in the last of the three cardinal planes, the GSM-YZ plane at X = 0, or the dawn-dusk meridian. This cross section is dominated by solar wind and magnetosheath flows. The magnetospheric lobes are prominent in both solar wind and polar wind proton pressure, though for solar protons, they are profoundly empty at the core, but with a mantle of low energy solar protons as well.

The polar wind protons populate the lobes, but at relatively low pressure. The magnetosheath contains an enhancement of polar wind proton pressure, but it is still a minor addition to solar wind pressure. For the solar wind, a dawn-dusk asymmetry between the magnetosheath and the inner region hints at the dominant entry pattern into the inner magnetosphere via the dawn flank, but the view of this is better in Plate 6.

The auroral wind O⁺ also shows the dawn-dusk asymmetry seen earlier in the XY plane, which extends into the lobes, where it has the opposite sense as in the inner magnetosphere (and the same sense as in the plasma sheet. This presumably reflects circulation asymmetries in convection, combined with the recirculation of auroral wind outflows reaching the dayside magnetopause.

In Plate 8, we display an array of velocity distributions for solar, polar and auroral wind particles. In the upper row, angular integrations are used to derive energy distributions by density contribution. We find that the solar wind contribution contains components at around 1 keV and at about 40 keV. The polar wind, in contrast, yields a population dominated by the lowest energies around 100 eV, declining steeply into the 10's of keV. The auroral wind has a 10's eV component evidently deriving from direct outflow from the dayside region, and a dominant 10's keV component that exceeds all others.

In the middle row, the full (2D) velocity distributions are given with color coding according to density contribution (a normalized phase space density). Here we see angular structure mainly in the polar wind core particles, which have a pancake distribution in this region. In the auroral wind particles, field alignment is seen in the lowest energy population, which originates from the night side auroral zone outflows. Additional details indicate the density, pressure and average energy for these populations.

In the lower row of panels, the same velocity distributions are coded with a color indicating the maximum excursion of each particle from Earth during its history prior to arrival in this spatial bin. The main point here for solar wind particles is that the energetic component has generally entered the inner magnetosphere without exceeding their original distance of 15 Re upstream. The low energy component has evidently traveled farther downstream, and these are the low energy particles of the incident solar wind that formed the mantle flow. The polar wind has an opposite trend with the most energetic component having traveled farthest down the tail before returning, with a sprinkling of exceptional particles. The lowest energy core of the polar wind particles has evidently been locally emitted from the ionosphere without passing through the plasma sheet. For auroral wind particles, there is a similar mixture of local low energy core

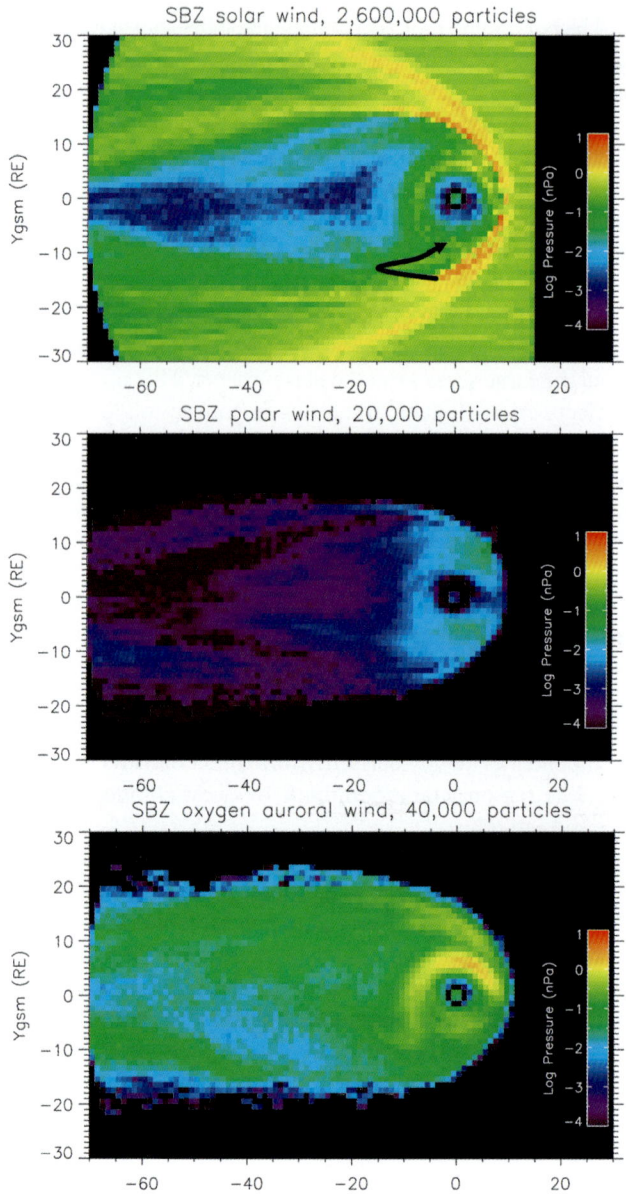

Plate 6. The plasma pressure is color contoured in the GSM-XY or equatorial plane. Solar wind particle pressures are shown in the upper panel; polar wind particle pressures are shown in the middle panel; and auroral wind pressure in the lower panel.

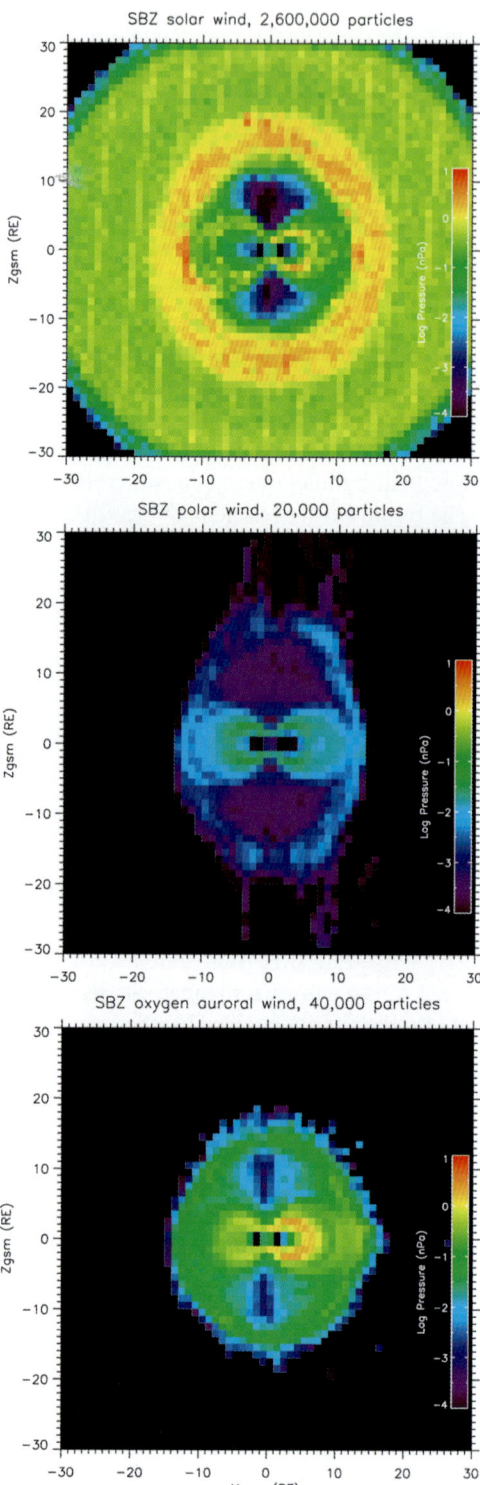

Plate 7. For the SBz case, the plasma pressure is color contoured in the GSM-YZ or dawn-dusk meridian plane. Solar wind particle pressures are shown in the upper panel; polar wind particle pressures are shown in the middle panel; and auroral wind pressures in the lower panel.

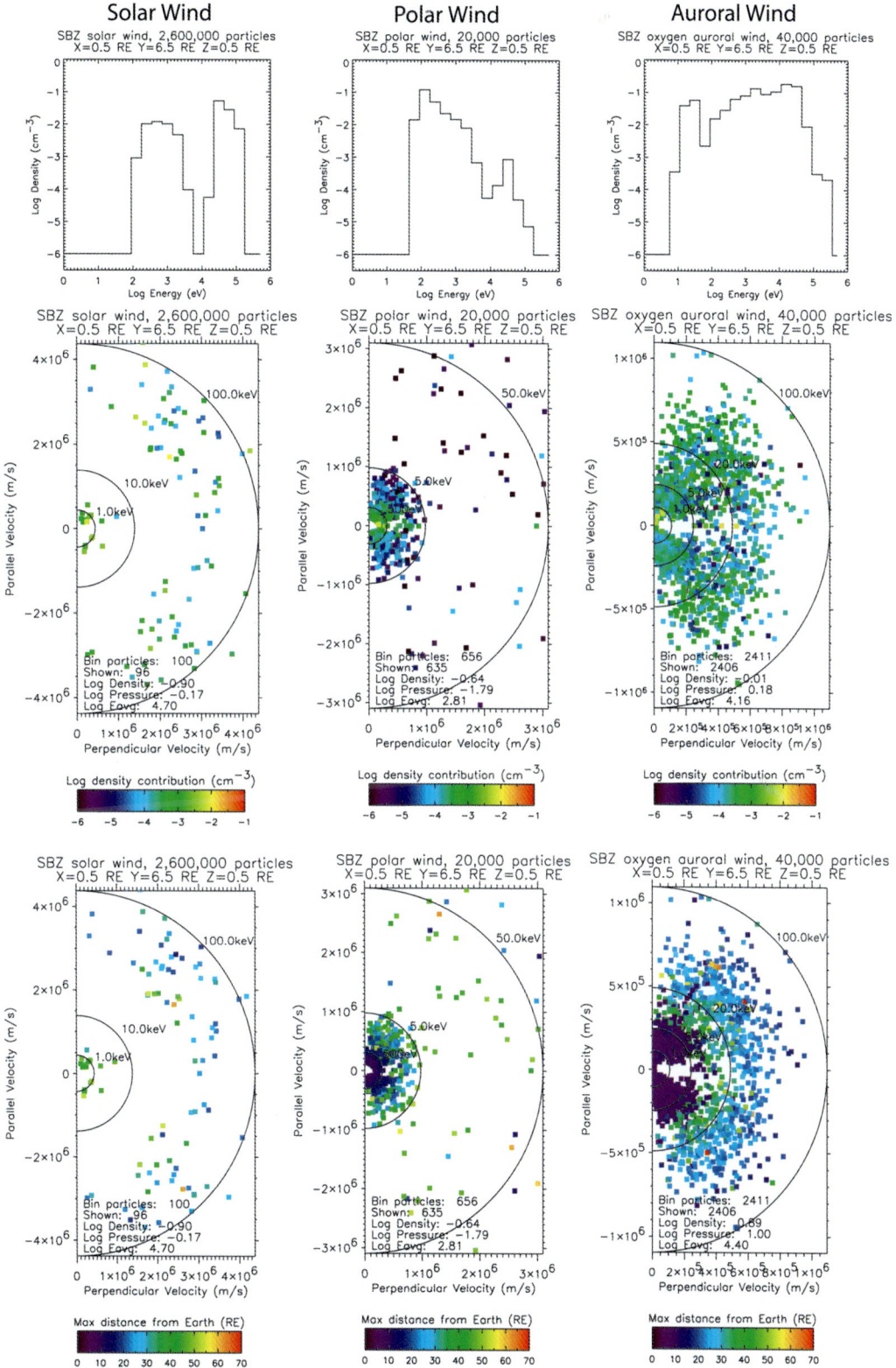

Plate 8. Energy, velocity, and maximum radius distributions, at the indicated dusk sector geosynchronous location, for solar wind, polar wind and auroral wind ions.

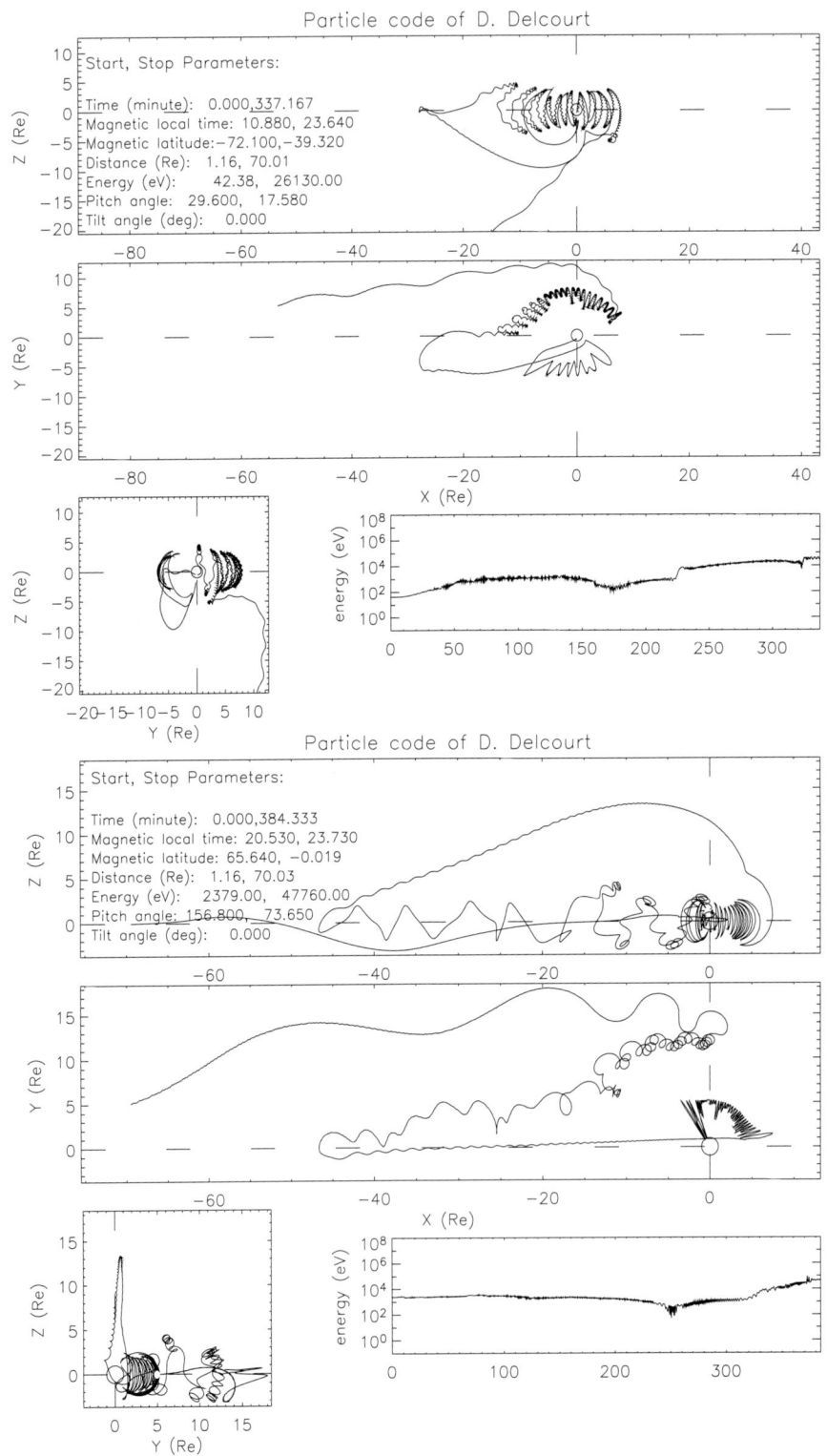

Figure 1. Example trajectories for the auroral wind O+ ions. The upper set of panels show three projections and the energy history for a dayside emitted auroral wind ion. The lower set of panels show three projections and the energy history for a nightside emitted auroral wind ion. Initial and final conditions are summarized in the inset text listings for each trajectory.

and energetic component that has traveled farther down tail, but then at the highest energies we again see that the origin becomes more local.

In Figure 1, we display a pair of O^+ trajectories typical of dayside and nightside outflows, illustrating some of the points made above. In particular, the upper panels show a dayside O+ ion, launched near 11 MLT with an initial energy of 42 eV, and initially convected tailward through the polar lobe. As a consequence of its low energy, the ion is deflected dawnward as it arrives in the near-Earth plasma sheet and convects sunward. When it reaches the dayside cusp region, it again travels tailward through the lobe, this time at higher altitude, and traveling farther down tail. It is significantly accelerated upon traversing the neutral sheet and transported duskward. Its energy is then high enough to drift through the dusk sector toward the dayside magnetopause, where it is again sent tailward, but this time with too much energy to be captured in the tail, approximately 25 keV.

The lower panels of Figure 1 illustrate a nightside auroral O^+ ion, launched at about 2100 MLT with an initial energy of 2.4 keV. This is a high enough energy that the ion travels sunward through the dusk sector on its first pass. The ion misses the stretched part of the plasma sheet and gains no significant energy in its first pass. It reaches the dayside and is again carried tailward through the lobes, reaching the neutral sheet at about -45 R_E and gaining substantial energy as it swings across the neutral sheet and is then convected earthward, but swings wide toward the dusk flank, where it escapes into the downstream magnetosheath. The final energy is similar to that of the dayside outflow O^+ ion, but whereas the dayside auroral wind ion is retained in the inner magnetosphere, the nightside auroral ion drifts duskward and fails to make a second pass through the inner magnetosphere.

5. SUMMARY DISCUSSION

We have previously suggested [*Moore et al.*, 2001] that much of the plasma sheet and ring current plasma may be provided by ionospheric sources, even though it is clear that the solar wind supplies the energy to power magnetospheric phenomena. Delcourt et al. [1992] looked at solar wind entry and concluded that it was exceedingly difficult to get solar wind protons arriving via the dayside cusp region to reach the inner magnetosphere through the plasma sheet. We failed at that time to consider entry along the low latitude boundary layers near the flanks of the magnetosphere, but others have suggested or pointed out that entry path [*Lennartsson*, 1994; *Richard et al.*, 2002, *Walker et al.*, 2003, *Peroomian*, 2003, *Winglee*, 2003; *Thomsen et al*, 2003]. On the other hand, the polar wind contribution to the plasma sheet has not been studied as carefully as the solar wind since Hill [1974]. We have taken a more comprehensive approach to both solar wind entry and polar wind circulation than we did earlier, by initiating particles in the boundary regions and tracking them through a realistic interaction from MHD simulations [*Moore et al.*, 2005]. To that we have now added the ionospheric source of O^+ from the auroral zones, both dayside and nightside, which we have termed "auroral wind".

To explore both the transport paths and kinetic behaviors of the plasma ions, we treated them using single particle motions in global fields. The fields derive from a snapshot of magnetohydrodynamic simulations and are therefore realistic in terms of bulk plasma and electrodynamic parameters, but frozen in time during the particle motions. No diffusive effects were included, but particles were introduced with randomized gyrophase and suffer some diffusive numerical effects. The fields are based only on a solar wind plasma source, revealing an implicit assumption that the ionospheric plasma does not alter the dynamics of the MHD simulation. The magnetosphere was considered at equinox with no dipole tilt off normal to the Sun-Earth line. These limitations have been introduced to make this work practical. We believe them to have relatively minor impacts on the conclusions drawn from this work, but plan to explore the dynamic effects of polar and auroral winds in future work using similar or multifluid simulations.

In agreement with results of others cited above, we find that the dawn low latitude boundary layer region is the most effective source of solar wind particles to the ring current region. This route supplies the inner ring current region independent of the path through the midnight plasma sheet near the reconnection region and dominates over that path, which is much lower in solar proton density and pressure, compared with the dawn entry route, but is well-populated with polar wind, which then flows into the ring current region along the traditional path. For the SBz conditions we studied, the solar wind contribution to the inner magnetospheric ring current region is dominant over the polar wind contribution in pressure and the two are comparable in density, with the solar wind energy distribution having a greater mean energy. This result supports the conclusions reached by Hill [1974].

The current results should be considered in the context of the geopause suggested by Moore and Delcourt [1995]. It is apparent that solar wind protons enjoy a special access route to the inner magnetosphere that is quite different from that taken by ionospheric outflows (with the possible exception of such outflows into the dawn flank region). This effect, which appears to result from magnetic gradient drifts of solar wind protons, leads to an effective "wormhole" or leak of solar protons from the dawn low latitude boundary layer to the ring current region. It produces considerable mixing of the two sources in the inner magnetosphere, for the conditions considered here and allows the solar wind plasma to dominate the interior ring current-like region.

In contrast with the solar wind, the auroral wind follows the polar wind path to the plasma sheet and ring current region, but has a substantially larger source flux and a lower velocity (dayside) or smaller convective separation (nightside) from the ring current region. As a result, the auroral wind is initially more confined to the inner magnetosphere and can for active conditions have a larger flux than the polar wind by an order of magnitude or more.

In contrast with the polar wind, the auroral wind remains slower, even after passage through the inner magnetosphere. When it reaches the dayside reconnection region, it subsequently acts more like solar wind mantle (slow) than solar wind core (fast). Therefore, it convects to the plasma sheet a second time and is recirculated to the inner magnetosphere a second time, again raising its energy and pressure. Together these make for substantially greater pressure of O^+ in the inner magnetosphere. Auroral O^+ is energized to a greater degree by its interactions with the plasma sheet and magnetospheric convection, and is successively energized during two or more passes through the system. Being slower at given energy, the auroral wind must reach higher energies before it is fast enough to escape downstream in the boundary layers. The net affect appears to be that heavy ion auroral wind is energized and concentrated within the magnetosphere to a greater degree than polar wind, and can also exceed the pressure of the solar wind proton component.

6. CONCLUSIONS

Solar wind ions enter principally along the dawn flank, bifurcating into a component that continues down the dawn flank, and another component that convects immediately into the inner magnetosphere and into the ring current without passing through the midnight plasma sheet. Polar wind fills the plasma sheet proper and supplies a plasma pressure contribution that is appreciable but minor, for the conditions we've studied in the ring current region. The density contributions of solar and polar wind protons are comparable, but the solar wind protons have a substantially higher mean energy and dominate the proton pressure.

In contrast, auroral wind O^+ has the capability to form the dominant component of the ring current pressure, particularly under active conditions that drive substantial dayside outflows [*Moore et al.*, 1999b; *Strangeway et al.*, 2000], which we have assumed to be present here. Moreover, the auroral wind follows the polar wind path to the ring current and is thus arguably subject to substorm effects as observed by Mitchell et al. [2003]. Relatively slow auroral wind heavy ions are uniquely capable of making multiple circuits around the global magnetospheric circulation flow, gaining energy from each cycle. This new finding suggests the possibility of cumulative effects on the ring current involving sustained magnetospheric convection, but needs additional investigation using a dynamic multifluid global simulation. Somewhat surprisingly, accelerated nightside auroral wind flows do not appear to contribute substantially to the ring current pressure here. Contributing factors include the following: we have assumed their flux to be substantially smaller; they do not initially travel as far down the tail; their greater initial energy means that they drift duskward and tend to escape downstream through the dusk side flank, rather than being recirculated through the inner magnetosphere.

Simulated velocity distribution features in the inner plasma sheet agree well with Polar observations of high Mach number lobal wind supply of the plasma sheet, with interpenetration of the polar wind streams from the two lobes and heating to form counterstreaming plasma sheet boundary layer populations and a hot isotropic central plasma sheet population. In the region observed by Polar, a hot solar wind proton distribution is found to occupy the current sheet region. These protons are on closed drift paths encircling the Earth for these conditions and may be considered part of the outer ring current. In the ring current-like region, the energy distributions of polar wind and solar wind protons agree very well with the results from AMPTE/CCE shown by Christon et al [1994] and support their identification of a soft component of polar wind origins, and an energetic peaked component of solar wind origins. The principal observed features of the inner plasma sheet are thus formed naturally from a combination of cold lobal wind outflows, combined with solar wind proton entry through the dawn flank region.

For conditions of moderate activity, magnetospheric transport of ionospheric and solar wind ions is very different. The solar wind entry path is via the dawn flank and thence rather directly into the inner plasma sheet and ring current regions. Consequently, the solar proton ring current should be less affected by substorms and other magnetotail phenomena, while the auroral wind outflows are expected to depend greatly on such processes. Thus, the behavior observed recently by Mitchell et al. [2003], with strong substorm modulation of oxygen fluxes in the ring current, but with relatively little modulation of proton fluxes, appears to follow from our results. However, to fully explore this will require trajectory simulations in dynamically evolving fields. Though this was beyond the scope of the current study, have developed the tools necessary for this and will report on it in the near future.

Acknowledgments. The authors acknowledge support from the NASA Polar Mission under UPN 370-08-43; from the NASA IMAGE Mission under UPN 370-28-20, from the NASA Geospace Sciences Program under UPN 344-42-01 for the Terrestrial Plasma Energization investigation, and from the NSF under grant ATM-0000251.

REFERENCES

Chandler, M. O., and T. E. Moore, Observations of the geopause at the equatorial magnetopause: Density and temperature, *Geophys. Res. Lett.*, SSC6, 15 Aug, 2003.

Chappell, C. R., T. E. Moore, and J. H. Waite, Jr., The ionosphere as a fully adequate source of plasma for the Earth's magnetosphere, *J. Geophys. Res.* 92, 5896, 1987.

Chen, S.-H., and T. E. Moore, Dayside flow bursts in the Earth's magnetosphere, *J. Geophys. Res.*, 109(A3), p.A03214, 18 March 2004.

Christon, S. P., G. Gloeckler, D. C. Hamilton, and F. M. Ipavich, A method for estimating the solar wind H+ contribution to magnetospheric plasma, in *Solar Terrestrial Energy Program, 5th COSPAR Colloquium*, D. N. Baker, V.O. Patitashvili and M.J. Teague, Eds., Pergammon Press, New York, 1994.

Cully, C.M., E.F. Donovan, A.W. Yau, and H.J. Opgenoorth, Supply of thermal ionospheric ions to the central plasma sheet *J. Geophys. Res*, 108(A2), 1092, 2003.

Daglis, I. A., G. Kasotakis, E. T. Sarris, Y. Kamide, S. Livi, and B. Wilken, Variations in the ion composition during an intense magnetic storm and their consequences, *Phys. Chem. Earth*, 24, p.229, 1999.

Delcourt, D. C., C. R. Chappell, T. E. Moore, and J. H. Waite, Jr., A three dimensional numerical model of ionospheric plasma in the magnetosphere, *J. Geophys. Res* 94, 11,893, 1989.

Delcourt, D. C., T. E. Moore, J. A. Sauvaud, and C. R. Chappell, "Non-Adiabatic Transport Features in the Outer Cusp Region," *J. Geophys. Res.*, 97(A11), 16,833–16,842, 1992.

Delcourt, D.C., J.A. Sauvaud, and T. E. Moore, Polar wind dynamics in the magnetotail, *J. Geophys. Res.*, 98(A6), 9155, 1993.

Delcourt, D. C., T. E. Moore, and C. R. Chappell, "Contributions of Low-Energy Ionospheric Protons to the Plasma Sheet," *J. Geophys. Res.,* 99(A4), 5681–5689, 1994.

Elphic, R. C., M. F. Thomsen, and J. E. Borovsky, The fate of the outer plasmasphere, *Geophys. Res. Lett.*, 24, 365, 1997.

Fedder, J. A., J. G. Lyon, S. P. Slinker, and C. M. Mobarry, Topological structure of the magnetotail as a function of interplanetary magnetic field direction, *J. Geophys. Res.*, 100, 3613–3621, 1995.

Foster, J., P. J. Erickson, A. J. Coster, J. Goldstein, and F. J. Rich, Ionospheric signatures of plasmaspheric tails, *Geophys. Res. Lett.*, v.29(13), p.1623, 2 July 2002.

Freeman, J. W., Jr., H. K. Hills, T. W. Hill, P. H. Reiff, and D. A. Hardy, Heavy ion circulation in the Earth's magnetosphere, *Geophys. Res. Lett.*, 4, 195, 1977.

Goldstein, J., *et al.,* IMF-driven Plasmasphere Erosion of 10 July 2000, *Geophys. Res. Lett.*, 30, doi:10.1029/2002GL016478, 2003.

Hamilton, D. C., G. Gloeckler, F. M. Ipavich, W. Studemann, B. Wilken, and G. Kremser, Ring current development during the great geomagnetic storm of February, 1986, J. Geophys. Res. 93, 14343, 1988.

Hill, T. W., 1974, Origin of the plasma sheet, Rev. Geophys. 12, 379.

Klumpar, D. M., Transversely accelerated ions: An ionospheric source of hot magnetospheric ions, *J. Geophys. Res.* 84, 4229, 1979.

Lennartsson, W, Tail lobe ion composition at energies of 0.1 to 16keV/e: Evidence for mass-dependent density gradients, *J. Geophys. Res.* 99(A2), p.2387, 1994.

Mitchell, D. G., Pontus C:Son Brandt, Edmond C. Roelof, Douglas C. Hamilton, Kyle C. Retterer, And Steven Mende, "Global Imaging Of O+ From Image/Hena," *Space Science Reviews* 0: 1–13, 2003.

Mobarry, C. M., J. A. Fedder, and J. G. Lyon, Equatorial plasma convection from global simulations of the Earth's magnetosphere, *J. Geophys. Res., 101*, 7859–7874, 1996.

Moore, T. E., and D.C. Delcourt, The Geopause, *Revs. Geophys.* 33(2), p. 175., 1995.

Moore, T. E., R. Lundin, et al., "Source processes in the high latitude ionosphere", Space Sci. Revs., 88(1–2), p.7; Chapter 2 of *Magnetospheric Plasma Sources and Losses*, ed. by B. Hultqvist and M. Oieroset, Kluwer, Dordrecht, Holland, 1999a.

Moore, T. E., W. K. Peterson, C. T. Russell, M. O. Chandler, M. R. Collier, H. L. Collin, P. D. Craven, R. Fitzenreiter, B. L. Giles, and C. J. Pollock, Ionospheric mass ejection in response to a CME, *Geophys. Res. Lett.*, 26(15), pp. 2339–2342, 1999b.

Moore, T. E., B. L. Giles, D. C. Delcourt, and M.-C. Fok, The plasma sheet source groove, *J. Atmos. Solar Terr. Phys.*, 62(6), p.505, 2000.

Moore, T. E., M.O. Chandler, M.-C. Fok, B.L. Giles, D.C. Delcourt, J.L. Horwitz, C.J. Pollock, Ring Currents and Internal Plasma Sources, *Space Science Reviews*, 95(1/2): 555–568, January, 2001.

Moore, T. E., M.-C. Fok, M. O. Chandler, C. R. Chappell, S. P. Christon, D. C. Delcourt, J. A. Fedder, M. Huddleston, M. W. Liemohn, W. K. Peterson, and S. Slinker, Plasma sheet and (non-storm) Ring Current Formation from Solar and Polar Wind Sources, J. Geophys. Res., in press, 2005.

Moore, T. E., M.-C. Fok, M. O. Chandler, C. R. Chappell, S. P. Christon, D. C. Delcourt, J. Fedder, M. Huddleston, M. Liemohn, W. K. Peterson, and S. Slinker, Plasma sheet and (nonstorm) ring current formation from solar and polar wind sources, *J. Geophys. Res.*, v.110, A02210, doi:10.1029/2004JA010563, 2005.

Peterson, W. K., R.D. Sharp, E.G. Shelley, R.G. Johnson, and H. Balsiger, Energetic Ion Composition of the Plasma Sheet, J. Geophys. Res. 86, 761, 1981.

Peterson, W.K., E.G. Shelley, G. Haerendel, and G. Paschmann, Energetic Ion Composition in the Subsolar Magnetopause and Boundary Layer, *J. Geophys. Res.* 87, 2139, 1982.

Peroomian V., The influence of the interplanetary magnetic field on the entry of solar wind ions into the magnetosphere, *Geophys. Res. Lett.*, 30 (7), 1407, doi: 10.1029/2002GL016627, 2003.

Richard, R. L., M. El-Alaoui, M. Ashour-Abdalla, and R. J. Walker, Interplanetary magnetic field control of the entry of solar energetic particles into the magnetosphere, *J. Geophys. Res.*, *107*(A8), 1184, 10.1029/2001JA000099, 2002.

Sandel, B. R., et al., Initial Results from the IMAGE Extreme Ultraviolet Imager, *Geophys. Res. Lett.*, 28(8), p. 1439, April 2001.

Sharp, R. D., R. G. Johnson, and E. G. Shelley, 1977, Observation of an ionospheric acceleration mechanism producing energetic (keV) ions primarily normal to the geomagnetic field direction, *J. Geophys. Res.* 82(22), 3324.

Sharp, R. D., W. Lennartsson, and R. J. Strangeway, The ionospheric contribution to the plasma environment in near-earth space, *Radio Sci.* 20, 456, 1985.

Shelley, E. G., R. G. Johnson, and R. D. Sharp, 1972, Satellite observations of energetic heavy ions during a geomagnetic storm, *J. Geophys. Res.* 77, 6104.

Slinker, S. P., J. A. Fedder, J. G. Lyon, Plasmoid formation and evolution in a numerical simulation of a substorm, *Geophys. Res. Lett.*, 22(7), 859–862, 10.1029/95GL00300, 1995.

Stern, D. P., The motion of a proton in the equatorial magnetosphere, *J. Geophys. Res.*, *80*, 595–599, 1975.

Strangeway, R. J., C. T. Russell, C. W. Carlson, J. P. McFadden, R. E. Ergun, M. Temerin, D. M. Klumpar, W. K. Peterson, T. E. Moore, Cusp Field-Aligned Currents and Ion Outflows, *J. Geophys. Res.* 105(A9), p.21129, 2000.

Su, Y.-J., J. L. Horwitz, T. E. Moore, M. O. Chandler, P. D. Craven, B. L. Giles, M. Hirahara, and C. J. Pollock, Polar wind survey with Thermal Ion Dynamics Experiment/Plasma Source Instrument suite aboard POLAR, *J. Geophys. Res.*, 103, 29,305, 1998.

Su, Y.-J., J. E. Borovsky, M. F. Thomsen, N. Dubouloz, M. O. Chandler, T. E. Moore, and M. Bouhram, Plasmaspheric material on high-latitude open field lines, *J. Geophys. Res.*, 106, 6085, 2001.

Thomsen, M. F., J. E. Borovsky, R. M. Skoug, and C. W. Smith, The delivery of cold, dense plasma sheet material into the near-earth region, *J. Geophys. Res.*, 108(A4), doi: 10.1029/ 2002JA009544, 2003.

Tsyganenko, N. A., A magnetospheric magnetic field model with a warped tail current sheet, *Planet. Space Sci.* 37, p.5, 1989.

Volland, H., A model of the magnetospheric convection electric field, *J. Geophys. Res.*,*83*, 2695–2705, 1978.

Walker, R. J., M. Ashour-Abdalla, T. Ogino, V. Peroomian, and R. L. Richard, Modeling magnetospheric sources, in *Earth's Low-Latitude Boundary Layer, Geophys. Monogr. Ser.*, vol. 133, edited by P. Newell and T. Onsager, p. 33, 2003.

Winglee, R. M., Circulation of ionospheric and solar wind particle populations during extended southward interplanetary magnetic field, *J. Geophys. Res.*, Vol. 108, No. A10, 1385, 10.1029/2002JA009819, 2003.

S. H Chen, NASA's Goddard Space Flight Center, LSSP Code 612.2, Greenbelt, MD 20771 USA

M. O. Chandler, NASA Marshall Space Flight Center, NSSTC, 320 Sparkman Dr., Huntsville, AL 35805 USA

S. P. Christon, NASA's Goddard Space Flight Center, LSSP Code 612.2, Greenbelt, MD 20771 USA

D. C. Delcourt, CETP, 4, Ave. de Neptune, St. Maur, 94107, France

J. Fedder, LET Inc., Washington DC USA

M. C. Fok, NASA's Goddard Space Flight Center, LSSP Code 612.2, Greenbelt, MD 20771 USA

M. Liemohn, University of Michigan, SPRL, 2455 Hayward St., Ann Arbor, MI 48109-2143 USA

T. E. Moore, NASA's Goddard Space Flight Center, LSSP Code 612.2, Greenbelt, MD 20771 USA

S. Slinker, Naval Research Laboratory, 4555 Overlook Ave. SW, Washington, DC 20375-5000 USA

Statistical Properties of Dayside Subauroral Proton Flashes Observed With IMAGE-FUV

Benoît Hubert[1], Jean-Claude Gérard[1], Stephen B. Mende[2], Stephen A. Fuselier[3]

The SI12 instrument of the FUV experiment onboard the IMAGE satellite is specifically devoted to the observation of the proton aurora. Transient subauroral proton aurora was detected with SI12 in response to a solar wind dynamic pressure increase. These Dayside Subauroral Proton Flashes (DSPF's) take place on field lines of L-Shell as low as 4, and possibly result of an increase of EMIC growth rate instability due to the compression of the dayside magnetosphere by the increased solar wind dynamic pressure. In this study, a set of 75 DSPF's observed with SI12 related with a solar wind dynamic pressure increase is studied. Statistical distributions of relevant quantitative and morphologic indicators of the DSPF's properties are computed. Correlations between these indicators and the solar wind properties are also studied. It is found that the solar wind dynamic pressure is the key parameter controlling the DSPF maximum power, maximum flux, magnetic latitude and extent in MLT. Also, DSPF's occur preferentially in the afternoon sector, where the plasma temperature anisotropy is higher, so that the EMIC instability threshold is more easily exceeded. Moreover, no correlation is found between the DSPF's characteristic decay time and the solar wind properties, suggesting that this parameter is internally controlled by the properties of the magnetospheric plasma. In this dataset, no correlation is found relating the IMF and the DSPF properties.

1. INTRODUCTION

Since the launch of the IMAGE spacecraft in March 2000 [Burch 2000], the Spectrographic Imager at 121,8 nm (SI12) instrument of the IMAGE-FUV experiment [Mende et al., 2000 a, b] has been widely used to image the Earth's proton aurora on a global scale. This experiment also includes two other far ultraviolet imagers, the Wideband Imaging Camera (WIC) and the Spectrographic Imager at 135.6 nm (SI13), providing images of the N_2-LBH band and OI-135.6 nm emissions respectively. These two emissions are mainly excited by electron impact, but they are also present in the proton aurora [*Hubert et al.*, 2001].

Recently, a new transient dayside subauroral feature was observed by *Hubert et al.* [2003]. These Dayside Subauroral Proton Flashes (DSPF) are connected to an increase of the solar wind dynamic pressure compressing the dayside magnetosphere. A comparison of the SI12, SI13 and WIC observations revealed that DSPF's are due to proton precipitations. Using in situ particles measurement, *Zhang et al.* [2003] confirmed that the precipitation causing these features is mostly composed of energetic protons. As shown by *Hubert et al.* [2003], the field lines threading the observed DSPF's map in the equatorial plane at distances as low as 4 R_E. It must be noted that DSPF's are possibly related with the subauroral emissions previously reported by *Liou et al.* [2002]

[1]Laboratoire de Physique Atmosphérique et Planétaire, Université de Liège, Belgium
[2]Space Sciences Laboratory, University of California, Berkeley, California, USA
[3]Lockheed Martin Advanced Technology Center, Palo Alto, California, USA

Inner Magnetosphere Interactions: New Perspectives from Imaging
Geophysical Monograph Series 159
Copyright 2005 by the American Geophysical Union.
10.1029/159GM15

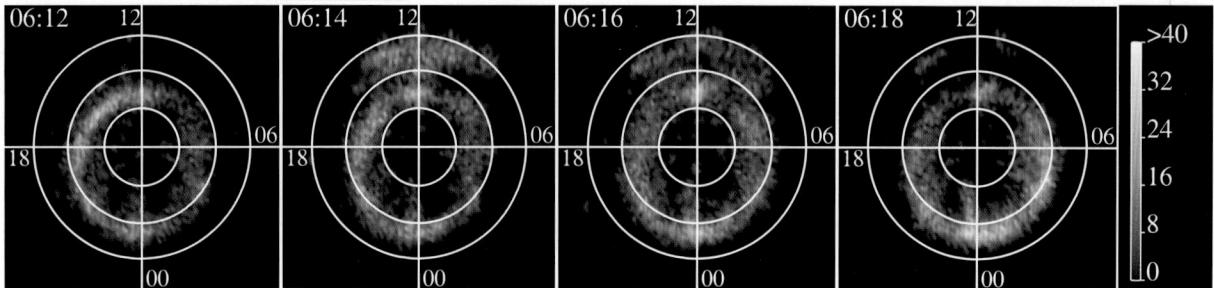

Figure 1. SI12 counts remapped in geomagnetic coordinates showing the subauroral proton flash of November 8 2000 at 0614 UT. The background has been removed. Concentric yellow circles are 10° MLAT apart, noon is at the top of each picture (MLT=12).

using POLAR-UVI data, without the ability to distinguish between electron and proton precipitations.

Fuselier et al. [2004] described a mechanism responsible for the proton precipitation in Dayside Subauroral Proton Flashes (DSPF): following compression of the dayside magnetosphere by the increased solar wind dynamic pressure, the Electromagnetic Ion Cyclotron (EMIC) growth rate turns unstable. This instability diverts protons into the loss cone along low L-shell field lines. Indeed, the stable/unstable issue of the EMIC growth rate is controlled by the β plasma parameter and the temperature anisotropy. Balance of the competing effects leads to an instability region corresponding to subauroral latitudes, eventually leading to a gap separating the auroral oval and the proton flash, as observed by *Hubert et al.* [2003].

In this paper, we use a set of 75 Dayside Subauroral Proton Flashes observed with IMAGE-FUV in order to determine their statistical properties. This set of events was built following a detailed inspection of the IMAGE-FUV dataset and of the solar wind properties measured with the ACE, WIND and GEOTAIL spacecrafts. In the present study, events were selected when a transient proton precipitation is seen in the SI12 images and is related with an increase of the solar wind dynamic pressure obtained from the ACE, GEOTAIL and/or WIND satellites (DSPF's appearing in the absence of a pressure increase are not included in this dataset and will be the subject of further studies). Not only CME-induced events were included in the set, but also weak proton flashes related with a moderate solar wind dynamic pressure increase.

2. POWER AND DECAY TIME

2.1. Statistical Distribution

The power precipitated into each proton flash was determined using the SI12 observations. Figure 1 presents the SI12 images obtained for the DSPF observed on November 8 2000 at 0614 UT, as already presented in Hubert et al. [2003] and reproduced here for convenience. The DSPF clearly appears on the dayside, well centered on the noon sector, and detached from the auroral oval. The power of the proton precipitation of the flash observed on November 8 2000 is calculated for each SI12 image [Hubert et al., 2002]. It is plotted versus time in Figure 2, with t = 0 corresponding to 0612 UT. The sudden increase of the power is conspicuous, and takes place on a time scale shorter than 2 minutes, i.e. less than the time resolution of the SI12 observations. The maximum power obtained from the SI12 data is 0.53 GW. After reaching its maximum value, the power decreases roughly exponentially, with a characteristic decay time of ~2 minutes. In this case, a second minor peak is observed some 12 minutes after the main peak. Considering that the solar wind dynamic pressure related with this event [Hubert et al., 2003], as deduced from the ACE satellite measurements, presents two successive ramps separated by ~12 minutes, it may be speculated that the main peak of the subauroral proton flash is related to the first pressure increase, whereas the second dynamic pressure

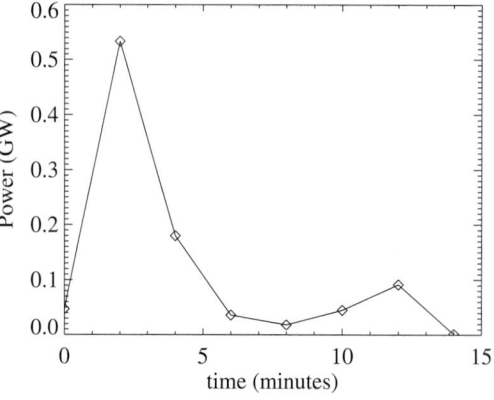

Figure 2. Proton power deduced from the SI12 observations of the Dayside Subauroral Proton Flash that occurred on November 8 2000 at 0614 UT, versus time. The time=0 mark corresponds to 0612 UT.

increase, though much more spectacular than the first one, is responsible for the minor peak of the observed DSPF, the dayside flux tubes having already been emptied of a significant part of their proton population.

Considering this example, it is natural to define two parameters describing the properties of a DSPF: the maximum power reached during the event, and its characteristic exponential decay time. The maximum power is an indicator of the brightness of the flash as a whole. The actual peak value can be larger, as the time resolution of 2 minutes could easily miss the peak value. Figure 3 shows the distribution of the maximum proton power for our set of selected DSPF's. This asymmetric distribution has an average value of 0.24 ± 0.003 GW, the standard deviation of the distribution being $\sigma = 0.26$ GW. The relation between the amplitude of the solar wind pressure pulse and the power of the proton precipitation of the resulting Dayside Subauroral Proton Flash will be investigated later. Figure 4 presents the distribution of the DSPF characteristic decay time obtained by fitting an exponential function to the proton power curve of each event. It is, on average ~199 ± 15 s. Anticipating on the following section, we note that no correlation could be found between the determined decay time and the solar wind properties. This suggests that the decay time is internally controlled by the properties of the magnetosphere. It must also be stressed that decay times smaller than 2 minutes are poorly estimated, because the FUV experiment has a time resolution of 2 minutes.

2.2. Correlation With the Solar Wind Parameters

In the present study, the criterion used to asses the correlation or uncorrelation between two parameters is the

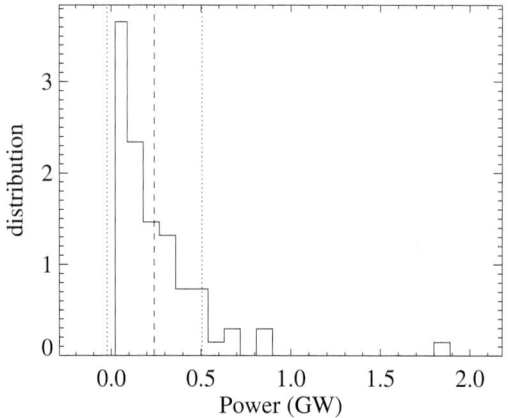

Figure 3. Distribution function of the maximum power reached during the observed DSPF's. The vertical dashed line represents the average value (m=0.24 GW), the vertical dotted lines are the average plus/minus one standard deviation (σ=0.26 GW). The uncertainty on m is thus ~.003 GW, that is ~13%.

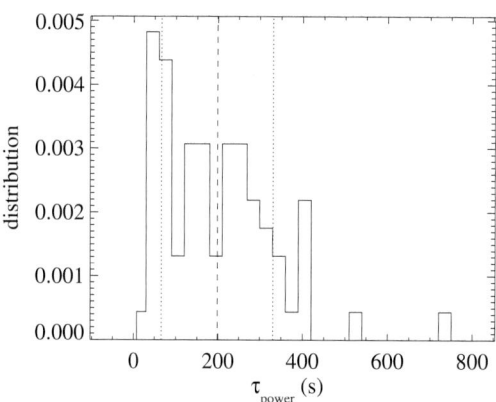

Figure 4. Distribution function of the characteristic decay time of the observed DSPF's. The average value is m=199 ± 15 s, and the standard deviation of the distribution is σ=132 s. The vertical dashed line represents m, the vertical dotted lines represent m ± σ.

uncorrelation criterion of Fisher [Press et al., 1989]. If r is the linear correlation coefficient, Z is defined as

$$Z = \tfrac{1}{2} \ln\left(\frac{1+r}{1-r}\right) \qquad (1)$$

Let also u_β be defined such that there is a probability β for a Gaussian random variable of mean 0 and standard deviation 1 to be smaller than u_β. Fisher's criterion then states that two variables are uncorrelated under a level of confidence α when the relation

$$\frac{-u_{1-\tfrac{\alpha}{2}}}{\sqrt{n-3}} \leq Z \leq \frac{u_{1-\tfrac{\alpha}{2}}}{\sqrt{n-3}} \qquad (2)$$

is verified, where n is the number of observations ($n \geq 10$). We will thus accept the hypothesis that two variables are correlated under the level of confidence 1-α when the relation (2) is not verified, that is when

$$|Z| > \frac{-u_{1-\tfrac{\alpha}{2}}}{\sqrt{n-3}} \qquad (3)$$

The larger |r|, the larger |Z|, so that one need |r| to be sufficiently large to accept the correlation hypothesis. As u_β increases with β, it is clear that the smaller α, the more constraining the constraint of relation (2) being not fulfilled. As a consequence, an increase of the level of confidence 1-α strengthens the requirements for Fisher's correlation test, as expected. If we select a confidence level of 0.9, i.e. $\alpha = 0.1$, it follows that $u_{1-\tfrac{\alpha}{2}}$=1.65, so that for n=75, the critical value for Z discriminating between uncorrelated and correlated variables corresponds to r ~ 0.2 only.

For those DSPF's apparently driven by a solar wind dynamics pressure increase, it is expected that the brightness, and hence the power of the observed DSPF's is

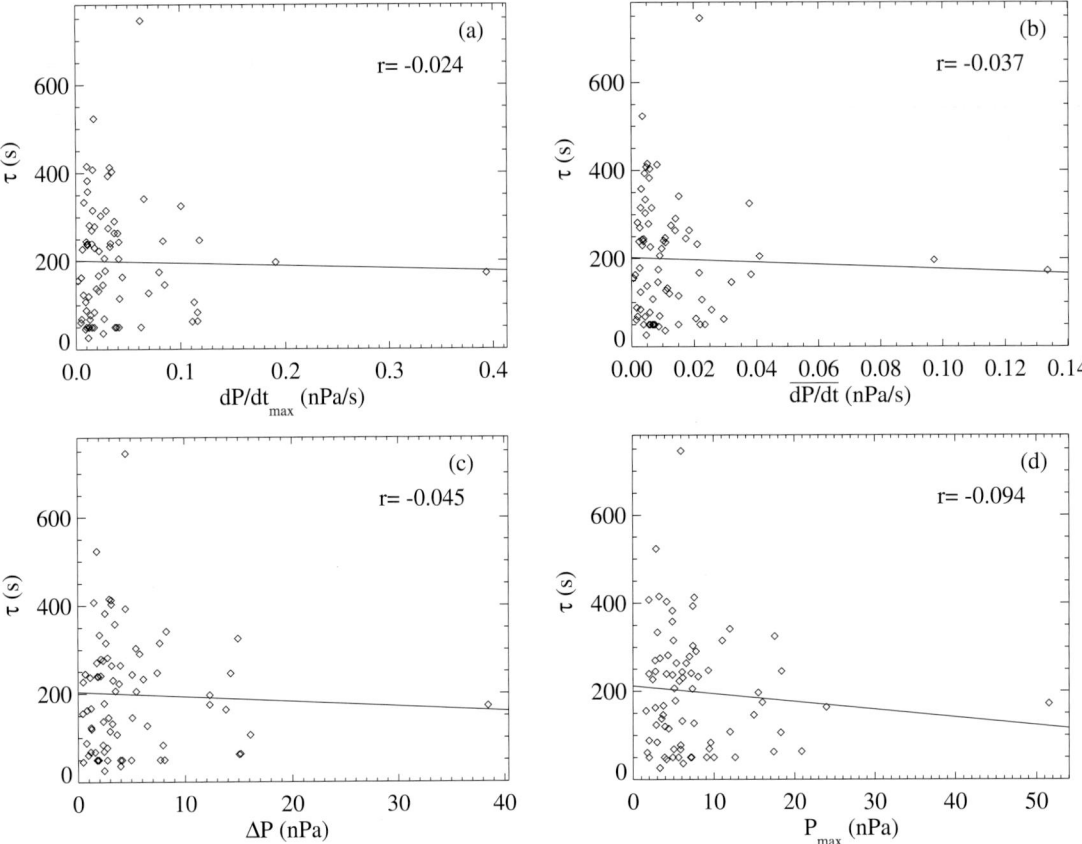

Figure 5. Correlation between the decay time of the proton power of the observed DSPF's and the corresponding maximum value of the temporal derivative of the solar wind dynamic pressure (a), the average value of the derivative (b), the solar wind dynamic pressure variation (c) and the maximum value of P_{dyn} (d), deduced from solar wind data of the ACE, WIND and GEOTAIL satellites. Each diamond represents an event, the solid line is a linear best fit to the observations. All four correlation coefficients are smaller than the threshold value of 0.2, so that the decay time is not correlated with these four quantities.

related in some way to the solar wind dynamic pressure P_{dyn} (assuming that the protons content of the disturbed flux tubes is sufficiently high). The time interval of the solar wind data related with an observed DSPF was individually identified for each case, accounting for the solar wind propagation time from the satellite to the front of the magnetosphere. Several quantities can then be defined to describe the dynamic pressure pulse (even in the case of a weak pulse) that is responsible for the DSPF proton precipitation. First, the maximum value reached by the temporal derivative of P_{dyn}, $\frac{dP}{dt}_{max}$ is an indicator of the strength of the dynamic pressure increase. However, this maximum value is only a punctual indicator, and a second indicator can be considered for the pressure ramp as a whole: the average temporal derivative of P_{dyn}, $\overline{\frac{dP}{dt}}$ computed on the time interval starting right prior to the pressure increase and ending when the dynamics pressure reaches its maximum.

Even if the average temporal derivative of P_{dyn} is large, the pressure increase may take place during such a short period of time that the pressure shock would actually be of small amplitude. Consequently, both the maximum pressure reached during the event, P_{max}, and the solar wind dynamic pressure variation ΔP, i.e. the solar wind dynamic pressure increase across the event, also appear as valuable indicators of the properties of the solar wind pressure pulse.

As already mentioned in a previous paragraph, no correlation could be found between the characteristic decay time of the power of the Dayside Subauroral Proton Flashes and the solar wind properties (Figure 5). Correlation coefficients of -0.024, -0.037, -0.045 and -0.094 were found with $\frac{dP}{dt}_{max}$, $\overline{\frac{dP}{dt}}$, ΔP and P_{max} respectively. The absence of correlation suggests that this parameter is controlled by the internal properties of the magnetosphere, which integrates over the longer term history of the system.

Figure 6a presents the correlation between the proton flash maximum power W_{p-max} and $\frac{dP}{dt}_{max}$. With a correlation coefficient of 0.78, larger than the threshold value of 0.2 determined before, these two quantities are significantly correlated. The outlier at $W_{p-max} \sim 1.8$ GW is a reliable point representing an event characterized by a large solar wind velocity of ~900 km/s. This point strongly constrains the regression. If ignored, the correlation coefficient is r = 0.54, still leading to the same conclusion regarding the correlation. The relations between W_{p-max} and the average temporal derivative $\frac{dP}{dt}$, the pressure variation ΔP and the maximum solar wind dynamic pressure P_{max} are shown as well and all present a statistically significant correlation. As the four correlation coefficients are roughly equal to each other, the four indicators proposed here to quantify the strength of the solar wind pressure pulse appear as equivalent. These correlations simply indicate that the stronger the solar wind disturbance, the stronger the response in power of the proton precipitation to the solar wind pressure pulse. However, the large dispersion of the data indicates the solar wind dynamic pressure is not the only parameter controlling the strength of the proton precipitation. One can indeed expect that the state of the magnetosphere also constrains the auroral response to the pressure increase. In addition, a correlation does not necessarily imply a causal link. From a physical standpoint, inferring such a causal link at the light of a correlation study only makes sense if an underlying precipitation mechanism driven by the pressure increase can be identified. Such a mechanism diverting protons into the loss cone was briefly proposed by Hubert et al. [2003] and thoroughly analyzed by Fuselier et al. [2004]. Two possible behaviors of the observed DSPF's may explain the dependence of W_{p-max} on the solar wind dynamic pressure: the precipitated proton flux can be stronger, or the spatial extent of the DSPF could be larger

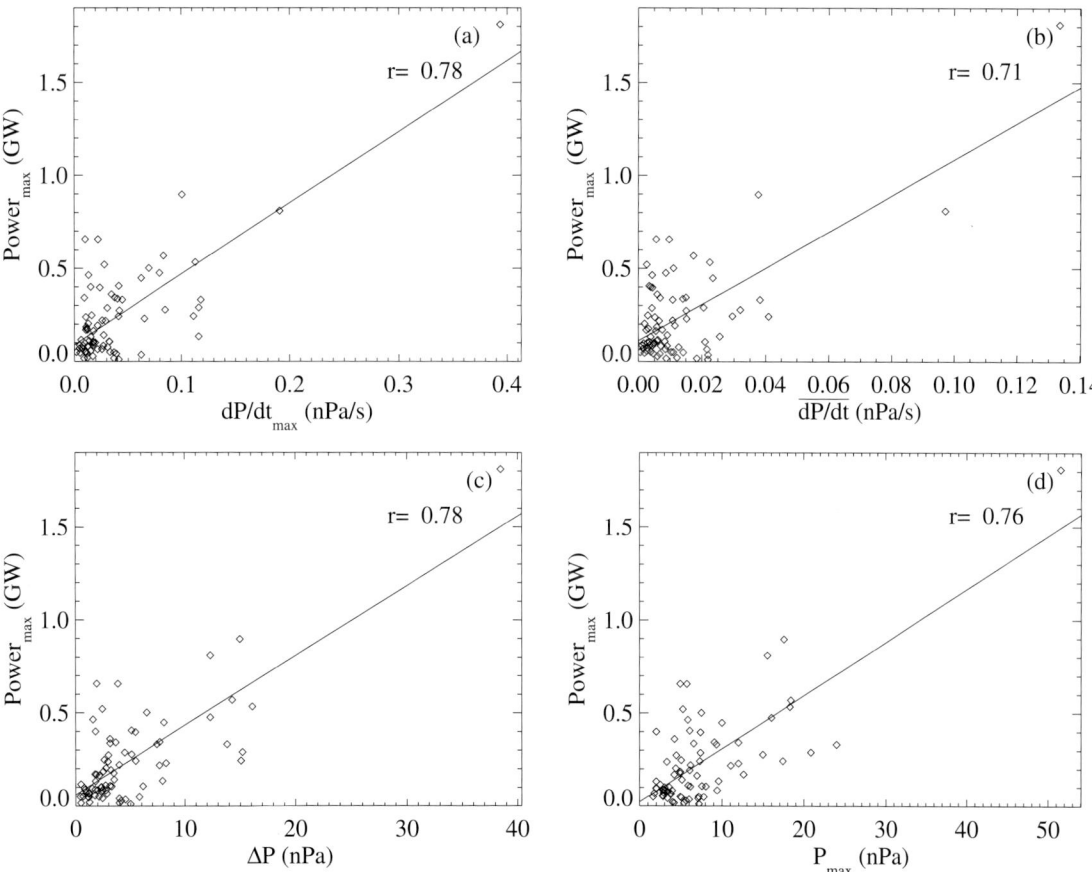

Figure 6. Correlation between the maximum power reached during the observed DSPF's and the corresponding maximum value of the temporal derivative of the solar wind dynamic pressure (a), the average value of the derivative (b), the solar wind dynamic pressure variation (c) and the maximum value of P_{dyn} (d), deduced from solar wind data of the ACE, WIND and GEOTAIL satellites. Each diamond represents an event, the solid line is a linear best fit to the observations. All four correlation coefficients are larger than the threshold value of 0.2, so that the maximum power is correlated with these four quantities.

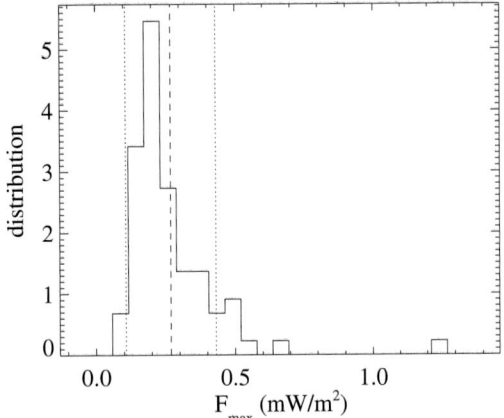

Figure 7. Statistical distribution of the maximum proton flux reached during the observed DSPF events. The average value is m = 0.27 ± 0.02 mW/m² and the standard deviation of the distribution is σ = 0.16 mW/m². The solid vertical line represents the average m, and the dotted vertical lines are at m ± σ.

(or both), when the solar wind pressure increase is stronger. We discuss these points in the next paragraphs.

3. PROTON FLUX

The maximum value reached by the proton flux during the event, F_{max}, can also be considered as an indicator of the brightness of the observed DSPF's. This indicator works at the local scale, in contrast with W_{p-max} that concerns the global scale. Figure 7 presents the statistical distribution of F_{max}. The predominance of rather weak events also appears in the asymmetry of the distribution. The average value is 0.27 ± 0.02 mW/m².

Figure 8 presents the correlation of F_{max} with $\frac{dP}{dt}_{max}$ (a), $\overline{\frac{dP}{dt}}$ (b), ΔP (c) and P_{max} (d). The correlation coefficients are 0.69, 0.66, 0.67 and 0.69 respectively. These results indicate that larger solar wind pressure pulses result in larger proton precipitation. This can be understood in terms of the mechanism proposed by *Fuselier et al.*

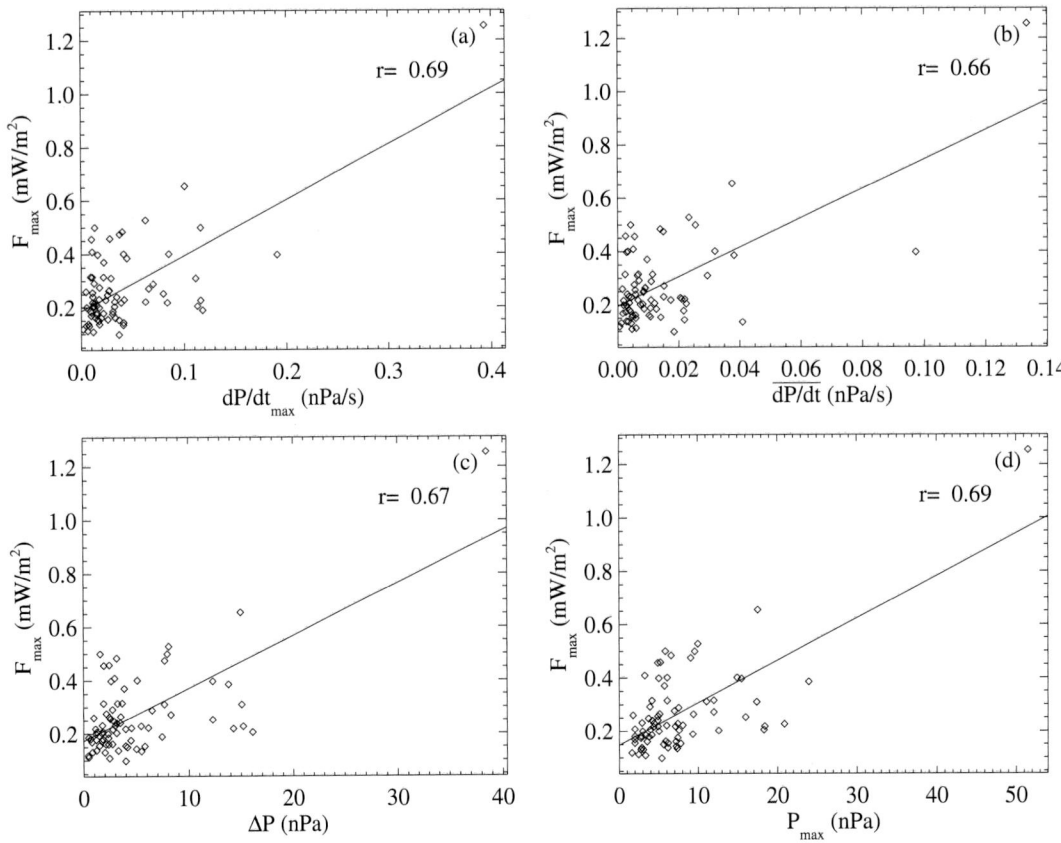

Figure 8. Correlation between the maximum proton flux reached during the observed DSPF's and the corresponding maximum value of the temporal derivative of the solar wind dynamic pressure (a), the average value of this derivative (b), the solar wind dynamic pressure variation (c) and the maximum value of P_{dyn} (d), deduced from solar wind data of the ACE, WIND and GEOTAIL satellites. Each diamond represents an event, the solid line is a linear best fit to the observations. The correlation coefficients are all larger than the threshold value of 0.2.

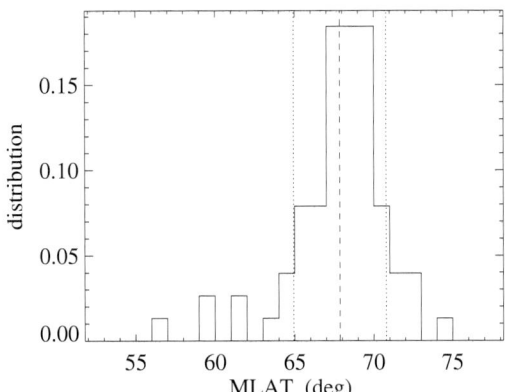

Figure 9. Distribution of the central MLAT of the observed DSPF's. The average MLAT is m = 68° ± 0.4° (dashed line), the standard deviation of the distribution is σ = 3°. The dotted lines are for m ± σ.

MLAT location of the observed DSPF is determined by the field lines along which the disturbance efficiently fills the loss cone, by establishing an unstable EMIC growth rate for example. The MLT extent of the DSPF quantifies the size of the magnetospheric region compressed by the solar wind pressure increase.

Figure 9 shows the distribution of the MLAT of the center of the dayside subauroral proton flash, defined as the average MLAT weighted by the proton flux. The distribution of MLAT is centered on 68° ± 0.3°, with a standard deviation of 3°. About 8% of the observed DSPF's occur at an average MLAT less than 65°. Figure 10 presents the distribution of the average magnetic local time (MLT) of the observed DSPF's and their MLT extent (ΔMLT). The DSPF's are seen preferentially in the afternoon sector (MLT = 1258 ± 0009 MLT on average). This may be related with the asymmetry of the temperature anisotropy observed in the dayside magnetosphere, this anisotropy being higher in the afternoon sector [*Thomsen M. F., personal communication, Anderson et al., 1996*]. A larger temperature anisotropy favors the EMIC instability thought to be responsible for the proton precipitation in DSPF's [*Fuselier et al., 2004*]. The average value of ΔMLT is 3.6 ± 0.18 MLT hours, the standard deviation of its distribution is 1.3 MLT hour.

[2004]: the stronger the compression of the dayside magnetosphere, the larger the disturbance of the inner geomagnetic field at dayside. Thus the EMIC growth rate will be larger and will turn more unstable.

4. MORPHOLOGICAL PARAMETERS

4.1. Statistical Distributions

We now focus on the magnetic latitude (MLAT) and the extent in magnetic local time (ΔMLT) of the proton precipitation of the observed DSPF events. A relation between these morphological parameters and the solar wind variation triggering the DSPF is expected. The

4.2. Correlation With Solar Wind Parameters

As shown in Figure 11, the magnetic latitude of the observed DSPF's appears statistically anticorrelated with the solar wind dynamic pressure variation and maximum value. The correlation coefficients with $\frac{dP}{dt}_{max}$, $\frac{dP}{dt}$, ΔP and

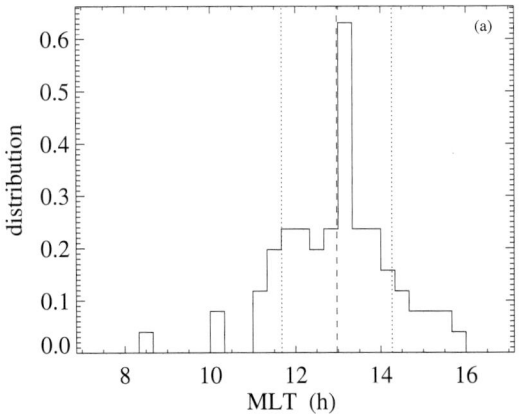
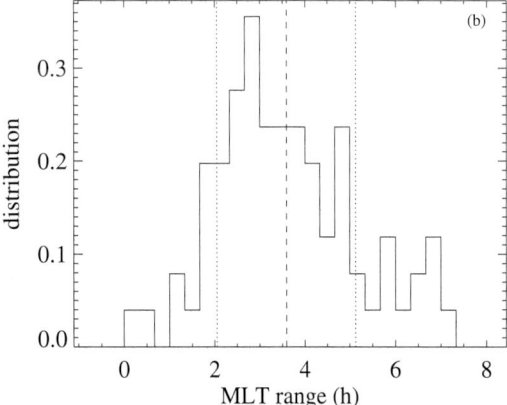

Figure 10. Statistical distribution of the average MLT of the observed DSPF's (a) and of their MLT extent (b). Vertical dashed lines indicate the average m of the distribution, dotted lines indicate m ± σ. The average MLT distribution is centered on 1258 ± 0009 MLT with a standard deviation of 0118 MLT hour. The MLT extent is 3.6 ± 0.18 MLT hours on average, the standard deviation of its distribution is 1.54 MLT hour.

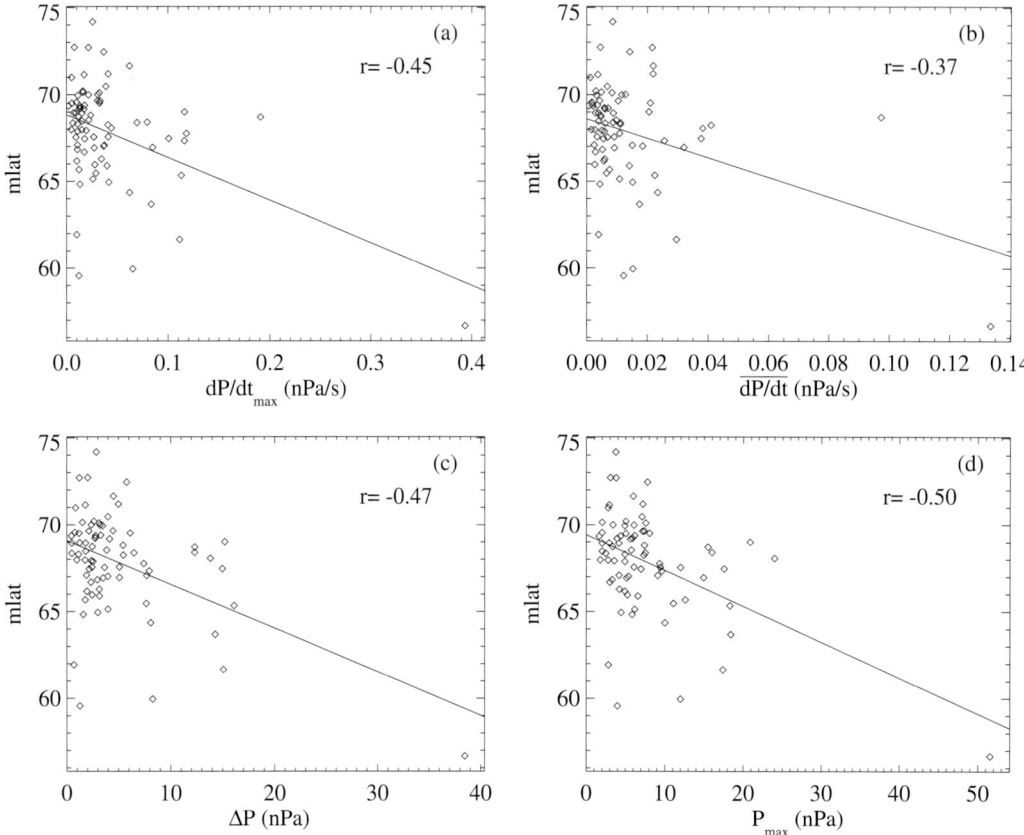

Figure 11. Correlation of the magnetic latitude of the observed DSPF's and the corresponding maximal value of the temporal derivative of the solar wind dynamic pressure (a), the average value of this derivative (b), the solar wind dynamic pressure variation (c) and the maximum value of P_{dyn} (d), deduced from solar wind data of the ACE, WIND and GEOTAIL satellites. Each diamond represents an event, the solid line is a linear best fit to the observations. The correlation coefficients are all larger (in absolute value) than the threshold of 0.2.

P_{max} are −0.45, -0.37, -0.47 and -0.50 respectively. This tendency is weakly pronounced, as the low correlation coefficients suggest, but it is nevertheless compatible with the EMIC mechanism proposed by *Fuselier et al.* [2004]: a stronger compression of the magnetosphere results in stronger disturbances at deeper L-shell, causing the instability threshold to be overcome on field lines threading regions of lower magnetic latitude.

Figure 12 examines the correlation of the MLT extent of the DSPF's and $\frac{dP}{dt}_{max}$, $\overline{\frac{dP}{dt}}$, ΔP and P_{max}. The correlation coefficients are 0.22, 0.13, 0.22 and 0.17 respectively. Only correlation coefficients larger than 0.2 can be considered as representing a correlation at a level of confidence of 0.9, as was discussed before. Nevertheless, rejecting the outlier having $\frac{dP}{dt}_{max} \approx 0.39$ nPa/s from the dataset leads to correlation coefficients of 0.32, 0.16, 0.30 and 0.23 respectively, so that the correlation hypothesis can be considered with all variables but $\overline{\frac{dP}{dt}}$, this case being disturbed by a second outlier at $\frac{dP}{dt} \approx 0.95$ nPa/s. The correlation of the MLT extent with the solar wind dynamic pressure indicators of the pressure pulse is not sharp at all. The morphology of the proton precipitation in DSPF's can also depend on other solar wind parameters, such as the orientation of the shock normal etc., but also on the state of the disturbed flux tubes, if we refer to the EMIC-based precipitation mechanism of *Fuselier et al.* [2004]. The magnitude of the solar wind pressure pulse is probably not the factor controlling the MLT extent of the DSPF precipitation.

5. INFLUENCE OF THE IMF

No correlation could be found between the IMF components and the morphological and quantitative properties of the observed DSPF's. Most of the correlation coefficients of the DSPF's properties and the IMF components were close to 0.2, and generally lower. The visual inspection of the few cases of correlation revealed that outliers were responsible for the alleged correlation. We also tested the correlation between the IMF components averaged over a few minutes to an hour before the DSPF events and found no correlation, so that no preconditioning of the magnetosphere by the IMF could be established based on our dataset. *Sonnerup and Cahill [1967]* proposed a method to determine the shock normal based on IMF measurements. We conducted a study to determine the possible relation between the shock normal orientation and the central MLT of the observed DSPF's. This study revealed inconclusive, but it must be noted that the concept of shock normal is loosely defined in the case of a small pressure increase.

6. DISCUSSION

Figure 6 suggests a correlation between the proton flash power (indicator of the global brightness of the observed dayside subauroral proton flashes) and the four dynamic pressure indicators used here, despite the scatter of the data. The EMIC mechanism is compatible with such a correlation, and the causal relation between the pressure increase and the proton flashes is demonstrated, as expected. A similar conclusion can be drawn concerning F_{max} (local indicator of the DSPF brightness) and its correlation with the solar wind dynamic pressure variation. Figure 8 shows the tendency: a stronger pressure increase leads to a more intense proton precipitation at dayside subauroral latitudes, with nevertheless some scatter of the data. Both the quantitative statistical criterion and the physical mechanism proposed to explain the proton precipitation are compatible with the conclusion of a correlation between the intensity of the pressure increase and the peak proton flux of the proton flash. The dispersion of the data is actually not surprising, as the compression of the dayside magnetosphere is not the only parameter controlling the precipitation mechanism. The variability of the plasma properties, in particular the magnitude of the trapped particle reservoir inside of the magnetosphere, probably play a role on the amount of precipitated proton flux and power.

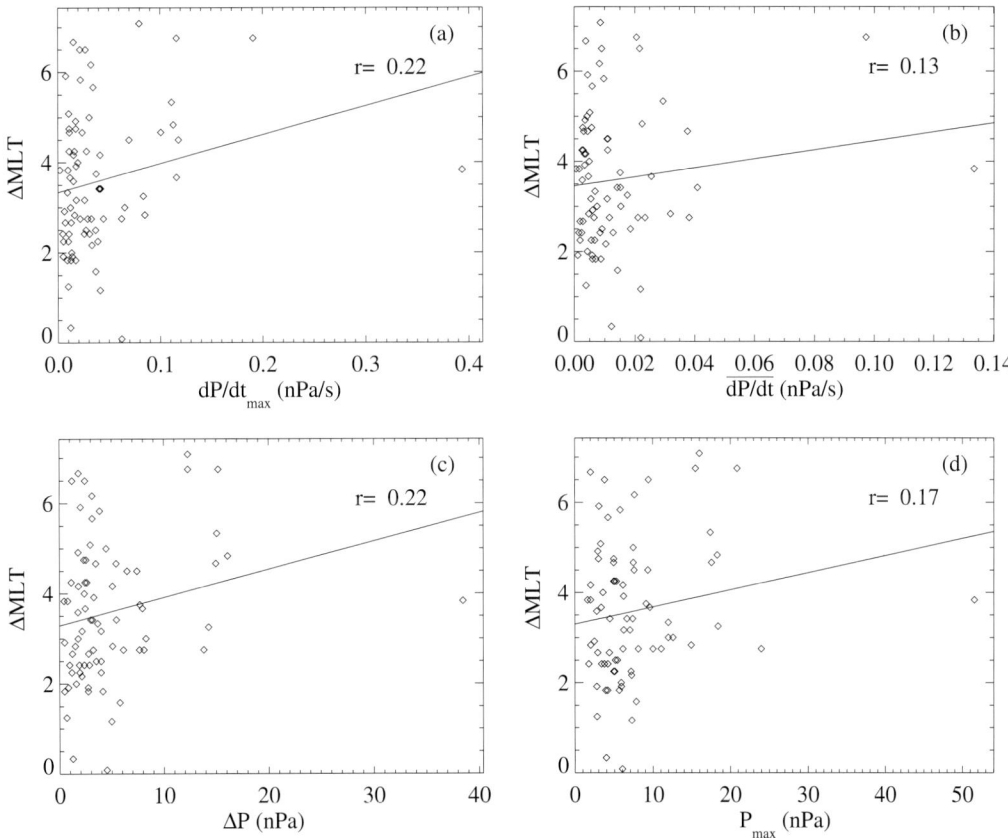

Figure 12. Correlation of the magnetic local time extent (ΔMLT) of the observed DSPF's and the corresponding maximal value of the temporal derivative of the solar wind dynamic pressure (a), the average value of this derivative (b), the solar wind dynamic pressure variation (c) and the maximum value of P_{dyn} (d), deduced from solar wind data of the ACE, WIND and GEOTAIL satellites. Each diamond represents an event, the solid line is a linear best fit to the observations. The correlation coefficients are 0.22 with $\frac{dP}{dt}_{max}$, 0.13 with $\overline{\frac{dP}{dt}}$, 0.22 with ΔP and 0.17 with P_{max}.

The correlation between the magnetic latitude of the observed DSPF's and the solar wind dynamic pressure variation also suffers a large scatter of the data, as shown in Figure 11. Visual inspection of the plots raises doubts concerning the relation between $\frac{dP}{dt}$ and MLAT (Figure 11 b). Nevertheless, the tendency remains apparent for the three other pressure indicators. Actually, a strict causal link is not expected between MLAT and the solar wind dynamic pressure increase, for the location of the proton precipitation is directly related with the field lines mapping to the region where the magnetospheric plasma has the required properties to allow the EMIC growth rate to turn unstable. The morphological properties of the proton flashes are thus expected to be only partly related with the solar wind dynamic pressure increase. The correlations found in this study, though low in the absolute sense, may be significant considering the complex mechanism relating the P_{dyn} increase and the final proton precipitation.

The lack of correlation of the characteristic decay time of the observed DSPF's and the solar wind properties suggests an internal magnetospheric control of the decay time. The absence of correlation between the observed proton flashes and the IMF is actually not unexpected considering that the precipitation mechanism is not directly related to a reconnection process.

The dataset presented in this study only includes cases of DSPF's observed in conjunction with a solar wind dynamic pressure increase. In addition, 47 weak DSPF cases were also found that developed in the absence of a solar wind dynamic pressure increase. These cases are not in contradiction with the causal link between the solar wind pressure and the subauroral proton precipitations, if one admits that any disturbance able to modify the EMIC growth rate up to the instability threshold can generate a dayside subauroral proton flash. We thus suspect that there exists at least one process other than dynamic pressure pulses able to trigger a DSPF precipitation. One such possibility is a directional discontinuity [*Burlaga,* 1971] causing a sudden change in the normal direction, so that, at the dayside magnetopause, local dynamic pressure variations generate local disturbances propagating to the inner magnetosphere and trigger subauroral proton precipitation.

7. CONCLUSIONS

In this study, we investigated the statistical morphology and the relation between the Dayside Subauroral Proton Flash phenomenon and the solar wind properties. A solar wind dynamic pressure increase is a driver able to trigger a DSPF, the intensity of which is dependent on the intensity of solar wind dynamic pressure increase, both at the global and local scales. This parameter also partly controls the morphology of the flash, its magnetic latitude and magnetic local time extent being weakly correlated with the magnitude of the pressure increase. The IMF does not appear as a factor controlling the precipitation mechanism, excluding the possibility of a mechanism dependent on a reconnection process between the magnetospheric field and the IMF. The characteristic decay time of the DSPF's does not depend on the solar wind conditions. We thus speculate that the decay time is internally controlled by the properties of the plasma of the inner magnetosphere. The dataset presented here all appear compatible with the mechanism based on the EMIC growth rate proposed by *Fuselier et al.* [2004]. Other triggering mechanisms than a solar wind dynamic pressure increase must also be considered, as DSPF's were also observed in the absence of such a pressure increase.

Acknowledgements. The success of the IMAGE mission is a tribute to the many dedicated scientists and engineers that have worked and continue to work on the project. The PI for the mission is Dr. J. L. Burch. Jean-Claude Gérard and Benoît Hubert are supported by the Belgian National Fund for Scientific Research (FNRS). This work was funded by the PRODEX program of the European Space Agency (ESA) and the Fund for Collective and Fundamental Research (FRFC grant 01-2.4569.01). Research at Lockheed Martin was supported by the IMAGE data analysis program through subcontract from the University of California, Berkeley. The IMAGE-FUV investigation was supported by NASA through SWRI subcontract number 83820 at the University of California, Berkeley, contract NAS5-96020. ACE level 2 data were provided by N.F. Ness (MFI) and D. J. McComas (SWEPAM), and the ACE Science Center. GEOTAIL data (L. Frank, U. Iowa) and WIND data (R. Lepping, NASA/GSFC) were obtained through the CDAweb site.

REFERENCES

Anderson, B. J., Denton R. E., Ho G., Hamilton D. C., Fuselier S. A., and Strangeway R. J., Observational test of local proton cyclotron instability in the Earth's magnetosphere, *J. Geophys. Res.*, 101, 21527, 1996.

Burch, J. L., IMAGE Mission overview, *Space Science Reviews*, 91, 1, 2000

Burlaga, L. F., Nature and origin of directional discontinuities in the solar wind, *J. Geophys. Res.*, 76, 4360, 1971.

Fuselier, S. A., , S. P. Gary, M. F. Thomsen, E. S. Claflin, B. Hubert, B. R. Sandel, and T. Immel, Generation of Transient Dayside Sub-Auroral Proton Precipitation, *J. Geophys. Res.*, 109, A1227, doi:10.1029/2004JA010393 , 2004.

Hubert, B., J.-C. Gérard, D. V. Bisikalo, V. I. Shematovich, and S. C. Solomon, The role of proton precipitation in the excitation of auroral FUV emissions, J. Geophys. Res., 106, 21475, 2001.

Hubert, B., J. C. Gérard, D. S. Evans, M. Meurant, S. B. Mende, H. U. Frey, and T. J. Immel, Total electron and proton energy input during auroral substorms: Remote sensing with IMAGE-FUV, *J. Geophys. Res.*, 107, doi: 10.1029/2001JA009229, 2002

Hubert, B., J. C. Gérard, S.A. Fuselier, and S. B. Mende, Observation of dayside subauroral proton flashes with the IMAGE-FUV imagers, *Geophys. Res. Lett.*, 30, doi: 10.1029/2002GL016464, 2003

Liou, K., C.-C. Wu, R. P. Lepping, P. T. Newell, and C.-I. Meng, Midday sub-auroral patches (MSPs) associated with interplanetary shocks, *Geophys. Res. Lett.*, 29, 1771, doi:10.1029/2001GL014182, 2002

Mende, S. B., H. Heetderks, H. U. Frey, M. Lampton, S. P. Geller, S. Habraken, E. Renotte, C. Jamar, P. Rochus, J. Spann, S. A. Fuselier, J. C. Gérard, G. R. Gladstone, S. Murphree, and L. Cogger, Far ultraviolet imaging from the IMAGE spacecraft: 1. System design, *Space Science Reviews*, 91, 243, 2000a.

Mende, S. B., H. Heetderks, H. U. Frey, J. M. Stock, M. Lampton, S. Geller, R. Abiad, O. Siegmund, S. Habraken, E. Renotte, C. Jamar, P. Rochus, J. C. Gérard, R. Sigler, and H. Lauche, Far ultraviolet imaging from the IMAGE spacecraft : 3. Spectral imaging of Lyman alpha and OI 135.6 nm, *Space Science Reviews*, 91, 287, 2000b.

Press, W. H., B. P. Flannery, S. A. Teukolsky, and W. T. Vetterling, Numerical recipes, the art of scientific computing, FORTRAN version, Cambridge University Press, New York, 1989.

Sonnerup, B. U. Ö, and L. J. Cahill, Jr., Magnetopause structure and attitude from Explorer 12 observations, *J. Geophys. Res.*, 72, 171, 1967.

Zhang, Y., L. J. Paxton, T. J. Immel, H. U. Frey, and S. B. Mende, Sudden solar wind dynamic pressure enhancements and dayside detached auroras: IMAGE and DMSP observations, *J. Geophys. Res.*, 108, doi:10.1029/2002JA009355, 2003.

Benoît Hubert and Jean-Claude Gérard, Institut d'Astrophysique et Géophysique, University of Liège, Allee du 6 Août, 17, Bât, B5c, 4000 Liège, Belgium, B.Hubert@ulg.ac.be

Stephen B. Mende, Space Sciences Laboratory, UC Berkely, 7 Gauss Way, Berkeley, CA 94720-7450, USA, mende@ssl.berkeley.edu

Stephen A. Fuselier, Lockheed Martin Advanced Technology Center, Palo Alto, CA 94304, USA, fuselier@star.spasci.com

Geospace Storm Processes Coupling the Ring Current, Radiation Belt and Plasmasphere

M.-C. Fok[1], Y. Ebihara[2], T. E. Moore[1], D. M. Ober[3], and K. A. Keller[4]

The plasmasphere/ring-current/radiation-belt are interacting systems. The magnetic field generated by the ring current changes the drift paths of energetic particles. Pressure gradients in the ring current produce the region 2 field aligned currents, which close in the ionosphere and create an electric field that acts to shield the lower-latitude region from the full force of convection. In turn, this shielding field alters the transport of the ring current and plasmaspheric plasmas. Furthermore, the anisotropy in the ring current plasmas excites waves that cause pitch-angle and energy diffusion of radiation belt and ring current particles. On the other hand, the precipitation of energetic electrons modifies the ionospheric conductances, and thus the electric field configuration in the magnetosphere-ionosphere system. A number of models of the plasmasphere, ring current and the radiation belt have been developed to study the behaviors of the inner magnetosphere during geospace storms. However, the majority of these models are designed to study a particular plasma population, without the consideration of interactions from others. In this paper, we briefly describe state-of-the-art models of the plasmasphere, ring current, and radiation belt, and present results from a preliminary coupling effort. The coupled models are shown to produce certain observed features of the inner magnetosphere: the post-midnight peak of storm main phase ring current ion flux; the plasmaspheric disturbance produced by impulsive substorm plasma injections, and the slow ramp-up of geosynchronous fluxes associated with energy diffusion. We conclude by presenting a framework on coupling these models together interactively to make significant progress toward a realistic plasmasphere/ring-current/radiation-belt interaction model.

1. INTRODUCTION

The inner magnetosphere is commonly defined in the vicinity above the topside ionosphere at ~ 1000 km to L ~ 8. In the inner magnetosphere, the particle magnetic drifts are strong and depends on particle energy and pitch angle. As a result, charged particles with different energies and pitch angles drift differently and they cannot be described as a single fluid. There are three major plasma populations in the inner magnetosphere: the plasmasphere, the ring current and the radiation belt. The plasmasphere consists of cold (~ 1 eV) electrons and ions with density on the order of 1000 cm^{-3} inside the sharp boundary

[1] NASA Goddard Space Flight Center, Greenbelt, Maryland, USA
[2] National Institute of Polar Research, Tokyo, Japan
[3] Mission Research Corporation, New Hampshire, USA
[4] S P Systems, NASA Goddard Space Flight Center, Greenbelt, Maryland, USA

called the plasmapause, whose location can vary from $L \sim 2$ to 6 depending on magnetic activities [*Carpenter and Anderson*, 1992]. The particle source of the plasmasphere is from the ionosphere. The shape of the plasmapause is controlled by convection, refilling rate from and loss rate to the ionosphere. The ring current is a population of hot electrons and ions with energies ranging from ~ 1 to 300 keV. It occupies from $L \sim 2$ to 8 with particle density in the range of 0.1–10s cm^{-3}. Even though the ring current is much less dense than the plasmasphere, it carries most of the plasma pressure (~ 10–100 nPa) in the inner magnetosphere. The main particle source of the ring current is the plasma sheet, which consists of particles from the solar wind and the ionosphere. During geospace storms, particles are injected and accelerated from the plasma sheet into the ring current region. Strong convection force can push the ring current deeply inward to $L \leq 2$ on the nightside. During the main phases of storms, the ring current is highly asymmetric in local time with the peak located on the nightside [*Le et al.*, 2004]. The ring current decays in the storm recovery and gradually becomes uniform in local-time. The radiation belt consists of relativistic electrons ($E > 30$ keV) and ions ($E > 20$ MeV). It occupies in the same region of the ring current. Radiation belt particles are originated in the solar wind, ionosphere and cosmic rays [*Walt*, 1996; *Baker et al.*, 1996]. They can be accelerated and enhanced in intensity during various events and processes: solar energetic proton events, interplanetary shocks, storm sudden commencements, storm and substorm injections [*Baker et al.*, 1996; *Hudson et al.*, 1996; *Summers et al.*, 1998; *Elkington et al.*, 1999]. Though radiation belt particles contribute insignificantly in plasma density and pressure in the inner magnetosphere, their sources, sinks and variabilities are subjects of great interest because of the possible radiation hazards to spacecraft electronics and humans in space.

The plasmasphere, the ring current and the radiation belts are not independent populations. They interact with each other in many different ways. The plasmasphere provides the environment for various plasma wave generation and propagation. These waves may interact with the hot plasmas, causing diffusion, degradation and acceleration in energy and pitch angle of the energetic particles. Ring current ions experience energy degradation by interacting with the plasmaspheric electrons and ions through Coulomb collisions. *Fok et al.* [1991] found that Coulomb interactions are more important than charge exchange in determining the decay lifetimes for ring current H$^+$ below a few keV and for ring current He$^+$ and O$^+$ below a few tens of keV. On the other hand, the energy transferred to the plasmaspheric ions and electrons through Coulomb collisions with the ring current ions is a source of plasmasphere heating. *Fok et al.* [1993] calculated the energy deposition rate to the plasmasphere electrons and the corresponding heat flux to the subauroral ionosphere. The calculated heat flux is sufficient to produce a subauroral electron temperature enhancement and stable auroral red (SAR) arc emissions that are consistent with observations during active periods. The effects of ring current heating to the plasmasphere ions through Coulomb interactions were also investigated by *Fok et al.* [1995]. They calculated the plasmaspheric ion temperature with the additional heat flux from the ring current and reproduced the high ion temperatures often seen at high altitudes during storm times.

Energetic particles in the ring current often have anisotropic phase space distribution functions. When the effective ion temperature anisotropy $A = T_\perp / T_\parallel - 1$ exceeds some positive threshold, these particles will provide the free energy source needed to generate electromagnetic ion cyclotron (EMIC) waves [*Cornwall*, 1964, 1965; *Kennel and Petsheck*, 1966; *Lyons and Williams*, 1984]. In turn, the EMIC waves provide a mechanism to control the ring current evolution and precipitation loss rate [*Cornwall et al.*, 1971]. During the storm main phase, the ring current decay rate due to resonant interaction with EMIC waves can be substantially faster than the decay rate due to charge exchange or Coulomb scattering. *Fok et al.* [1993] and *Kozyra et al.* [1998] investigated the decay of the ring current ions using simple magnetic storm models considering loss mechanisms such as charge exchange and Coulomb collisions. They found additional loss processes, possibly wave-particle interactions, have to be included to account for the observed decay rates. Ring current particles also play important roles on the global convection in the inner magnetosphere and the subauroral ionosphere. A non-zero divergence in the ring current produces field aligned currents that flow out or into the ionosphere, where currents are closed through ionospheric currents [*Wolf*, 1983]. The resultant electric field generated in the ionosphere often opposes the original convection electric field and provides shielding of the inner magnetosphere from the full force of convection [*Wolf* 1995]. Another signature of the ring current is the self-generated magnetic field. During the main phase of a storm, the magnetic field produced by the ring current can significantly deplete the main field and alter the drift paths of plasmas, especially radiation belt particles. In order to conserve the third adiabatic invariant, energetic charged particles drift outward when the magnetic field is reduced. The particle energy thus decreases, and some of them may be lost from the trapped region if their drift paths encounter the magnetopause boundary. This is the well-known Dst effect of the radiation belt that is often seen as a flux decrease during storm main phases [*Dessler and Karplus*, 1961; *Kim and Chan*, 1997].

Similar to the ring current ions, anisotropic pitch-angle distribution in energetic electrons (10–100 keV) excite whistler-mode waves in the magnetosphere [*Kennel and Petschek*, 1996; *Lyons et al.*, 1972]. Wave-particle interactions play important roles on the development of radiation belt plasmas. *Lyons et al.* [1972] suggested that pitch-angle diffusion of radiation belt electrons resulting from resonant interactions with plasmaspheric whistler-mode waves (~ 300–1000 Hz) is responsible for the formation of the quiet-time electron slot region. *Summers and Ma* [2000] derived the energy spectra of relativistic electrons (> 1 MeV) in the inner magnetosphere by solving an energy diffusion equation, with the diffusion coefficient calculated based on gyroresonant electron-whistler mode wave interaction and parallel wave propagation. They found this stochastic acceleration of electrons is a viable mechanism for generating killer electrons during geomagnetic storms. This diffusive interaction is strong just outside the plasmapause, where the cold plasma density is low and the magnetic field strength is still relatively high [*Summers et al.*, 1998]. As the shape of the plasmasphere evolves during a storm, the region of strong energy diffusion will vary accordingly. This is another ample example of plasmasphere/radiation-belt coupling. Furthermore, energetic electrons have influences on the electric coupling in the global magnetosphere-ionosphere (M-I) system. Enhanced electron precipitations during geospace storms increase the ionospheric conductances and change the convection electric field, which, in turn, modifies the transport of the ring current and plasmasphere.

A number of physics-based models of the plasmasphere, ring current and the radiation belt have been developed to study the behavior of the inner magnetosphere during geospace storms [*Rasmussen et al.*, 1993; *Ober et al.*, 1997; *Chen et al.*, 1994; *Fok and Moore*, 1997; *Fok et al.*, 2001b, *Toffoletto et al.*, 2003; *Jordanova et al.*, 1997; 2001; *Bourdarie et al.*, 1996; 1997; *Fok et al.*, 2001a; *Zheng et al.*, 2003]. However, the majority of these models are designed to study a particular plasma population, without the vigorous consideration of interactions from others. In this paper, we will briefly describe state-of-the-art models of the plasmasphere, ring current, and radiation belt. Only one model from each plasma population will be discussed as well as results from a preliminary coupling effort. We conclude by presenting a framework on coupling all these plasma populations together interactively to make significant progress toward a realistic plasmasphere/ring-current/radiation-belt interaction model.

2. STATE-OF-THE-ART MODELS OF THE INNER MAGNETOSPHERE

2.1 The Radiation Belt Environment (RBE) Model

The Radiation Belt Environment (RBE) model is a kinetic model that calculates the temporal variation of the phase space density of energetic electrons by solving the following bounce-averaged Boltzmann transport equation [*Fok et al.*, 2001a; *Zheng et al.*, 2003]:

$$\frac{\partial f_s}{\partial t} + \langle \dot{\lambda}_i \rangle \frac{\partial f_s}{\partial \lambda_i} + \langle \dot{\phi}_i \rangle \frac{\partial f_s}{\partial \phi_i} = \frac{1}{\sqrt{M}} \frac{\partial}{\partial M} \left(\sqrt{M} D_{MM} \frac{\partial f_s}{\partial M} \right) + \frac{1}{T(y) \sin 2\alpha_o} \frac{\partial}{\partial \alpha_o} \left(T(y) \sin 2\alpha_o D_{\alpha_o \alpha_o} \frac{\partial f_s}{\partial \alpha_o} \right) - \left(\frac{f_s}{0.5 \tau_b} \right)_{\text{loss cone}} \quad (1)$$

where $f_s = f_s(t, \lambda_i, \phi_i, M, K)$, is the average distribution function on the field line between mirror points. λ_i and ϕ_i are the magnetic latitude and local time, respectively, at the ionosphere foot point of the geomagnetic field line. M is the relativistic magnetic moment and $K = J/\sqrt{8 m_o M}$, where J is the second adiabatic invariant. The motion of the particles is described by their drifts across field lines which are labeled by their ionospheric foot points. The M range is chosen to well-represent the energy ranges of electrons from 10 keV to 4 MeV. The K range is chosen to cover the loss cone so that particle precipitations can be estimated as well.

The left hand side of (1) represents the drifts of the particle population and the terms on the right hand side of (1) refer to diffusion and loss. The calculation of the bounce-averaged drift velocities, $\langle \dot{\lambda}_i \rangle$ and $\langle \dot{\phi}_i \rangle$, were described in detail in *Fok and Moore* [1997]. These drifts include gradient and curvature drift, and E×B drift from convection and corotation electric fields. The effects of inductive electric field due to time-varying magnetic field are also taken into account implicitly in the model. For this purpose, we have assumed that field lines are rooted at the ionosphere, so that the inductive electric field there is zero. However, the shapes of field lines at higher altitudes vary as a function of time according to the magnetic field model. If field lines are perfect conductors, the field line motion at high altitudes, e.g., at the equator, will generate an induction electric field of the form,

$$\mathbf{E}_{\text{ind}} = -\mathbf{v}_o \times \mathbf{B}_o \quad (2)$$

where $\mathbf{v}_o$ and $\mathbf{B}_o$ are the field line velocity and magnetic field at the equator.

The first term on the right hand side of (1) represents particle diffusion in M as a result of energy diffusion due to interactions with plasma waves. The relation between energy diffusion coefficient (D_{EE}) and the corresponding coefficient in M (D_{MM}) is given as,

$$D_{MM} = D_{EE} \left(\frac{\partial M}{\partial E} \right)^2 = D_{EE} \left(\frac{E_o + E}{E_o B_m} \right)^2 \quad (3)$$

where E_o is the electron rest energy and B_m is the magnetic field at the mirror point. The second term on the right hand side of (1) represents pitch-angle diffusion from interacting with waves, where α_o is the equatorial pitch angle. For pure pitch-angle diffusion (E unchanged) in the (M, K) coordinates, we first map the particle phase space density from the (M, K) to (E, α_o) coordinates, perform diffusion in α_o, and then map the updated distribution back to the (M, K) coordinates [*Fok et al.*, 1996]. The diffusion terms are followed by the loss term of the loss cone, the boundary of which is assumed to correspond to mirror height of 120 km. Particles in the loss cone are assumed to have a lifetime of one half bounce period (0.5 τ_b).

Eq. (1) includes multiple terms of different processes. We use the method of fractional step to decompose the equation and solve only one term at a fractional step [*Fok et al.*, 1993]. To solve (1), we have to specify the electric, magnetic fields and the particle distribution on the nightside boundary, which is set at 10 Earth radii (R_E). We have been using empirical models such as Tsyganenko 1996 model [*Tsyganenko* 1995] for the magnetic field and Weimer model [*Weimer* 2001] for electric field [*Zheng et al.*, 2003]. We have also run the RBE model with the magnetic and electric fields output from the Block-Adaptive-Tree-Solarwind-Roe-Upwind-Scheme (batsrus) MHD model [*Groth et al.*, 2000; *Gombosi et al.*, 2003]. Both the electric and magnetic fields are updated every 5 minutes or less. The effect of radial diffusion is integrated in these time-varying fields.

The RBE model is almost the unique existing model that provides predictions of energetic electron distributions covering the entire radiation belt region and energy, with the considering of realistic and time-varying magnetic and electric fields. *Zheng et al.* [2003] showed that the RBE model gives reasonably well agreements with the observed energetic (50 keV–1.5 MeV) electron fluxes at the geosynchronous orbit. The model-data agreements are better when time-varying magnetic field was employed, indicating the importance of magnetic field configuration in controlling the transport of radiation belt particles.

2.2 The Comprehensive Ring Current Model (CRCM)

The Comprehensive Ring Current Model (CRCM) combines the Rice Convection Model (RCM) [*Harel et al.*, 1981] and the Fok ring current model [*Fok et al.*, 2001b]. The Fok ring current is similar to the RBE model described above, except the M range is chosen to cover the ring current energy (~ 1–300 keV) and charge exchange loss is also included. To couple with the Fok model, the RCM algorithm for calculating Birkeland currents has been generalized to arbitrary pitch angle distribution [*Fok et al.*, 2001b]. Plate 1 shows the model logic of the CRCM. Given an initial ring current distribution (f_s), the RCM component of the CRCM computes the ionospheric electric current:

$$J_{\|i} = \frac{B_i}{B} \sum_j \mathbf{b} \cdot \nabla \eta_j \times \nabla E_j \quad (4)$$

where $J_{\|i}$ is the current per unit area parallel to $\mathbf{B}_i$; positive current is down into the ionosphere, B_i is the magnetic field strength in the ionosphere, η_j is the number of particles of type j per unit magnetic flux in the range $\Delta M \Delta K$, and E_j is the kinetic energy. The η_j associated with a range $\Delta M \Delta K$ is related to the distribution function, f_s, by [*Fok et al.*, 2001b]:

$$\eta_j = f_s 4\sqrt{2}\pi\ m_o^{3/2} M_j^{1/2} \Delta M_j \Delta K_j \quad (5)$$

The summation in (4) sums all the contributions from different particle type (M and K) to the field aligned current. Given a specification of ionospheric conductance, the ionospheric potential, Φ, is calculated by solving

$$\nabla \cdot \left(-\bar{\bar{\Sigma}} \cdot \nabla \Phi \right) = J_{\|i} \sin I \quad (6)$$

where $\bar{\bar{\Sigma}}$ is a conductance tensor, and I is the magnetic dip angle. The Fok model then advances the plasma distribution using the electric field computed by (6) and at the same time calculates particle losses along drift paths. The updated distributions are then returned to the RCM to complete the computation cycle.

Input models required for running the CRCM include a magnetic field model, the electric potential at high-latitude ionosphere boundary (near the polar cap), an ionospheric conductance model, and the plasma sheet distribution function at the equator at the CRCM outer boundary. Currently the Tsyganenko 1996 model [*Tsyganenko* 1995] is used for the magnetic field configuration. The electric potential at the polar boundary is modeled by the Weimer 2000 model [*Weimer* 2001] or by the Boyle model [*Boyle et al.*, 1997]. The CRCM conductance model superimposes a *Hardy et al.* [1987] auroral enhancement on a background conductance based on the MSIS neutral atmosphere [*Hedin*, 1991], the IRI-90 ionospheric model [*Bilitza et al.*, 1993], and collision-frequency expressions given by *Riley* [1994]. The plasma sheet distribution at 10 R_E is assumed to be a Maxwellian

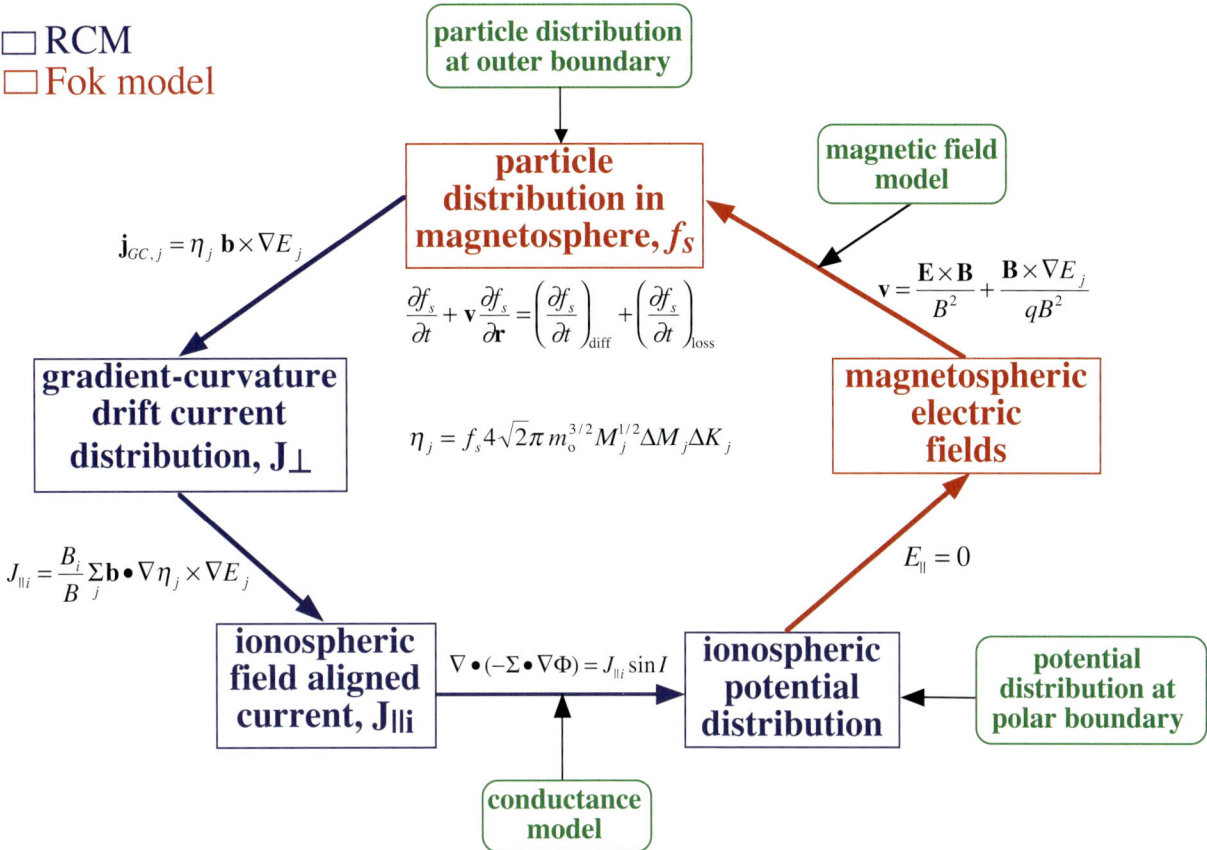

Plate 1. The model logic of the Comprehensive Ring Current Model. Blue boxes are the Rice Convection Model, red boxes the Fok model, and green boxes input models.

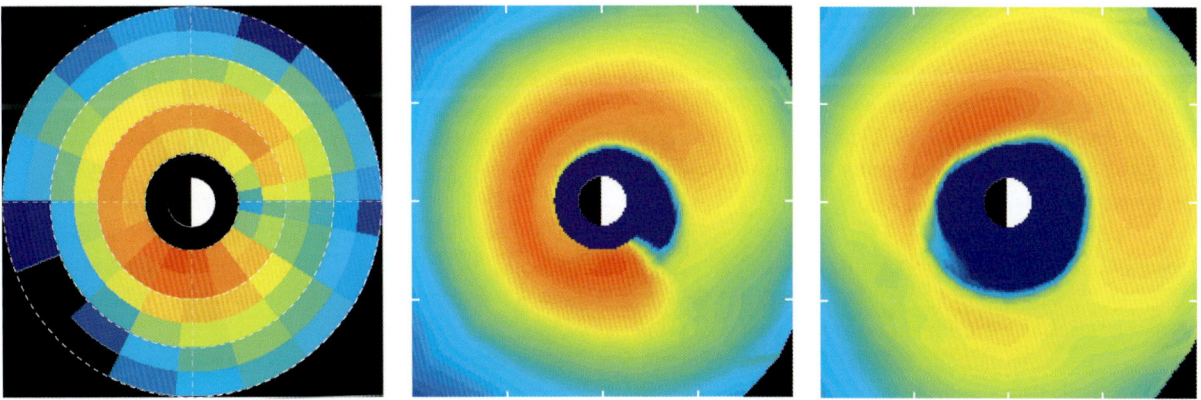

Plate 2. Left panel: equatorial flux of 32 keV H$^+$ inverted from the HENA image at 09:00 UT on 12 August 2000. Middle panel: simulated flux at the same energy and time calculated by the CRCM. Right panel: simulated flux using the Weimer electric field model.

with constant density of 0.5 cm^{-3} and temperature 5 keV. In some studies, we varied the boundary temperature and density according to the upstream solar wind conditions [*Fok et al.*, 2003].

The CRCM is the first ring current model that self-consistently solves the magnetospheric plasma distribution and ionospheric current and potential, with the consideration of arbitrary pitch-angle distribution in the ring current plasmas. The CRCM has been very successful in reproducing observable features of the storm-time ring current. In modeling the storm on 2 May 1986, near the peak of the storm when convection was strong, the model developed a region of strong outward electric field at $L \sim 3$ in the dusk-midnight sector [*Fok et al.*, 2001b]. This appears to be the same feature that was noted by *Rowland and Wygant* [1998] and *Burke et al.* [1998] in CRRES electric data for major storms. In the ionosphere, this signature corresponds to a strong poleward field in the subauroral region that has very similar characteristics as the subauroral polarization stream (SAPS) identified by *Foster and Vo* [2002]. The CRCM also reproduced and provided possible explanations to the post-midnight enhancements in storm-time ring current seen by the High Energy Neutral Atom (HENA) imager on the Imager for Magnetopause-to-Aurora Global Exploration (IMAGE) mission [*C:son Brandt et al.*, 2002]. *Fok et al.* [2003] found that the post-midnight peak is a combined effect of the irregularities in the ionosphere conductance and the strong shielding field generated by the ring current ions. Plate 2 shows the equatorial flux of 32 keV H$^+$ inverted from the HENA image [*Roelof and Skinner*, 2000] at 09:00 UT on 12 August 2000 (left panel), showing a flux maximum near dawn. The middle panel plots the simulated flux at the same energy and time calculated by the CRCM. It can be seen that the CRCM produces a very similar local-time distribution. In contrast, the simulation using the empirical electric field model of *Weimer* [1995] gives the peak flux at the dusk-midnight sector (right panel). This comparison strongly illustrates the superiority of self-consistent electric field over empirical models.

2.3 Dynamic Global Core Plasma Model (DGCPM)

A Dynamic Global Core Plasma Model (DGCPM) has been developed to calculate the plasma flux tube contents and equatorial plasma density distribution versus time throughout the magnetosphere, including the influences of convection on the flux tube volumes, as well as daytime refilling and nighttime draining of plasma [*Ober et al.*, 1997]. The model solves the following continuity equation of the total ion content of a magnetic flux tube:

$$\frac{D_\perp N}{Dt} = \frac{F_N + F_S}{B_i} \quad (7)$$

where D/Dt is the convective derivative in the $E \times B$ frame of the flux tube, N is the total ion content per unit magnetic flux, F_N and F_S are the ionospheric fluxes in or out of the flux tube at northern and southern ionospheres, and B_i is the magnetic field at the ionospheric foot points of the flux tube. The equatorial plasma density is assumed to be equal to the average density in the flux tube.

The net flux of plasmas in or out of a flux tube depends on the instantaneous content of the flux tube. The particle flux on the dayside, F_d, is given by:

$$F_d = \frac{n_{sat} - n}{n_{sat}} F_{max} \quad (8)$$

where n_{sat} is the saturation density [*Carpenter and Anderson*, 1992], n is the plasma density in the flux tube, and F_{max} is the limiting flux from the ionosphere [*Chen and Wolf*, 1972]. The nightside flux, F_n, is approximated by:

$$F_n = -\frac{N B_i}{\tau} \quad (9)$$

where τ is the downward diffusion lifetime on the nightside, which is assumed to be 10 days.

Observable features in the plasmasphere are reproduced by the DGCPM. A subauroral ion drift (SAID) event was modeled by the DGCPM [*Ober et al.*, 1997]. They found that imposing a SAID event in the dusk-evening sector of 30 minutes leads to the formation of a narrow embedded plasma density troughs generally resemble plasmasphere density profiles observed from DE 1 measurements. The DGCPM has the feasibility to allow user provided convection field and magnetic field (flux tube volume), and is thus easy to couple with other models.

3. PRELIMINARY RESULTS ON MODEL COUPLING

We have started coupling the DGCPM, CRCM, and RBE in various ways. We have incorporated the DGCPM inside the CRCM, driving the plasmasphere model with the electric field output from the CRCM. The storm on 17 April 2002 is studied. At 10:20 UT on 17 April 2002, an interplanetary shock was recorded by the Advanced Composition Explorer (ACE) satellite. About 50 minutes later at $\sim$ 11:10 UT, the shock arrived at the Earth, strongly compressed the dayside magnetopause and produced a sharp jump in the symH index. A magnetic storm commenced after the compression and symH attained a value of –90 nT at $\sim$19:00 UT. From 19:00–20:40 UT, IMAGE was ascending from the apogee. The EUV imager [*Sandel et al.*, 2000] on IMAGE captured clear images of the plasmasphere during this period of time.

FOK ET AL. 213

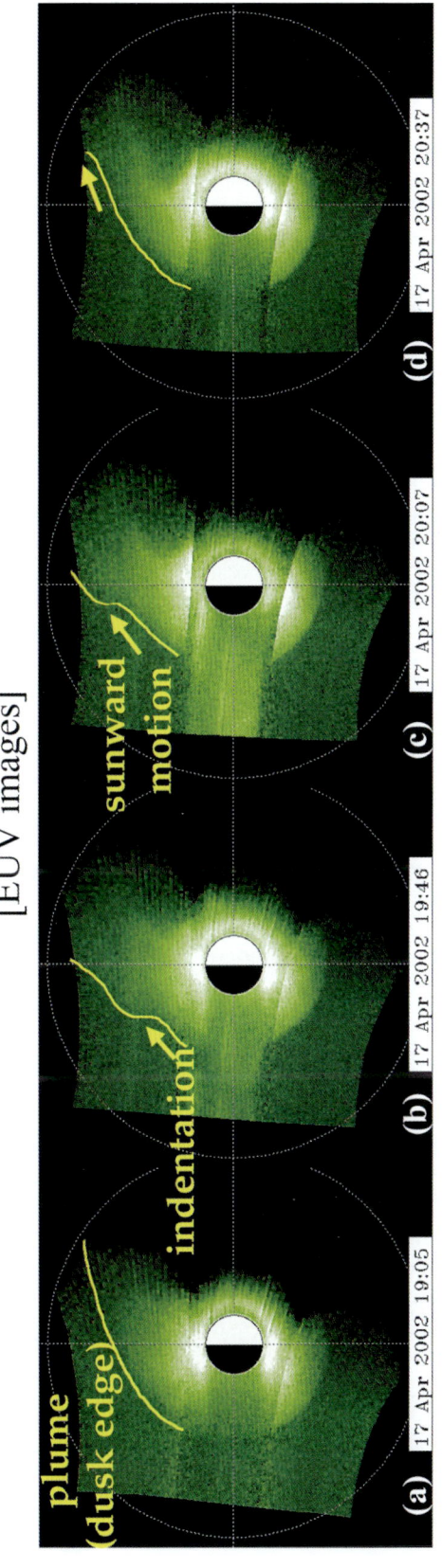

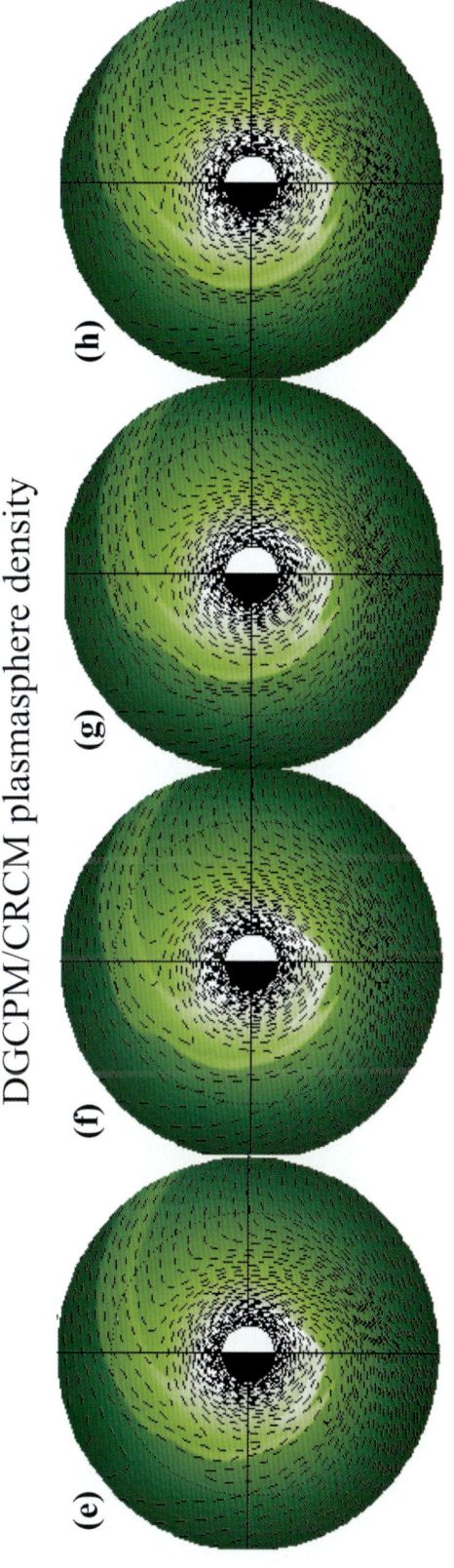

Plate 3. Top panels: plasmasphere undulation seen by IMAGE/EUV imager from 19:05–20:37 UT on 17 April 2002. Bottom panels: DGCPM/CRCM simulated images of the undulation event.

The upper panels of Plate 3 show four EUV images from 19:05–20:37 UT. The dim area on the nightside is the shadow of the Earth. As shown in Plate 3a, a plasma plume was seen on the dusk-side at 19:05 UT. At this moment, a substorm onset was detected by the IMAGE/FUV SI13 aurora images and IMAGE/HENA saw ring current injection after the onset (Jerry Goldstien, private communication). At 19:46 UT, an indentation of the plasmapause was seen in the EUV image (Plate 3b). This notch feature then moved westward to the dayside in the following hour and disappeared at ~ 20:37 UT (Plate 3c, d).

The lower panels of Plate 3 display the calculated plasmasphere density by the DGCPM, which drives the flux tube motion by the CRCM electric field. The black dashed lines are convection potential contours with co-rotation. As shown in Plate 3e–h, the combined DGCPM/CRCM reproduces the plasmasphere undulation seen by IMAGE/EUV data. By comparing the calculated electric potentials during and prior the undulation event, we found that the model predicts a strong shielding field produced by ring current injection during the undulation. The weak convection field at the plasmapause at 18–21 MLT causes outward motion or undulation of the plasmapause at this local time region (Plate 3f). The notch is then striped westward by the stronger convection later in time (Plate 3g–h). This scenario is consistent with the HENA data, which detect a substorm injection and enhanced ring current pressure during the plasmasphere undulation at ~ 19–20 UT. The field aligned current generated from the freshly injected ring current ions shields (or even over-shields) the plasmasphere region from the convection field and produces this wavy structure of the plasmasphere. *Liemohn et al.* [2004] simulated the plasmasphere shape during this storm using three electric field models: the McIlwain analytical model, the Weimer statistical model, and a self-consistent model. They also found the self-consistent model is the best in reproducing the plasmapause locations observed by the IMAGE/EUV data.

An accurate specification of the plasmasphere is necessary in order to precisely calculate the pitch-angle and energy diffusion of radiation belt particles due to wave-particle interactions. We have also integrated the DGCPM with the RBE model, driving both the transport of the plasmasphere and radiation belt plasmas with the same electric and magnetic fields. We have used the combined DGCPM/RBE model to simulate the energetic electron fluxes during the space storm on May 2–6, 1998. A coronal mass injection (CME) and magnetic cloud were observed by the ACE satellite on May 2–4. At the end of the CME on May 4, a high-speed stream was observed with large increases in solar wind speed, temperature and magnetic field [*Skoug et al.*, 1999]. This stream hit the magnetosphere on early May 4 and triggered a geospace storm with a minimum Dst of –200 nT. The recovery phase was long and jagged as seen in the Dst index.

We calculated the energetic electron fluxes on May 2–6, 1998 using the RBE model with the electric and magnetic fields output from the batsrus MHD model. The magnetic and electric field is updated every 4 minutes. The plasma sheet distribution at the nightside boundary of the RBE model at 10 R_E is assumed to be a kappa function with density (N_{ps}) and characteristic energy (E_{ps}) modeled by linear relations with the upstream solar wind conditions [*Zheng et al.*, 2003]:

$$N_{ps}(t) = [0.02 * N_{sw}(t-3\text{hr}) + 0.316] * \sqrt{\text{amu}}$$
$$E_{ps}(t) = 0.016 * V_{sw}(t-3\text{hr}) - 2.4 \quad (10)$$

where N_{ps} is in cm^{-3}, N_{sw} is the solar wind density in the same unit, amu is the atomic mass unit of electron, E_{ps} is in keV, and V_{sw} is solar wind velocity in km s^{-1}. Note that there is a 3-hour time lag between the plasma sheet condition and solar wind condition at the dayside magnetopause.

We have performed two model runs for this event. In the first one we consider particle drift and loss-cone loss only. In the second run, we include the effect of energy diffusion due to interacting with the whistler mode waves. *Summers and Ma* [2000] have derived a simple expression for the energy diffusion coefficient based on gyroresonant electron-whistler mode wave interaction and parallel wave propagation. They expressed the energy diffusion coefficient, D_{EE}, as:

$$D_{EE} = D_o [E_n(E_n+2)]^{1/2}/(E_n+1)$$
$$D_o = \frac{\pi(\gamma-1)}{8} \frac{e}{\sqrt{m_p}} \left(\frac{\Delta B \, B_o}{m_e}\right)^2 \left(\frac{\varepsilon_o}{n_e}\right)^{1.5} \quad (11)$$

where E_n is electron kinetic energy normalized by electron rest mass energy, γ is the turbulence spectral index which is assumed to equal to 5/3, ΔB is the whistler amplitude, B_o is the magnetic field at the equator, n_e is the plasmasphere density, and the rest of the symbols represent their commonly used definitions. In order for electrons to resonate with whistler waves, the kinetic energy of the electron must exceed a critical value [*Summers and Ma*, 2000]:

$$E_n \geq E_c = \left(1 + \frac{m_p v_A^2}{m_e c^2}\right)^{1/2} - 1 \quad (12)$$

where E_c is the normalized critical energy and v_A is the Alfven speed. As shown in (11), strong energy diffusion

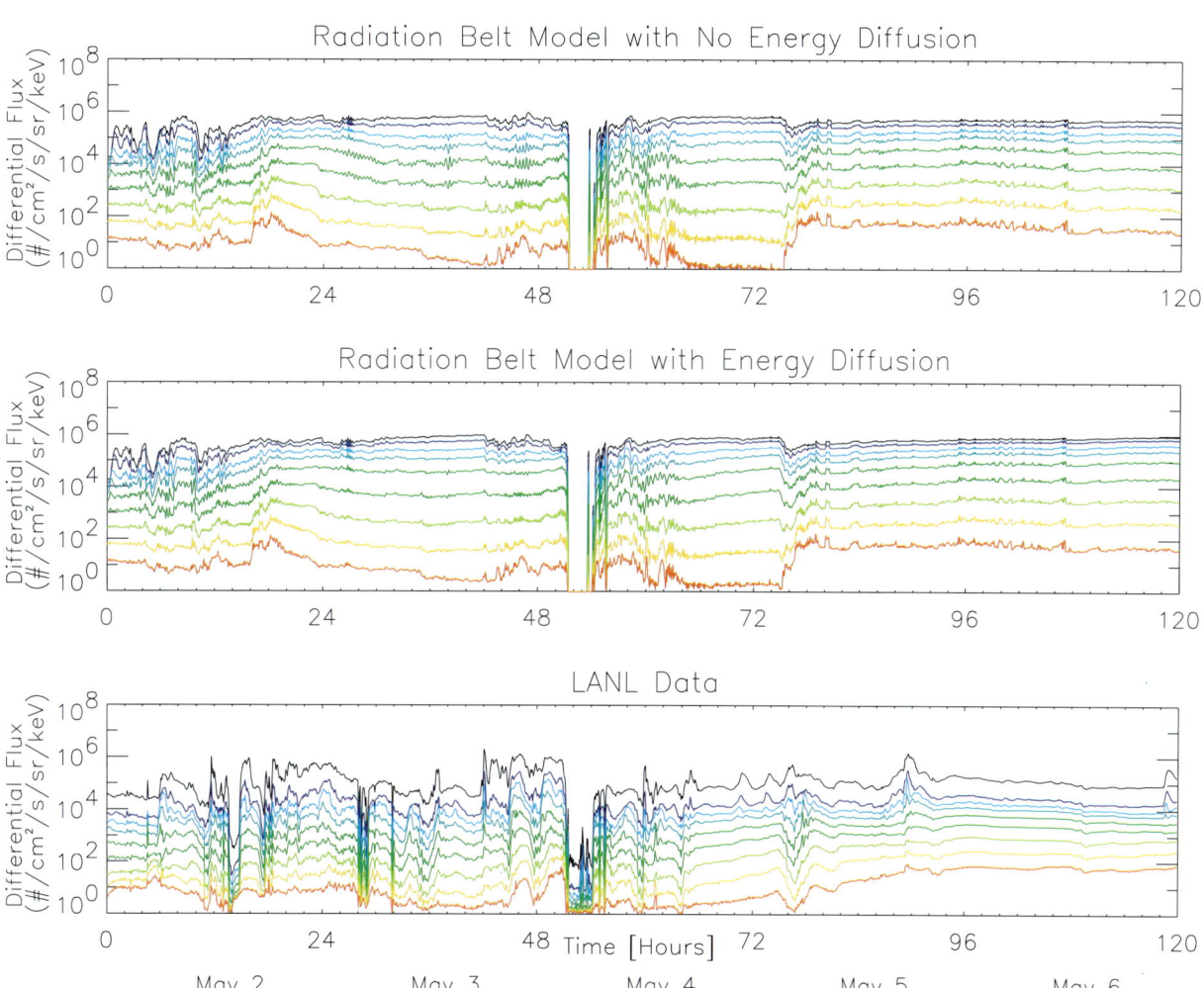

Plate 4. Top panel: simulated geosynchronous electron fluxes at 104° longitude on May 2-7, 1998, without energy diffusion. Middle panel: same as the top panel except with energy diffusion. Bottom panel: energetic electron data from LANL satellite 1994-084.

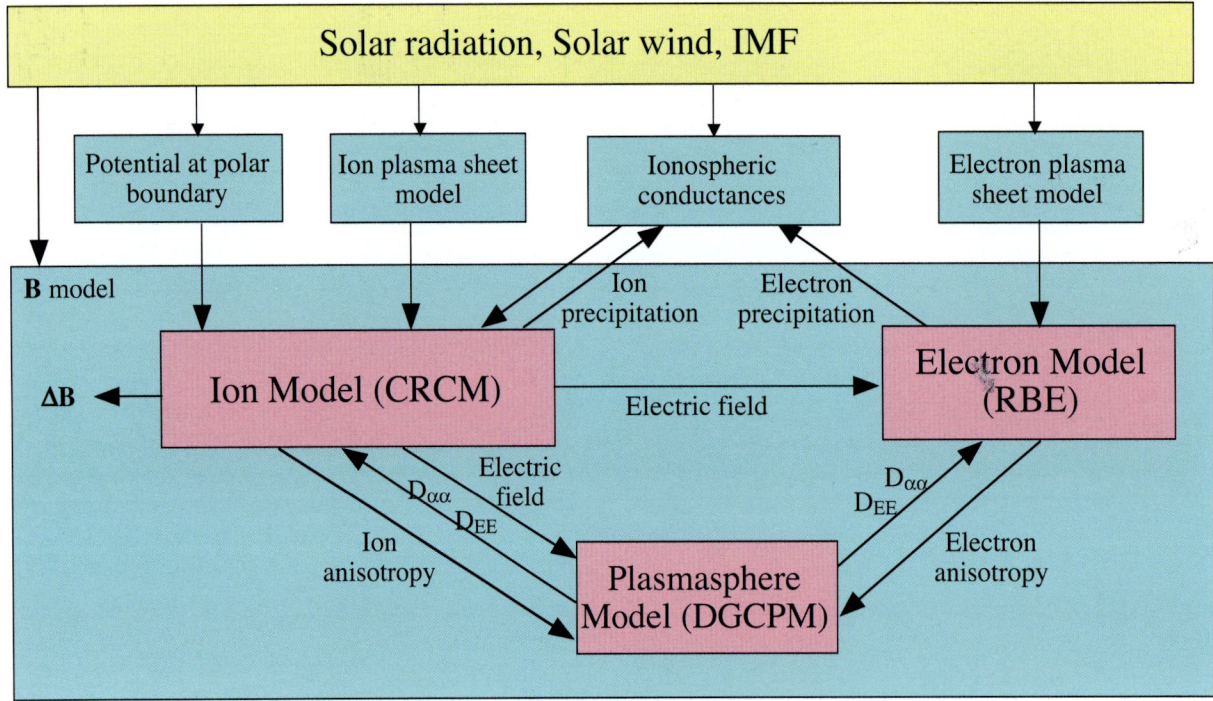

Plate 5. The model logic of the plasmasphere/ring-current/radiation-belt interaction model (PRRIM).

happens just outside the plasmapause, where B_o is strong and n_e is low. While the shape of the plasmasphere is changing during a storm, the region of strong energy diffusion varies accordingly. We calculate the energy diffusion coefficient using (11) with n_e given by the DGCPM imbedded in the RBE model. The whistler wave amplitude (ΔB) is assumed to be a constant of 50 pT throughout the storm.

Plate 4 shows the simulated and observed electron differential fluxes at energies from 50 keV to 1.3 MeV at the geosynchronous orbit at the longitude of the Los Alamos National Laboratory (LANL) satellite 1994-084. The top and middle panels are calculated electron fluxes without and with energy diffusion, respectively. The LANL Synchronous Orbit Particle Analyzer (SOPA) data are plotted in the bottom panel. As shown in the figure, the calculated fluxes generally agree well with the LANL data. The sharp flux drop out at 04:00–06:00 UT on May 4 is seen in both the observed and calculated data. This flux decrease is due to the incursion of the dayside magnetopause inside the geosynchronous orbit. The batsrus MHD model is able to reproduce the shape of this severely-compressed magnetosphere. During the recovery phase of the storm from late May 4 to May 7, the observed fluxes with energies $\gtrsim$ 300 keV slowly increased except at 04:00–06:00 UT on May 5. The simulated fluxes show a jump also at 04:00–06:00 UT on May 5 when the interplanetary magnetic field (IMF) was turning northward (Plate 4, top two panels). However, other than this fluctuation, the calculated fluxes are pretty steady during the storm recovery. With the inclusion of energy diffusion (middle panel), the calculated fluxes are higher than those without energy diffusion and agree better with the LANL data. In particular, on late May 4 to early May 5, the calculated electron distributions with energy diffusion show a slow increase in fluxes, consistent with the LANL data.

4. A PLASMASPHERE/RING-CURRENT/RADIATION-BELT INTERACTION MODEL

The simulation results from coupling the ionosphere-magnetosphere and hot-cold plasmas presented in the previous sections are very encouraging. However, there is still much room for improvement. The CRCM ion flux shown in Plate 2 is calculated using an empirical model of the height integrated ionospheric conductance. *Khazanov et al.* [2003a] calculated the ionosphere conductance according to the precipitated energetic electron and ion fluxes output from their self-consistent ring current model during the storm on 2 May 1986. They found deeper penetration of the convection electric field when conductances were calculated self-consistently. In order to estimate the effects of energy diffusion on radiation belt enhancement, we have applied a constant wave amplitude. In fact, it is well known that the rate of wave growth and thus wave amplitude depends on the temperature anisotropy of energetic electrons and ions [e.g., *Kennel and Petsheck*, 1966].

Plate 5 presents our design of a comprehensive plasmasphere/ring-current/radiation-belt interaction model (PRRIM). Many of the important processes linking the ionosphere, ring current, plasmasphere and radiation belt are included. In PRRIM, the geospace system is solely driven by the solar wind, IMF and solar radiation. The magnetic field model, cross polar cap potential, ion and electron plasma sheet distributions are determined by the IMF, solar wind density and temperature in the subsolar region. The background conductance varies with the solar F10.7 flux. The simulated ion and electron precipitations from the CRCM and RBE model are utilized to calculate the auroral conductance. The energy and pitch-angle diffusion of energetic ions and electrons are computed consistent with the ion and electron anisotropies [*Jordanova et al.*, 2001; *Khazanov et al.*, 2003b; *Horne et al.*, 2003], in the core plasma environment given by the DGCPM.

The CRCM, RBE and DGCPM are driven by the same magnetic field model and the electric field output from the CRCM. In the recent Tsyganenko models [*Tsyganenko* 1995; 2002], Dst is used to model the strength of the symmetric and partial ring current. We propose a more self-consistent way to simulate the ring current effect on the global magnetic configuration, that is to modify the symmetric and partial components of the ring current in Tsyganenko models by the actual ion currents calculated in the CRCM. In the case of MHD provided magnetic and electric fields, implementing the ring current contribution to the magnetic field would not be so straight forward. However, better description of the subauroral electric field can be obtained by replacing the MHD electric field by the field output from the CRCM. The MHD model only provides the potential at the CRCM high-latitude boundary.

5. DISCUSSIONS AND SUMMARY

The main purpose of this paper is to present our design of a plasmasphere/ring-current/radiation-belt interaction model. With all the components having been developed in certain degree of maturity, it is timely to couple and integrate these models together interactively. The PRRIM outlined in Plate 5 does not represent a perfect and fully completed model of the inner magnetosphere. One obvious weakness of the PRRIM is its crude or lack of self-consistency in handling the magnetic field. A vigorous way to treat the magnetic coupling self-consistently requires a large amount of computer time and careful schemes to assure numerical stability [*DeZeeuw et al.*, 2001; *Gombosi et al.*, 2003]. We have chosen a simple approach (modifying the ring current component in Tsyganenko models) to implement the effect of ring current on the magnetic field because we aim to develop a precise but efficient model of the inner magnetosphere. The scheme presented in Plate 5 already represents a major progress in modeling the M-I system. In summary, we have

(1) given brief descriptions of three state-of-the-art models of the ring current (CRCM), radiation belt (RBE) and plasmasphere (DGCPM).

(2) Coupling results of these models produce certain observed features of the inner magnetosphere: the post-midnight peak of storm main phase ring current ion flux; the plasmaspheric disturbance produced by substorm plasma injections, and the slow ramp-up of geosynchronous fluxes associated with energy diffusion.

(3) We have outlined a framework on coupling these models interactively to develop a comprehensive plasmasphere/ring-current/radiation-belt interaction model.

Acknowledgments. We thank Richard Wolf for many insightful discussions. The IMAGE/EUV data are provided by Jerry Goldstein and LANL/SPOA data by Geoff Reeves. The MHD output for the May 1998 event is provided by the Community Coordinated Modeling Center. This work is supported by the NASA Office of Space Science Sun-Earth Connection Guest Investigator Program under RTOP Grant 370-16-00-11 and IMAGE mission under UPN 370-28-12.

REFERENCES

Baker, D. N., S. G. Kanekal, M. D. Looper, J. B. Blake, and R. A. Mewaldt, Jovian, Solar, and other possible sources of radiation belt particles, in *Radiation Belts: Models and Standards, Geophys. Monogr. Ser.*, vol. 97, edited by J. F. Lemaire et al., pp. 49–55, AGU, Washington, D. C., 1996.

Bilitza, D., K. Rawer, L. Bossy, and T. Gulyaeva, International reference ionosphere–Past, present, future, *Adv. Space Res.*, *13*(3), 3–23, 1993.

Bourdarie, S., D. Boscher, T. Beutier, and J. Sauvaud, Magnetic storm modeling in the Earth's electron belt by the Salammbô code, *J. Geophys. Res.*, *101*, 27,171–27,176, 1996.

Bourdarie, S., D. Boscher, T. Beutier, J.-A. Sauvaud, and M. Blanc, Electron and proton radiation belt dynamic simulations during storm periods: A new asymmetric convection-diffusion model, *J. Geophys. Res.*, *102*, 17,541–17,552, 1997.

Boyle, C. B., P. H. Reiff, and M. R. Hairston, Empirical polar cap potential, *J. Geophys. Res.*, *102*, 111–125, 1997.

Burke, W. J., N. C. Maynard, M. P. Hagan, R. A. Wolf, G. R. Wilson, L. C. Gentile, M. S. Gussenhoven, C. Y. Huang, T. W. Garner, and F. J. Rich, Electrodynamics of the inner magnetosphere observed in the dusk sector by CRRES and DMSP during the magnetic storm of June 4–6, 1991, *J. Geophys. Res.*, *103*, 29,399–29,418, 1998.

Carpenter, D. L. and R. R. Anderson, An ISEE/whistler model of equatorial electron density in the magnetosphere, *J. Geophys. Res.*, *97*, 1097–1108, 1992.

Chen, A. J., and R. A. Wolf, Effects on the plasmasphere of a time-varying convection electric field, *Planet. Space Sci.*, *20*, 483–509, 1972.

Chen, M. W., L. R. Lyons, and M. Schulz, Simulations of phase space distributions of storm time proton ring current, *J. Geophys. Res.*, *99*, 5745–5759, 1994.

Cornwall, J. M., Cyclotron instabilities and electromagnetic emission generation, *J. Geophys. Res.*, *69*, 4515, 1964.

Cornwall, J. M., Cyclotron instabilities and electromagnetic emission in the ultra low frequency and very low frequency ranges, *J. Geophys. Res.*, *70*, 61, 1965.

Cornwall, J. M., F. V. Coroniti, and R. M. Thorne, Unified theory of SAR arc formation at the plasmapause, *J. Geophys. Res.*, *76*, 4428, 1971.

C:son Brandt, P., S. Ohtani, D. G. Mitchell, M. -C. Fok, E. C. Roelof, and R. Demajistre, Global ENA observations of the storm mainphase ring current: Implications for skewed electric fields in the inner magnetosphere, *Geophys. Res. Lett.*, *29(20)*, 1954, doi:10.1029/2002GL015160, 2002.

Dessler, A. J., and R. Karplus, Some effects of diamagnetic ring currents on Van Allen radiation, *J. Geophys. Res.*, *66*, 2289–2295, 1961.

DeZeeuw D., A. Ridley, G. Toth, T. Gombosi, K. Powell, and R. Wolf, Inner magnetosphere simulations–Coupling the Michigan MHD model with the Rice Convection Model, *Eos. Trans. AGU*, *82*(47), Fall Meet. Suppl., Abstract SM42A-0830, 2001.

Elkington, S. R., M. K. Hudson, and A. A. Chan, Acceleration of relativistic electrons via drift-resonant interaction with toroidal-mode Pc-5 ULF oscillations, *Geophys. Res. Lett.*, *26*, 3273–3276, 1999.

Fok, M.-C., and T. E. Moore, Ring current modeling in a realistic magnetic field configuration, *Geophys. Res. Lett.*, *24*, 1775–1778, 1997.

Fok, M.-C., J. U. Kozyra, A. F. Nagy, and T. E. Cravens, Lifetime of ring current particles due to Coulomb collisions in the plasmasphere, *J. Geophys. Res.*, *96*, 7861–7867, 1991.

Fok, M.-C., J. U. Kozyra, A. F. Nagy, C. E. Rasmussen, and G. V. Khazanov, Decay of equatorial ring current ions and associated aeronomical consequences, *J. Geophys. Res.*, *98*, 19381–19393, 1993.

Fok, M.-C., P. D. Craven, T. E. Moore, and P. G. Richards, Ring current—plasmasphere coupling through Coulomb collisions, in *Cross-Scale Coupling in Space Plasmas, Geophys. Monogr. Ser.*, vol. 93, edited by J. L. Horwitz, N. Singh, and J. L. Burch, pp. 161–171, AGU, Washington, D. C., 1995.

Fok, M.-C., T. E. Moore, and M. E. Greenspan, Ring current development during storm main phase, *J. Geophys. Res.*, *101*, 15,311–15,322, 1996.

Fok, M.-C., T. E. Moore, and W. N. Spjeldvik, Rapid enhancement of radiation belt electron fluxes due to substorm dipolarization of the geomagnetic field, *J. Geophys. Res.*, *106*, 3873–3881, 2001a.

Fok, M.-C., R. A. Wolf, R. W. Spiro, and T. E. Moore, Comprehensive computational model of the Earth's ring current, *J. Geophys. Res.*, *106*, 8417–8424, 2001b.

Fok, M.-C., T. E. Moore, G. R. Wilson, J. D. Perez, X. X. Zhang, P. C:son Brandt, D. G. Mitchell, E. C. Roelof, J.-M. Jahn, C. J. Pollock, and R. A. Wolf, Global ENA IMAGE simulations, *Space Sci. Rev.*, *109*, 77–103, 2003.

Foster, J. C. and H. B. Vo, Average characteristics and activity dependence of the subauroral polarization stream, *J. Geophys. Res*, *107*(A12), 1475, doi: 10.1029/2002JA009409, 2002.

Gombosi, T. I., D. L. DeZeeuw, K. G. Powell, A. J. Ridley, I. V. Sokolov, Q. F. Stout, and G. Toth, Adaptive mesh refinement for global magnetohydrodynamic simulation, in *Space Plasma Simulation*, edited by J. Buchner, C. T. Dum, and M. Scholer, pp. 247–274, Springer-Verlag Berlin Heidelberg, 2003.

Groth, C. P. T., D. L. Zeeuw, T. I. Gombosi, and K. G. Powell, Global three-dimensional MHD simulation of a space weather event: CME formation, interplanetary propagation, and interaction with the magnetosphere, *J. Geophys. Res.*, *105*, 25053–25078, 2000.

Hardy, D. A., M. S. Gussenhoven, R. Raistrick, and W. J. McNeil, Statistical and functional representations of the pattern of auroral energy flux, number flux, and conductivity, *J. Geophys. Res.*, *92*, 12,275–12,294, 1987.

Harel, M., R. A. Wolf, P. H. Reiff, R. W. Spiro, W. J. Burke, F. J. Rich, and M. Smiddy, Quantitative simulation of a magnetospheric substorm, 1, Model logic and overview, *J. Geophys. Res.*, *86*, 2217–2241, 1981.

Hedin, A. E., Extension of the MSIS thermospheric model into the middle and lower atmosphere, *J. Geophys. Res.*, *96*, 1159–1172, 1991.

Horne R. B., N. P. Meredith, R, M Thorne, D. Heynderickx, R. H. A. Iles, and R. R. Anderson, Evolution of energetic electron pitch angle distributions during storm time electron acceleration to MeV energies, *J. Geophys. Res.*, *108*(A1), 1016, 2003.

Hudson, M. K., S. R. Elkington, J. G. Lyon, V. A. Marchenko, I. Roth, M. Temerin, and M. S. Gussenhoven, MHD/particle simulations of radiation belt formation during a storm sudden commencement, in *Radiation Belts: Models and Standards, Geophys. Monogr. Ser.*, vol. 97, edited by J. F. Lemaire, D. Heynderickx, and D. N. Baker, pp. 57–62, AGU, Washington, D. C., 1996.

Jordanova, V. K., J. U. Kozyra, A. F. Nagy, and G. V. Khazanov, Kinetic model of the ring current-atmosphere interactions, *J. Geophys. Res.*, *102*, 14,279–14,291, 1997.

Jordanova, V. K., C. J. Farrugia, R. M. Thorne, G. V. Khazanov, G. D. Reeves, and M. F. Thomsen, Modeling ring current proton precipitation by electromagnetic ion cyclotron waves during the May 14–16, 1997, storm, *J. Geophys. Res.*, *106*, 7–22, 2001.

Kennel, C. F., and H. E. Petscheck, Limit on stably trapped particles fluxes, *J. Geophys. Res.*, *71*, 1, 1966.

Khazanov, G. V., M. W. Liemohn, T. S. Newman, M.-C. Fok, and R. W. Spiro, Self-consistent magnetosphere-ionosphere coupling: Theoretical studies, *J. Geophys. Res.*, *108*(A3), 1122, doi:10.1029/2002JA009624, 2003a.

Khazanov, G. V., K. V. Gamayunov, and V. K. Jordanova, Self-consistent model of magnetospheric ring current and electromagnetic ion cyclotron waves: The 2–7 May 1998 storm, *J. Geophys. Res.*, *108*(A12), 1419, doi:10.1029/2003JA009856, 2003b.

Kim, H.-J., and A. A. Chan, Fully relativistic changes in storm time relativistic electron fluxes, *J. Geophys. Res.*, *102*, 22107–22116, 1997.

Kozyra, J. U., M.-C. Fok, E. R. Sanchez, D. S. Evans, D. C. Hamilton, and A. F. Nagy, The role of precipitation losses in producing the rapid early recovery phase of the Great Magnetic Storm of February 1986, *J. Geophys. Res.*, *103*, 6801–6814, 1998.

Le, G., C. T. Russell, and K. Takahashi, Morphology of the ring current derived from magnetic field observations, *Ann. Geophys.*, *22*, 1267–1295, 2004.

Liemohm, M. W., A. J. Ridley, D. L. Gallagher, D. M. Ober, and J. U. Kozyra, Dependence of plasmaspheric morphology on the electric field description during the recovery phase of the 17 April 2002 magnetic storm, *J. Geophys. Res.*, *109*, A03209, 2004.

Lyons, L. R., and D. J. Williams, *Quantitative Aspects of Magnetospheric Physics*, D. Reidel, Norwell, Mass. 1984.

Lyons, L. R., R. M. Thorne, and C. F. Kennel, Pitch-angle diffusion of radiation belt electrons with the plasmasphere, *J. Geophys. Res.*, *77*, 3455–3474, 1972.

Ober, D. M., J. L. Horwitz, and D. L. Gallagher, Formation of density troughs embedded in the outer plasmasphere by subauroral ion drift events, *J. Geophys. Res.*, *102*, 14,595–14,602, 1997.

Rasmussen, C. E., S. M. Guiter, and S. G. Thomas, Two-dimensional model of the plasmasphere: refilling time constants, *Planet. Space Sci.*, *41*, 35–43, 1993.

Riley, P., Electrodynamics of the low latitude ionosphere, Ph. D. thesis, Rice Univ., Houston, Tex., 1994.

Roelof, E. C., and A. J. Skinner, Extraction of distributions from magnetospheric ENA and EUV images, *Space Sci. Rev.*, *91*, 437–459, 2000.

Rowland, D. E., and J. R. Wygant, Dependence of the large-scale, inner magnetospheric electric field on geomagnetic activity, *J. Geophys. Res.*, *103*, 14,959–14,964, 1998.

Sandel, B. R. et al., The extreme ultraviolet imager investigation for the IMAGE mission, *Space Sci. Rev.*, *91*, 197–242, 2000.

Skoug, R. M. et al., A prolonged He^+ enhancement within a coronal mass injection ejection in the solar wind, *Geophys. Res. Lett.*, *26*, 161–164, 1999.

Summers, D., and C.-Y. Ma, A model for generating relativistic electrons in the Earth's inner magnetosphere based on gyroresonant wave-particle interactions, *J. Geophys. Res.*, *105*, 2625–2639, 2000.

Summers, D., R. M. Thorne, and F. Xiao, Relativistic theory of wave-particle resonant diffusion with application to electron acceleration in the magnetosphere, *J. Geophys. Res.*, *103*, 20,487–20,500, 1998.

Toffoletto, F., S. Sazykin, R. Spiro, and R. Wolf, Inner magnetospheric modeling with the Rice Convection Model, *Space Sci. Rev.*, *107*, 175–196, 2003.

Tsyganenko, N. A., Modeling the Earth's magnetospheric magnetic field confined within a realistic magnetopause, *J. Geophys. Res., 100*, 5599–5612, 1995.

Tsyganenko, N. A., A model of the near magnetosphere with a dawn-dusk asymmetry 2. Parameterization and fitting to observations, *J. Geophys. Res., 107* (A8), 10.1029/2001JA000220, 2002.

Walt, M, Source and loss processes for radiation belt particles, in *Radiation Belts: Models and Standards, Geophys. Monogr. Ser.,* vol. 97, edited by J. F. Lemaire et al., pp. 1–13, AGU, Washington, D. C., 1996.

Weimer, D. R., Models of high-latitude electric potentials derived with a least error fit of spherical harmonic coefficients, *J. Geophys. Res., 100*, 19595–19607, 1995.

Weimer, D. R., An improved model of ionospheric electric potentials including substorm perturbations and applications to the Geospace Environment Modeling November 24, 1996, event, *J. Geophys. Res., 106*, 407–416, 2001.

Wolf, R. A., The quasi-static (slow-flow) region of the magnetosphere, in *Solar Terrestrial Physics*, edited by R. L. Carovillano and J. M. Forbes, pp. 303–368, D. Reidel, Norwell, Mass., 1983.

Wolf, R. A., Magnetospheric Configuration, in *Introduction to Space Physics*, edited by M. G. Kivelson and C. T, Russell, chap. 10, Cambridge University Press, New York, 1995.

Zheng, Y., M.-C. Fok, and G. V. Khazanov, A radiation belt—ring current forecasting model, *Space Weather, 1*(3), 1013, 2003.

Y. Ebihara, National Institute of Polar Research, 1-9-10 Kaga, Itabashi-ku, Tokyo 173-8515, Japan.

M.-C. Fok, and T. E. Moore, NASA Goddard Space Flight Center, Code 612.2, Greenbelt, MD 20771, USA.

K. A. Keller, S P System, NASA Goddard Space Flight Center, Code 612.3, Greenbelt, MD 20771, USA.

D. M. Ober, Mission Research Corporation, 589 West Hollis Street Suite 201, Nashua, NH 03062, USA

Toward Understanding Radiation Belt Dynamics, Nuclear Explosion-Produced Artificial Belts, and Active Radiation Belt Remediation: Producing a Radiation Belt Data Assimilation Model

Geoffrey D. Reeves[1], Reiner H. W. Friedel[1], Sebastien Bourdarie[2], Michelle F. Thomsen[1], Sorin Zaharia[1], Michael G. Henderson[1], Yue Chen[1], Vania K. Jordanova[3], Brian J. Albright[4], and Dan Winske[4]

The space radiation environment presents serious challenges to spacecraft design and operations: adding costs or compromising capability. Our understanding of radiation belt dynamics has changed dramatically as a result of new observations. Relativistic electron fluxes change rapidly, on time scales less than a day, in response to geomagnetic activity. However, the magnitude, and even the sign, of the change appears uncorrelated with common geomagnetic indices. Additionally, observations of peaks in radial phase space density are not readily explained by diffusion processes. These observations lead to a complex picture of acceleration and loss process all acting on top of adiabatic changes due to the storm-time magnetic field. Of even greater practical concern for national security applications is the threat posed by artificial radiation belts produced by high altitude nuclear explosions (HANE). The HANE-produced environment, like the natural environment, is subject to global transport, acceleration, and losses. Radiation belt remediation programs aim to exploit our knowledge of natural loss processes to artificially enhance the removal of particles from the radiation belts. The need to open up new orbits and new capabilities has raised questions about the space environment that, up to this time, we have been unable to fully answer. Here we describe the development of a next-generation model for specifying natural and HANE-produced radiation belts using data-assimilation based modeling. We exploit the convergence of inexpensive high-performance parallel computing, new physical understanding, and an unprecedented set of satellite measurements to improve national capability to model, predict, and control the space environment.

[1.] Space Sciences and Applications, Los Alamos National Laboratory, Los Alamos New Mexico
[2.] CERT-ONERA/DERTS, Toulouse, France
[3.] Institute for the Study of Earth, Oceans, and Space Science Center, University of New Hampshire, Durham New Hampshire
[4.] Plasma Physics, Los Alamos National Laboratory, Los Alamos New Mexico

Inner Magnetosphere Interactions: New Perspectives from Imaging
Geophysical Monograph Series 159
Copyright 2005 by the American Geophysical Union.
10.1029/159GM17

NATURAL VARIABILITY OF THE EARTH'S ELECTRON RADIATION BELTS

The discovery of the Earth's radiation belts was one of the first of the space age. Since that time many measurements of the radiation belts have been made and, as recently as ten years ago, the radiation belts and the processes affecting them were considered to be relatively well-understood. Text books still teach that radiation belt dynamics are primarily controlled by radial and pitch angle diffusion as described by *Schulz and Lanzerotti,* [1974]. However, observations from a variety of satellite programs (such as CRRES, geosynchronous, GPS, HEO, SAMPEX, POLAR, Akebono, and others) have revealed fundamental holes in our understanding of radiation belt structure and dynamics (plate 1). The shift of radiation belt physics from a sleepy backwater of space physics to a cutting edge research topic [*Friedel et al.*, 2002; *Kintner et al.*, 2002] and a national science priority [*NRC Space Studies Board*, 2002; *National Security Space Architect*, 1997] can be traced to the March 1991 CRRES satellite observation that an entirely new belt of >13 MeV electrons was produced in a matter of minutes through the interaction of an interplanetary shock with the Earth's magnetosphere [*Blake et al.*, 1992].

To date most studies have focused on the radiation belt electron flux increases seen at geosynchronous orbit. Those studies showed that the peak fluxes are typically observed one to three days after the storm main phase, in the middle of the ring current recovery phase [*Baker et al.*, 1990]. The delayed response was originally explained by the "recirculation" model of Fujimoto and *Nishida* [1990]. More recently, multi-spacecraft observations revealed that this delay is primarily a characteristic of the outer edges of the radiation belts near geosynchronous orbit, while in the heart of the radiation belts the enhancement can occur in a matter of hours, too fast for classical radial diffusion or recirculation [*Reeves et al.*, 1998; *Li et al.*, 1999]. New theories are being developed that account for enhanced diffusion through, for example, enhanced ULF drift resonance [e.g. *Elkington et al.*, 1999; 2003] but other observations are even more of a challenge to the "diffusion-only" scenario. *Green and Kivelson* [2004] have published observations of peaks in the radial profile of phase space density that provide strong evidence that local stochastic acceleration and/or radially-localized pitch angle scattering from wave particle interactions may dominate over diffusive processes. Those proposed stochastic processes have led to new theoretical studies of relativistic wave-particle interactions [e.g. *Horne and Thorne*, 2003]. Another challenge for theory and models came from the discovery that enhanced geomagnetic activity could produce either large increases or large decreases in relativistic electron flux suggesting a delicate balance between enhanced electron acceleration and enhanced storm-time losses [*Reeves et al*, 2003]. *Summers and Ma* [2000] proposed a framework which included the combined processes of radial diffusion and interactions between electrons and both whistler and EMIC waves. The combined effect of the different interactions could produce both enhanced energization and enhanced precipitation operating simultaneously.

All of these proposed processes (and others not discussed here—See *Friedel et al.*, [2002] for a review) are still somewhat speculative or poorly quantified because the observational evidence has not been combined with global physics-based models in a way that can definitively prove or rule out competing scenarios. One reason for this is that understanding acceleration, transport, and losses requires simultaneous multi-point measurements of phase space densities at fixed values of the three invariants of the particle motion, which in turn requires knowledge of the global, storm-time magnetic field—knowledge that can only come from global models. In order to enable a future space weather capability these models will need to be time-dependent and data-driven. But, they must also apply over sufficiently long time scales to enable reliable and cost-effective spacecraft design.

MAN-MADE BELTS FROM HIGH ALTITUDE NUCLEAR EXPLOSIONS (HANE)

Like the March 1991 interplanetary shock event, high altitude nuclear explosions (HANE) are known to produce sudden, intense, and long-lived radiation belts (plate 2). Therefore the acceleration, transport, and loss processes that apply to the natural radiation environment are equally important for the HANE environment. However, understanding and mitigating the threat to space-based systems from man-made belts presents additional challenges to modeling. One of the biggest challenges to modeling is that we must apply those models to conditions that have never been observed or have been poorly observed. This requires a high degree of confidence in the physical understanding of key processes that are encoded in the models' algorithms, a high degree of validation based on the variability of the natural environment, and event-specific scenarios that incorporate the full set of space observations.

The impact of a single high altitude nuclear explosion can be severe. It is estimated that the STARFISH explosion (1.4 Mt at 400 km altitude) set off in July 1962 produced about 10^{26} fission electrons with MeV energies[*Brown et al.*, 1963; *Van Allen et al.*, 1963]. No measurable high energy protons were produced. The belt was relatively narrow, being centered at L = 1.2 with a peak flux of ~ 10^9/cm2-s, with the electron density reduced to 10% at L = 1.8 and 1% at L = 2.2 [*Hess*, 1963]. However, some fission electrons were detected as far out as L ~ 5–6, implying some outward radial transport. (Similar cases have been proposed for the natural environment by *Reeves et al.* [1998] and *Green and Kivelson*[2004].) Most of the initial electrons were in the range 1–4 MeV, consistent with fission (where electrons occur up to about 10 MeV), and the artificial enhancement of the radiation belts was observed to have a lifetime of years. However, at low L, the calculated lifetimes (based on pitch angle diffusion and atmospheric precipitation) were

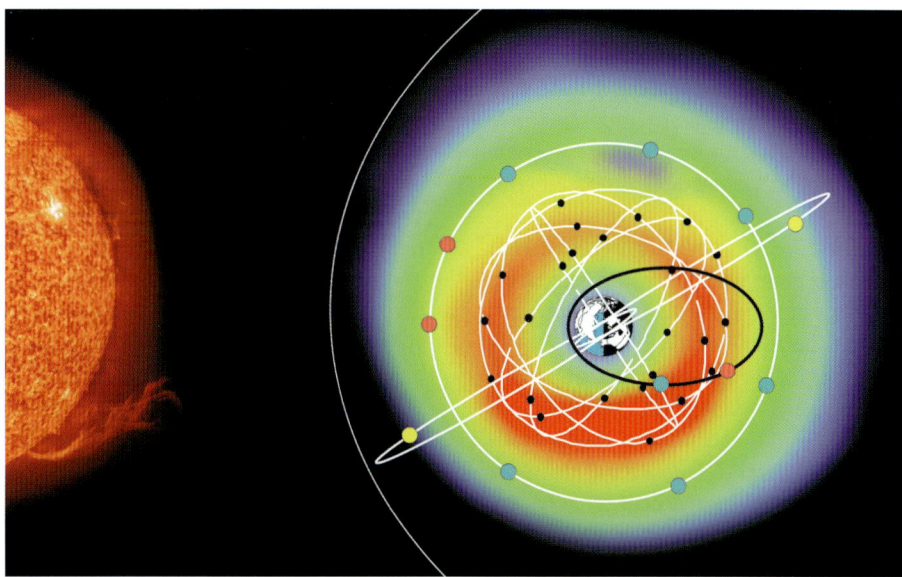

Plate 1. This schematic illustration shows some of the resources available for global data-assimilation based models of the Earth's radiation belts. Illustrated are the color-coded MeV electron fluxes as viewed from above the equatorial plane. Also illustrated are the relative positions of satellites expected to be operational during the coming years: geosynchronous orbit (in which currently 2 GOES and 6 LANL satellites are currently operational), the GPS orbits (24 satellites with 4 in each of 6 orbits), Molniya orbits (with 2 polar highly elliptically orbiting satellites), and the 2 Radiation Belt Storm Probes (RBSP) that are one component of NASA's Living With A Star program in an equatorial elliptical orbit shown in black.

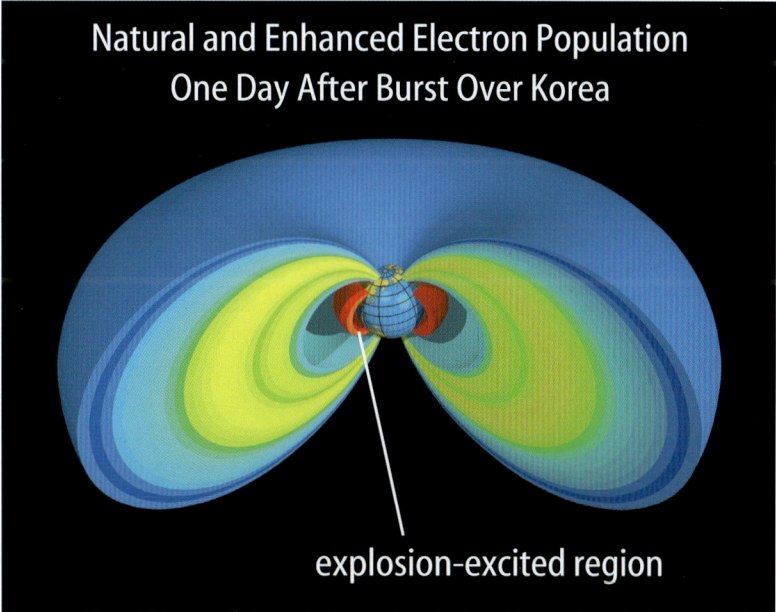

Plate 2. A schematic illustration of the fluxes of natural radiation belts and those produced by a hypothetical high altitude nuclear explosion (HANE) of 1 Mt over Korea. The artificial, HANE belt is expected to be relatively narrow but have peak fluxes 100–1000 times higher than the average at the peak of the outer belt. Predicting the evolution and the potential effects of an artificial belt requires the same physical understanding and modeling that is currently being applied to the natural belts. This is particularly true when one considers that the observations and models must be extrapolated to conditions that have never been observed or have been only poorly observed.

longer than observed by an order of magnitude, implying an unknown mechanism source of strong diffusion over short periods of time. Similar results have been seen for interplanetary shock-produced radiation belts by the SAMPEX satellite, but possible mechanisms for enhanced wave-particle scattering have not yet been modeled or compared with observations.

Since the enactment of the Limited Nuclear Test Ban Treaty in 1963, the likelihood that a country would detonate a weapon in space has been a matter of policy debate and dependent on the current global political conditions. Currently, several scenarios have emerged as having the greatest risk to systems that are space-based or depend on critical space-based components (e.g. communications and navigation) [*Murch*, 2001]. The scenario currently considered the most likely is collateral damage from regional nuclear conflict. A high altitude nuclear explosion could be used as a nuclear warning shot in an escalating regional conflict or a deliberate effort to damage adversary forces and infrastructure through a nuclear-generated electromagnetic pulse (EMP). A second scenario is detonation of a salvage-fused warhead during an attempted exoatmospheric intercept. A third scenario has been described as a 'Space Pearl Harbor'—a deliberate effort to cause economic damage and decreased military capability through asymmetric attack. Such a strategy could be used by a rogue state facing economic strangulation or imminent military defeat and could occur over their own sovereign territory. Increasingly, risk scenarios must include terrorist actions that seek to pose large-scale economic and cultural impact with lower risk of nuclear retaliation.

While there is ample room for debate over the probability that any of these scenarios would actually occur, the potential consequences are sufficiently severe that it would be unwise not to develop modern models that could better predict the creation and evolution of man-made radiation belts. Studies by the Defense Threat Reduction Agency (DTRA) [*Murch*, 2001] concluded that "One low-yield (10–20 kt), high-altitude (125–300 km) nuclear explosion could disable (in weeks to months) all LEO satellites not specifically hardened to withstand radiation generated by that explosion." Satellites at risk include communications, imaging-mapping, and manned spaceflight. Replacement cost alone is estimated at over $50 billion. Reduced lifetimes of even a few critical Department of Defense (DoD) satellite systems would have a significant detrimental effect in conducting military campaigns [*Metz and Babcock*, 2004].

RADIATION BELT REMEDIATION

In space as on the ground the old joke still applies: "Everybody talks about the weather but nobody does anything about it." Increasingly though, the civilian and military space community is moving from a strategy that has been described as "cope and avoid" to one that is characterized as "predict and control". Improved physical understanding of the natural loss processes in the radiation belts has the potential to enable systems that can exploit those processes and reduce or remediate the threat from natural or man-made radiation belt electron fluxes.

Relativistic electron fluxes are depleted either by loss through the magnetopause, precipitation into the atmosphere, or possibly through de-energization—removing electrons from the system or reversing the initial acceleration processes. Comparisons between satellites such as SAMPEX and POLAR show that the fluxes in the drift loss cone (which are precipitated in one drift period or less) track the fluxes at high altitude closely [*Kanekal et al.*, 1999]. This shows that relativistic electron precipitation occurs nearly continuously. However, in contrast to the quasi-steady "drizzle" of electrons from the radiation belts, strong geomagnetic activity during storms can produce very rapid rates of precipitation leading to permanent and dramatic reductions in the trapped electron fluxes [*Onsager et al.*, 2002; *Reeves et al.*, 2003]. *Green et al.*, [2004] have reviewed the possible causes of this loss, evaluated those mechanisms against observations, and concluded the most probable mechanism for loss is electron precipitation through enhanced pitch angle scattering through wave-particle interactions [e.g. *Horne and Thorne*, 2003]. Low-altitude satellite measurements of electron precipitation "bands" and "microbursts" yield electron loss rates that could completely remove all relativistic electrons from the radiation belts in a matter of days. (See e.g. *Nakamura et al.*, [1995, 2000]; *Lorentzen et al.*, [2000, 2001]; *Millan et al.*, [2002]; *Blake et al.*, [1995] for further discussion.)

While some radiation belt remediation schemes are, to say the least, impractical (for example, wrapping the Earth's equator with a solenoid to cancel out the geomagnetic dipole field) others aim to exploit our knowledge of natural processes to artificially enhance the rates of electron precipitation at specific times, energies, or altitudes to mitigate radiation hazards from natural or man-made events. One promising method of radiation belt remediation under current investigation involves enhancing the electron pitch-angle scattering rate via cyclotron-resonant wave-particle interactions. VLF radio waves can be injected either from space or from ground-based sources (such as ionospheric heaters that modulate ionospheric conductivity). Properly coupled, the wave-induced scattering will reduce the magnetic mirror altitude of trapped electrons, increase atmospheric collisions, and dramatically increase precipitation losses [e.g. *Inan et al.*, 2003]. Other methods to increase the pitch-angle scattering rate being investigated include electrostatic and magnetostatic processes implemented by space based tethers or DC magnets.

In order to evaluate any of the proposed radiation belt remediation techniques, or to optimize the effectiveness of any particular technique, one must be able to accurately and quantitatively predict the effects on a global scale. To achieve this level of understanding and predictive capability requires improved understanding of natural processes, targeted active experiments in space, and the global, data-driven physical models that we describe in further detail below.

DATA ASSIMILATION FOR RADIATION BELT MODELING

To address the needs and solve the questions posed in the preceding sections requires a focused international effort with three components: (1) a targeted, multi-satellite observational campaign such as the NASA Living With A Star (LWS) Radiation Belt Storm Probes [Kintner et al., 2002] to fill in holes in our knowledge of radiation belt dynamics; (2) a strong program to develop improved theoretical descriptions of key processes such as wave-particle interactions and multi-dimensional relativistic electron diffusion; and (3) development of global, time-dependent, data-driven but physics-based models of the radiation belts. Here we address the elements that would enable the successful execution of the third component—a next-generation radiation belt model.

To be useful for the applications described above, a next-generation radiation belt model must have several features. It must have high fidelity to the known physical equations governing the particles and fields in the inner magnetosphere—not just for physical understanding, but also to be useful in extrapolating to conditions or scenarios that have not yet been observed. (However, this does not necessarily require a first-principles model, such as global MHD models, that start with conditions in the solar wind or at the solar surface.) At the same time, the model must use all available observations in order to accurately represent the dynamic changes that occur during any specific individual event. The model must also be able to accurately represent the changes in the global magnetic field during geomagnetic storms. This is particularly critical for transforming spacecraft observations from a spatial coordinate system to a magnetic coordinate system where data from multiple satellites can be properly compared and physical equations can be solved consistently. While purely empirical and purely first-principle physics models are most appropriate for some applications, the requirements discussed here lead us toward data assimilation models that use physical equations together with all relevant observations to produce a "best fit" description of the dynamics of the radiation belts.

Data assimilation techniques are ideally suited for combining the data and models in such a way that the limitations of one component are balanced by the strengths of another component. Data assimilation models have been used extensively in other fields such as meteorology and climate modeling [e.g. Ghil et al., 1997] but, except in the area of ionospheric physics models, they have not been extensively applied to space. Radiation belt dynamics are well-suited to the methods of data assimilation—more so than other problems in space plasma physics. Compared to other regions of the magnetosphere, the inner magnetosphere is relatively well-ordered by the geomagnetic field, the physical equations governing the majority of particle dynamics relatively well-known, and there is a relatively large number of satellites (tens) covering the volume of the system.

There are a number of well-established techniques for data assimilation but among the most powerful and widely-used is Kalman filtering [Kalman, 1960] (and here we use the term to included extended Kalman filtering). Kalman filtering is a technique to simultaneously incorporate data (with specified errors) and adjust physical parameters within the model using a recursive solution to the discrete-data linear filtering problem. The Kalman filter is a set of mathematical equations that provides an efficient computational (recursive) means to estimate the state of a process, in a way that minimizes the mean of the squared error. The filter is very powerful in several aspects: it supports estimations of past, present, and even future states, and it can do so even when the precise nature of the modeled system is unknown [Welsh and Bishop, 1995]. Since the technique was first described in the 1960's the method has been the subject of extensive research and application due in large part to advances in digital computing that allow for the solution of highly-coupled systems like the storm-time inner magnetosphere.

Irrespective of the specific data assimilation technique used, there are several components that need to be combined to solve the coupled system of ring current, radiation belts, electric potentials, magnetic fields and waves in the inner magnetosphere. A realistic model of radiation belt dynamics that is valid for geomagnetic storm times must also include a self-consistent calculation of the storm-time ring current (carried by keV protons), a sophisticated description of diffusion in energy, pitch angle and L-shell (including off-diagonal matrix elements), a specification of the spatial and temporal distribution of whistler and EMIC wave fields, and a calculation of the stochastic effects of wave particle interactions. All the necessary components of such a model now exist.

To be realistic, this model must also be consistent with all the available data sources—measurements of radiation belt particles, ring current particles, wave fields, local magnetic fields, and solar wind inputs. An unprecedented set of all these measurements has now been collected and critical new measurements will be added by the LWS Radiation Belt Storm Probes (and possibly, by other proposed missions such as ORBITALS and COMPASS).

A comprehensive program is needed to coordinate all the aspects of theory, modeling, data validation, and application of data assimilation techniques. Such a comprehensive program is a significant, but highly valuable, endeavor. We now outline the components of such a program and the initial steps that have been taken to bring these pieces together.

DIRECT DATA INSERTION USING THE SALAMMBO CODE

One example of the value of even very simple data assimilation is provided by a three-satellite study of one month of storms using the Salammbo code [*Boscher et al.*, 1996; *Bourdarie et al.*, 1996]. The Salammbo model is a diffusion model that solves the Fokker Planck equation in three dimensions: L-shell, Energy, and pitch angle. Salammbo currently uses the simplifying assumption that the magnetic field is a pure dipole and therefore not time-dependent and not azimuthally asymmetric. The model uses statistical relationships between solar wind parameters and indices of geomagnetic activity (such as Kp) to parameterize processes such as diffusion rates or wave-particle interactions where the radiation belts overlap plasmasphere.

Plate 3 shows the period of September 21–October 20, 1998: one of the intervals selected for study by the NSF Geospace Environment Modeling (GEM) working group on radiation belt dynamics. Plate 3-a shows the results of a run in which only a single geosynchronous spacecraft was used as input into the simulation. The top panel in plate 3-a shows the fluxes measured by the Aerospace Corporation's instruments on a highly elliptical orbit (HEO) satellite. Electron flux above ≈ 4 MeV is plotted as a function of L-shell and time. The bottom panel shows the Dst index. There are three clear enhancements of fluxes (of diminishing intensity) in the range $L \approx 3-8$ in response to three intervals of enhanced geomagnetic activity. The second panel of plate 3-a shows the output of the Salammbo model using the geosynchronous data as input and the next panel shows the ratio of the measured and model fluxes on a log scale and in the same format of L vs. time. In the plot of ratios, black represents a value of 1 or perfect agreement; red represents measured fluxes that are ten times higher than the model predicts and dark blue represents measured fluxes that are up to ten times lower.

One, known, problem with the location of plasmapause in the Salammbo model produces a pair of bright red bands near L=3. More troubling is the relative lack of dynamic changes in the Salammbo model compared to the observations. This is particularly true in regions above L=5.5. One might expect that region to be more well-specified since the input from geosynchronous is at L=6.6. However, the HEO satellite, which is the basis for the observational comparison, crosses L=6.6 at high magnetic latitudes and is therefore quite sensitive to the assumed (isotropic) pitch angle distribution at geosynchronous orbit. Therefore, even at L=6.6 the agreement between model and measurement is poor.

Plate 3-b is identical to plate 3-a except that in this run two satellites were used as input into the model LANL-GEO 1994-084 and GPS NS-33. The data were incorporated through simple 'direct data insertion' which involves adjusting the values at certain grid points to exactly fit the observations when the satellite was at that location and then allowing the solution to propagate from those points throughout the system as defined by the equations of motion. Once again the HEO fluxes were used as an independent comparison and not as an input to the model so the top panel of plate 3-b is the same as previously. In the second and third panels of plate 3-b we can see that the inclusion of GPS makes a significant difference in the output from the model. While the GPS data (with coverage only above L>4.2) do not change the problems at very low L-shells, the addition of those data make the three intervals of enhanced electron flux much more apparent in the model. The ratio plot shows significantly better quantitative agreement with more regions of black throughout the plot. The substantially better agreement in the vicinity of geosynchronous orbit is due to the fact that GPS crosses L=6.6 at high magnetic latitudes (similar to HEO) and the two point measurements on the same L-shell allow a direct (but limited) determination of the pitch angle distribution.

This relatively simple illustration shows the value of incorporating multiple data sets but does not yet show the full potential of data assimilation. With even as few as two satellites it is possible to allow adjustable, time-dependent diffusion rates in the model—a technique known as adaptive data assimilation. For example, *Koller and Friedel,* [2005] have recently demonstrated that we can use this technique to not only determine radial diffusion rates that best reproduce the observations, but actually to determine the time dependence of those diffusion rates. The result is not just a better match to the HEO data (after all we could just use the HEO observations for that) but rather an improved understanding of the time-dependence of the diffusion rates and their rela-

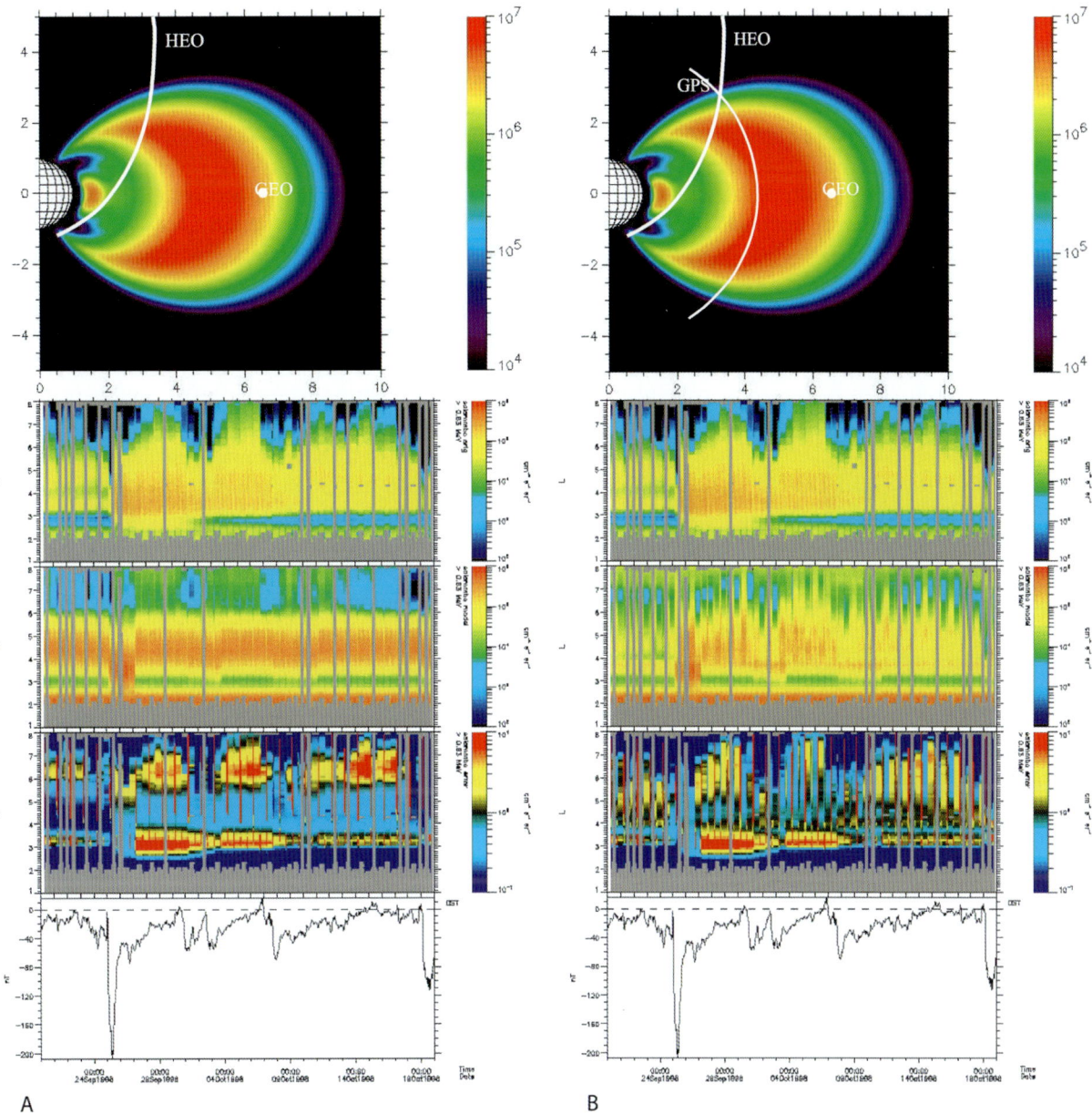

Plate 3 shows observations and models of a storm on September 21–October 20, 1998. Column A shows the results of a run in which only a single geosynchronous spacecraft was used as input into the simulation. The top panel in plate 3-a shows the fluxes measured by the Aerospace Corporation's instruments on a highly elliptical orbit (HEO) satellite. Electron flux above ≈4 MeV is plotted as a function of L-shell and time. The bottom panel shows the Dst index. The second panel of plate 3-a shows the output of the Salammbo model using the geosynchronous data as input and the next panel shows the ratio of the measured and model fluxes on a log scale and in the same format of L vs. time. In the plot of ratios, black represents a value of 1 or perfect agreement; red represents measured fluxes that are ten times higher than the model predicts and dark blue represents measured fluxes that are up to ten times lower. Column B is identical to Column A except that in this run two satellites were used as input into the model LANL-GEO 1994-084 and GPS NS-33. Again the HEO fluxes were used as an independent test and not as an input to the model. In the second and third panels of plate 3-b we can see that the inclusion of GPS makes a significant difference in the output from the model.

tionship to geomagnetic activity. Improved understanding provides improved confidence in our ability to extrapolate other energies, locations, and conditions.

As noted above, Salammbo uses a relatively simple dipole description of the magnetic field. Recently, though, the model has been extended to better incorporate non-dipole fields. This is done by transforming the spacecraft observations from spatial coordinates (latitude, longitude, radius) into magnetic coordinates (most importantly pitch angle and L*). All calculations and spacecraft intercomparisons are done in magnetic coordinates. To compare against an independent set of observations such as HEO, the transformation is reversed. This improvement allows much more accurate and consistent treatment of the data which is especially important as additional data from Geo, GPS, POLAR, SAMPEX, HEO, and Akebono are added.

STORM-TIME RING CURRENT MODELING

While diffusion is certainly one of the most important processes affecting the radiation belts, full understanding of the structure and dynamics of the radiation belts requires a more physically realistic description that self-consistently represents a variety of interacting populations and processes. Those include, time-dependent convection electric fields, the development of the storm-time plasma sheet and ring current, the interaction of the ring current with the plasmasphere to produce EMIC waves, the interaction of EMIC and storm-time whistler waves with the radiation belt electrons, the inflation of the geomagnetic field due to the ring current, and the adiabatic response of the radiation belt electrons to the changing geomagnetic field.

Several models of this type have been developed, originating with work done at the University of Michigan [e.g. *Jordanova et al.*, 1994; *Liemohn* 2001; *Fok*, 2001]. We have been using the UNH version of the RAM code which incorporates all the above-mentioned processes and has already been used with considerable success to reproduce realistically the storm-time evolution of the near-Earth plasma sheet and ring current [e.g., *Jordanova et al.*, 2003a,b].

The model numerically solves the bounce-averaged kinetic equation for the distribution function of charged particles in specified global electric and magnetic fields. The model treats ions (H+, O+, and He+) with kinetic energies from 15 eV to 400 keV and has recently been extended up to relativistic energies for electrons [*Jordanova et al.*, 2005]. Like Salammbo, the present version of UNH-RAM represents the magnetic field of the Earth as a dipole. However, the code is capable of solving the equations of motion in an arbitrary magnetic field. The model has been updated to use any electric field specification including new models with high temporal and spatial resolution like the AMIE model [*Richmond and Kamide*, 1988], the *Weimer* [2001] model, or real-time data-driven descriptions that are under development.

All major loss processes of magnetospheric particles are included in the UNH-RAM model, including charge exchange, Coulomb collisions, wave-particle interactions, and loss due to atmospheric precipitation (see *Jordanova et al.* [1996; 1997] for more details). For example, the convective growth rates are obtained from the dispersion relation, which is coupled and solved simultaneously with drift transport in order to treat the process of wave-particle interactions self-consistently.

STORM-TIME MAGNETIC FIELD SPECIFICATION

As discussed above, a major challenge in understanding important radiation belt processes such as diffusion, stochastic acceleration, or particle precipitation is to accurately calculate the phase space density at fixed values of the adiabatic invariants. While the first invariant (defined by electron gyromotion) can be calculated based on local measurements of particle pitch angle distributions and magnetic field strength, the second (bounce) invariant requires an integral along a magnetic field line and the third (drift) invariant requires an integral around the entire drift orbit of the electron. Therefore, highly accurate, time-dependent, and event-specific magnetic field models are required.

The requirements on the global magnetic field model are most stringent during the main phase of geomagnetic storms when acceleration and loss processes may both be most intense but when the perturbation to the magnetic field is also the most dramatic. The diamagnetic effects of the ring current decrease the field, moving particles outward and decreasing their energy in order to conserve the third invariant [*Kim and Chan*, 1997]. This can result in flux changes of 2–3 orders of magnitude during a storm main phase—which is exactly the time when the other processes have their largest effects. In order to calculate radiation belt dynamics correctly we need to separate the nonadiabatic acceleration and loss processes from the adiabatic effects of the ring current (the "Dst effect"). Equally important is including the full local time asymmetries of the inflated magnetic field (the asymmetric ring current) in order to correctly map spacecraft at different local times to the same magnetic coordinate system.

Two classes of storm-time magnetic field models currently exist. One provides a statistical representation of the average storm time field [*Tsyganenko et al.*, 2002] but is not event specific and is only parameterized by solar wind and geomagnetic conditions. The other class is represented by the UNH-RAM code which calculates the perturbation field produced by the ring current but solves the electron and ion motion in a dipole field. We have investigated two approaches to solving this problem which can ultimately be combined together.

Self-Consistent Magnetic Field-Ring Current Calculations

No ring current models currently calculate ring current dynamics in a self-consistent magnetic field. Particle dynamics are calculated in a dipole field and then used to determine the perturbation field produced by those particle distributions. In reality, though any perturbation to the field perturbs the trajectory of the ring current particles themselves and hence, to be realistic, the particle trajectories and global magnetic field must be calculated self-consistently.

This adds several layers of complexity. First, the codes must be generalized to be able to solve the equations of motion in a non-dipole field, which requires more numerical integration (and computation time) but is not conceptually complex. In an arbitrary field, bounce-averaging of the general gradient-curvature drift is necessary [*Shukhtina*, 1992]. In order to study the process of wave-particle interactions and address questions related to acceleration and loss of energetic particles by plasma waves, the full pitch angle dependence of the distribution function in the equatorial plane must be retained. Incorporating a self-consistent field specification therefore has impacts on pressure distributions, wave growth, wave particle interactions, and particle trajectories as well as the large adiabatic effects discussed above. We have recently implemented a parallel computing version of the UNH-RAM code using the Message Passing Interface standard to implement domain decomposition, making the added computations easily feasible.

The second problem is that the complex 3-dimensional magnetic field that would be in force balance with the particle (plasma) population needs to be found and can no longer be represented analytically. Additionally, in the inner magnetosphere large temperature anisotropies are common and equilibrium solutions must accommodate those anisotropies. There are several ways to solve the problem of the magnetic field/plasma equilibrium in 3 dimensions. Among the most promising, and the method we have chosen, is an iterative solution [*Zaharia et al.*, 2004] that uses an Euler Potential specification of the field and finds the magnetic configuration in force balance with a prescribed pressure distribution (plate 4). The pressure only needs to be prescribed at one location along each field line (e.g. on the equatorial plane), as mapping along the field line provides it everywhere else.

These two changes to the numerical calculations will enable a fully self-consistent calculation of the ring current and the perturbed global, storm-time magnetic field. To do so, first we calculate the plasma pressure distributions as already specified by UNH-RAM code. We then calculate a 3-dimensional magnetic configuration in force balance with those pressure distributions, using our Euler potential equilibrium code. Once the field and the electric currents have been found, we need to replace the dipole field in UNH-RAM with the new, more realistic field. This cycle will be repeated iteratively until the solution converges. The first step has been successfully demonstrated by *Zaharia et al.* [2005]. The resulting field can be used to calculate the dynamics of radiation belt electrons directly since they do not carry sufficient currents to further perturb the magnetic field.

Empirical Magnetic Field Specification

Another, complementary, technique to specify the global magnetic field is to exploit the magnetic drift motion of energetic particles and Liouville's theorem, which states that particle phase space density is conserved along a dynamic trajectory. The technique is illustrated in plate 5 [*Reeves et al.*, 1997]. For simplicity, imagine particles with 90 degree equatorial pitch angles that drift along contours of constant equatorial magnetic field strength. Two satellites that cross the same contour of constant equatorial field strength at nearly the same time should, according to Liouville's theorem, measure the same phase space density provided that there has been no appreciable acceleration or loss of particles as they drift from one satellite location to the other. Furthermore this must be true for all energies and all pitch angles (which follow slightly different trajectories in a non-dipole field). Thus since all particles move in the same magnetic field, each energy and each pitch angle measured can essentially provide an independent constraint on the configuration of the large-scale magnetic field. Those constraints can be used either to verify the accuracy of the magnetic field calculated by other means (e.g. through self-consistent ring current modeling) or can be used to specify where the phase space densities should match and hence how the magnetic field model should be modified to ensure that match.

Another strength of this technique is that it can be applied to observations for which only particle measurements, and not magnetic field measurements, are available. In that case, phase space density can be calculated based on the model magnetic field. Again phase space densities at two (or more) satellite locations can be compared and adjustments made to the field as needed. Of course this modifies the phase space density based on the model field calculation so the process must be repeated iteratively. For spacecraft with magnetometer measurements the field is

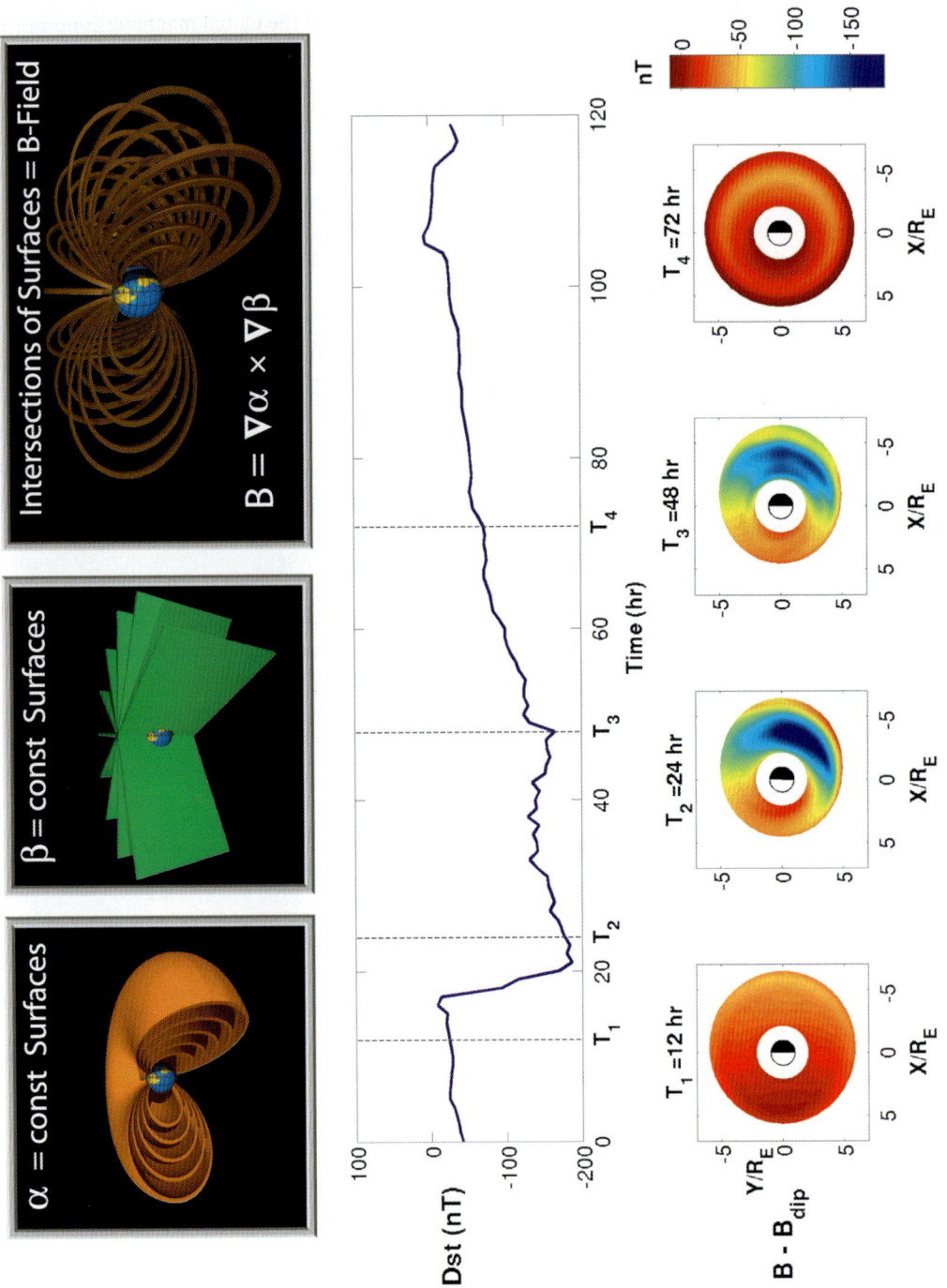

Plate 4 illustrates the methods we have applied to the problem of determining the global storm-time magnetic field using representations that (a) are self consistent with the ring current particle distributions and (b) can be efficiently used in numerical calculations. The top panel illustrates the use of Euler potential specification of the geomagnetic field. In the Euler specification two scalars are used the magnetic field vector at any point is given by the cross product of the gradient of those scalars. Added benefits are that the field is, by design, divergence free, and that magnetic field lines are simply the intersection of those two iso-potential contours. In the bottom panel we show a first step in determining a self consistent calculation of the ring current and storm-time magnetic field using the UNH RAM code (after *Zaharia et al.,* [2005]). Zaharia et al. calculated the plasma pressure distributions in the UNH-RAM code (using a dipole field), then calculated the 3-dimensional magnetic configuration in force balance with those pressure distributions, using an Euler potential equilibrium code. The bottom panel shows the difference in the field calculated using this procedure and the original dipole field. Values of B-Bdip over 100 nT during the main phase illustrate the need to calculate particle motions in a self-consistent magnetic field model. The middle panel shows Dst for this storm (beginning October 21, 2001) with four selected times marked.

known and those known values of the field need to be used in the process of optimizing the magnetic field model so that any adjustments to the field do not produce contradictory values of the field. Practically this is done through the use of an error function that one minimizes in order to get the best agreement between phase space densities and measured vs model field values.

At low energies electric field drifts dominate over magnetic gradient-curvature drifts and at intermediate energies both electric and magnetic drifts contribute. Thus, in principle, with a sufficiently dense set of measurements over the right range of energies the electric field can be constrained in the same manner as the magnetic field (although validation of the electric field is more straight-forward than specification of the field).

Whether magnetic and electric fields or just magnetic fields are being specified, the more satellite measurements that are available the more powerful this technique can be. Currently there are six geosynchronous satellites carrying LANL particle instruments, two GOES satellites carrying SEC particle and field instruments, POLAR, HEO, and Akebono in inclined orbits, SAMPEX and numerous other low-Earth-orbit satellites, and the GPS constellation. Within the next six years we expect the full constellation of 24 GPS satellites to be equipped with energetic particle detectors. In about the same time frame NASA also plans to launch the two Radiation Belt Storm Probes into CRRES-like equatorial elliptical orbits. This represents a sufficiently dense set of satellites that many and frequent conjunctions occur over a broad range of L-shells.

This technique has already been demonstrated for the field in the vicinity of geosynchronous orbit by *Onasger et al.*, [2004] and *Chen et al.*, [2005a, b]. *Onsager et al.*, [2004] used the fact that two geosynchronous satellites at different longitudes are also at different latitudes and therefore trace out different trajectories through L* (which is a common representation of the drift invariant [*Roederer*, 1970]). At two points on the orbit the satellites cross the same L* where the phase space densities should match. (See plate 6.). At other locations the two satellites measure the instantaneous local radial gradient in phase space density. *Chen et al.*, [2005a, b] extended this technique using both the GOES and LANL measurements. They calculated phase space density for a variety of equatorial pitch angles which all follow slightly different contours of L* and cross in slightly different locations. They found that, at times the standard empirical magnetic field models (e.g. *Olsen Pfitzer* [1974]; *Tsyganenko* [2002]; etc.) gave excellent agreement over a very broad range of adiabatic invariants. *Chen et al.*, [2005b] then went beyond verification to use the particle measurements to specify the magnetic field. They implemented 7 different magnetic field models and constructed an error function based on matching both the magnetic field (vector or unit vector) and particle phase space densities. By calculating the error function in each field they could determine which empirical field best matched the global magnetic conditions near geosynchronous orbit at any given time and then switch, dynamically, between models to determine the evolution of the magnetic field throughout a storm.

So far this technique has been limited by the choice of spacecraft and field models. The next logical steps are to extend it in L-shell by using other spacecraft and to use a continuously deformable magnetic field to remove the current temporal discontinuities caused by instantaneously switching magnetic field models. Ultimately, the two techniques described in this section—empirical and self-consistent calculations of the magnetic field—can be combined using the standard, but powerful, techniques of data assimilation that we have discussed briefly here.

SUMMARY

The Earth's radiation belts provide a rich field of study for basic physical processes as well as an important topic for the design and operation of space-based technology systems. Recent observations and newly-emerging theories have made significant advances in our understanding but still leave many important questions unanswered. Better understanding and quantitative prediction of changes in Earth's natural radiation belts have high value for spacecraft design, systems operations, and manned exploration programs. These issues are generally well-known within the space physics community. Less well-known are the threats posed by high altitude nuclear explosions (HANE) or the possible steps that could be taken to mitigate those threats through active radiation belt remediation (RBR). Evaluating the production and dynamics of artificial radiation belts, understanding their potential impact on national space-based infrastructure, and quantitatively evaluating the effectiveness of mitigation strategies requires the same ability to understand and model physical processes as does study of the natural variation of the Earth's belts.

Developing physical understanding of the key transport, acceleration, and loss processes requires a three-pronged approach: (1) a targeted, multi-satellite observational campaign to fill in holes in our knowledge of radiation belt dynamics; (2) a strong program to develop improved theoretical descriptions of key processes; and (3) development of global, time-dependent, data-driven but physics-based models. We have described one promising approach to provide the third of these critical components, a next-generation data assimilation model of the radiation belts. Our approach combines data-driven but physics-based modeling of the storm-

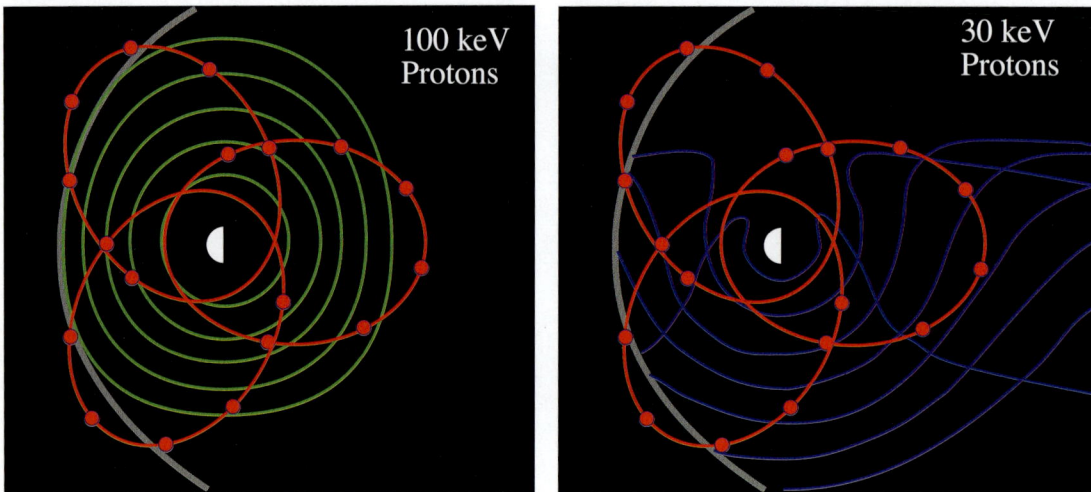

Plate 5 shows how particle measurements can be used to further improve the accuracy of global magnetic field models. Shown, schematically, are the drift paths of 100 keV and 30 keV protons along with a proposed configuration for a multi-satellite inner magnetosphere mission [*Reeves et al.*, 1997]. Particle measurements can be used along with field models to calculate phase space density at each observation point. Liouville's theorem specifies that (in the absence of non-adiabatic effects) the phase space density must be constant where satellites measure particles on the same dynamic trajectory. This must be true at all energies and all pitch angles independently (but for the same global field). Thus Liouville's theorem can be used with multi-point particle measurements to adjust the global field (and recalculate the adiabatic invariants) until optimal matching of phase space densities is obtained.

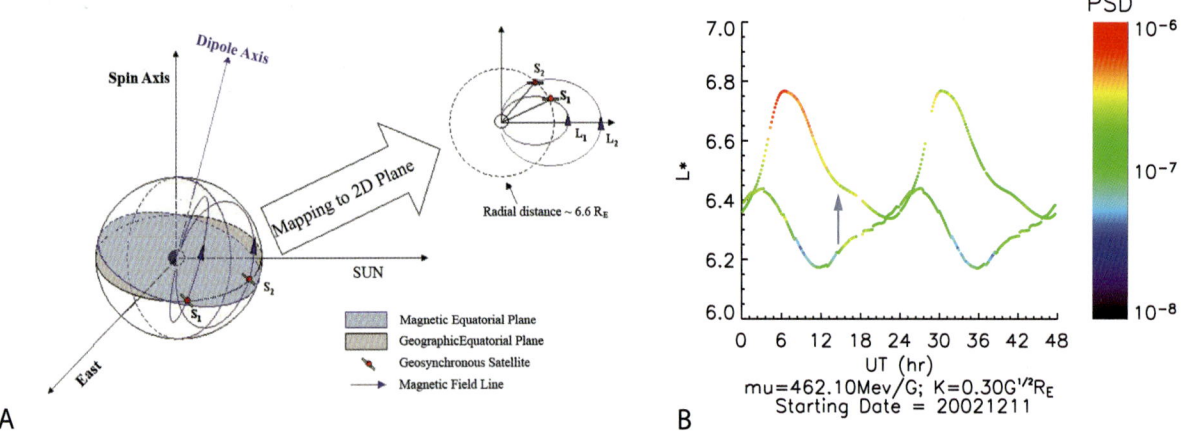

Plate 6 shows the application of phase space density matching to determination of the optimal global magnetic field configuration in the vicinity of geosynchronous orbit. Panel A illustrates the technique. Because of the Earth's dipole tilt geosynchronous satellites at different longitudes are at slightly different magnetic latitudes and therefore trace out different paths in L* as they complete their 24-hour orbit around the Earth. Panel B shows the phase space density color coded as a function of location (in L*) and time for two geosynchronous satellites. Each satellite traces a different path in L*. Where those paths cross the phase space densities should match (as shown by the same color dots). Where satellites sample different L* the phase space density gradient over that range of L* can be measured [*Chen et al.*, 2005a, b]. (This plot shows phase space density calculated by optimally fitting 7 different magnetic field models to measurements of the geosynchronous magnetic field made by GOES. Some discontinuities in L* appear where the optimal magnetic field model changes.)

time ring current, a self-consistent specification of the global magnetic field, and powerful data assimilation techniques that are well-developed in other fields but only recently applied to the radiation belt problem. One of the promises of this proposed approach is that many of the limitations of the current (extensive) set of spacecraft observations can be overcome through assimilation with physics-based models. For example the lack of magnetometers, pitch angle measurements, or the specification of adiabatic invariants can be compensated for if one applies a magnetic field model that is physically realistic. At the same time the observations can be used not only as input or as a consistency check on the models but can actually modify the models to have the accuracy and realistic dynamics required.

The space physics community has begun to take critical steps to implement the program outlined in this paper. We have verified that the key techniques do work, that critical observations are available and can be assimilated with physical models, that the needed improvements in numerical methods and implementation are possible: in short, the problem can be solved if the right resources are applied. Clearly the program outlined here is not trivial and certainly requires effort significantly larger than commonly provided by individual research grants. However, those challenges do not seem insurmountable—especially when compared to the potential reward in scientific understanding, in real cost savings to satellite designers and operations, and to improved reliability and risk assessment for commercial and military programs that depend on systems in space.

Acknowledgements. We gratefully acknowledge R. A. Christensen, E. Noveroske, T. E. Cayton, J. B. Blake, J. F. Fennell, T. G. Onsager, and many colleagues from the NSF Geospace Environment Modeling (GEM) workshops for providing data used in this study and for many helpful and inspiring discussions. We thank the US Department of Energy, the NSF GEM program, and NASA's Living With A Star (LWS) program for financial support of this study.

REFERENCES

Baker, D. N., R. L. McPherron, T. E. Cayton, and R. W. Klebesadel, Linear prediction filter analysis of relativistic electron properties at 6.6 Re, *J. Geophys. Res., 95,* 15,133, 1990.

Blake, J. B., W. A. Kolasinski, R. W. Fillius, and E. G. Mullen, Injection of electrons and protons with energies of tens of MeV into L<3 on 24 March 1991., *Geophys. Res. Lett., 19,* 821, 1992.

Blake, J. B., M. D. Looper, D. N. Baker, R. Nakamura, B. Klecker, and D. Hovestadt, New high temporal and spatial resolution measurements by SAMPEX of the precipitation of relativistic electrons, *Adv. Space. Res., 18,* 171–186, 1995.

Boscher, D., S. Bourdarie, R. M. Thorne, and B. Abel, Influence of the wave characteristics on the electron radiation belt distribution, *Adv. Space Res. 26,* 163–166, 2000.

Bourdarie, S., D. Boscher, T. Beutier, J.-A. Sauvaud, and M. Blanc, Magnetic storm modeling in the earth's electron belt by the Salammbo code, *J. Geophys. Res. 101,* 27171, 1996.

Brown, W. L., et al. Collected papers on the artificial radiation belt from the July 9, 1962, nuclear detonation, *J. Geophys. Res., 68,* 605–606, 1963.

Chen, Y., R. H. W. Friedel, G. D. Reeves, T. G. Onsager, and M. F. Thomsen, Multi-satellite Determination of the Relativistic Electron Phase Space Density at Geosynchronous Orbit: I. Methodology and Initial Results During Geomagnetic Quiet Times, *J. Geophys. Res., in press*, 2005a.

Chen, Y, R. H. W. Friedel, and G. D. Reeves, Multi-satellite Determination of the Relativistic Electron Phase Space Density at Geosynchronous Orbit: II.Application to Geomagnetic Storm Times, *submitted to J. Geophys. Res.,* 2005b.

Elkington, S. R., M. K. Hudson, and A. A. Chan, Acceleration of relativistic electrons via drift-resonant interaction with toroidal-mode Pc-5 ULF oscillation, *Geophys. Res. Lett. 26,* 3273–3276, 1999.

Elkington, S. R., M. K. Hudson, and A. A. Chan, Resonant acceleration and diffusion of outer zone electrons in an asymmetric geomagnetic field, *J. Geophys. Res. Vol. 108* No. A3, 10.1029/2001JA009202, 2003.

Fok, M.-C., R. A. Wolf, R. W. Spiro, and T. E. Moore, Comprehensive computational model of Earth's ring current, *J. Geophys. Res., 106(A5),* 8417–8424, 10.1029/2000JA000235, 2001.

Friedel, R. H. W., G. D. Reeves, and T. Obara, Relativistic electron dynamics in the inner magnetosphere: A review, *J. Atmos. Solar-Terrestrial Phys., 64,* 265–282, 2002.

Fujimoto, M., and A. Nishida, Energization and isotropization of energetic electrons in the Earth's radiation belt by the recirculation process, *J. Geophys. Res., 95,* 4625, 1990.

Ghil, M., K. Ide, A. F. Bennett, P. Courtier, M. Kimoto, and N. Sato, eds., *Data Assimilation in Meteorology and Oceanography: Theory and Practice,* Meteorological Society of Japan and Universal Academy Press, 1997.

Green, J. C., and M. G. Kivelson, Relativistic electrons in the outer radiation belt: Differentiating between acceleration mechanisms, *J. Geophys. Res., 109,* A03213, doi:10.1029/2003JA010153, 2004.

Green, J. C., T. G. Onsager, and T. P. O'Brien, Testing loss mechanisms capable of rapidly depleting relativistic electron flux in the Earth's outer radiation belt, *J. Geophys. Res., 109,* A12211, doi:10.1029/2004JA010579, 2004.

Hess, W. N., The artificial radiation belt made on July 9, 1962, *J. Geophys. Res., 68,* 667–683, 1963.

Horne, R. B., and R. M. Thorne, Relativistic electron acceleration and precipitation during resonant interactions with whistler-mode chorus, *Geophys. Res. Lett. Vol. 30* No. 10 10.1029/2003GL016973, 2003.

Inan, U. S., T. F. Bell, J. Bortnik, and J. M. Albert, Controlled precipitation of radiation belt electrons, *J. Geophys. Res., 108(A5),* 1186, doi:10.1029/2002JA009580., 2003.

Jordanova, V. K., J. U. Kozyra, G. V. Khazanov, A. F. Nagy, C. E. Rasmussen, and M.-C. Fok, A bounce-averaged kinetic-model of the ring current ion population, *Geophys. Res. Lett., 21,* 25, 2785, 1994.

Jordanova, V. K., L. M. Kistler, J. U. Kozyra, G. V. Khazanov, and A. F. Nagy, Collisional losses of ring current ions, *J. Geophys. Res., 101(A1),* 111–126, 10.1029/95JA02000, 1996.

Jordanova, V. K., J. U. Kozyra, A. F. Nagy, and G. V. Khazanov, Kinetic model of the ring current-atmosphere interactions, *J. Geophys. Res., 102(A7),* 14279–14292, 10.1029/96JA03699, 1997.

Jordanova, V. K., L. M. Kistler, M. F. Thomsen, and C. G. Mouikis, Effects of plasma sheet variability on the fast initial ring current decay, *Geophys. Res. Lett., 30,* 6, 1311, 2003b.

Jordanova, V. K., New insights on geomagnetic storms from model simulations using multi-spacecraft data, *Space Sci. Rev., 107,* 1–2, 157, 2003a.

Jordanova, V. K., Sources, transport, and losses of energetic particles during geomagnetic storms, *Physics and Modeling of the Inner Magnetosphere*, AGU Monograph, this volume, 2005.

Kalman, R. E., A New Approach to Linear Filtering and Prediction Problems, *Transaction of the ASME—Journal of Basic Engineering,* 35–45, March, 1960.

Kanekal, S. G., R. H. W. Friedel, G. D. Reeves, D. N. Baker, and J. B. Blake, Relativistic electron events in 2002: Studies of pitch angle isotropization, *J. Geophys. Res., submitted,* 2005.

Kim, H.-J., and A. A. Chan, Fully-adiabatic changes in storm-time relativistic electron fluxes, *J. Geophys. Res., 102,* 22,107–22,116, 1997.

Kintner, P. M., et al., *The LWS Geospace Storm Investigations: Exploring the extremes of Space Weather,* NASA/TM-2002–21613, September, 2002.

Koller, J., and R. H. W. Friedel, Radiation Belt Data Assimilation and Parameter Estimation, *J. Geophys. Res., submitted,* 2005.

Li, X., D. N. Baker, M. Teremin, T. E. Cayton, G. D. Reeves, R. S. Selesnick, J. B. Blake, G. Lu, S. G. Kanekal, and H. J. Singer, Rapid enhancements of relativistic electrons deep in the magnetosphere during the May 15, 1997, magnetic storm, *J. Geophys. Res., 104,* 4467–4476, 1999.

Liemohn, M. W., J. U. Kozyra, M. F. Thomsen, J. L. Roeder, G. Lu, J. E. Borovsky, and T. E. Cayton, Dominant role of the asymmetric ring current in producing the stormtime Dst*, *J. Geophys. Res., 106(A6),* 10883–10904, 10.1029/2000JA000326, 2001.

Lorentzen, K. R., M. P. McCarthy, G. K. Parks, J. E. Foat, R. M. Millan, D. M. Smith, R. P. Lin, and J. P. Treilhou, Precipitation of relativistic electrons by interaction with electromagnetic ion cyclotron waves, *J. Geophys. Res., 105,* 5381–5390, 10.1029/1999JA000283, 2000.

Lorentzen, K. R., J. B. Blake, U. S. Inan, and J. Bortnik, Observations of relativistic electron microbursts in association with VLF wave activity, *J. Geophys. Res., 106,* 6017–6027, 2001.

Metz, A.P. and R.R. Babcock, *Military Utility Analysis: Radiation Belt Remediation Technology,* AFRL Technical Report, AFRL-VS-PS TR-2004-1033, AFRL/VS, Kirtland AFB, NM, Mar 2004.

Millan, R. M., R. P. Lin, D. M. Smith, K. R. Lorentzen, and M. P. McCarthy, X-ray observations of MeV electron precipitation with a balloon-borne germanium spectrometer, *Geophys. Res. Lett., 29(24),* 2194, doi:10.1029/2002GL015922, 2002.

Murch, R.S., *High Altitude Nuclear Detonations (HAND) Against Low Earth Orbit Satellites (HALEOS)*, DTRA Advanced Concept Office briefing, Apr 2001.

Nakamura, R., D. N. Baker, J. B. Blake, S. Kanekal, B. Klecker, and D. Hovestadt, Relativistic electron-precipitation enhancements near the outer edge of the radiation belt, *Geophys. Res. Lett. 22,* 1129–1132, 1995.

Nakamura, R., M. Isowa, Y. Kamide, D. N. Baker J. B. Blake, and M. Looper, SAMPEX observations of precipitation bursts in the outer radiation belt, *J. Geophys. Res., 105,* 15,875–15,885, 2000.

National Security Space Architect, *An Executive Overview for FY 1998—2003*, March 1997.

NRC Space Studies Board, *The Sun to the Earth—and Beyond: A Decadal Research Strategy in Solar and Space Physics* THE NATIONAL ACADEMIES PRESS Washington, D.C., 2002

Olsen, W. P., and K. A. Pfitzer, A quantitative model of the magnetospheric magnetic field, *J. Geophys. Res., 79,* 3739, 1974.

Onsager, T. G., A. A. Chan, Y. Fei, S. R. Elkington, J. C. Green, and J. J. Singer, The radial gradient of relativistic electrons at geosynchronous orbit, *J. Geophys. Res., 109,* A05, 221, doi:10.1029/2003JA010368., 2004.

Onsager, T. G., A. A. Chan, Y. Fei, S. R. Elkington, J. C. Green, and J. J. Singer, The radial gradient of relativistic electrons at geosynchronous orbit, *J. Geophys. Res., 109,* A05221, doi:10.1029/2003JA010368., 2004.

Reeves, G. D., R. D. Belian, T. E. Cayton, R. H. W. Friedel, M. G. Henderson, D. N. Baker, and H. E. Spence, Energetic Particle Contributions to a Magnetospheric Constellation Mission, *EOS Trans. AGU,* Fall Meeting, San Francisco, CA, 8–12 December, 1997.

Reeves, G. D., D. N. Baker, R. D. Belian, J. B. Blake, T. E. Cayton, J. F. Fennell, R. H. W. Friedel, M. G. Henderson, S. Kanekal, X. Li, M. M. Meier, T. Onsager, R. S. Selesnick, and H. E. Spence, The Global Response of Relativistic Radiation Belt Electrons to the January 1997 Magnetic Cloud, *Geophys. Res. Lett., 17,* 3265–3268, 1998.

Reeves, G. D., K. L. McAdams, R. H. W. Friedel, and T. P. O'Brien, Acceleration and loss of relativistic electrons during geomagnetic storms, *Geophys. Res. Lett.,30 (10),* 1529, doi: 10.1029/2002GL016513, 2003.

Richmond, A. D., and Y. Kamide, Mapping electrodynamic features of the high-latitude ionosphere from localized observations: Technique, *J. Geophys. Res., 93(A6),* 5741–5759, 10.1029/88JA01165, 1988.

Roederer, J.G., *Dynamics of geomagnetically trapped radiation*, Springer-Verlag, New York, 1970.

Schulz, M., and L. J. Lanzerotti, *Particle Diffusion in the Radiation Belts*, Springer-Verlag, New York., 1974.

Shukhtina, M. A., and V. A. Sergeyev, Modeling of the drift of energetic particles in the magnetosphere near the geosynchronous orbit Modeling of the drift of energetic particles in the magnetosphere near the geosynchronous orbit, *Geomagn. Aeron., 31(5),* 627–631, 10.1029/92GA00395, 1992.

Summers, D., and C. Ma, A model for generating relativistic electrons in the earth's inner magnetosphere based on gyroresonant wave-particle interactions., *J. Geophys. Res. 105*, 2625–2639, 2000.

Tsyganenko N. A., A model of the magnetosphere with a dawn-dusk asymmetry, 1, Mathematical structure, *J. Geophys. Res., 107 (A8)*, doi:10.1029/2001JA000219, 2002.

Van Allen, J. A., et al., Satellite observations of the artificial radiation belt of July 1962, *J. Geophys. Res., 68*, 619–627, 1963.

Weimer, D. R., An improved model of ionospheric electric potentials including substorm perturbations and application to the Geospace Environment Modeling November 24, 1996, event, *J. Geophys. Res., 106(A1)*, 407–416, 10.1029/2000JA000604, 2001.

Welsh, G., and G. Bishop, *An introduction to the Kalman filter*, University of North Carolina technical report TR 95-014, 1995.

Zaharia, S., C. Z. Cheng, and K. Maezawa, 3-D force-balanced magnetospheric configurations, *Annales Geophysicae 22*: 251–265, 2004.

Zaharia, S., M. F. Thomsen, J. Birn, M. H. Denton, V. K. Jordanova, and C. Z. Cheng, Effect of Storm-Time Plasma Pressure on the Magnetic Field in the Inner Magnetosphere, *Geophys. Res. Lett., 32*, L03102, doi:10.1029/2004GL021491, 2005.

Geoffrey D. Reeves, Reiner H. W. Friedel, Michelle F. Thomsen, Sorin Zaharia, Michael G. Henderson, and Yue Chen, Space Sciences and Applications, Los Alamos National Laboratory, Mail Stop D-644, Los Alamos NM, 87545

Sebastien Bourdarie, CERT-ONERA/DERTS, 2 Avenue Edouard Belin, PO Box 4025, 31055 Toulouse, France

Vania K. Jordanova, Institute for the Study of Earth, Oceans, and Space Science Center, University of New Hampshire, Durham NH, 03824

Brian J. Albright, and Dan Winske, Plasma Physics, Los Alamos National Laboratory, Mail Stop B-259, Los Alamos NM, 87545

Simulated Stormtime Ring-Current Magnetic Field Produced by Ions and Electrons

Margaret W. Chen[1], Michael Schulz[2], Shuxiang Liu[3], Gang Lu[4],
Larry R. Lyons[3], Mostafa El-Alaoui[5], and Michelle Thomsen[6]

Using drift-loss simulations, we investigate the stormtime ring-current magnetic field produced by protons and electrons during the 19 October 1998 storm. We compute the guiding-center drift motion of equatorially-mirroring protons and electrons in a model magnetosphere and consider H charge exchange loss and electron precipitation due to pitch-angle scattering. We simulate perpendicular particle pressure distributions and the ring-current magnetic field in the equatorial plane. Our quiet-time proton ring current model reproduces a radial perpendicular pressure profile that is similar to statistical observations. Both the proton and electron ring current form asymmetrically early in the storm main phase and become stronger and more nearly symmetric later in the main phase. We find that the quiet-time electron ring-current magnetic field is negligible compared to the quiet-time proton ring-current magnetic field. However, during storms, electrons contribute ~ 15–20% of protons to the total *Dst*. Thus, the electrons contribute significantly to the total ring current during storms. There are large asymmetric stormtime ring-current magnetic field depressions at $r_0 \sim 3.5-6\ R_E$. The simulated large magnetic field depressions agree qualitatively with statistical observations. These magnetic field depressions are a significant fraction of the geomagnetic field, indicating a need for a magnetically self-consistent treatment of the particle transport. The large ring-current magnetic depressions should be included in global descriptions of the inner magnetosphere.

1. INTRODUCTION

The defining feature of a magnetic storm is the formation of a ring current that perturbs the Earth's geomagnetic field. This magnetic perturbation contributes significantly to the widely used *Dst* index that represents the longitude-averaged magnetic field perturbation at low-latitudes on the surface of the Earth due to all current systems. Thus, the strength of the ring current and magnetic storms has been conveniently associated with the *Dst* index. Large depressions (~ –100 to –200 nT) in the magnetic field due to the ring current has also been measured in-situ at *L* of 3–4 during the main phase of storm events [*Cahill*, 1966]. Several recent statistical analyses of large databases of in-situ magnetic field measurements of the inner magnetosphere [*Tsyganenko*

---

[1] Space Sciences Applications Laboratory, The Aerospace Corporation, El Segundo, California
[2] Lockheed Martin Advanced Technology Center, Palo Alto, California
[3] Department of Atmospheric Sciences, University of California, Los Angeles, California
[4] High Altitude Observatory, Boulder, Colorado
[5] Institute of Geophysics and Planetary Physics, University of California, Los Angeles, California
[6] Los Alamos National Laboratory, Los Alamos, New Mexico

Inner Magnetosphere Interactions: New Perspectives from Imaging
Geophysical Monograph Series 159
Copyright 2005 by the American Geophysical Union.
10.1029/159GM18

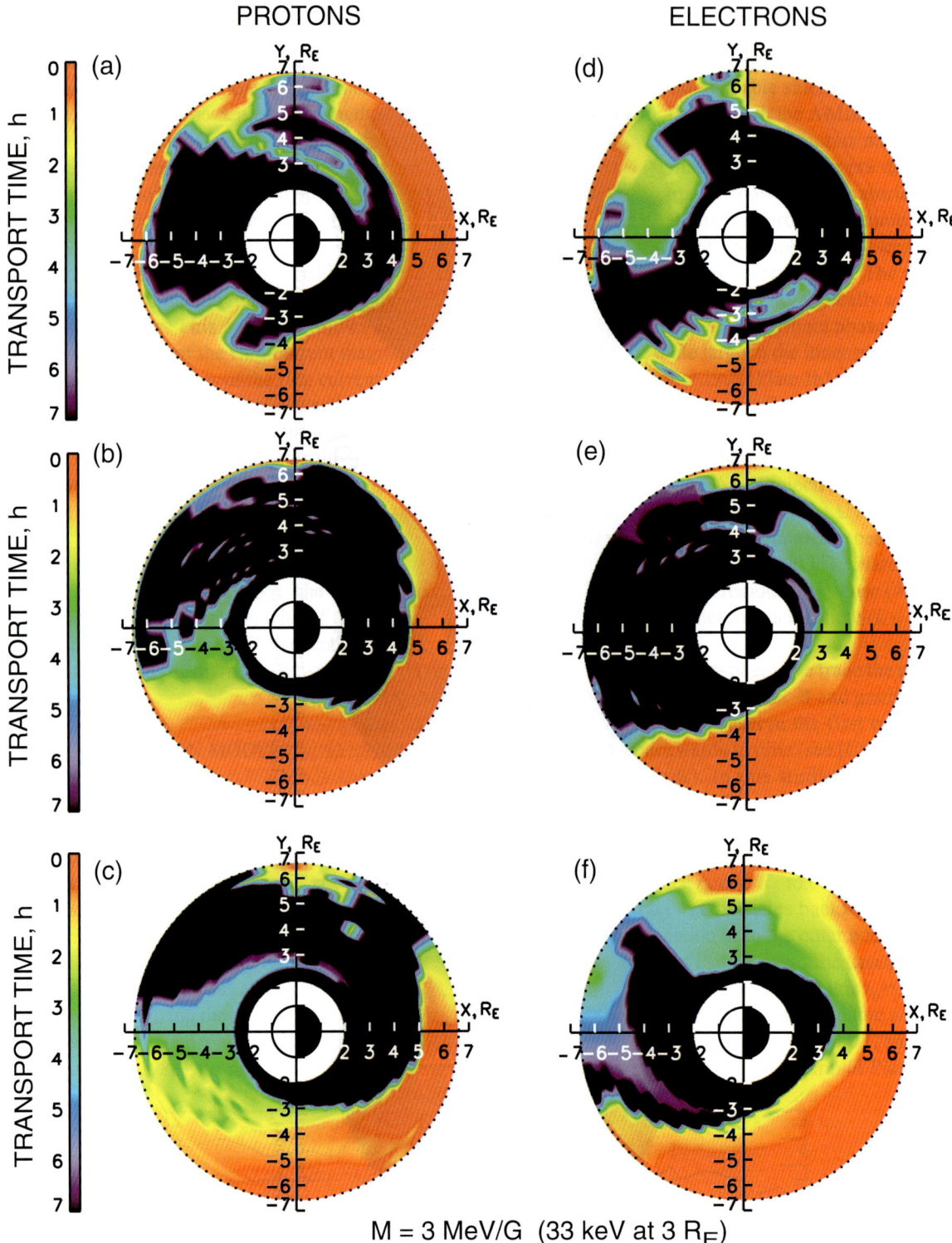

Plate 2. Drift times required for plasma sheet protons (left column) and electrons (right column) with $M = 3$ MeV/G (33 keV at 3 R_E) to reach the equatorial inner magnetosphere at (a,d) 0500 UT, (b, e) 0800 UT, and (c, f) 1130 UT on 19 October 1998.

the corresponding transport times of electrons with $M = 3$ MeV/G. About one hour into the main phase at 0500 UT, the region of very rapid transport (< 1 hr) is roughly the same for both protons and electrons. However, protons with $M = 3$ MeV/G drift westward while electrons drift eastward. Thus, for times later in the main phase the drift times in the afternoon quadrant are shorter for protons than for electrons and vice-versa in the morning quadrant.

The boundary conditions for both protons and electrons during the storm are based on fluxes obtained from the MPA instruments on LANL satellites at geosynchronous altitude. The MPA instrument measures total ion counts. However, for this study we have assumed the measured total ion flux is the proton flux. To obtain the longitudinal variation of the boundary fluxes, we performed a least-squares fit of the Kp-dependent averaged LANL data of *Korth et al.* [1999] to MLT for each Kp values. Then, we renormalized this MLT-dependence to the LANL real time data for the associated Kp value to the real time data.

We obtain initial conditions for the trapped (on closed drift trajectories) proton phase-space density distribution by solving the steady-state transport equation that balances quiet-time radial diffusion against proton charge exchange, the dominant collisional loss process for the protons. Details about this solution can be found in [*Chen et al.*, 1994]. Outside of the trapping region, the proton phase-space density distribution is assigned to be same as that at the outer boundary of our model which represents the plasma sheet. We use the LANL ion data for $Kp = 0$ of *Korth et al.* [1999] as our quiet-time outer boundary conditions. For each M value of interest, we find the drift separatrix under the pre-storm AMIE electric field to demarcate the boundary between the quiet-time trapped and untrapped (on open drift trajectories) populations.

The electron initial conditions is obtained by using the numerical solutions of the steady-state transport equation that balances radial diffusion with loss due to electron precipitation in the limit of weak pitch-angle diffusion [*Albert*, 1994]. The solutions provide the phase-space density distributions within an assumed plasmasphere of $r_0 = 5.5\ R_E$. We extrapolate the phase-space distribution from 5.5 R_E to geosynchronous orbit and renormalize the initial conditions using the quiet-time boundary conditions based on the quiet-time *Korth et al.* [1999] LANL electron data. Details on the electron ring-current model can be found in *Liu et al.* [2003].

We use the particle trajectory simulation results and invoke Liouville's theorem of conservation of phase-space density modified by loss to obtain the stormtime phase-space density distributions. As the particles drift, we take account of their loss. For protons we consider their loss against charge exchange with neutral H. Plate 3a shows a plot of the lifetime against charge exchange for protons with $M = 3$ MeV/G. Protons with $M = 3$ MeV/G have an energy of 3.2 keV at $r_0 = 6\ R_E$ and have a lifetime against charge exchange of approximately 2 days. For electrons we consider their loss due to pitch-angle scattering due to waves. Plate3b shows the lifetime against pitch-angle scattering for electrons with $M = 3$ MeV/G. We use the theoretical weak-diffusion lifetimes due to plasmaspheric hiss of *Albert* [1994] earthward of the plasmapause. Outside of the plasmapause, the electron lifetimes are based on the theoretical calculations of diffusion against electrostatic cyclotron harmonic waves of *Lyons* [1974] or strong diffusion [*Schulz*, 1998]. Electrons with $M = 3$ MeV/G at $r_0 = 6\ R_E$ have a lifetime against pitch-angle scattering of approximately 2.4 hours. Thus, the lifetime of ring-current electrons can be significantly shorter than those of ring-current protons. The electron lifetimes are on the order of the storm main phase. See [*Liu et al.*, 2003] for further details on the ring current electron model.

3. SIMULATION RESULTS

3.1 Perpendicular Pressure Distributions

We map the stormtime phase distributions for both protons and electrons at selected times of interest during the storm main phase: 0500 UT, 0800 UT, and 1130 UT (see Figure 1). We consider M values from 1 to 100 MeV/G and extrapolate the phase-space density spectra beyond 100 MeV/G as a power law. For $M < 1$ MeV/G, we consider that the phase-space spectrum is equal to the boundary spectrum. From the phase-space density distributions, we compute the perpendicular particle pressure $P_\perp$ over all energies or first invariant values. The derivation of the perpendicular pressure equation for our field model can be found in *Chen et al.* [1994]. Plate 4 shows examples of the simulated perpendicular pressure for protons (left column) and electrons (right column) at different times during the 19 October 1998 storm. Only about one hour into the main phase of the storm (0500 UT; Plate 4b), $P_\perp$ is the most intense on the duskside at roughly $r_0 = 3.5$–5 R_E. One can see that the perpendicular pressure distribution is longitudinally asymmetric. Later in the main phase (1130 UT; Plate 4c), the distribution of $P_\perp$ looks more nearly azimuthally symmetric as more of the protons have been transported to the morning side and the cross polar cap potential has weakened.

In Figure 2, radial profiles of the model proton perpendicular pressure at 2100 MLT for pre-storm (dotted curve), 0500 UT (solid curve), 0800 UT (dashed-dotted-dotted curve), and 1130 UT (dash-dotted curve) are plotted. The dashed curve shows the average proton perpendicular pres-

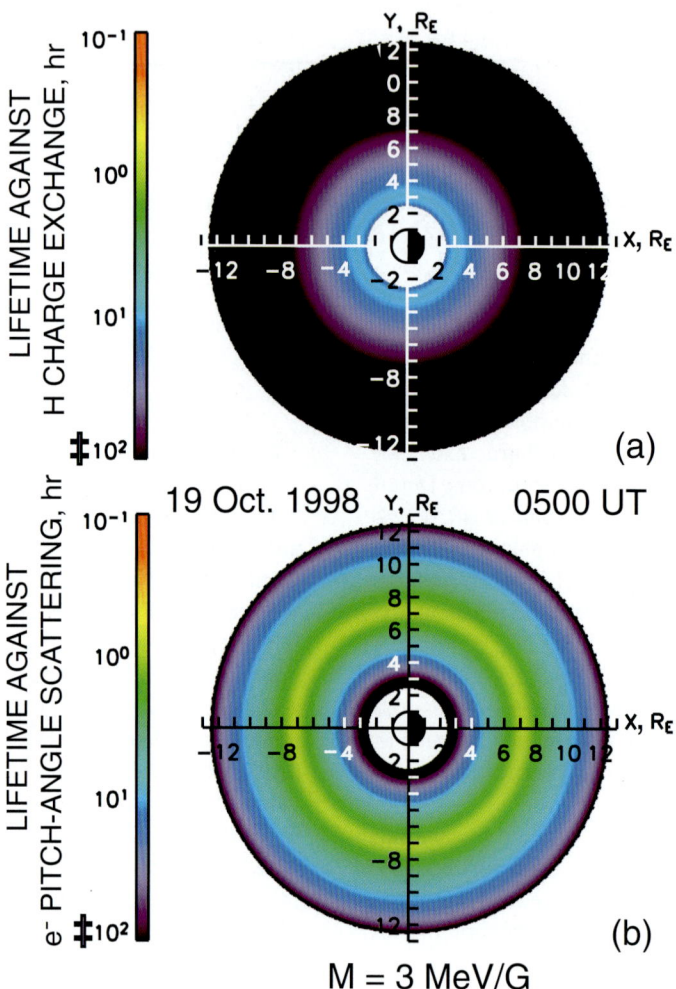

Plate 3. (a) Equatorial lifetimes against charge exchange between protons with 3 MeV/G (3.2 keV at 6 R_E) and neutral H. (b) Equatorial lifetimes against electron pitch-angle scattering for electrons with 3 MeV/G.

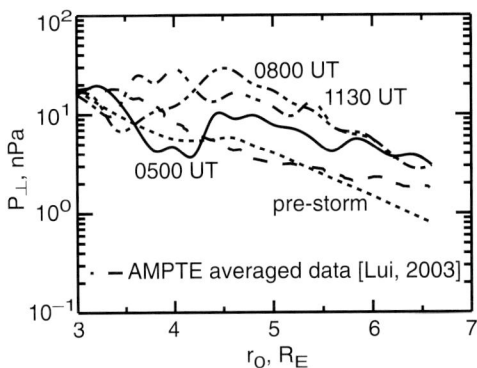

Figure 2. Radial profiles of the proton perpendicular pressure at 2100 MLT. The dotted black curve corresponds to the model pre-storm and the dashed black curve to averaged AMPTE quiet-time data of [*Lui*, 2003]. The solid, dash-dotted-dotted, and dash-dotted curves correspond to the model profiles at 0500 UT, 0800 UT, and 1130 UT on 19 October 1998, respectively.

sure from AMPTE/CCE observations of *Lui* [2003]. The AMPTE data was averaged over $Kp < 2$, longitudes of 2100–2200 MLT, and energies of 1 keV to 4 MeV. This large range of energy encompasses the most significant contributions to the total ring current pressure. Thus, it is fair to compare our total model pre-storm proton $P_\perp$ profile (black solid) to the observed proton $P_\perp$ (black dashed) profile. We find that the model pre-storm $P_\perp$ looks quite similar to the quiet-time averaged $P_\perp$ from AMPTE observations. Deviations between the model and observed $P_\perp$ get bigger at large r_0, possibly due to our use of a very simplified model magnetic field that is less realistic at farther equatorial radial distances. Nevertheless, the similarity between the model and observed pre-storm perpendicular pressure profiles provides confidence in our quiet-time model. At 2100 MLT the simulated stormtime proton $P_\perp$ peaks progressively inward as the main phase develops. At 1130 UT on 19 October 1998 the peak at 2100 MLT is $P_\perp \sim 30$ nPa at $r_0 \sim 4\ R_E$. *El-Alaoui et al.* [2004] have simulated the same 19 October 1998 event using MHD simulations. At 2100 MLT, the MHD peak total proton pressure is ~1.5 nPa to 2 nPa at $L \sim 5\ R_E$ during the main phase of the storm. The peak MHD proton pressure is considerably smaller than the stormtime peak pressure from kinetic simulations. The MHD peak pressure also occurs farther out than what is found from the kinetic simulations. In future work, we will modify the MHD model to more explicitly account for particle drifts to realistically include a ring current.

Distributions of the electron perpendicular pressure are shown in the right column of Plate 4. Overall, the pre-storm electron $P_\perp$ (Plate 4d) is relatively much smaller than the pre-storm proton $P_\perp$ (Plate 4a). This is because the pre-storm conditions are based on a long-term steady-state solution in which the electrons have suffered large losses due to precipitation. During the storm main phase, there can be quite large electron pressures in localized regions. Early in the main phase (0500 UT; Plate 4e), the electron $P_\perp$ has intensified in the evening quadrant at $r_0 \sim 4$–$5\ R_E$ and on the dayside around $r_0 \sim 3\ R_E$. The latter occurs because there were electrons that had been injected to this region before the main phase. Although not shown, as the main phase progresses, the pressure intensifies on the morning side and the electron $P_\perp$ distribution looks azimuthally asymmetric. By 1130 UT (not shown), the electron perpendicular pressure distribution looks more nearly symmetric and is relatively intense at $r_0 \sim 2.5$–$3\ R_E$.

3.2 Current Distributions

Knowing the perpendicular pressure distributions, we can compute the particle current distributions. The gradient-drift current is given by

$$\mathbf{J}_g = \frac{P_\perp}{B^2} \hat{\mathbf{B}} \times \nabla \mathbf{B} \qquad (1)$$

Plate 5a shows an example of the simulated proton gradient current per unit equatorial radial distance dI_g/dR corresponding to 0500 UT on 19 October 1998. At this time the proton gradient-drift current is intense on the dusk and dayside where the proton perpendicular pressure is intense. We close the partial gradient-drift currents with field-aligned currents. To explain this closure we consider a simple current loop with twice as much current on the dusk side as the dawn side. The partial current loop bifurcates at the equator at both noon and midnight with half of the parallel current going up a field line to and through the ionosphere and then back toward the equator. The other half of the parallel current goes down a field line to and through the ionosphere and back toward the equator. Closing of the current is necessary to satisfy Ampere's Law. Plate 5b shows an example of the field-aligned proton current needed to close the gradient-drift current shown in Plate 5a. In this example the field-aligned current per unit equatorial radial distance $dI_\parallel/dR$ is large at dusk, but is relatively small in absolute magnitude compared to the gradient-drift current.

In addition to the gradient-drift and field-aligned currents there is a magnetization current associated with particle gyration. The magnetization current is included in the expression for the total equatorial current density

$$\mathbf{J}_\perp = \left(\frac{c}{B^2}\right) \mathbf{B} \times \nabla P_\perp \qquad (2)$$

[*Schulz and Lanzerotti*, 1974, pp. 26–27].

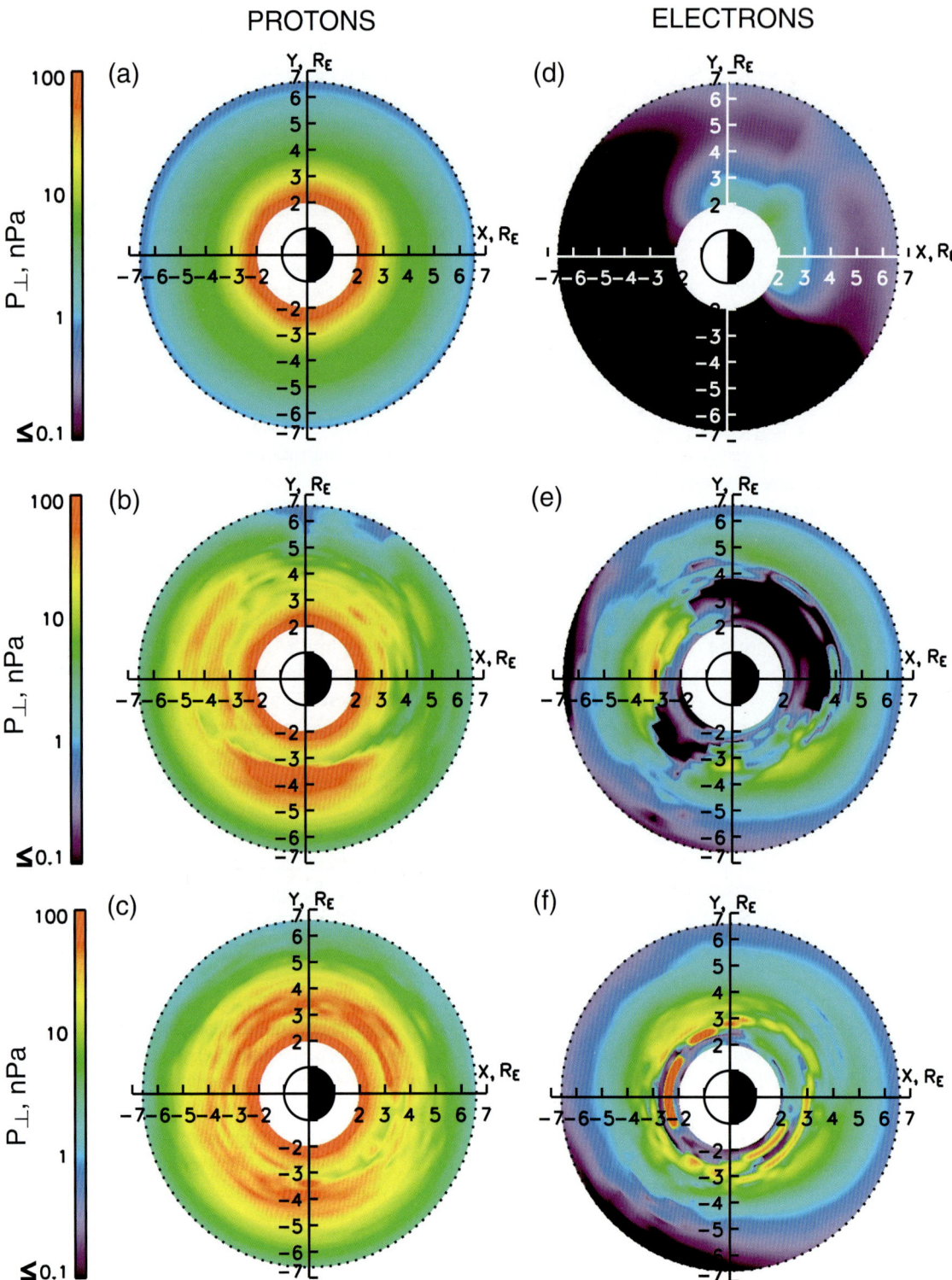

Plate 4. Simulated perpendicular pressure of equatorially-mirroring protons (left column) and electrons (right column) at (a,d) pre-storm, (b,e) 0500 UT, and (c, f) 1130 UT on 19 October 1998.

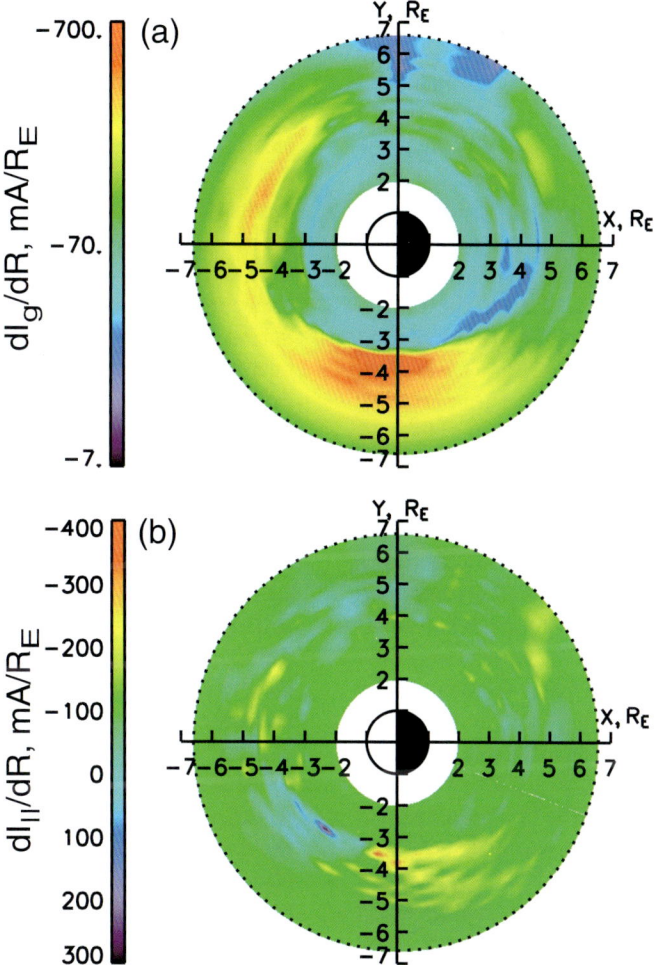

Plate 5. (a) Simulated proton gradient current per unit equatorial radial distance and (b) field-aligned proton current needed to close (a). Both panels correspond to 0500 UT on 19 October 1998.

3.3 Ring-Current Magnetic Field

In the past we computed the drift-averaged ring-current magnetic field for trapped protons [*Chen et al.*, 1994] in terms of the gradient of the perpendicular pressure from (2), but did not consider partial ring currents as we do here. In this work, we find it illustrative to compute the magnetic field-contributions from the gradient-drift, field-aligned, and magnetization currents separately. To compute the magnetic field perturbation from the gradient-drift current, we apply the Biot-Savart law. Plate 6a shows an example of the simulated magnetic intensity perturbation from the proton gradient-current drift ΔB_g corresponding to 0500 UT on 19 October 1998. The largest depressions ($-\Delta B_g$) at this time occurred near dusk at $r_0 \sim 3.5\ R_E$, where the gradient-drift current per unit equatorial radial distance is intense (see Plate 5a). The calculated depression might be somewhat smoothed out if we include particles that mirror off the equator in our simulations as we plan to do in the future. Similarly, to compute the magnetic field contribution from the field-aligned current $\Delta B_\parallel$, we apply the Biot-Savart law. An example of the simulated magnetic intensity perturbation due to field-aligned proton current at 0500 UT on 19 October 1998 is shown in Plate 6b. The largest $\Delta B_\parallel$ occurs at dusk around $r_0 \sim 3-4\ R_E$, where the field-aligned current is the largest (see Plate 5b). Noting the different color bar scales in Plates 6a and 6b, the magnetic intensity perturbation due to field-aligned currents is very small compared to the magnetic intensity perturbation due to the gradient-drift current. Plate 6c shows an example of the simulated proton magnetic intensity perturbation due to magnetization at 0500 UT on 19 October 1998. The ΔB_{mag} is large and negative where there is plasma and small and positive where there is not much plasma.

The total simulated proton ring-current magnetic field at 0500 UT on 19 October 1998 is shown in Plate 6d. At this time there are very large ($\Delta B \lesssim -400$ nT) asymmetric magnetic field depressions from $r_0 \sim 3.5-5\ R_E$. Comparison of the panels in Plate 6 reveals that the magnetic field due to magnetization and the gradient-drift are significant contributors to the large depressions in the proton ring-current magnetic field. The magnetic field due to field-aligned currents is almost negligible.

The large stormtime magnetic field depressions in the inner magnetosphere that we find in our simulations agree qualitatively with stormtime empirical models and statistical observations during active times. *Tsyganenko et al.* [2003] have found magnetic field depressions as large as ~ -400 nT at $L \sim 2-4$ from the afternoon to pre-dawn in their empirical model during a different storm event on 6 April 2000. Based on statistical analysis of AMPTE magnetic field data, *Lui* [2003] reported that there are large magnetic field depressions in the dusk-midnight sector at $L \sim 3.5-6$ for high magnetic activity ($Kp > 3$). Our present simulations likely overestimate the magnetic field depressions because of the lack of a magnetically self-consistent treatment of the particle transport. In fact, the simulated stormtime magnetic field depressions can be a significant fraction of the geomagnetic field, indicating a need for a magnetically self-consistent treatment of the particle transport in inner magnetospheric models. We plan to develop a self-consistent ring current model in the near future.

Thus far, we have discussed only the proton ring-current magnetic field. Next, we present some results of the simulated electron ring-current magnetic field and its contribution to the overall ring-current magnetic field. Plate 7a shows the model pre-storm proton ring-current magnetic field. The magnetic field depressions are relatively small during quiet times, occurring at $r_0 \sim 4-5.5\ R_E$. The proton magnetic field intensity perturbation at the center of the Earth is -78.3 nT during pre-storm. In contrast, the pre-storm electron ring-current magnetic field intensity (Plate 7b) is clearly much smaller in absolute value than the proton ring-current magnetic field intensity. The electron magnetic field intensity perturbation at the center of the Earth is -2.69 nT. Plate 7c shows the sum of the proton and the electron pre-storm ring-current magnetic field. During pre-storm or quiet times the perturbed electron ring-current magnetic field is small compared to the proton ring-current magnetic field. However, we find that electrons can significantly contribute to the ring-current magnetic field during storms. Plate 7d shows the proton ring-current magnetic field at 0800 UT on 19 October 1998. The magnetic field intensity perturbation at the center of the Earth is -140.8 nT. The electron ring-current magnetic field at this time is shown in Plate 7e. There are fairly large dawn side magnetic field depressions from slightly pre-midnight to noon at $r_0 \sim 3-3.5\ R_E$, where the electron perpendicular pressure is large. The electron ring current looks more asymmetric than the proton ring current (Plate 7d) at this time. The electron magnetic field intensity perturbation at the center of the Earth at this time is -21.6 nT. Plate 7f shows the sum of the proton and electron ring-current magnetic field at 1130 UT. With inclusion of the electrons, there are larger magnetic field depressions from midnight to pre-dawn at $r_0 \sim 3-3.5\ R_E$. The magnetic field intensity perturbation at the center of the Earth is -162.4 nT. Thus, for this example, the electrons contribute about 13% of what protons contribute to the total *Dst* at this time. At 0500 UT and 1130 UT on 19 October 1998, the electrons contribute 14% and 18%, respectively, of the proton contributions to the total *Dst*. Thus, the ratio we find is $\sim 15-20\%$. This is somewhat less than the estimate of $\sim 25\%$ that *Frank* [1967] reported based on analysis of the energy content of electrons

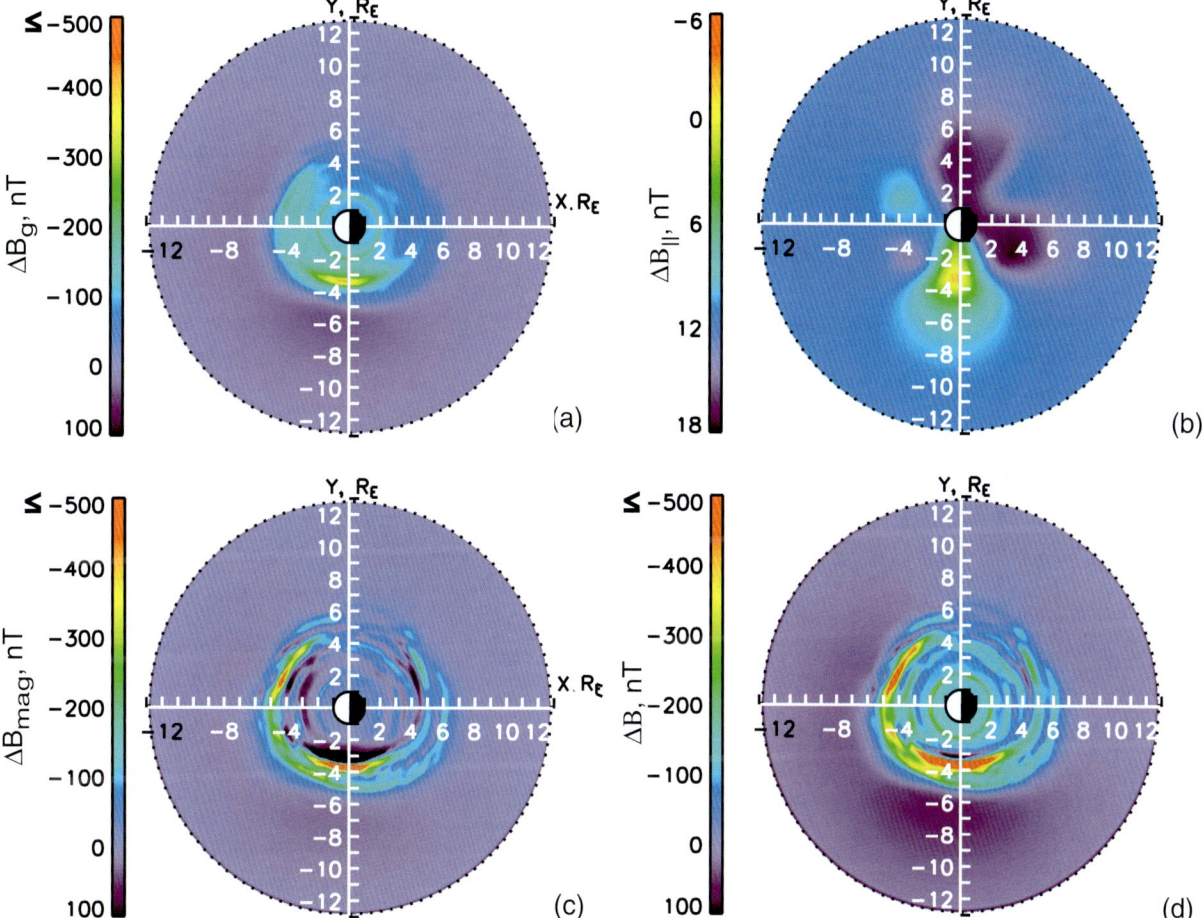

Plate 6. The simulated magnetic field perturbations due to the proton (a) gradient-drift, (b) field-aligned, and (c) magnetization current at 0500 UT on 19 October 1998. (d) The total simulated proton ring-current magnetic field at the same time.

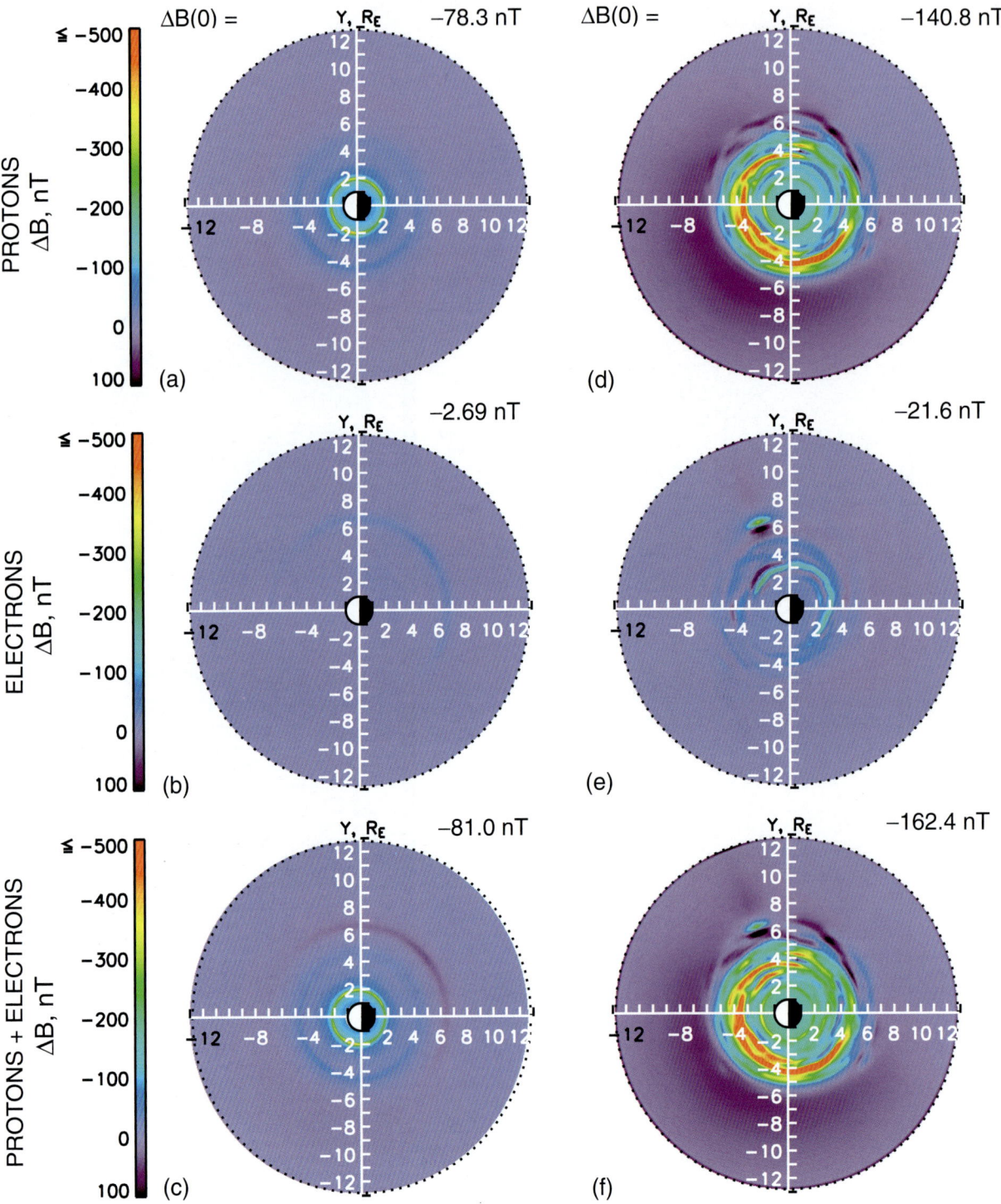

Plate 7. The left column shows the simulated equatorial ring-current magnetic field for (a) protons only, (b) electrons only, and (c) protons and electrons during pre-storm. The right column shows a similar display of the ring-current magnetic field for 0800 UT on 19 October 1998. The ring-current magnetic field at the center of the Earth $\Delta B(0)$ is labeled in each panel.

and protons of energies for a small storm. Nonetheless, our simulations show that during storms, electrons can significantly contribute to the ring-current magnetic field, especially in the morning sector.

4. SUMMARY AND CONCLUSIONS

In summary, we have simulated the perpendicular pressure distribution and ring-current magnetic field for both protons and electrons during the main phase of the 19 October 1998 storm. Below is a summary of the important findings.

Our quiet-time proton ring current model reproduces a radial perpendicular pressure profile that is similar to the AMPTE/CCE observations of *Lui* [2003].

Both the proton and electron ring current form asymmetrically early in the storm main phase. However, the electron ring current tends to be concentrated on the dawnside because electrons drift westward whereas the proton ring current tends to be more concentrated on the duskside because ions drift eastward. The proton and electron ring current becomes stronger and more nearly symmetric later in the main phase as the cross polar cap potential weakens.

The magnetic field intensity perturbation from magnetization current is very important where there is plasma. The magnetic field intensity perturbation from field-aligned currents that are needed to close the partial gradient-drift current is small compared to both the magnetic field intensity perturbations from the gradient-drift and magnetization currents.

The quiet-time electron ring-current magnetic field is negligible compared to the quiet-time proton ring-current magnetic field. However, during storms, electrons significantly contribute to the ring-current magnetic field, especially on the dawnside. From our simulations the electrons contribute ~ 15–20% of protons to the total *Dst*.

There are large asymmetric stormtime ring-current magnetic field depressions from $r_0 \sim 3.5-6\ R_E$. The simulated large magnetic field depressions agree qualitatively with statistical observations of *Lui* [2003] and the empirical model of *Tsyganenko et al.* [2003].

This simulation study shows that the stormtime ring-current magnetic field significantly distorts the geomagnetic field. This indicates a need for magnetically self-consistent models of the ring current. We are beginning to develop such a self-consistent model. The large stormtime depressions in the ring-current magnetic field also need to be taken account of in global descriptions of the inner magnetosphere.

Acknowledgments. The authors thank A. T. Lui for helpful discussions of his statistical analysis of the AMPTE/CCE data. The work of M. W. Chen, S. Liu, and L. R. Lyons was supported in part by NSF grant NSF-ATM-0202108. The work of M. W. Chen and M. El-Alaoui was supported by NASA NAG5-12106 under UCLA subaward 2090GCC340. M. W. Chen's work was also supported by NSF Grant NSF-ATM-0207160 and by The Aerospace Corporation's Independent Research and Development Program. The work of M. Schulz was supported by NASA contract NAS5-30372, by NSF grant ATM-0000340, and by the Independent Research and Development Program of Lockheed Martin Space Systems Company. Computing resources were provided by the Maui High Performance Computing Center.

REFERENCES

Albert, J. M., Quasi-linear pitch angle diffusion coefficients: Retaining high harmonics, *J. Geophys. Res., 99(A12)*, 23,741–23,745, 1994.

Cahill, L. J., Inflation of the inner magnetosphere during a magnetic storm, *J. Geophys. Res., 71*, 4505–4519, 1996.

Chen, M. W., M. Schulz, L. R. Lyons, and D. J. Gorney, Stormtime transport of ring current and radiation-belt ions, *J. Geophys. Res., 98*, 3835–3849, 1993.

Chen, M. W., M. Schulz, and L. R. Lyons, Simulations of phase space distributions of stormtime proton ring current, *J. Geophys. Res., 99*, 5745–5759, 1994.

Chen, M. W., M. Schulz, J. L. Roeder, J. F. Fennell, and L. R. Lyons, Simulations of ring current proton pitch-angle distributions, *J. Geophys. Res., 103*, 165–178, 1998.

Chen, M. W., M. Schulz, G. Lu, and L. R. Lyons, Quasi-steady drift paths in a model magnetosphere with AMIE electric field: Implications for ring current formation, *J. Geophys. Res., 108(A5)*, 1180, doi: 10.1029/2002/JA009584, 2003.

Dungey, J. W., Interplanetary magnetic field and the auroral zones, *Phys. Rev. Lett., 6*, 47–78, 1961.

El-Alauoi, M., R. L. Richard, M. Ashour-Abdalla, and M. W. Chen, Low Mach number bow shock locations during a magnetic cloud event: Observations and magnetohydrodynamic simulations, *Geophys. Res. Lett., 31*, L03813, doi:10.1029/2003GL018788, 2004.

Frank, L. A., On the extraterrestrial ring current during geomagnetic storms, *J. Geophys. Res., 72(15)*, 3753–3767, 1967.

Korth, H., M. F. Thomsen, J. E. Borovsky, and D. J. McComas, Plasma sheet access to geosynchronous orbit, *J. Geophys. Res., 104*, 25,047–25,061, 1999.

Le, G., C. T. Russell, K. Takahashi, Morphology of the ring current derived from magnetic field observations, *Annal. Geophys., 22*, 1267–1295, 2004.

Liu, S., M. W. Chen, L. R. Lyons, H. Korth, J. M. Albert, J. L. Roeder, and P. C. Anderson, Contribution of convective transport to stormtime ring current electron injection *J. Geophys. Res., 108(A10)*, 1372, doi:10.1029/2003JA010004, 2003.

Lui, A. T. Y., Inner magnetospheric plasma pressure distribution and its local time asymmetry, *Geophys. Res. Lett., 30 (16)*, 1846, doi:10.1029/2003GL017596, 2003.

Lyons, L. R., Electron diffusion driven by magnetospheric electrostatic waves, *J. Geophys. Res., 79(4)*, 575–580, 1974.

Ostapenko, A. A, and Yu. P. Maltsev, LT and Dst dependence of the ring current, *Physics of Auroral Phenomena*, Proc. XXVI Annual Seminar, Apatity, Russian Academy of Science, pp. 83–86, 2003.

Richmond, A. D., and Y. Kamide, Mapping electrodynamic features of the high-latitude ionosphere from localized observations: Technique, *J. Geophys. Res., 93*, 5471–5759, 1988.

Schulz, M., Particle drift and loss rates under strong pitch angle diffusion in Dungey's model magnetosphere, *J. Geophys. Res., 103*, 61–67, 1998.

Tsyganenko, N. A., H. J. Singer, J. C. Kasper, Storm-time distortion of the inner magnetosphere: How severe can it get?, *J. Geophys. Res., 108(A5)*, 1209, doi: 10.1029/2002JA009808, 2003.

Margaret W. Chen, Space Sciences Applications Laboratory, The Aerospace Corporation, El Segundo, California, USA.

Michael Schulz, Lockheed Martin Advanced Technology Center, Palo Alto, California, USA.

Shuxiang Liu, Department of Atmospheric Sciences, University of California, Los Angeles, California, USA.

Gang Lu, High Altitude Observatory, Boulder, Colorado, USA.

Larry R. Lyons, Department of Atmospheric Sciences, University of California, Los Angeles, California, USA.

Mostafa El-Alaoui, Institute of Geophysics and Planetary Physics, University of California, Los Angeles, California, USA.

Michelle Thomsen, Los Alamos National Laboratory, Los Alamos, New Mexico, USA.

Radiation Belt Responses to the Solar Events of October–November 2003

D. N. Baker[1], S. G. Kanekal[1], J. B. Blake[2], J. H. Allen[3]

The solar disturbances of October–November 2003 produced large enhancements of the energetic particles deep within the Earth's magnetosphere. Most notably, there were acceleration and redistribution processes within the radiation belts leading to a complete filling of the "slot" region around L~2.5 (which is usually devoid of highly relativistic electrons). The radiation belts were enhanced in intensity for more than 40 days following the so-called "Halloween storm" event. The solar and magnetospheric particle radiation appeared to be related to a variety of spacecraft anomalies and other space weather effects. The observed solar particles and radiation belt changes in this extended interval are here placed into the context of the historical record of SAMPEX observations and are related to plasmaspheric changes seen by IMAGE. It is concluded that the solar drivers and the magnetospheric responses were amongst the largest that have been recorded in the spacecraft data record.

I. INTRODUCTION

The Van Allen belts have been studied continuously for over 12 years with the Solar, Anomalous, and Magnetospheric Particle Explorer (SAMPEX) spacecraft (see Baker et al., 1993 for satellite description]. In the SAMPEX lifetime there have been several major enhancements and dropouts of the radiation belt particle populations. It is found that the center of the outer Van Allen belt usually is about 4 R_E (Earth radii) away from Earth's center. During October–November 2003, the Van Allen radiation belt electron population was powerfully accelerated and was reconfigured inward toward Earth's surface to a degree not observed before [*Baker et al., 2004*]. From November 1 to November 10, the outer belt had its center only at about 2.5 R_E geocentric distance. This normally is a place where there are almost no energetic electrons at all [*Lyons and Williams,* 1975; *McIlwain,* 1996].

The magnetospheric responses in late October and early November of 2003 can be traced to sunspot group number 484 (that appeared on the east limb of the Sun's disk on 18 October 2003) and to concurrent sunspot groups 486 and 488. They rotated across the solar disk during this time period and the Sun produced large enhancements of X-rays, solar energetic particles, and some of the strongest geomagnetic storms of this solar cycle [*Lopez et al., 2004*]. Figure 1 portrays several of the sunspot features and solar flares as seen in late October. The interplanetary shock waves and coronal mass ejections launched by the Sun from these active regions reached Earth's magnetosphere in time periods as short as one day. These high-speed solar disturbances compressed, distorted, and enhanced the Earth's Van Allen belts. It is shown here that there is good observational evidence for powerful associated changes produced in Earth's radiation belts. It is discussed why these effects persisted for many weeks and months following the geomagnetic storms themselves. It is also noted that important practical consequences—"space weather" effects—were produced by such a remarkable solar-terrestrial sequence of events.

---

[1] Laboratory for Atmospheric and Space Physics, University of Colorado, Boulder, Colorado
[2] The Aerospace Corporation, El Segundo, California
[3] SCOSTEP c/o NOAA/NGDC, Boulder, Colorado

Inner Magnetosphere Interactions: New Perspectives from Imaging
Geophysical Monograph Series 159
Copyright 2005 by the American Geophysical Union.
10.1029/159GM19

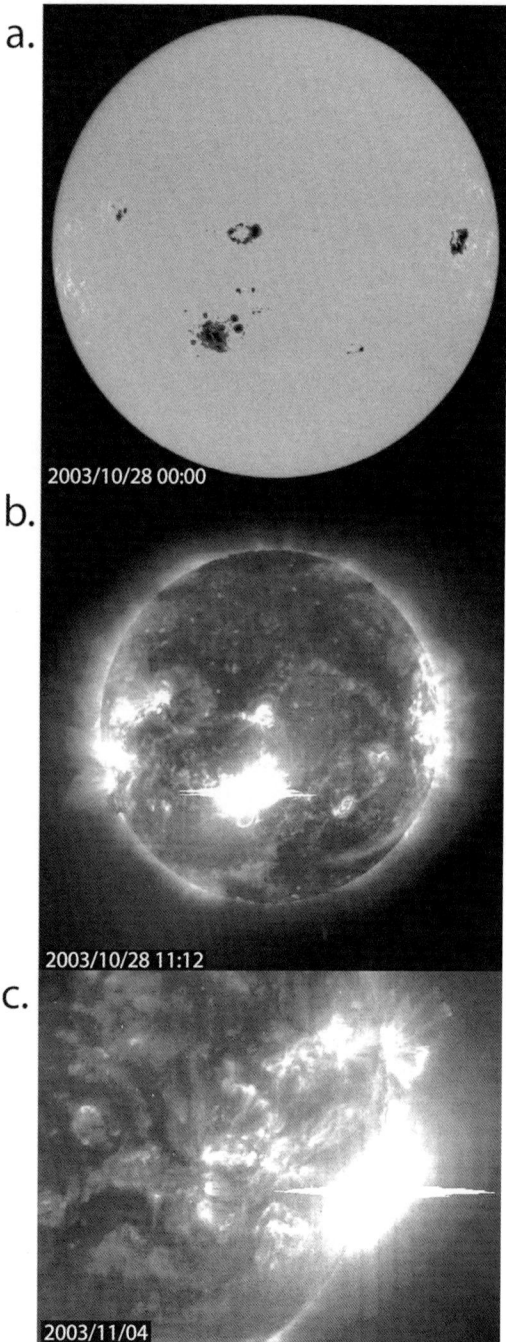

Figure 1. Selected data for the late-October 2003 period showing solar features that ultimately led to the "Halloween Storms" of 2003: (a) Sunspot groups 484, 486, and 488; (b) SOHO EIT image showing solar flare on 28 October; (c) similar to (b) for 4 November 2003.

2. OBSERVED RADIATION BELT CHANGES

A long-term view of the structure and variability of the Earth's electron radiation belts is shown in Plate 1 using data from one particular energy channel (2–6 MeV) from the SAMPEX spacecraft. SAMPEX is in a low-altitude (~600 km), high-inclination (82°) orbit [*Baker et al.*, 1993]. For each 98-min-long orbit, data are sorted according to spatial location. The resulting sorted data were averaged for each day. Data were corrected for dead-time, pulse pileup, and mode-change artifacts.

Plate 1 shows the color-coded fluxes of 2–6 MeV electrons measured by SAMPEX from launch (July 1992) to March 2004. The individual years are demarcated as are the months that are shown by minor ticks. Data are plotted in terms of the magnetic L-shell parameter (which roughly speaking is a measure in Earth radii of where a magnetic dipole field line crosses the magnetic equatorial plane). The homogeneous data set shown in Plate 1 delineates very well the outer Van Allen belt ($3 \leq L \leq 7$), the "slot" region ($2 \leq L \leq 3$), and the inner Van Allen zone ($L \leq 2$). There is a high degree of electron flux variability on daily and monthly timescales. We have compared the electron flux patterns with the average solar sunspot number (SSN) and the average solar wind speed (V_{SW}) measured upstream of the Earth [*Baker et al.*, 2004]. It is found that the high-energy electrons in the outer Van Allen zone generally are strongly correlated with high solar wind speeds ($V_{SW} \geq 500$ km/s). It is also clear that the energetic electrons do not maximize in intensity at the time of highest sunspot number (2000–2001), but rather they peak in the declining phase (1994–1995 and again in 2002–2004) of solar activity [see *Baker et al.*, 1999].

The electron flux profiles shown in Plate 1 demonstrate intense, 27-day recurrent electron acceleration events (due to high-speed solar wind streams) in 1993–1995 for ($3 \leq L \leq 5$). The L=3–5 band is almost solidly red from mid-1993 through 1994. On the other hand, during sunspot minimum (1996) there were significant electron flux enhancements only briefly around the spring and fall equinoxes [*Baker et al.*, 1999]. Intense, but occasional, events associated with (non-recurrent) coronal mass ejection storm episodes were seen in May 1997 and again in May 1998. Several intense events of this sort were also seen in the latter part of 1998 and in later years. Plate 1 also shows a weak, but persistent, feature in the inner zone ($1 \leq L \leq 2$) that died away throughout 1992 and 1993; it intensified in 1994 and then gradually died away again throughout 1995 and 1996. This may, in part, be related to the "new" radiation belt created during the March 1991 storm [*Looper et al.*, 1994; *Li et al.*, 1993].

Plate 2(a) shows a blowup of the L-versus-time data of Plate 1 for the period 2003 and early 2004. This plot reveals

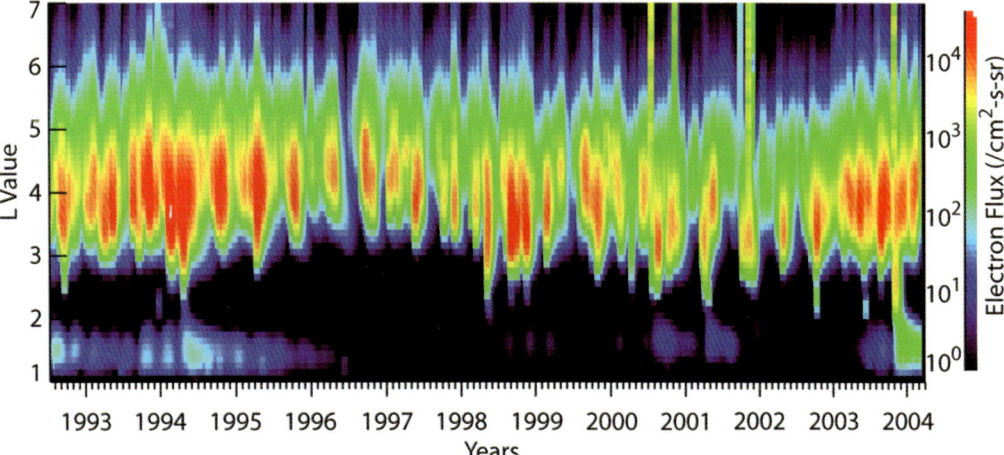

Plate 1. Color-coded flux levels of 2–5 MeV electrons measured by the SAMPEX spacecraft from July 1992 to March 2004. The vertical axis is the magnetic L-shell parameter which is effectively the distance in Earth radii at which a magnetic field line crosses the magnetic equatorial plane and the horizontal axis is time.

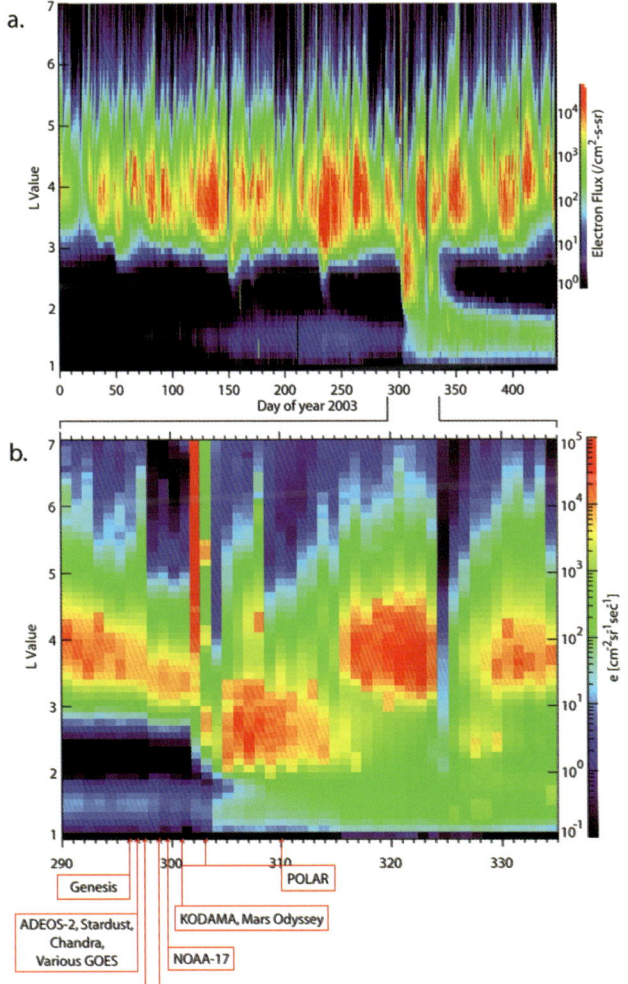

Plate 2. (a) A blowup of SAMPEX electron data similar to Fig. 2 for 2003 and the beginning of 2004. The Halloween Storm effects of interest commenced shortly after Day 300 of 2003; (b) A detailed plot similar to Fig. 2 for DOY 290 to DOY 335 of 2003. Also shown below is a timeline of important spacecraft anomaly events that occurred in October and November of 2003.

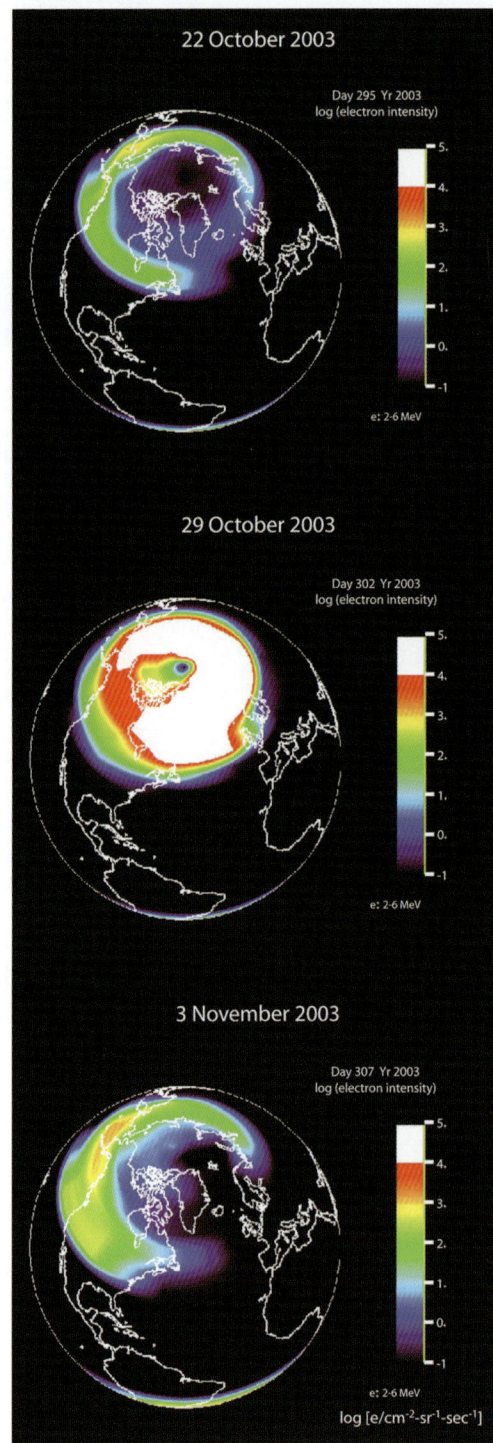

Plate 3. (a) A diagram showing the Earth and a projection of the outer radiation belt for 22 October 2003. (b) Similar to (a), but showing the high polar cap fluxes and greatly distorted radiation belt location on 29 October 2003. (c) Similar to (a) showing the expanded radiation belts seen after the Halloween storms on 3 November 2003.

the remarkable degree to which the 2–6 MeV electrons were displaced inward toward the Earth after Day of Year (DOY) ~300 following the Halloween storms. The data show that the normal peak of electron fluxes around L=4.0 that was seen before (and well after) the Halloween storm period was displaced inward to L~2.5 for several weeks. The normal radiation belt slot region—typically devoid of high-energy electrons—was filled with electrons from DOY~300 to DOY~340. Such an intense slot-filling event was not previously seen in the long run of SAMPEX data (Plate 1). Note, also, in Plate 2(a) that the inner zone ($1 \leq L \leq 2$) was filled to a similarly unprecedented degree with high-energy electrons and this has persisted since the Halloween storm onsets. The further expanded time scale plot of the data in Plate 2(b) shows details of the radiation belt changes seen between DOYs 290 and 385 of 2003.

Plate 3 shows the radiation belt and polar cap flux levels measured by SAMPEX as projected onto a global map. We show three selected days: (a) 22 October 2003, well before the Halloween storm period when the radiation belts were in a "normal" configuration with moderate electron intensities; (b) 29 October 2003, when extremely high electron intensities were detected even over the entire polar cap region (due to solar electrons penetrating on open field lines down to SAMPEX altitudes); and (c) 3 November 2003, when the radiation belt projections were higher in intensity and much broader in latitudinal extent than was seen prior to the Halloween storms. Obviously, the Halloween storms had both intense, prompt effects over broad geographic regions as well as longer-lasting consequences.

3. IMAGE PLASMASPHERE DATA

We have found evidence that the distortion and displacement of the outer radiation belt during and following the Halloween storm was closely associated with a major reaction of the Earth's plasmasphere [*Baker et al.*, 2004]. The plasmasphere is the region of cold, dense ionized gas (mostly protons and helium ions) that resides on the magnetic field lines close to the Earth [*Carpenter and Park*, 1973]. The plasmasphere is threaded by magnetic flux tubes that are persistently 'closed' so that plasma from the Earth's ionosphere has filled the flux tubes and reached a near-equilibrium state along the entire field line [*Grebowsky*, 1970]. The outer boundary of the plasmasphere—the plasmapause—can act to refract, and trap, electromagnetic waves propagating in the whistler mode [*Lyons et al.*, 1972]. Such confined waves in the plasmasphere strongly scatter trapped electrons into the atmospheric loss cone, causing the electrons to be precipitated into the Earth's upper atmosphere [*Walt*, 1996]. Such precipitated electrons are removed from the magnetosphere permanently. It is generally believed that the radiation belt slot region (described above) is due to such strong scattering and loss of electrons near the plasmapause caused by strong wave-particle interactions [*Lyons et al.*, 1972; *Walt*, 1996].

The Imager for Magnetopause-to-Aurora Global Exploration (IMAGE) spacecraft [*Burch et al.*, 2001] makes it possible to view directly the physical extent of the plasmasphere by imaging the extreme ultraviolet [EUV] emissions from He+ (at 30.4 nm wavelength) within the plasmasphere [*Sandel et al.*, 2003]. With its high inclination (90°) and highly elliptical (1000 x 46000 km altitude) orbit, IMAGE is able to view the Earth's plasma environs for extended periods [*Burch et al.*, 2001].

The effects of the Halloween storms on the plasmasphere have been reported by *Baker et al.* [2004] and an adaptation of those data are shown here as Fig. 2 [courtesy of *J. Goldstein and J. Burch*]. Before the storm, on 28 October, the plasmapause radius was at a geocentric distance between about 4 and 5 R_E (Fig. 5a). During the storm, on 31 October, the plasmapause radius was inside 2 R_E, and at some longitudes was at 1.5 R_E (Fig. 2b). Such an extremely small plasmasphere only occurs during the strongest geomagnetic storms [*Sandel et al.*, 2003], and the plasmapause shrinkage of the Halloween 2003 storm was the most pronounced yet identified [*Baker et al.*, 2004]. In the several days after the storm the plasmasphere recovered from its extreme confinement (Fig. 2c) as plasma from the Earth's ionosphere again gradually leaked out into space and repopulated the plasmasphere flux tubes [see *Baker et al.*, 2004].

A detailed SAMPEX L-versus-time plot of 2–6 MeV electron fluxes was made by *Baker et al.* [2004] for the period 1 September 2003 to 15 March 2004. In a general sense, it is known that there is a correspondence between the inward extent of the radiation belt and the modeled plasmapause location [*Li et al.*, 2003]. However, we have found by detailed consideration during the Halloween storm period that much of the time the model estimates of the plasmapause location do not agree very well with the actual inner edge of the electron radiation belt. However, using available IMAGE data, *Baker et al.*, [2004] determined the plasmapause locations throughout late October and into mid-November 2003. The minimum L-value for this boundary from the IMAGE-EUV data for the plasmapause tracked quite well with the inner extent of the electron radiation belt population measured by SAMPEX. In other words, the IMAGE-EUV data have shown a good and remarkably close correspondence for the period analyzed in detail between the inner edge of the outer Van Allen belt and the measured plasmapause location.

Under normal circumstances, the outer radiation belt extends quite far from the Earth with its maximum intensity

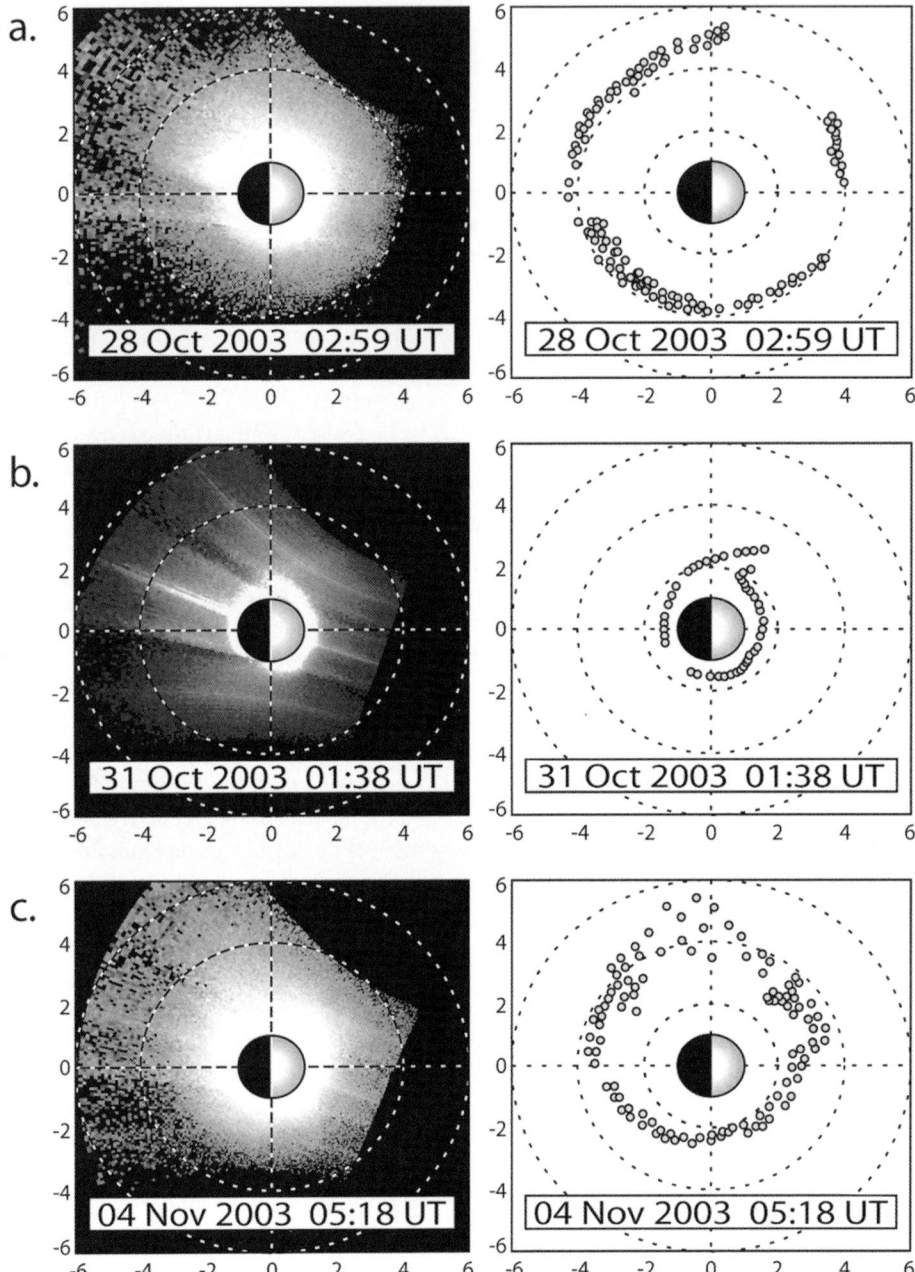

Figure 2. Selected IMAGE EUV data as describe in the text [adapted from *Baker et al.*, 2004].

occurring at ~4–5 R_E geocentric distance. The plasmasphere extends outward to 3–4 R_E geocentric instance. We have found during (and following) the Halloween 2003 storm that the plasmasphere was greatly contracted inward toward the Earth and remained in a very reduced state for many days. Concurrently, the Van Allen belt was also reformed very far inward and the highest electron intensities were actually seen at 2–3 R_E geocentric distance (see Plates 1 and 2).

4. ACCELERATION AND LOSS PROCESSES

It is widely accepted that the inner edge of the outer Van Allen belt should correspond rather closely to the plasmapause location [*Lyons et al.*, 1972]. Our data for the (post) Halloween storm period show this to very much be true. As shown by the analysis described above, as the plasmasphere and the magnetic field lines were displaced inward (to a nearly unprecedented degree) in October–November 2003, the whole radiation belt structure was completely transformed. The slot region, normally devoid of particles, became the location of highest radiation belt particle intensities. Such a powerful deformation of the entire inner magnetosphere has not previously been observed in detail by a complete suite of sensors.

An event of such magnitude and extent as the October–November 2003 solar disturbances provides a unique opportunity to study the acceleration, transport, and loss of energetic particles in Earth's magnetosphere. In many ways, the whole state of the Earth's radiation belt was altered in a very short time by the Halloween storms. An important point to note from SAMPEX data is that the immediate consequence of solar-generated disturbances hitting Earth's magnetosphere (on ~31 October 2003) was to deplete the radiation belt at almost all L-values. Then, in the subsequent day or two the radiation belt electron fluxes were regenerated to quite high levels. However, this regenerated population of electrons was formed in an unexpected location, i.e., in the normal slot region were few electrons ordinarily can survive for an extended period. The powerful acceleration of electrons that had to occur over DOY ~305–315 was associated with very strong wave activity measured both in space by the Polar spacecraft [*Cartwright et al.*, 2004] and on the Earth's surface as determined by ground-based magnetometers [*I. Mann*, private communication, 2004]. Thus, we would expect that waves in the ultra-low frequency (ULF) range could drive rapid radial diffusion of electrons (hence accelerating the particles) [*Liu et al.*, 1998]. However, it is also possible that strong 'chorus' emissions (in the whistler mode) or even electromagnetic ion cyclotron (EMIC) waves could locally heat and accelerate electrons to very high energies [*Summers and Ma*, 2000] in the L=2–3 range under these exceptional conditions.

We note with emphasis here that losses of the radiation belt electrons can be well studied by such a sharply defined set of events as the Halloween storm. As we have discussed, the inner extent of the 'new' radiation belt electron population at $2 \leq L \leq 3$ corresponds quite closely to the outward extent of the plasmasphere. Thus, even under the conditions of extreme distortion of the inner magnetosphere, it seems that strong wave-particle scattering and loss near the plasmapause must determine where the outer belt electrons can persist [e.g., *Lyons et al.*, 1972].

Finally, we note that many electrons made it through the normal slot-region "barrier" following the Halloween storm and entered the inner zone ($1 \leq L \leq 2$) region (see Plate 2 (a) and (b)). These electrons now constitute a new, intense population of relativistic electrons in the inner zone that have not been present there since the remnants of the March 1991 storm died away [*Looper et al.*, 1994].

5. SOLAR PARTICLES AND THE SPACE STATION

Besides the magnetospheric electron population described above, another particle population important for space weather during the October–November 2003 period was solar energetic protons (ions). It is interesting to note that numerous spacecraft suffered operational anomalies during the late October solar particle event [*Webb and Allen*, 2004]. In particular, many sensor systems had high trigger rates and other such problems on 28–30 October. Although the operational problems were not fatal in most instances, they were annoying and detrimental to space operations.

Another effect of solar storms in late October was the production of relatively low-latitude magnetic cutoffs that existed for large parts of the Halloween storm interval. Figure 3 summarizes the normal region of polar cap access of 8–15 MeV/nucleon helium (shown by the lighter-shaded ellipses for the (a) northern hemisphere and (b) southern hemisphere). The measurements were made with the MAST sensor system onboard SAMPEX [courtesy *R. Leske*, Caltech]. The darker-shaded extended tracks in each panel show the much lower latitude cutoffs measured for many orbits during the Halloween storm. The smooth curved lines encircling the Earth show the tracks (footprints) of the International Space Station (ISS) for representative orbits. It is seen that solar energetic ions (equivalent in rigidity to 30–60 MeV protons) often had access to the ISS orbit during the Halloween storm(s).

The panel (c) of Figure 3 shows equivalent results to those in panels (a) and (b) but for the "Bastille Day" event of 15 July 2000. Note that the October 2003 storms produced broader and more extensive solar proton (ion) access to the polar cap and the ISS than did the much-publicized Bastille Day event.

6. OTHER SPACE WEATHER IMPACTS

The presence of very energetic electrons in the Earth's magnetosphere constitutes an important 'space weather' hazard to spacecraft operating in near-Earth space [e.g., *Baker*, 2002]. As discussed above, the Halloween storms generated large fluxes of solar energetic particles as well as the high fluxes of radiation belt electrons we have reported here. The combination of solar and magnetospheric particles produced a wide variety of spacecraft and ground "anomalies" in human technological systems during October–November 2003 [*Webb and Allen*, 2004]. As illustrated in the lower portion of Plate 3b, a number of serious spacecraft failures and other significant losses were directly attributable to this spate of space weather.

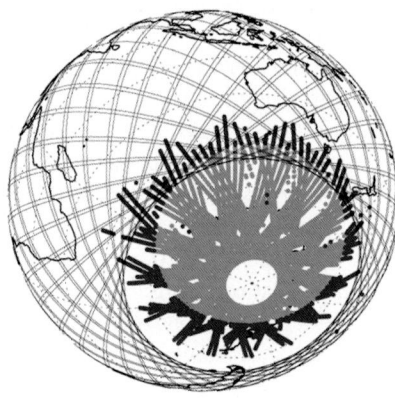

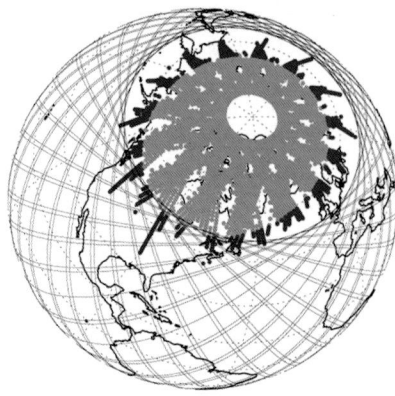

Figure 3. Solar energetic ion access to the polar caps for October 2003. (a) Northern hemisphere access region, (b) Southern hemisphere access region. (c) Northern access region for the Bastille Day event of July 2000.

We particularly point out that many spacecraft systems operate, or contemplate operation, in the middle Earth orbit (MEO) altitude range to take advantage of the normally benign radiation environment associated with the slot region. A set of events as occurred with the Halloween storm changed substantially the radiation conditions at L ≤ 3 and there suddenly developed a hostile space weather environment in a region that normally is quiescent. Future models and design strategies must certainly take account of possible extreme events as occurred in October–November 2003.

7. SUMMARY AND CONCLUSIONS

The October–November 2003 solar activity produced a wide variety of enhancements of solar energetic particles as well as numerous CMEs and solar flares. The CMEs produced interplanetary clouds and associated shock waves that impacted Earth's magnetosphere in several key instances. As shown in this paper, the solar outputs produced remarkable changes in the Earth's radiation belts and inner magnetosphere [see also *Baker et al.*, 2004].

We particularly emphasize that the radiation belt slot region was filled with E>2 MeV electrons for over a month following the Halloween storm. SAMPEX, Polar, and other spacecraft have given a rather clear global picture of the radiation belt enhancement and the decay characteristics. We have noted here several of the spacecraft operational anomalies that occurred in October–November. The interested reader is referred to *Webb and Allen* [2004] for more details.

Acknowledgments. This work was supported by NASA. The authors thank colleagues from the SAMPEX, Polar, IMAGE, and SOHO missions for valuable contributions to this work.

REFERENCES

Baker, D. N., How to cope with space weather, *Science*, 297, 30 August 2002.

Baker, D. N., et al., An overview of the SAMPEX mission, *IEEE Trans. Geosci. Elec.*, 31, 531, 1993.

Baker, D. N., S. G. Kanekal, T. I. Pulkkinen, and J. B. Blake, Equinoctial and solstitial averages of magnetospheric relativistic electrons: A strong semiannual modulation, *Geophys. Res. Lett.*, 26, 3193–3196, 1999.

Baker, D. N., S. G. Kanekal, X. Li, S. P. Monk, J. Goldstein, and J. L. Burch, An extreme distortion of the Van Allen belt arising from the 'Hallowe'en' solar storm in 2003, *Nature*, 432, 878–881, doi:10.1038/nature03116, 2004.

Blake, J. B., et al., New high temporal and spatial resolution measurements by SAMPEX of the precipitation of relativistic electrons, *Adv. Space Res.*, 18(8), 171–186, 1996.

Burch, J. L., et al., Views of the Earth's magnetosphere with the IMAGE satellite, *Science*, 291, 619–624, 2001.

Carpenter, D. L., and C. G. Park, What ionospheric workers should know about the plasmapause/plasmasphere, *Rev. Geophys.*, 11, 133–154, 1973.

Cartwright, J., et al., Solar wind control of the compressional fluctuation level in the magnetosphere: How the solar wind controls the radiation belts? Presented in GEM, Snowmass, CO, 22 June 2004.

Grebowsky, J. M., Model study of plasmapause motion, *J. Geophys. Res.*, 765, 4329–4333, 1970.

Li, X., et al., The predictability of the magnetosphere and space weather, *Eos*, 84, 361+369–370, 2003.

Li, et al., Simulation of the prompt energization and transport of radiation particles during the March 23, 1991 SSC, *Geophys. Res. Lett.*, 20, 2423–26, 1993.

Liu, W. W., G. Rostoker, and D. N. Baker, Internal acceleration of relativistic electrons by large-amplitude ULF pulsations, *J. Geophys. Res.*, 104, 17391–17407, 1998.

Looper, M. D., et al., Observations of remnants of the ultrarelativistic electrons injected by the strong SSC of 24 March 1991, *Geophys. Res. Letters*, 21, 2079–2082, 1994.

Lopez, R. E., D. N. Baker, and J. H. Allen, Sun unleashes Halloween Storm, *Eos*, 85, No. 11, 105,108, 2004.

Lyons, L. R., and D. J. Williams, The quiet time structure of energetic (35–560 keV) radiation belt electrons, *J. Geophys. Res.*, 80, 943–949, 1975.

Lyons, L. R., R. M. Thorne, and C. F. Kennel, Pitch-angle diffusion of radiation belt electrons within the plasmasphere, *J. Geophys. Res.*, 77, 3455–3474, 1972.

McIlwain, C. E., Processes acting upon outer zone electrons, in Radiation Belts; Models and Standards, 15–26, Geophys. Monograph 97, Amer. Geophys. Un., Washington, DC, 1996.

O'Brien, T. P., and M. B. Moldwin, Empirical plasmapause models from magnetic indices, *Geophys. Res. Lett.*, 30(4), 1152, doi:10.1029/2002GL016007, 2003.

Sandel, B. R., J. Goldstein, D. L. Gallagher, and M. Spasojevic, Extreme ultraviolet imager observations of the structure and dynamics of the plasmasphere, *Space Sci. Rev.*, 198, 25–46, 2003.

Summers, D., and C. Ma, A model for generating relativistic electrons in the Earth's magnetosphere based on gyro-resonant wave-particle interactions, *J. Geophys. Res.*, 105, 2625–2639, 2000.

Walt, M., Source and loss processes for radiation belt particles, in Radiation Belts: Models and Standards, 1–13, *Geophys. Monograph* 97, Amer. Geophys. Un., Washington DC, 1996.

Webb, D. F., and J. H. Allen, Spacecraft and ground anomalies related to the October–November 2003 solar activity, *Space Weather*, 2, SO3009, doi: 10.1029/2004SW000075, 2004.

D. N. Baker, Laboratory for Atmospheric and Space Physics, University of Colorado, Boulder, 1234 Innovation Drive, Boulder, CO 80303-7814; daniel.baker@lasp.colorado.edu

S. G. Kanekal, Laboratory for Atmospheric and Space Physics, University of Colorado, Boulder, 1234 Innovation Drive, Boulder, CO 80303-78142

J. B. Blake, The Aerospace Corporation, 2350 E. El Segundo Blvd., El Segundo, CA 90245-4691

J. H. Allen, SCOSTEP c/o NOAA/NGDC, 325 Broadway, Boulder, CO 80305

Hemispheric Daytime Ionospheric Response To Intense Solar Wind Forcing

A. J. Mannucci[1], Bruce T. Tsurutani[1], Byron Iijima[1], Attila Komjathy[1], Brian Wilson[1], Xiaoqing Pi[1], Lawrence Sparks[1], George Hajj[1], Lukas Mandrake[1], Walter D. Gonzalez[2], Janet Kozyra[3], K. Yumoto[4], M. Swisdak[5], J. D. Huba[6], and R. Skoug[7]

We investigate the ionospheric response to events where the z-component of the interplanetary magnetic field, B_z, becomes large and negative for several hours, associated with the largest geomagnetic storms over the prior solar maximum period (2000-2004). We compute the average vertical total electron content (TEC) in the broad region covering 1200–1600 local time and ±40 degrees geomagnetic latitude (dipole), using data from the global network of Global Positioning System (GPS) receivers. In several cases, we find approximately a two-fold increase in total electron content within 2–3 hours of the time when the southward-B_z solar wind impinged on the magnetopause. We also analyze daytime super-satellite TEC data from the GPS receiver on the CHAMP satellite orbiting at approximately 400 km altitude, and find that for the October 30, 2003 storm at mid-latitudes the TEC increase is nearly one order of magnitude relative to the TEC just prior to the B_z southward onset. The geomagnetic storm-time phenomenon of prompt penetration electric fields into the ionosphere from enhanced magnetospheric convection is the most likely cause of these TEC increases, at least for certain of the events, resulting in eastward directed electric fields at the equator. The resulting dayside vertical **ExB** drift of plasma to higher altitudes, while solar photons create more plasma at lower altitudes, results in a "daytime super-fountain" effect that rapidly changes the plasma structure of the entire dayside ionosphere. This phenomenon has major practical space weather implications.

---

[1]Jet Propulsion Laboratory, California Institute of Technology, Pasadena, California
[2]Instituto Nacional de Pesquisas Espaciais, São José dos Campos, Brazil
[3]University of Michigan, Ann Arbor, Michigan
[4]Space Environment Research Center, Kyushu University, Fukuoka, Japan
[5]Icarus Research, Inc., Bethesda, Maryland
[6]Plasma Physics Division, Naval Research Laboratory, Washington, DC
[7]Los Alamos National Laboratory, Los Alamos, New Mexico

Inner Magnetosphere Interactions: New Perspectives from Imaging
Geophysical Monograph Series 159
Copyright 2005 by the American Geophysical Union.
10.1029/159GM20

1. INTRODUCTION

It is well known that the interaction between southward interplanetary magnetic fields (IMFs) and the Earth's magnetic field leads to strong dawn-to-dusk electric fields in the magnetosphere and an overall increase in magnetospheric convection. This convection, in turn, causes intense ring current buildup and magnetic storms [*Gonzalez and Tsurutani*, 1987; *Gonzalez et al.*, 1994; *Kozyra et al.*, 1997; *Kamide et al.*, 1998]. This rapid change in magnetospheric conditions can have dramatic effects on the Earth's ionosphere. If electric fields can penetrate to the low latitude ionosphere before shielding builds up [*Tanaka and Hirao*, 1973], they can modify equatorial ionospheric electrodynamics. Resulting eastward electric fields in the daytime will cause upward **ExB** drift in the equatorial ionosphere during the day. This paper concerns the ionospheric effects of electric fields associated with geomagnetic disturbances.

In this paper we analyze data from several intense interplanetary events characterized by strong southward (negative B_z in GSM coordinates) interplanetary magnetic field, and measure the ionospheric consequences during the initial phase of the magnetic storm. We report that at the equator and middle latitudes, significant increases in ionospheric total electron content (TEC) occur, as measured by dual-frequency Global Positioning System (GPS) receivers (*Mannucci et al.*, 1998; *Mannucci et al.*, 1999). Little is presently reported about the full hemispheric dayside response to intense magnetic storms.

Significant changes in TEC can be produced soon after event onset by intense disturbance-related electric fields originating from the magnetosphere-ionosphere interaction. Disturbance electric fields at low latitudes have been identified principally from two causes, as: (a) prompt penetration zonal electric field often observed in the equatorial latitudes [*Sastri*, 1988; *Fejer and Scherliess*, 1995; *Abdu et al.*, 1995; *Sobral et al.*, 1997; *Sobral et al.*, 2001; *Fejer*, 2002; *Sastri et al.*, 2002; *Kelley et al.*, 2003; *Vlasov et al.*, 2003] and/or (b) delayed electric fields produced by the disturbance dynamo driven by modified thermospheric winds due to energy input at high latitudes [*Blanc and Richmond*, 1980; *Richmond and Lu*, 2000; *Scherliess and Fejer*, 1997]. Zonal electric fields from these two causes, depending on their polarity and duration, could cause large uplifts or downdrafts of the ionospheric plasma leading to large-scale local-time dependent enhancements or decreases of the vertical TEC.

Electrodynamics that occurs in daytime is especially important because the ionospheric plasma is highly responsive to such disturbance electric fields, and the daytime is where the quantity of terrestrial plasma is largest. GPS observations suggest that large quantities of plasma are generated rapidly (within 2–3 hours) in the equatorial region through mid-latitudes, possibly affecting the amount of plasma entering higher latitudes. A suggestive picture of full hemispheric response is shown in Plate 1. A global network of GPS receivers has been used to estimate vertical total electron content along the lines of sight between each receiver and several GPS satellites in view simultaneously (typically 6–10). The map shows the measured average TEC, obtained at a 30-second cadence, between UT 2100 and 2115 for October 29, 2003. Over North America, an enhanced narrow plume of plasma can be seen stretching from middle to high latitudes in the mid-afternoon sector. Vertical TEC data from a low-Earth orbiting GPS receiver, which has no data gaps over the oceans, suggests that in the early afternoon large TEC increases have occurred spanning the entire low-to-middle latitude ionosphere. The existence of this large "plasma pool" may contribute to the large TEC gradients associated with the plume.

These hemispheric-scale plasma increases have important practical consequences. There are many applications for which large TEC gradients, and associated instabilities, are major concerns, including to radar, communications and GPS-based navigation systems, including those used by the Federal Aviation Administration (FAA) for civil aircraft navigation. The fact that plasma gradients at mid-latitudes can be enhanced by plasma uplift spanning the low-to-midlatitude dayside hemisphere is important for predictive purposes and for understanding the magnitude and temporal evolution of significant gradients in TEC at mid-latitudes. The performance of the FAA's system is limited by incomplete scientific understanding of low-through-middle latitude plasma response.

In this paper we will address the problem of the dayside ionospheric response to intense solar wind forcing, using measurements provided by the Global Positioning System (GPS) from ground and space-borne receivers. A companion paper [*Tsurutani et al.*, 2004b] analyzed the November 6, 2001 event in detail using multiple observations. In this paper, we explore the general phenomenon of prompt TEC increases after B_z southward turnings in the solar wind for a number of events. The data tend to support the hypothesis that, in the early phases of intense geomagnetic storms, B_z southward turning is often associated with significant daytime TEC increases. Whether these are all related to a similar cause is a matter of future research.

2. DATA ANALYSIS AND METHOD

Our measurements of the column density of electrons, referred to as total electron content (TEC), were obtained from the Global Positioning System (GPS) satellite signals as received by both ground-based receivers (~200 distributed around the globe) and a receiver onboard the CHAMP satellite in low-Earth orbit at approximately 400 km altitude. The locations of both the satellites and receivers are known to a few decimeters or better, and since straight line propagation is essentially correct for the signals, the electron content is measured along a geographically well-located line-of-sight between GPS satellite and receiver.

A schematic of the TEC measurement geometry is shown in Figure 1. There are approximately 28 GPS satellites located in circular Earth orbit at an altitude of 20,200 km. For simplicity, only one GPS satellite is illustrated in the figure. Each receiver simultaneously tracks multiple

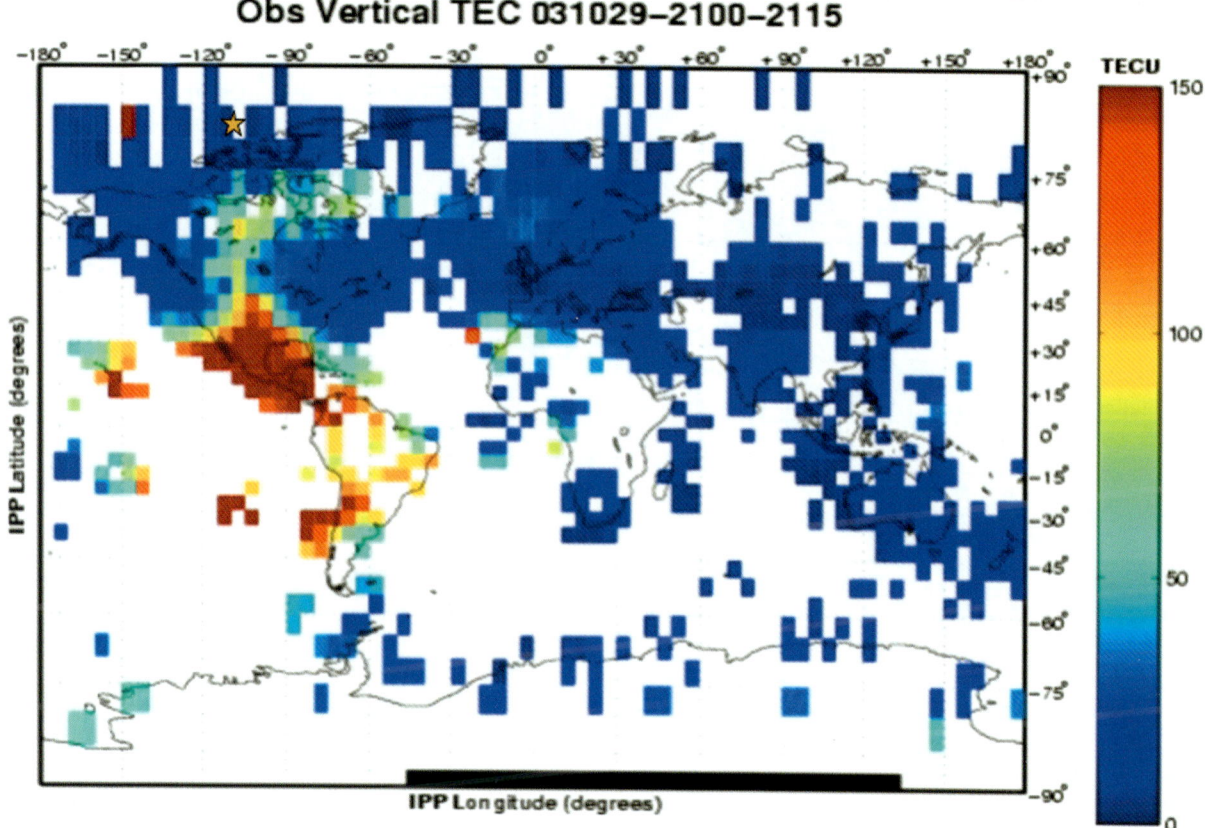

Plate 1. A global "snapshot" of vertical total electron as measured by a global network of GPS receivers on October 29, 2003 from 2100–2115 UT when a geomagnetic storm was in progress. A plasma tongue structure extending to high latitudes is visible in North America, which may be dependent on enhanced plasma generation at low latitudes due to disturbance electric fields. A star indicates the geomagnetic pole location. Units: 1 TECU = 10^{16} electrons/m^2.

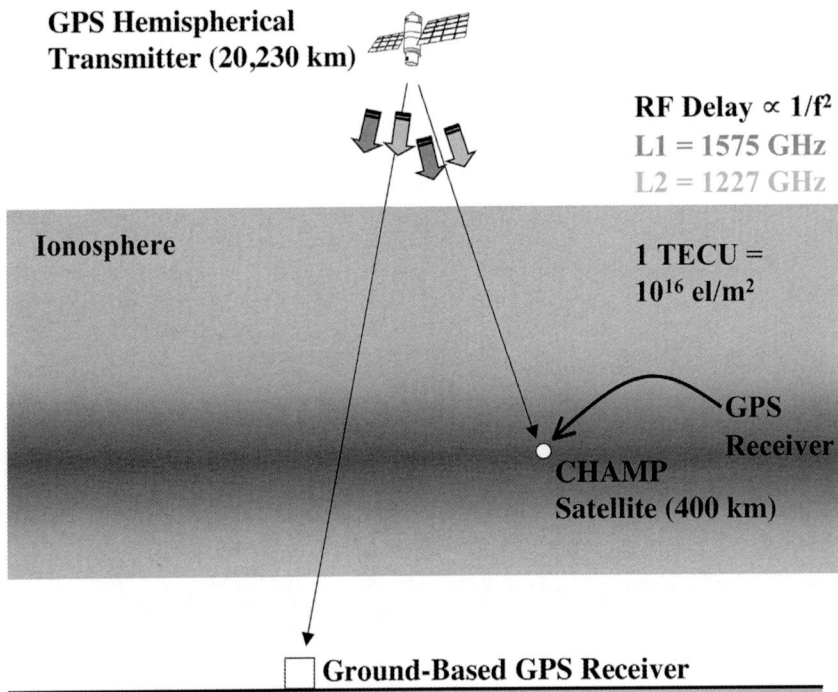

Figure 1. A schematic of a GPS satellite transmitter and low altitude satellite and ground-based GPS receivers. The GPS satellites broadcast continuous transmissions at two frequencies f (L1 and L2). Signal processing is used to extract ionospheric electron density structure (total electron content) towards multiple satellites in view simultaneously. The shaded area represents the ionosphere.

(4–10) GPS satellite signals. The relative delay between the two GPS transmission frequencies L1 (1575 GHz) and L2 (1227 GHz) is directly related to the column density of electrons (total number of electrons per unit area) along the line of sight [*Mannucci et al.*, 1999; *Mannucci et al.*, 1998].

The CHAMP satellite (*Reigber et al.*, 2002) possesses a GPS receiver and zenith viewing antenna that tracks all GPS satellites in view at positive elevation angles. CHAMP is in a polar orbit (87° inclination) at a slowly declining altitude of ~400–430 km over the period of this study, with an orbital period of approximately 100 minutes. Data for elevation angles within 70° of the vertical, or higher, were used exclusively in this study. Otherwise no other data deletions have been performed. To normalize the measurements obtained at multiple elevation angles, the slant TEC data have been used to estimate the vertical TEC directly above the low-Earth orbiter (LEO), assuming a simple geometrical factor that accounts for the difference between slant and vertical TEC. We assume the vertical distribution of density is a spherical shell ionosphere of uniform (horizontally stratified) density, 700 km-thick, above the CHAMP altitude. For elevation angles greater than 30° as used in this study, the error from this simplifying assumption is not a significant factor for our analysis, as suggested in a later section.

The ground-based GPS data set used is composed of ~100 stations from the International GPS Service data centers [*Moore*, 2001]. The obliquity function used to estimate the vertical TEC from the slant observations is computed modeling the ionosphere as a spherical slab of uniform electron density between 450 and 650 km altitude. The latitude and longitude at which the ground-to-satellite line-of-sight intersects the ionosphere is computed using a spherical shell at 450 km altitude. Detailed discussion concerning the removal of instrumental offsets for both ground-based receivers and satellite-based receivers is beyond the scope of this paper. If the reader is interested in further information on the topic of extracting TEC measurements from ground-based GPS receivers, see *Mannucci et al.* [1999].

We use Level 2 data from the ACE satellite upstream of the Earth (GSM position x,y,z = 1.4×10^6 km, 1.2×10^5 km, -2.0×10^5 km) to provide the estimates of the B_z southward turning onset times that initially cause the geomagnetic disturbance, and hourly *Dst* geomagnetic data from the National Geophysical Data Center to monitor the total intensity of the resulting geomagnetic storm.

3. NOVEMBER 6, 2001 STORM

The event of November 5–6 2001 is analyzed first, following the detailed discussion in Tsurutani et al., 2004b. Measured parameters pertaining to the event are shown in Figure 2. The top two panels are interplanetary parameters taken by the ACE spacecraft located at 1.4×10^6 km upstream of the Earth. They are: the solar wind speed and z-component of the magnetic field B_z (GSM coordinates). Using the measured solar wind speed of ~700 km/s, a convection delay time of ~34 minutes from ACE to the magnetosphere is estimated. The solar wind data has therefore been shifted by this time in the Figure to match the ground based AE and Dst index data (the bottom two panels).

The dashed vertical line labeled "C" indicates the start of a magnetic cloud [Klein and Burlaga, 1982]. The speed is ~420 km/s and is thus a "slow" cloud (see Tsurutani et al. [2004a] for discussion of slow cloud properties). It is identified by quiet magnetic fields with the general absence of large amplitude Alfvén waves [Tsurutani et al., 1988] and very low proton temperatures, $~2.5 \times 10^4$ K. The plasma density is ~18 cm^{-3} and |B| ~18 nT.

The solid vertical line labeled "S" is a fast forward shock. The shock occurred at ~0120 UT at ACE. It is identified by an abrupt solar wind speed increase from ~420 km/s to ~700 km/s, a proton temperature increase to $~8 \times 10^5$ K, a density increase to ~48 cm^{-3}, and a magnetic field increase to ~70 nT. The magnetic field magnitude reached a maximum value of 80 nT ~1 hour and 40 minutes after the shock passage. The timing and large magnitude of the B_z-southward turning are of most significance here.

The shock interaction with the upstream slow magnetic cloud has a profound effect on the resultant geomagnetic activity at Earth. The cloud has a steady $B_z = -7$ nT (southward) field. The interplanetary shock compresses this pre-existing interplanetary negative B_z [Tsurutani et al., 1988] to $B_z = -48$ nT. At the peak field strength, ~1 hour and 40 minutes after the shock, the B_z component reached –78 nT. Two hours after the shock, B_z was ~ -65 nT. B_z remained at large negative values until ~3+ hrs after the shock. This

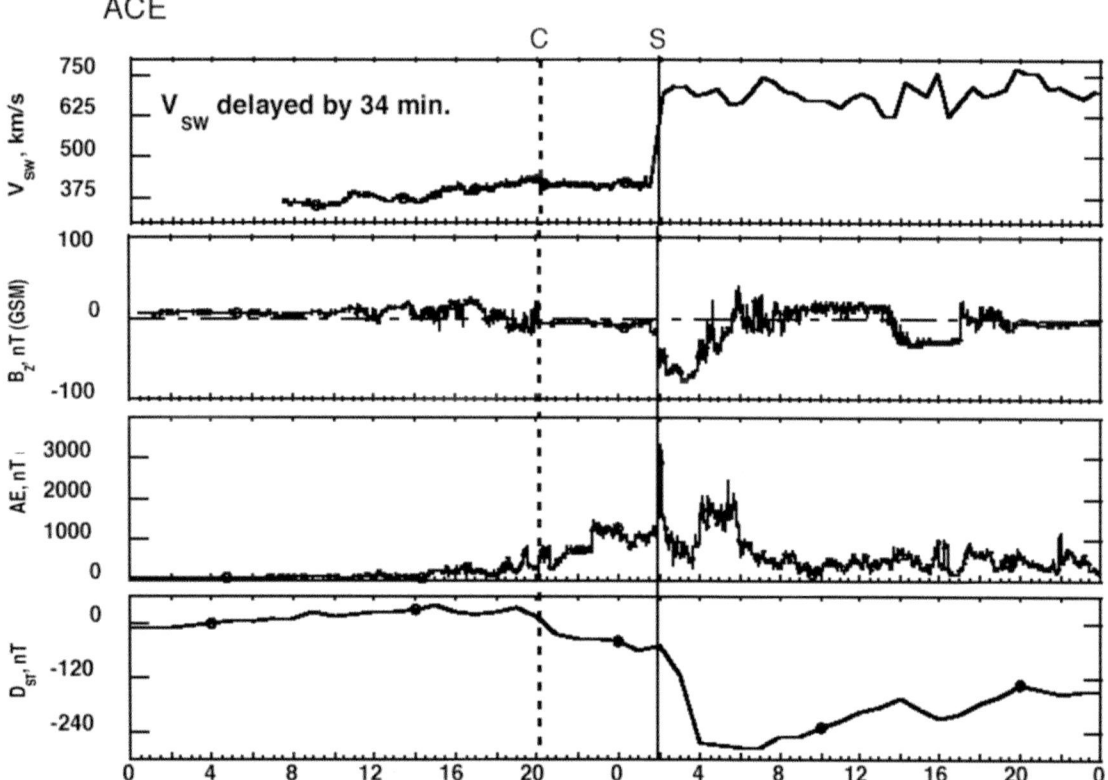

Figure 2. The interplanetary event of 5–6 November 2001. The interplanetary data is taken by the ACE spacecraft. The magnetic field is plotted in GSM coordinates. The time delay of the solar wind and magnetic field convection from ACE to the magnetopause is ~34 min. Thus the interplanetary shock should impinge upon the magnetosphere at ~0154 UT. A strong dawn-to-dusk electric field is imposed on the magnetosphere at this time. An AE ~ 3000 nT substorm onset and a magnetic storm reaching a magnitude of Dst = -275 nT also start at the time of the shock arrival

intensely negative IMF B_z feature of several-hour duration is the cause of the main phase of the magnetic storm reaching $D_{ST} \cong -275$ nT. Such intense IMF B_z events lasting of order hours are always present during major magnetic storms [*Gonzalez and Tsurutani*, 1987; *Gonzalez et al.*, 1994].

For the remainder of the paper, we will discuss the impact of sudden intense interplanetary electric field (IEF) on the Earth's ionosphere (IEF = $V_{SW} \times B_z$). This electric field onset occurs at ~0120 UT at ACE and assuming a time shift of 34 minutes, it should have been imposed on the magnetosphere at ~0154 UT. It should also be remembered that there is a small but important "precursor" interplanetary electric field associated with the negative B_z (due to the slow magnetic cloud ahead of the shock) and a concomitant moderate storm, occurring prior to the shock electric field event. For other events discussed in this paper, the onset time of the southward B_z turnings, shifted in time according to the measured solar wind speed, are used to assess the correspondence between changes in magnetospheric convection and ionospheric modification.

A synoptic view of the dayside TEC response to the interplanetary shock is available from ground-based TEC measurements as shown in Figure 3. The data are plotted as a function of local time and magnetic latitude. Figure 3a (top panel) shows November 4, 2001 quiet-time "baseline" data from 0409 to 0456 UT two days before the interplanetary event. Each gray-scale point in the figure represents an estimate of vertical TEC from a link between a GPS ground receiver and a GPS satellite. About ten data epochs are recorded along each satellite-receiver link during the 45-minute period of the map. Each receiver tracks numerous satellites, typically a number between four and ten, at every epoch. Estimates of the TEC above the CHAMP satellite track from an altitude of 430 km, at approximately 1900–2000 local time (LT) over most latitudes, are superposed.

Figure 3a shows a typical pattern of global TEC, that reaches a maximum on the dayside due to solar UV and X-ray irradiation, and is centered at approximately 1400 local time. The area of TEC values near to or exceeding 100 TECU, the lightest-shaded area, extends from –34° MLAT to +20° MLAT. The bias towards southern latitudes is typical for northern wintertime (November) due to the influence of the prevailing circulation from south-to-north.

Figure 3b shows the TEC distribution after the shock event on 6 November 2001, from 0414 to 0500 UT (~2 to 3 hours after the shock). The post-shock TEC distribution is markedly different from the quiet-time distribution: the region of enhanced TEC (> 100 TECU) is much larger ranging from +48° MLAT to -40 MLAT. The westward and eastward extent of the red areas is approximately the same pre- and post-shock: ~0900 to ~2100 LT.

From a detailed comparison (not shown) of Figure 3a and Figure 3b at 14 LT, where the maximum TEC occurs at low latitudes, it is noted that on 6 November the ground-based TEC has increased from ~145 TECU (4 November) to ~170–180 TECU. Thus, there is an absolute increase in TEC at this local time by ~21%. Figures 3a and 3b also contain the CHAMP upward viewing TEC data obtained at dusk. The TEC values above CHAMP are the same as those above the ground within an uncertainty of 10%. Thus, at ~1900 LT, nearly all of the ionospheric plasma is above ~430 km altitude.

TEC increases can be caused by eastward-directed electric fields causing plasma uplift to altitudes of reduced recombination rate (see *Tsurutani et al.* 2004b, and references therein; also *Tanaka and Hirao*, 1973). The relatively early ionospheric response for this event suggests that the mechanism for the electric fields is prompt penetration, as opposed to thermospheric dynamo. Data that tends to corroborate the prompt electric field mechanism is shown in Figure 4, an estimate of the equatorial electrojet (EEJ) current strength using magnetometer data. The east-west flowing EEJ is an indirect measure of eastward-directed electric fields during daytime, since the intensity of the EEJ depends on vertical **ExB** drift velocity through the intermediate mechanism of vertically-directed polarization fields—the larger the vertical **ExB** drift (due to eastward-directed fields), the larger the polarization that develops in the vertical direction (Anderson et al. 2002; Rastogi and Klobuchar, 1990). Figure 4 shows the equatorial electrojet current intensity over the Pacific equatorial station Yap (9.3° N, 138.5° E, dip angle: -0.6°), which was obtained by subtracting the diurnal range of the horizontal (northward) component of magnetic field intensity (_H) over a non-EEJ station Guam (13.58° N, 144.87° E, dip angle: 9°) from that of Yap. The medium-gray trace shows the storm-time EEJ estimate; compare to the quiet-time trace in lightest gray. Although the EEJ estimate cannot be compared directly to the IEF magnitudes (plotted as the black trace), it is clear that excellent agreement exists for the temporal trends comparing the storm-time EEJ estimate with that of the IEF estimate from ACE data.

During the interplanetary shock event and consequential storm, Yap was in the midday sector. The large increase of the EEJ intensity over the Pacific sector (medium-gray trace) coincident with the shock arrival time ("S" in the figure) is consistent with a large disturbance penetration electric field of eastward polarity produced on the dayside by the shock event. The interplanetary electric field is superposed on this figure (black trace, calculated from the data in Figure 2 but at a cadence of 30 minutes) and shows a pattern similar to that measured at the equator (for another example of EEJ correlation to solar wind, see Kelley et al., 2003).

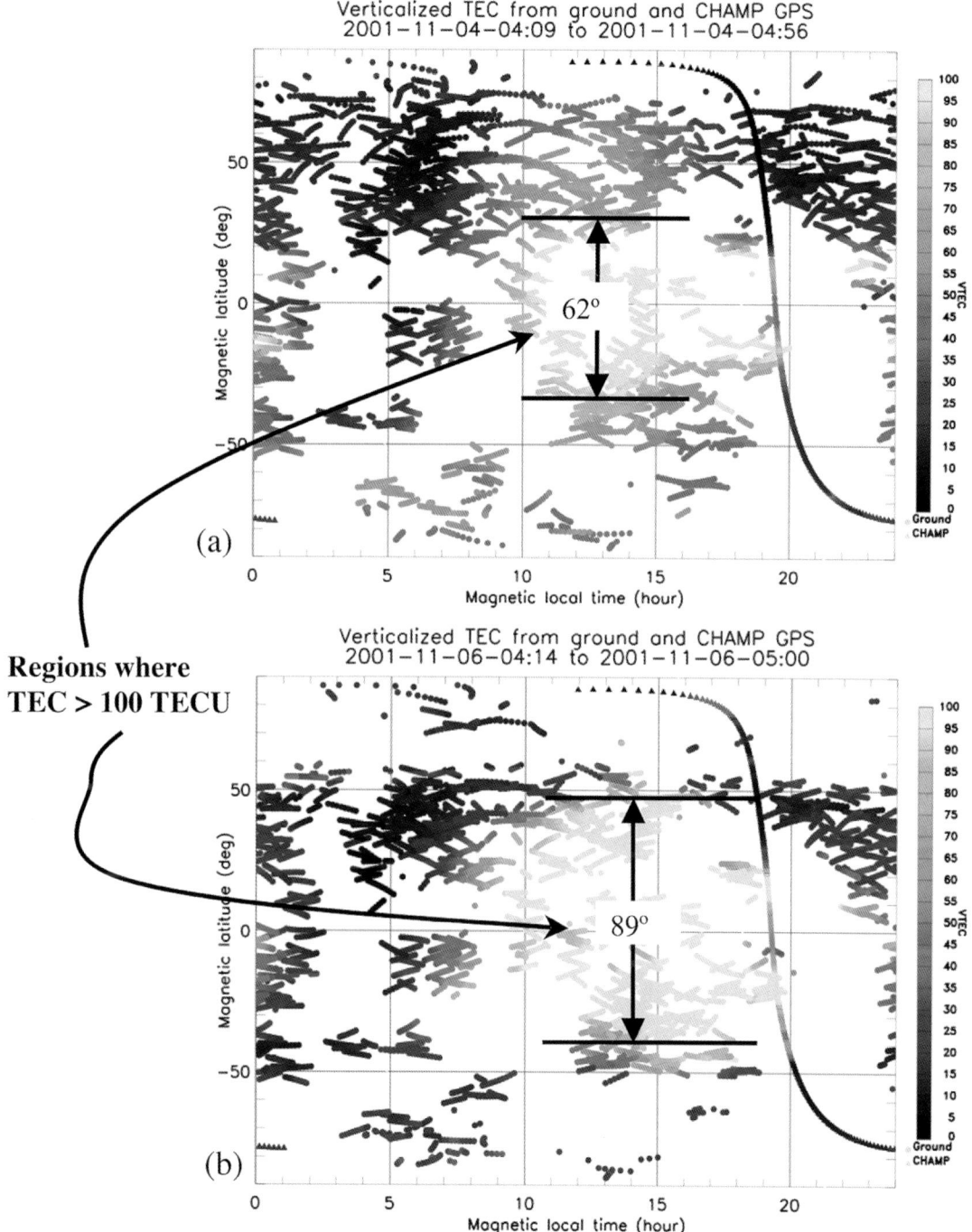

Figure 3. The ~100-station ground-based TEC data for 4 November from 0409 to 0456 UT (background) in panel (a) and for 6 November from 0414 to 0500 UT (post shock event) in panel (b). Integrated electron content data above CHAMP altitude of 430 km are also shown (to the same gray scale). The dayside post-shock region of enhanced TEC (>~ 100 TECU) is much broader in latitudinal extent, ranging from +48 ° MLAT to -40 ° MLAT, compared to the range 28° → –34° for the region where TEC > ~ 100 TECU on the quiet day.

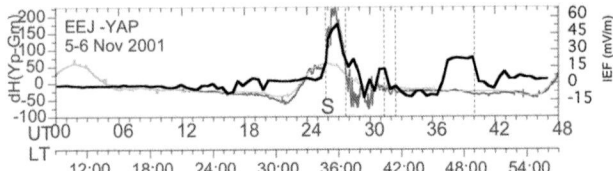

Figure 4. The Equatorial Electrojet –EEJ current intensity over the Yap magnetometer station in the midday sector is shown (medium gray). The reference day curve for Yap (light gray) is taken from 5 Nov. The starting time in the figure is 00 UT 5 Nov. The LT at Yap (UT plus 9 hrs) is shown at the bottom. Vertical line 1 indicates the onset of the shock. Shown in black is the interplanetary electric field ($V_x B_z$) time shifted at each point according to the velocity measured at ACE and assuming a distance to ACE of 1.4×10^6 km. The adjusted curve was subsequently shifted as a whole to line up with the curve for November 6 (medium gray).

An eastward disturbance electric field at daytime through dusk is consistent with *Nopper and Corovillano* [1978], which contains a calculation of equatorial electric field direction when Region 1 field aligned currents are much larger than Region 2 currents (under-shielding) assuming an average ionospheric conductance pattern. (See also *Senior and Blanc*, 1984; *Spiro et al.*, 1988; *Fejer et al.*, 1990). It is clear from Figure 4 that a large electric field, as inferred via magnetometer data, tracks the pattern of IEF onset and decay. This is suggestive of penetration electric fields, as opposed to other causes (see also Kelley et al., 2003). The analysis of multiple events in the next section, and the fact that TEC increases are always observed following soon after the large B_z southward event, suggests that the prompt penetration mechanism is a major contributor to the hemispherical dayside plasma redistribution.

4. MULTI-EVENT ANALYSIS

The November 6, 2001 event was studied in great detail in Tsurutani et al., 2004b. A conclusion of that work was that the TEC changes were due to eastward-directed electric fields at the equator that appeared immediately after the shock reached the magnetopause. The **ExB** plasma drift associated with this eastward electric field will lift the ionospheric plasma to higher altitudes where recombination rates are low, thereby reducing the average loss (recombination) rate of ionospheric plasma. Solar UV radiation will form new electron-ion pairs at lower altitudes, leading to major increases in TEC that is the major focus of this paper. In this section, we will look at a number of interplanetary events of similar structure to assess the dayside TEC response.

We are interested in the temporal behavior of daytime TEC before, during and after a number of candidate events leading to intense geomagnetic storms. We compute a time series of average TEC in a fixed local-time/latitude region using data from a global network of ~100 GPS receivers. The coverage of the network is extensive enough that some data exists in the prescribed local time region at all universal times, although the number of points (coverage) varies with time, as does the latitude/longitude distribution of sites. For this reason, we plot the averages for two days preceding the storm day to reveal the quiet time variability solely due to changing coverage and natural ionospheric variability. As in the previous section, an obliquity function is used to estimate the vertical TEC from the slant TEC measurements obtained above 10 degrees elevation angle. The ionospheric pierce point location is used to select data included in the average.

Figure 5 shows the TEC averages for the local time range 1200–1600 LT and magnetic latitude range ±40 degrees (dipole) every 30-minutes of Universal Time (UT). The vertical line(s) in each panel indicate the time at which B_z turns southward. For all these events, there was an extended southward turning of large magnitude (> 20 nT). In some cases, there were other southward turnings (often minor), hence two or three lines in some cases. The times are shifted by our best estimate of the velocity of the solar wind protons, as measured by the SWEPAM instrument on ACE (see for example Skoug et al., 2004), plus one hour. The uncertainty of these times should be viewed as ±30 minutes.

4.1 Event Discussion

All the events included here led to significant geomagnetic storms as indicated by the minimum *Dst* value reached over the storm period (column 3, table 1), except for the storm on July 17, 2004 which is interesting as a "control" case and also because major impact in the North American continent was reported (this impact is not discussed further). Except for July 2004, these storms represent the largest Dst events in the period 2000–2004. Despite the approximations inherent in the analysis due to non-uniform coverage of the ground stations, the following pattern and properties are discernable from Figure 5:

1. In all these events, a TEC increase is clearly visible above the background variability in the preceding quiet days.
2. The TEC increases always begin *after* an identifiable B_z southward turning, in some cases immediately (within one hour) of the B_z southward event, and in other cases several hours elapse.
3. A TEC decrease is often discernable several hours (10–20) after the B_z southward turning. The decrease will not be addressed further here.

These data suggest that B_z southward events in the solar wind, when associated with major geomagnetic storms, are

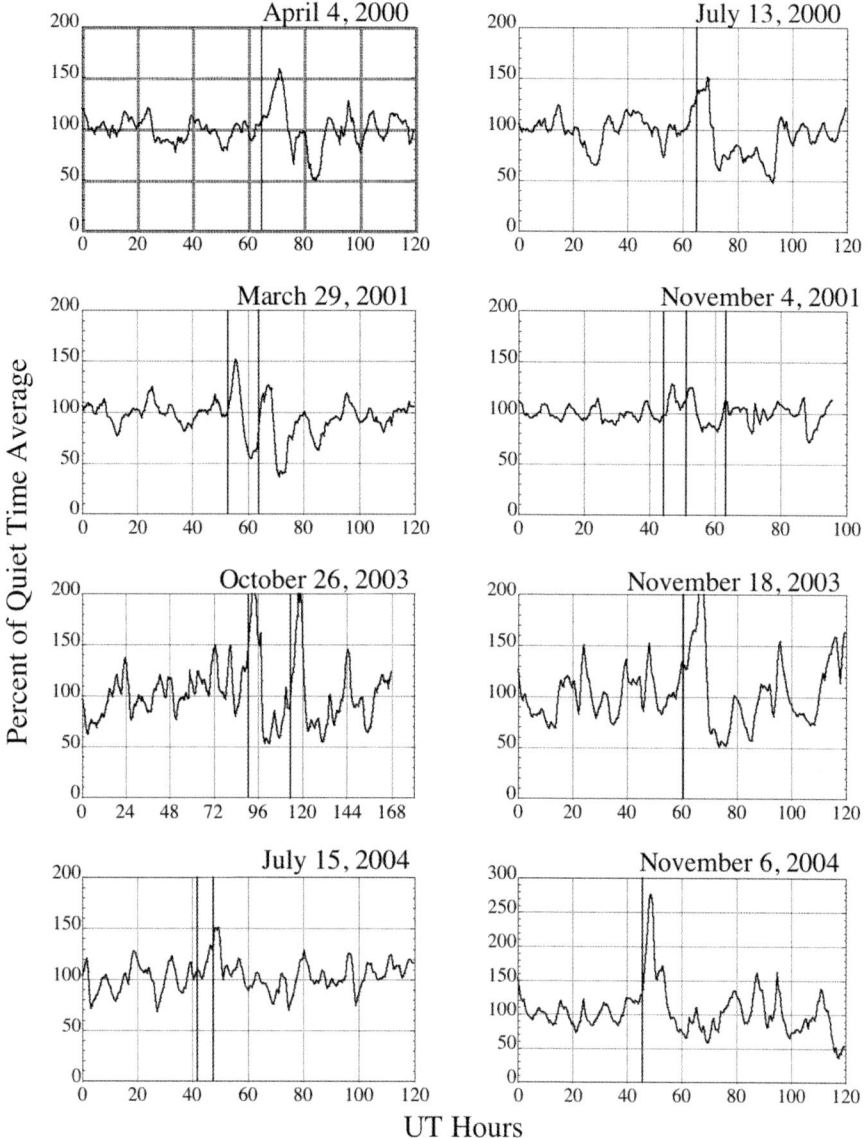

Figure 5. Average TEC change as a function of Universal Time for several events for the local time/latitude region 1200–1400 LT and ±40 degrees geomagnetic latitude, normalized to the quiet time average TEC of the 2–3 days preceding the storm event (100 = quiet time average). The vertical lines indicate times when B_z becomes negative, as measured by the ACE spacecraft, time shifted according to the delay at which the solar wind reaches the magnetopause from ACE.

a reliable predictor of dayside TEC increases, in some cases very significant. In Figure 6 we examine the correlation between the magnitude of the geomagnetic storm and the magnitude of the fractional TEC increase. The peak Dst value reached for the storm is plotted against the peak TEC increase relative to quiet-time average over the preceding two days. There does appear to be a general correspondence between the largest TEC increases and the largest Dst values, although with significant case-to-case variability. For example, Dst values in the range 360-380 are associated with peak TEC increases in a broad range of 63%–163%. The largest geomagnetic storm (Dst reaches –451 nT) is not associated with the largest TEC fractional increase, which occurred for the November 8, 2004 storm. We note that the TEC averages reported in this study must be regarded as an approximate quantity, as the true TEC average in the region studied differs from the measured values because the distribution of receivers in the region varies with Universal Time.

Table 1. Event dates and times for Figure 6.

Start Date	B_z South Date	Minimum Dst Value (nT)	B_z SOUTH TIME UT HOURS	B_z South Time (2)	B_z South Time (3)
4/4/00	4/6/00	-290	17.0		
7/13/00	7/15/00	-290	16.8		
11/4/01	11/8/01	-265	20.2	3.1	15.2
3/29/01	3/31/01	-358	4.7	15.7	
10/26/03	10/29/03	-363	18.2		
10/26/03	10/30/04	-400	17.6		
11/18/03	11/20/03	-451	12.6		
7/15/04	7/17/04	-79	17.5	23.5	
11/6/04	11/7/04	-380	20.6		

We expect the inherent "noise" in the analysis technique is a significant contributing factor to the results in Figures 5 and 6. Nevertheless, it does appear that significant dayside TEC increases are a common feature of superstorm events.

These storms occurred at varying phases of the solar cycle, so the fractional increases occurred over highly varying quiet-time backgrounds, because of the variable ionizing solar flux which declines with solar cycle. Analysis of these events in terms of absolute TEC increase may be worth further study.

There may be several physical effects leading to the general patterns observed in Figure 5. Following the interpretation of *Tsurutani et al.* (2004b), we expect that promptly penetrating electric fields from enhanced magnetospheric convection are responsible for at least some of the observed TEC increase caused by the uplift/reduced recombination mechanism mentioned earlier, particularly those that occur rapidly after B_z southward turning. The detailed relationship between enhanced convection and magnitude of TEC increase, including pre-conditioning factors such as the ionospheric conductivity pattern possibly modulated by auroral precipitation, and the effectiveness of inner-magnetosphere shielding on the dayside, is a complex subject requiring further study and modeling.

For the October 2003 events, we have examined electric field data from the DMSP satellite F13 to assess whether these data reflect disturbed conditions at low latitudes. The data are shown in Figure 7 and are only available from dusk or dawn local time; we used dusk data. In Figure 7a (left panel), the electric field values within ±10 degrees geomagnetic latitude measured by DMSP (SSIES) are shown for each pass, for several days starting with October 25, 2003, shaded according to direction (black is westward, gray is eastward). The electric field is nearly always westward until October 29 and 30, when larger eastward values appear. Figure 7b is an expanded plot centered on the storm days, showing the appearance of large eastward directed fields shortly after the B_z south events listed in Table 1, as indicated by vertical lines. Although these electric field data are only available at dusk and not daytime, the appearance of eastward-directed fields appearing when they do is suggestive of low-latitude electrodynamic disturbance connected to solar wind conditions. (The earlier eastward field event on October 29, at ~102 hours in Figure 7a, is nearly coincident with a large but short-lived B_z southward event at ~4 UT on October 29, which is not considered in this analysis of persistent B_z south events).

5. OBSERVATIONS FROM SPACE-BORNE GPS

The analysis of average regional TEC in the previous section suggests that TEC increases are a common feature associated with persistent B_z southward turnings and the resulting superstorms. The detailed spatial structure of the TEC increase is not revealed by this analysis. In this section, we use GPS receivers on a low-Earth orbiting platform

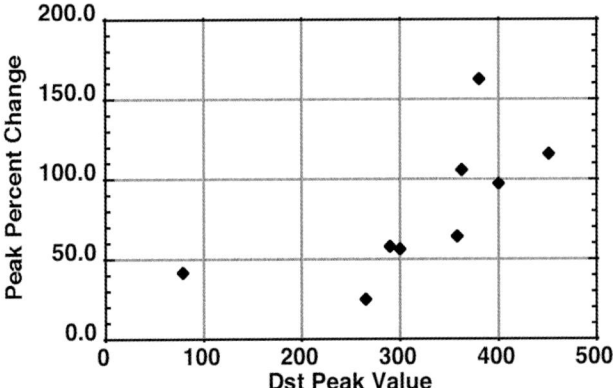

Figure 6. Peak percent TEC change in the disturbed period compared to quiet-time average of the preceding two days, for the events considered in Figure 5, versus peak *Dst* excursion magnitude for the storm.

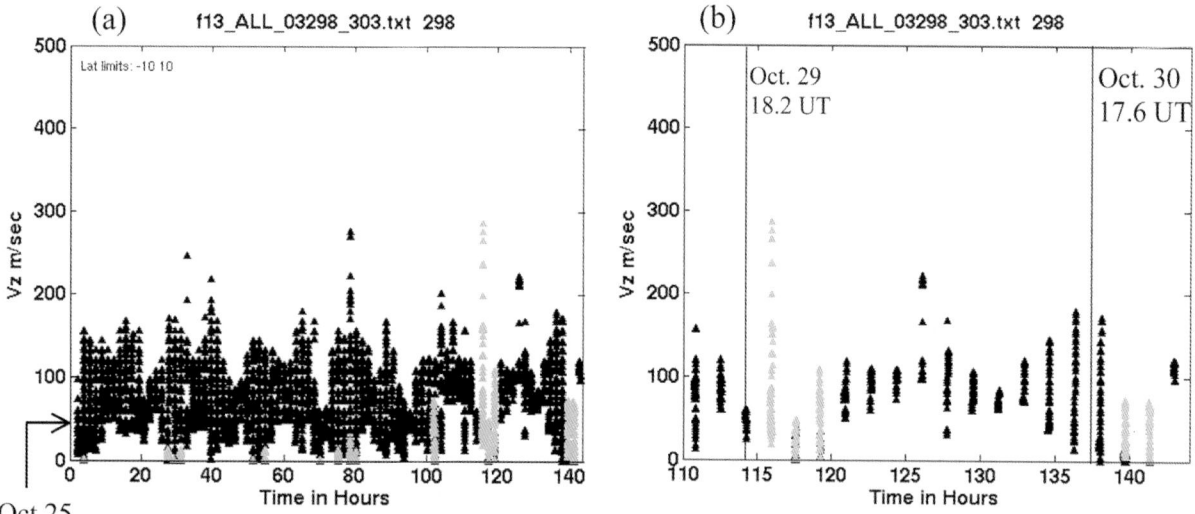

Figure 7. Electric field data from the SSIES instrument onboard the F13 DMSP satellite when the satellite was within ±10 degrees geomagnetic latitude at dusk (1745 LT). Black indicates westward fields and gray eastward. The left panel shows several days of data starting with October 25, 2003, and the right panel is focused on the October 29–30, 2003 storm period. Vertical lines correspond to B_z south event times.

(CHAMP) to provide more information on the structure of the TEC increases as a function of latitude, taking advantage of the satellite orbit to measure TEC versus latitude at the fixed local time of the orbit.

Daytime observations of the TEC above the CHAMP satellite altitude of 400 km on Oct 30 were available from the upward-viewing GPS antenna, plotted in Figure 8 (1300 local time). Slant measurements are used to estimate vertical TEC directly above the satellite as mentioned earlier. There are generally several points at a given epoch (seen as multiple traces for a single color) because there is more than one GPS satellite in view of CHAMP at elevations greater than 40 degrees. The separate traces generally agree, except for the red traces between 40 and 60 degrees latitude, which differ because of the azimuth of the raypaths to the GPS satellites: the lower TEC values are associated with a satellite being tracked in the northwest direction versus northeast/east-directed azimuths for the higher TEC values. These traces differ because of unusually large horizontal TEC gradients in the vicinity of the satellite.

Three daytime ascending passes are shown in Figure 8, the first starting within one hour of the interplanetary B_z south event (lightest gray), and the two subsequent passes. The first latitudinal profile shows the typical "two-peak" structure characteristic of the quiet-time equatorial anomaly [*Hanson and Moffett*, 1966].

The next pass (medium gray), starting at 2012 UT, measures a vastly increased TEC above CHAMP, up to 200 TECU, with the peaks farther apart. This trace begins approximately 2.75 hours after the onset of the IMF southward-B_z event at 1736 UT. The TEC increases still further to ~350 TECU in the southern hemisphere at approximately 4 hours after the onset of the IMF southward-B_z event (second pass after southward B_z, shown in black). The twin peak features previously identified as an *equatorial* anomaly now appear at mid-latitudes (± 28 degrees). The ionospheric TEC above CHAMP altitudes has increased by nearly an order of magnitude (900%) at mid-latitudes (e.g. -30° geomagnetic).

In Figure 8, we also plot estimated vertical TEC obtained from ground-based GPS receivers in North America (light and medium gray dots at 38 and 39 degrees geomagnetic latitude), for measurements located within ±6 minutes and within ±3 degrees longitude of the CHAMP location projected to the Earth's surface. Prior to the interplanetary event's impact (lightest gray dots), the electron content above the CHAMP altitude of approximately 400 km is about half of the total electron content measured from the ground. After the interplanetary event (medium gray dots), the fraction of electron content above CHAMP has increased significantly, suggesting more plasma resides at higher altitudes where recombination rates are reduced. This shift in the vertical plasma distribution is consistent with upward plasma drift on the dayside causing TEC increases.

Preliminary modeling results computing low to mid-latitude TEC using the NRL first-principles ionospheric model SAMI2 [*Huba et al.*, 2000] are presented, based on a simulation study comparing observations obtained at

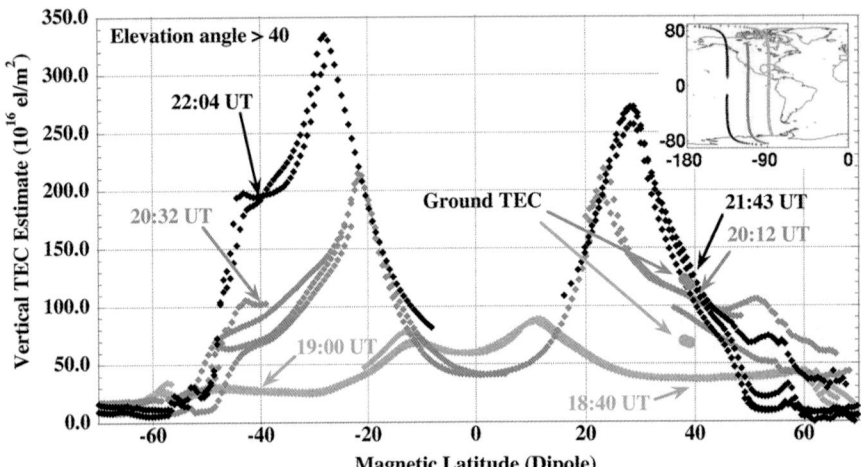

Figure 8. The super-satellite integrated electron content (IEC) as measured by the CHAMP spacecraft is shown at (lightest gray) and after (medium gray and black traces) the onset of the interplanetary event of October 30. The local time of the CHAMP orbit ranges from 1230–1330 LT for latitudes within ±60 degrees. Points missing near the anomaly trough are due to the elevation angle cut-off. The universal times of the –40 and +40 degree latitude cross-over points are shown for each trace. Also shown in the upper right are the geographical location of the traces. Total electron content averages from ground data near to the CHAMP ground track are shown as round dots.

Millstone Hill with model results during storm time conditions [Swisdak et al., 2005]. The results presented here are in qualitative agreement with the results presented in Figure 8 (assuming that the structure of electron content above CHAMP altitude is similar to total electron content structure). The purpose of the simulations is to illustrate the impact of a strong, eastward electric field and a strong equatorward neutral wind on the structure of the low- to mid-latitude ionosphere. Three simulations were performed: (1) quiet day as a benchmark, (2) strong eastward electric field and a quiet day wind, and (3) strong eastward electric field and a strong equatorward neutral wind. A simple sinusoid model of the vertical **E×B** drift at the magnetic equator is used; the peak drift speed was 40 m/s on the quiet day and 120 m/s on the storm day. These drifts correspond to peak electric fields of roughly 1.0 mV/m and 3.0 mV/m, respectively. The strong wind case imposed a strong equatorward wind ($Vn = 250$ m/s) at the time of peak **E×B** drift. The simulation longitude is chosen to pass over Millstone Hill (288.5° E longitude). The vertical TEC for these runs is shown in Figure 9 for 1315 LT. The curve with square symbols denotes the quiet time simulation, the curve with triangular symbols the strong eastward electric field, and the curve with asterisk symbols the combined strong eastward electric field and strong equatorward wind. The imposition of a strong electric field substantially increases the vertical TEC and broadens the width of the ionization crests. Adding the strong equatorward neutral wind largely enhances the TEC but does not broaden the ionization crests significantly. The structures that appear in these modeling results are generally consistent with the data presented in Figure 8. The simulation results are suggestive of the physical processes responsible for the observed increase in TEC and widening of the ionization crests; in a future work we intend to perform simulation studies to be directly comparable to data.

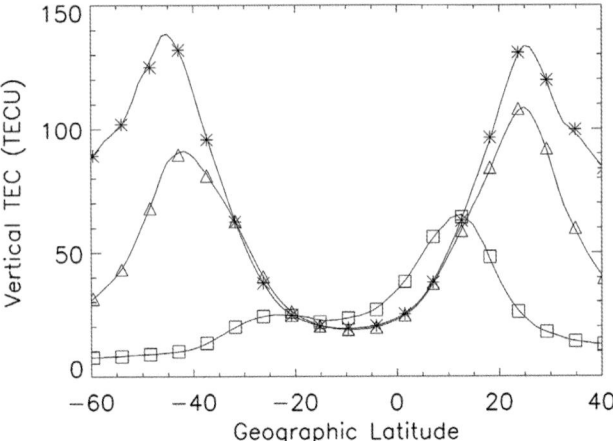

Figure 9. Modeling results comparing TEC under quiet time conditions (□), an applied sinusoidal electric field model with peak eastward field of 3 mV/m (△), and an additional strong equatorward neutral wind ($Vn = 250$ m/s) applied at the time of peak **E×B** drift (∗).

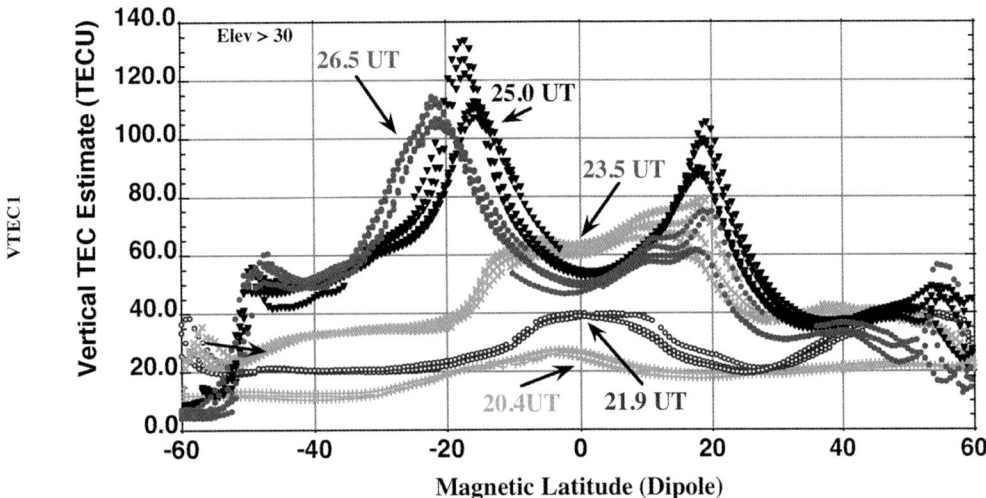

Figure 10. CHAMP observations of TEC above 400 km altitude, similar to Figure 8, for the event on November 7, 2004.

The TEC above the CHAMP satellite for the November 7, 2004 storm is shown in Figure 10, obtained when CHAMP was in a 1400 local time orbit. The trace at 20.4 UT (equatorial crossing time) was obtained just prior to the B_z southward turning and the traces at 21.9 UT and 23.5 UT are the two passes immediately following the interplanetary event after it reached the magnetopause. Additional changes are seen for this event in the next two passes (25.0 and 26.5 UT which occurred on November 8). Before the interplanetary event, the absence of the twin-peak anomaly structure suggests that tidally generated quiet-time eastward electric fields were weak at CHAMP longitudes prior to onset. Immediately after IMF onset, TEC increases are observed without the equatorial anomaly structure. Eventually, the twin-peak anomaly structures form, resulting in widely separate peaks and large increases in mid-latitude plasma content, qualitatively similar to the October 30 event.

Plasma uplift may be creating the large TEC increases observed here, but the detailed temporal evolution of the increase appears qualitatively different than that observed on October 30, 2003. We note that the measured solar flux at 10.7 cm wavelength (UV radiance proxy) is 271.4 for October 30, 2003 versus less than half that value (124.1) for November 8, 2003. Less UV production of plasma may reduce the intensity of the daytime super-fountain, which may be one factor contributing to the differences observed between the November 8, 2004 and October 30, 2003 events. The relatively long period over which electric fields appear to be acting raises questions about the effectiveness of dayside shielding for this event, if the fields originate from magnetospheric convection as we expect (see *Tsurutani et al.*, 2004b).

5. CONCLUSIONS

A multi-event analysis of ionospheric total electron content at low to mid-latitudes suggests that significant increases are a common response to the onset of increased interplanetary electric fields leading to superstorms. The most plausible physical mechanism for the TEC increase is prompt penetration electric fields resulting from increased magnetospheric convection and large under-shielded Region 1 currents entering the high-latitude ionosphere, causing dayside eastward directed electric fields and plasma uplift. Raising the plasma to higher altitudes decreases the average recombination rate of the plasma, while plasma production at lower altitudes due to solar photons continues unabated. The net result is a significant overall increase in daytime TEC at low to high latitudes, as a result of this "dayside super-fountain" effect. Peak TEC increases over the daytime region between ±40 degrees latitude are commonly in the range 50-200%, whereas TEC increases above 400 km altitude of nearly an order of magnitude have been observed at mid-latitudes for the Halloween storms of 2003.

The dramatic hemispheric changes reported here have major practical consequences because the overall TEC increases may intensify spatial TEC gradients and irregularity formation due to mechanisms such as Subauroral Polarization Streams (SAPS) electric fields (*Foster et al.*, 2002b), which also become intense during geomagnetic storms. Therefore, it is important to understand in more detail the physics of the daytime super-fountain and its dependence on: solar wind parameters, ionospheric and magnetosphere conditions and preconditions, and the effectiveness of inner magnetospheric shielding.

Acknowledgements. The research for this paper was performed at the Jet Propulsion Laboratory, California Institute of Technology under contract with NASA. We gratefully acknowledge the Center for Space Sciences at the University of Texas at Dallas and the US Air Force for making available the DMSP drift velocity data.

REFERENCES

Abdu, M. A., I. S. Batista, G. O. Walker, J. H. A. Sobral, N. B. Trivedi, and E. R. de Paula, Equatorial ionospheric electric field during magnetospheric disturbances: local time/longitude dependences from recent EITS campaigns, *J. Atmos. Terr. Physics,* 57, p. 1065, 1995.

Anderson, D., A. Anghel, K. Yumoto, M. Ishitsuka, and E. Kudeki, Estimating daytime vertical ExB drift velocities in the equatorial F-region using ground-based magnetometer observations, Geophysical Research Letters, 29 (12), pp. 1596–1599, 2002.

Blanc, M., and A. D. Richmond, The ionospheric disturbance dynamo, *J. Geophys. Res.,* 85, 1669, 1980.

Fejer, B. G., R. W. Spiro, R. A. Wolf, and J. C. Foster, Latitudinal variation of perturbation electric fields during magnetically disturbed periods: 1986 SUNDIAL observations and model results, *Ann. Geophys.,* **8**, 441, 1990.

Fejer, B. G., and L. Scherliess, Time dependent response of equatorial ionospheric electric fields to magnetospheric disturbances, *Geophys. Res. Letts.,* 22, 851, 1995.

Fejer, B. G., Low latitude storm time ionospheric electrodynamics, *Journal of Atmospheric and Solar-Terrestrial Physics,* **64**, pp. 1401–1408, 2002.

Foster, J. C., P. J. Erickson, A. J. Coster, J. Goldstein and F. J. Rich, Ionospheric signatures of plasmaspheric tails, *Geophys. Res. Lett., 29(13),* 10.1029/2002GL015067, 2002a.

Foster, J. C. and H. B. Vo, Average characteristics and activity dependence of the subauroral polarization stream, *J. Geophys. Res., 107 (A12),* 1475, doi:10.1029/2002JA009409, 2002b.

Gonzalez, W. D. and B. T. Tsurutani, Criteria of interplanetary parameters causing intense magnetic storms (Dst < -100nT). *Planetary Space Science,* 35(9): 1101, 1987.

Gonzalez, W. D.; Joselyn, J. A.; Kamide, Y.; Kroehl, H. W.; Rostoker, G.; Tsurutani, B. T.; Vasyliunas, V. M. What is a geomagnetic storm? *J. Geophys. Res., 99(A4):* 5771, 1994.

Hanson, W.B., and R.J. Moffett, Ionization Transport Effects in Equatorial F Region, *Journal of Geophysical Research,* **71** (23), p. 5559, 1966.

Huba, J.D., G. Joyce, and J.A. Fedder, Sami2 is Another Model of the Ionosphere (SAMI2): A new low-latitude ionosphere model, *Journal of Geophysical Research-Space Physics,* **105** (A10), pp 23035–23053, 2000.

Kamide, Y., W. Baumjohann, I.A. Daglis, W.D. Gonzalez, M. Grande, J.A. Joselyn, R.L. McPherron, J.L. Phillips, E.G.D. Reeves, G. Rostoker, A.S. Sharma, H.J. Singer, B.T. Tsurutani, and V.M. Vasyliunas, Current understanding of magnetic storms: Storm-substorm relationships, *Journal of Geophysical Research-Space Physics,* **103** (A8), pp. 17705–17728, 1998.

Kelley, M. C., J. J. Makela, J. L. Chau, and M. J. Nicolls, Penetration of the solar wind electric field into the magnetosphere/ionosphere system, *Geophys. Res. Letts.,* 30, 1158, 2003.

Klein, L.W., and L.F. Burlaga, Inter-planetary magnetic clouds at 1-AU, *Journal of Geophysical Research-Space Physics,* **87** (NA2), 613, 1982.

Kozyra, J. U., V. K. Jordanova, R. B. Horne, and R. M. Thorne, Modeling the contribution of the electromagnetic ion cyclotron (EMIC) waves to stormtime ring current erosion, in Magnetic Storms, *Geophys. Monogr. Ser.,* volume 98, edited by B. T. Tsurutani et al., pp. 187 – 202, AGU, Washington, D. C., 1997.

Mannucci, A.J., B.D. Wilson, D.N. Yuan, C. M. Ho, U.J. Lindqwister, T. F., Runge. A global mapping technique for GPS-derived ionospheric total electron content measurements. *Radio Science* 33 (3), 565, 1998.

Mannucci, A.J., Iijima, B. A., Lindqwister, U. J., Pi, X., Sparks, L., Wilson, B.D., GPS and Ionosphere, published in *URSI Reviews of Radio Science, 1996–1999,* Oxford University Press, August 1999.

Moore A.W., A review of currently available IGS network summaries, Phys. And Chem. of the Earth – Part A, **26** (6–8), 591–594, 2001.

Nopper, R.W., and R.L. Carovillano, Polar-Equatorial Coupling During Magnetically Active Periods, *Geophysical Research Letters,* 5 (8), 699–702, 1978.

Rastogi, R.G., and J.A. Klobuchar, Ionospheric Electron-Content within the Equatorial F2 Layer Anomaly Belt, *Journal of Geophysical Research-Space Physics,* **95** (A11), pp. 19045–19052, 1990.

Reigber, C; Luhr, H; Schwintzer, P, CHAMP mission status, *Advances In Space Research,* **30** (2), pp. 129–134 2002.

Richmond, A. D., and G. Lu, Upper-atmospheric effects of magnetic storms: a brief tutorial, *J. Atmos. Solar-Terrest. Phys.,* 62, 1115, 2000.

Sastri, J. H., Equatorial electric fields of the disturbance dynamo origin, *Annales Geophysicae* 6, 635, 1988.

Sastri, J. H., K. Niranjan, and K.S.V Subbarao, Response of the equatorial ionosphere in the Indian (midnight) sector to the severe magnetic storm of July 15, 2000, *Geophys. Res. Lett.,* 29, No. 13, 10.1029/2002GL015133, 2002.

Scherliess, L., and B. G. Fejer, Storm time dependence of equatorial disturbance dynamo zonal electric field, *J. Geophys. Res.,* 102, 24037, 1997.

Senior, C., and M. Blanc, On the control of magnetospheric convection by the spatial distribution of ionospheric conductivities, *J. Geophys. Res.,* 89, p. 261 (1984).

Skoug, R. M., J. T. Gosling, J. T. Steinberg, D. J. McComas, C. W. Smith, N. F. Ness, Q. Hu, and L. F. Burlaga, Extremely high speed solar wind: 29–30 October 2003, J. Geophys. Res., 109, A09102, doi:10.1029/2004JA010494, 2004.

Sobral, J. H. A., M. A. Abdu, W. D. Gonzalez, I. Batista, A. L. Clua de Gonzalez, Low-latitude ionospheric response during intense magnetic storms at solar maximum, *J. Geophys. Res.* 102, 14305, 1997.

Sobral, J. H. A., M. A. Abdu, W. D. Gonzalez, C. S. Yamashita, A. L. Clua de Gonzalez, , I. Batista and C. J. Zamlutti, Responses of the low latitude ionosphere to very intense geomagnetic storms, *J. Atmos. Solar Terr. Phys.*, **63**, 965, 2001.

Spiro, R. W., R. A. Wolf, and B. G. Fejer, Penetration of high latitude electric field effects to low latitudes during SUNDIAL 1984, *Ann. Geophys.*, **6**, 39, 1988.

Swisdak, M., J.D. Huba, G. Joyce, and C.S-Huang, Simulation study of a positive storm phase observed at Millstone Hill, submitted to Geophys. Res. Lett., 2005.

Tanaka, T. and K. Hirao, Effects of an electric field on the dynamical behavior of the ionospheres and its application to the storm time disturbances of the F-layer, *J. Atmos. Terr. Phys.*, **35**, 1443, 1973.

Tsurutani, B.T., W.D. Gonzalez, F. Tang, S.I. Akasofu, and E.J. Smith, Origin of Interplanetary Southward Magnetic-Fields Responsible for Major Magnetic Storms near Solar Maximum (1978–1979), *Journal of Geophysical Research—Space Physics*, **93** (A8), 8519, 1988.

Tsurutani, B. T., W. D. Gonzalez, X. –Y. Zhou, R. P. Lepping, and V. Bothmer, Properties of slow magnetic clouds, *J. Atmosph. Solar Terr. Physics*, **66**, 147, 2004a.

Tsurutani, Bruce; Mannucci, Anthony; Iijima, Byron; Abdu, Mangalathayil Ali; Sobral, Jose Humberto A.; Gonzalez, Walter; Guarnieri, Fernando; Tsuda, Toshitaka; Saito, Akinori; Yumoto, Kiyohumi; Fejer, Bela; Fuller-Rowell, Timothy J.; Kozyra, Janet; Foster, John C.; Coster, Anthea; Vasyliunas, Vytenis M., Global dayside ionospheric uplift and enhancement associated with interplanetary electric fields, *J. Geophys. Res.*, **109** (A8), p. A08302, doi 10.1029/2003JA010342, 2004b.

Vlasov, M., M.C. Kelley, and H. Kil, Analysis of ground-based and satellite observations of F-region behavior during the great magnetic storm of July 15, 2000, *Journal of Atmospheric and Solar-Terrestrial Physics*, **65** (11–13), 1223–1234, 2003.

Walter D. Gonzalez, Instituto Nacional de Pesquisas Espaciais, CP 515, 12245-970, São José dos Campos, Sao Paulo, Brazil.

J.D. Huba, Code 6790, Plasma Physics Division, Naval Research Laboratory, Washington, DC 20375– 5320, USA.

Janet Kozyra, Department of Atmospheric, Oceanic, and Space Sciences, University of Michigan, 1414 Space Research Building, Ann Arbor, Michigan, 48109-2143, USA.

A. J. Mannucci, Byron Iijima, Attila Komjathy, Brian Wilson, Xiaoqing Pi, Lawrence Sparks, George Hajj, Lukas Mandrake, Jet Propulsion Laboratory, 4800 Oak Grove Drive, MS 138-308, Pasadena, CA 91109, USA.

R. Skoug, Los Alamos National Laboratory, MS D466, Los Alamos, New Mexico, 87545, USA.

M. Swisdak, Icarus Research, Inc., PO Box 30780 Bethesda, Maryland, 20824-0780, USA.

B. T. Tsurutani, Jet Propulsion Laboratory, 4800 Oak Grove Drive, MS 169-506, Pasadena, CA 91109, USA.

K. Yumoto, Space Environment Research Center, Kyushu University, 6-10-1 Hakozaki, Fukuoka 812-8581, Japan.

Redistribution of the Stormtime Ionosphere and the Formation of a Plasmaspheric Bulge

John C. Foster[1], Anthea J. Coster[1], Philip J. Erickson[1], William Rideout[1]
Frederick J. Rich[2], Thomas J. Immel[3], and Bill R. Sandel[4]

Plasmasphere drainage plumes resulting from the erosion of the plasmasphere boundary layer by disturbance electric fields have been identified from both ground and space. Here we describe a localized enhancement of total electron content (TEC) seen at the base of the erosion plume, on field lines mapping into the outer plasmasphere. Observations suggest that this enhanced TEC results from a poleward redistribution of post-noon sector low latitude ionospheric plasma during the early stages of a strong geomagnetic disturbance. Ground based and low-altitude observations with GPS TEC, incoherent scatter radar, and DMSP in situ observations provide details and a temporal history of the evolution of such events. Seen from space by IMAGE EUV, the region of enhanced TEC appears as a pronounced brightening in the inner plasmasphere. IMAGE FUV provides complementary images at lower altitude of this inner-plasmasphere feature, showing that it is associated with localized enhancement in the vicinity of the equatorial anomaly peak. These effects are especially pronounced over the Americas, and we suggest that this results from a strengthening of the equatorial ion fountain due to undershielded (penetrating) electric fields in the vicinity of the South Atlantic magnetic anomaly. The enhanced low-latitude features, seen both from the ground and from space, corotate with the Earth once they are formed. The high-TEC plasma in these regions contributes to the intensity of the erosion plumes arising in the American sector during strong disturbance events.

INTRODUCTION

Carpenter and Lemaire [2004] have coined the term plasmasphere boundary layer (PBL) to describe the region of dynamic interaction between the plasmas of the inner and outer magnetosphere at the outer extent of the plasmasphere. The thermal plasma in the vicinity of the PBL participates in a variety of processes associated with electric field penetration and shielding, and the dynamic redistribution of plasma during disturbed conditions (c.f. *Carpenter et al.* [1993]). The MIT Millstone Hill incoherent scatter radar, located at 55°Λ (invariant latitude) near the ionospheric projection of the plasmapause and the PBL, regularly observes plumes of storm enhanced density (SED) which stream from the pre-midnight sub-auroral ionosphere toward the noontime cusp during the early stages of magnetic storms [*Foster*, 1993]. Recent observations using both ground and space-based thermal plasma imaging techniques have revealed such

[1]Massachusetts Institute of Technology, Haystack Observatory, Westford, Massachusetts
[2]Air Force Research Laboratory, Hanscom AFM, Massachusetts
[3]University of California, Berkeley, Berkeley, California
[4]University of Arizona, Lunar and Planetary Laboratory, Tucson, Arizona

Inner Magnetosphere Interactions: New Perspectives from Imaging
Geophysical Monograph Series 159
Copyright 2005 by the American Geophysical Union.
10.1029/159GM21

ionospheric SED events to be the low-altitude signature of the plasmasphere drainage plumes recently observed from space by the IMAGE EUV imager [Sandel et al., 2001]. Using radar and GPS observations of total electron content (TEC) to produce two-dimensional snapshots, [Foster et al, 2002] found that the SED/TEC plumes identified at low altitude map directly to the magnetospheric boundaries of the plasmapause and plasma-spheric erosion plume determined by IMAGE EUV.

Here we describe a localized enhancement of total electron content (TEC) observed at the base of the erosion plume, on field lines mapping into the outer plasmasphere. Observations suggest that this enhanced TEC results from a poleward redistribution of post-noon sector low-latitude ionospheric plasma during the early stages of a strong geomagnetic disturbance. After its formation, this feature corotates with the Earth and is seen from space as an enhanced-density plasmaspheric bulge. At ionospheric heights, our observations show that the equatorward extent of the sub-auroral polarization stream electric field (SAPS [Foster and Vo, 2002]) overlaps this region of enhanced TEC, drawing it noonward and poleward along the SAPS convection channel, and producing intense plumes of SED. We use ground-based and low-altitude DMSP observations to probe the spatial extent and temporal evolution of these features and IMAGE EUV and FUV space-based imagery to place the observations in the context of the global plasmasphere-ionosphere system.

OBSERVATIONS OF AN ENHANCED PLASMA SOURCE FOR THE EROSION PLUME

Plate 1a (left) presents in polar projection a GPS TEC map of the spatial extent of the strong plume of storm enhanced density seen during the March 31, 2001 event discussed by Foster et al [2002]. GPS TEC measures the integrated column content of cold electrons through the ionosphere and overlying plasmasphere to an altitude of ~20,000 km (~4 Re) [Coster et al., 2003]. The SED plume stretches from a region of enhanced TEC in the SE USA, poleward to the limit of the GPS observations near the noontime cusp over north central Canada. In Plate 1b (right) we have used the Tsyganenko [2002] magnetic field model to project the ionospheric footprint of the TEC observations into the magnetosphere equatorial plane. The SED plume maps into a narrow drainage plume reaching sunward from the greatly eroded plasmapause position near L=2 to the dayside magnetopause near noon (noon is at the right of the figure).

The IMAGE extreme ultraviolet (EUV) imager [Sandel et al., 2001] directly observes the plasmasphere by detecting 30.4 nm sunlight resonantly scattered by plasmaspheric He^+ ions. In Plate 2 we present the IMAGE EUV observation of the plasmasphere and drainage plume taken at the time of the TEC observations of Plate 1. In the dusk sector, a narrow erosion plume is seen on the outer edge of a broader region which bulges sunward from the post-noon plasmasphere. Goldstein et al [2003a] have discussed the complementary roles of the convection electric field and SAPS in forming these features of the post-noon plasmasphere. The lower-threshold sensitivity for the EUV images is 40 electrons cm^{-3} (~ 4–8 He^+ cm^{-3}) [Goldstein et al, 2003b]. During this event the erosion plume is seen only weakly, but here we concentrate on the distribution of plasma equatorward of L=2, inside the plasmasphere where significant large-scale structure is evident. The azimuthal distribution of EUV intensity at two radial distances, 1.7 Re and 2.2 Re, is presented in Figure 1 plotted versus magnetic local time. The Earth's shadow suppresses the intensity across the nightside. In the post-noon sector there is a 50%–100% enhancement in intensity centered near 15 LT at 1.7 Re and 2.2 Re, respectively. This intensification lies at the base of the erosion plume which extends sunward from the post-noon plasmasphere. The sequence of frames taken by the EUV camera indicates that this enhancement approximately corotates into the dusk sector during this event.

We now turn our attention to the ionosphere and in Figure 2 present the temporal/spatial evolution of the TEC enhancement seen at the base of the SED plume in Plate 1. North American TEC maps similar that shown in Plate 1a have been generated at 5-min intervals. We present the time variation of the longitude / UT distribution of TEC observed in a 2-degree width latitude band centered at 32 N latitude across the American sector. For the duration of the event, this latitude lies equatorward of the sharp gradient in TEC which denotes the plasmapause position. Time runs from bottom to top of the figure and a 11-hour interval is shown. A strong enhancement of TEC (>100 TECu; 1 TECu = an integrated column density of 10^{16} m^{-2}) begins near 16 UT across longitudes from 70 W to 85 W. This enhancement persists for 6 hours and remains centered near 75 W longitude.

To indicate the relationship of the enhanced-TEC region to the inner-plasmasphere structure described above, we have mapped its approximate center point at 19:30 UT (32 N, 75 W) to the equatorial plane (to a radial position of 1.75 Re at 15 MLT), where it coincides with the bright inner-plasmasphere enhancement observed by IMAGE EUV. Figure 2 indicates that the TEC enhancement at the base of the SED plume forms rapidly (on the order of 1 hour) and then approximately corotates with the Earth. Mechanisms involved in the formation of the high-TEC inner-plasmasphere enhancement and the erosion plume are discussed in the following section.

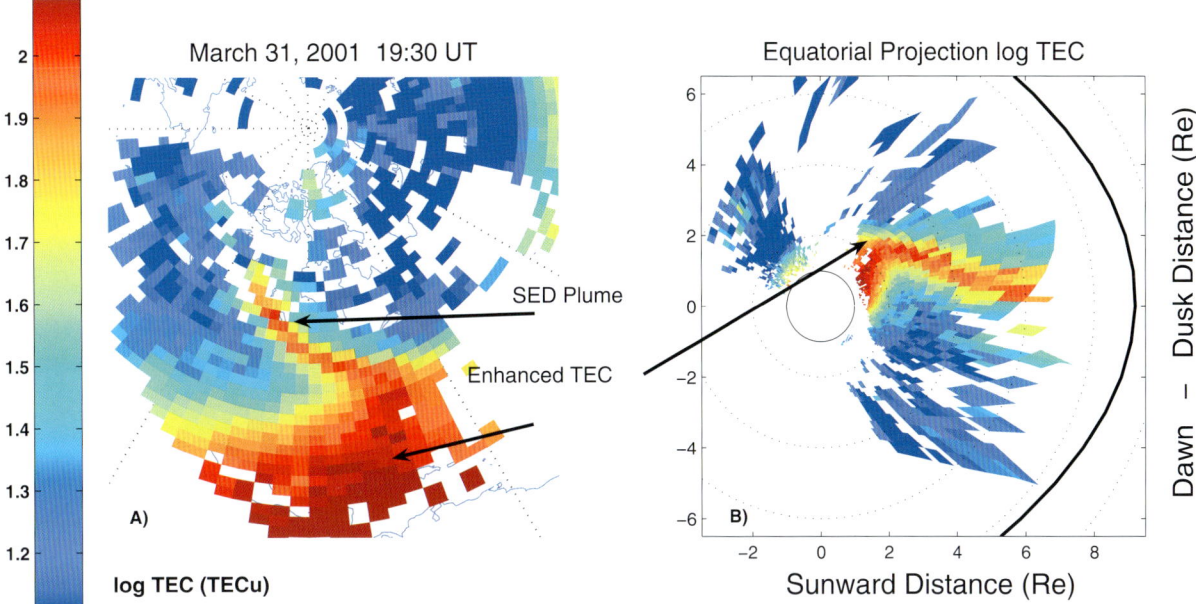

Plate 1. (A) A region of enhanced GPS TEC was observed at the base of the plume of storm enhanced density seen over North America during the March 31, 2001 event. (B) Projecting the GPS TEC observations into the magnetospheric equatorial plane using Tysganenko mapping (with the sun at the right), indicates that the enhancement at the base of the plume is field lines threading the outer plasmasphere.

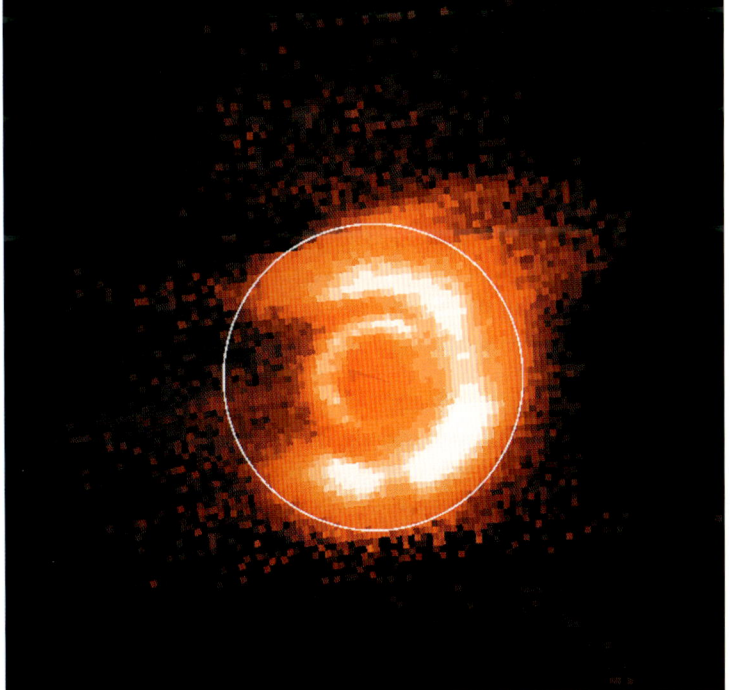

Plate 2. IMAGE EUV observations of plasmasphere structure at the time of the TEC observations of Plate 1 show significant structure in the outer plasmasphere, including a bright enhancement near 15 LT at the base of the erosion plume (the Sun is to the right). A circle is shown at 2 Re and the northern aurora is seen in the center of the image.

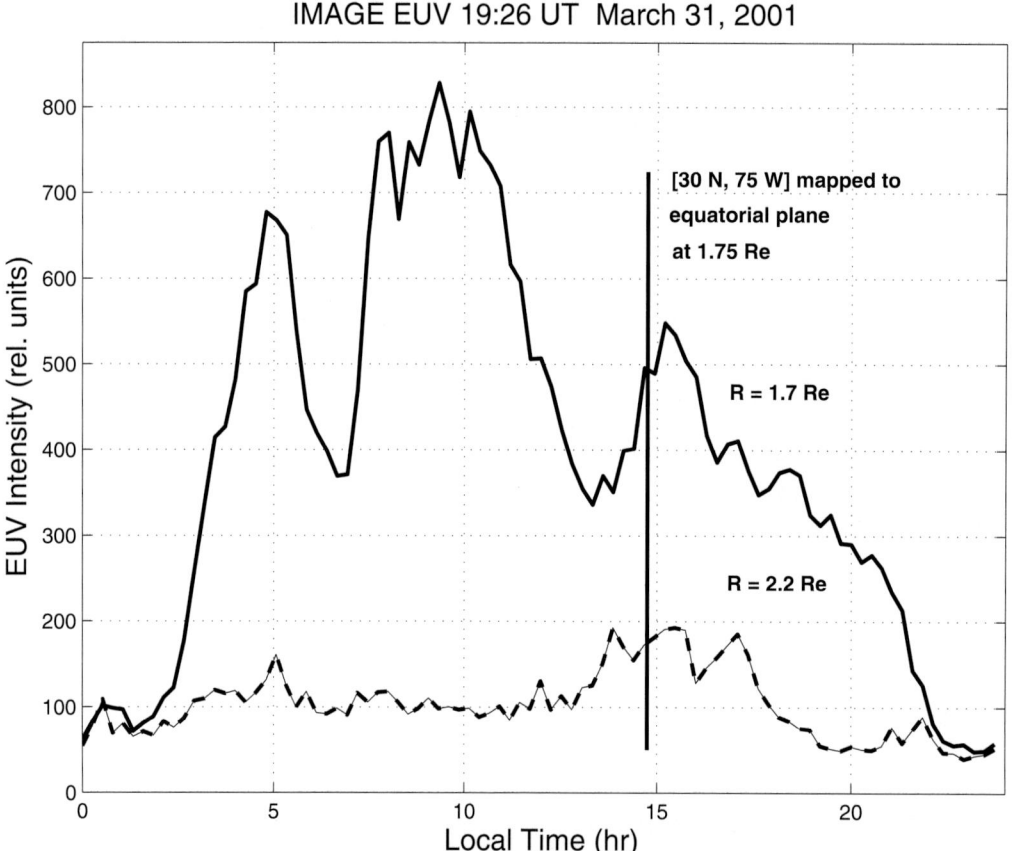

Figure 1. The azimuthal distribution of EUV intensity is shown at two radial distances distances for the image shown in Plate 2. The Earth's shadow suppresses the intensity across the nightside. In the post-noon sector there is a 50%–100% enhancement in intensity centered near 15 LT.

STORMTIME DISTURBANCE ELECTRIC FIELDS

During the early phase of a geomagnetic storm, enhanced cross-tail electric fields drive plasmasheet particles inward. There is little shielding in place at the time, and the fields penetrate deep into the inner magnetosphere. Penetrating eastward electric fields are observed at mid and low latitudes in the post-noon sector (e.g. *Foster and Rich* [1978]), driving the *F*-region plasma upward and poleward in the **ExB** direction. At equatorial latitudes, the ionosphere is lifted and spreads poleward along the field lines in both hemispheres under the influence of gravity and plasma pressure to form the equatorial ion fountain and the enhanced plasma of the equatorial ionospheric anomaly (EIA) peaks. During extreme uplift events (e.g. March 1989 [*Greenspan et al.,* 1991] or July 2000 [*Basu et al,* 2001]) the bottomside of the equatorial *F* region can rise above the 830-km altitude orbit of the DMSP satellites. (Such drastic uplifting is seen only at longitudes near the South Atlantic magnetic anomaly (SAA) and most often in the dusk sector [*Sultan and Rich,* unpublished manuscript]).

Plate 3 presents a cartoon schematic of the dual action of stormtime electric fields associated with inner-magnetosphere cold plasma redistribution. An initial strong equatorial upwelling redistributes the low-latitude ionospheric plasma to higher-latitude flux tubes within the plasmasphere, resulting in enhanced levels of density on flux tubes in the PBL near the plasmapause. As the event progresses, a shielding layer is set up where the freshly-injected ring-current particles abut the plasmapause. The inward extent of the energetic ring-current ions lies equatorward of the plasmasheet electrons, and Region II currents are driven into the sub auroral ionosphere. There a strong poleward electric field is needed to drive poleward-directed Pedersen closure currents in the low-conductivity ionosphere equatorward of the precipitating auroral electrons. This sub auroral polarization stream electric field (SAPS [*Foster and Vo,* 2002]) is strongest in the dusk/pre-midnight sector where it lies equatorward of and often distinct from the region of sunward ionospheric convection at auroral latitudes. It is the SAPS electric field which overlaps the PBL at the base of the erosion plumes, resulting in the erosion of the enhanced outer layer of the stormtime plasmasphere (see discussion below, also *Foster et al.*, [2002; 2004]). Nearer to noon, the SAPS convection merges into the equatorward extent of the auroral convection [*Foster and Vo*, 2002], bringing the SED plume at ionospheric heights into the dayside cusp [*Foster et al.*, 2004], as seen in Plate 1.

OBSERVATIONS OF THE REDISTRIBUTION OF LOW-LATITUDE COLD PLASMA

Plate 4 presents simultaneous GPS TEC and DMSP in situ plasma density observations which illustrate the features of the plasma-redistribution event shown schematically in Plate 3. The strong disturbance of May 29/30, 2003 produced a plasma-redistribution event very similar to those reported by *Greenspan et al* [1991] and *Basu et al* [2001]. In the center frame we present vertical TEC derived from the available observing sites in North and South America. Red lines show the orbital tracks of two DMSP spacecraft which crossed the region near the time of the GPS observations. In situ observations of topside (830 km) plasma density for each DMSP are shown at the sides of the TEC map. DMSP F15 observed a deep depletion over the magnetic equator in eastern Brazil, coincident with a localized region of very-low of ionosphere-plasmasphere TEC imaged by the GPS technique. A much lesser depletion was seen at equatorial latitudes by DMSP F14, 15° longitude to the west. A region of greatly-elevated TEC (>100 TECu) is seen near 285 E longitude off the coast of Florida and this enhancement extends poleward to the sharp gradient of the plasmapause seen near 40° N latitude. A SED erosion plume stretches NW across the US from this enhancement. Such an extensive region of elevated TEC at the base of the plume in the American sector is a characteristic occurrence during major storms, and contributed to the strong SED plumes observed in the May 30, 2003, March 31, 2001, April 11, 2001, and July 15, 2000 events.

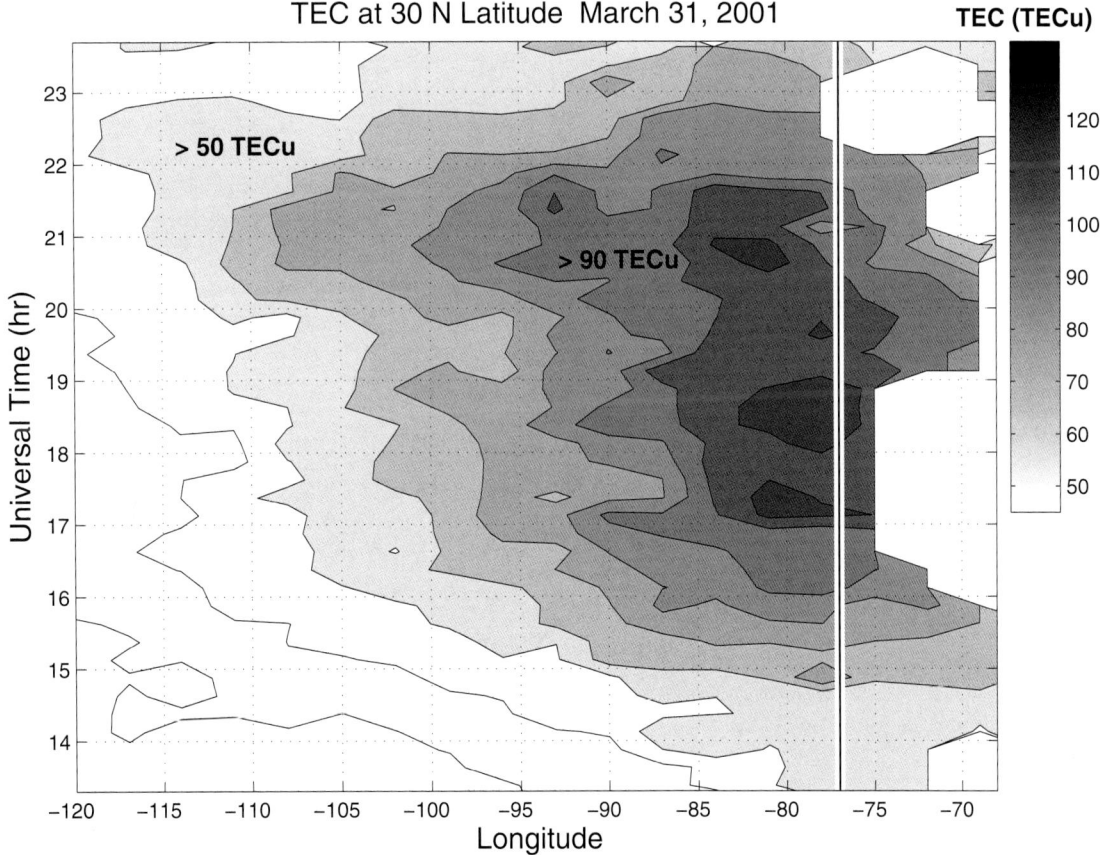

Figure 2. The temporal/spatial evolution of the TEC enhancement seen at the base of the SED plume in Plate 1 is determined in a 2- degree width latitude band centered at 30 N latitude across the American sector. Time runs from bottom to top of the figure. A strong enhancement of TEC (>100 TECu) begins near 16 UT centered near 75 W longitude.

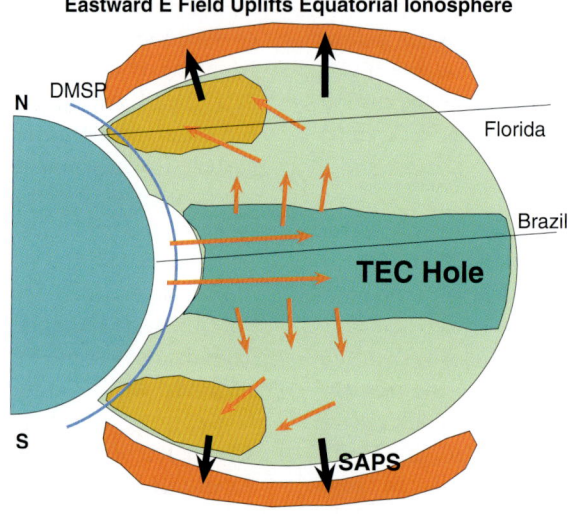

Plate 3. The dual effects of disturbance electric fields are presented schematically. Undershielded penetration electric fields uplift the equatorial ionosphere redistributing equatorial plasma poleward, while SAPS electric fields strip away the enhanced outer layers of the plasmasphere.

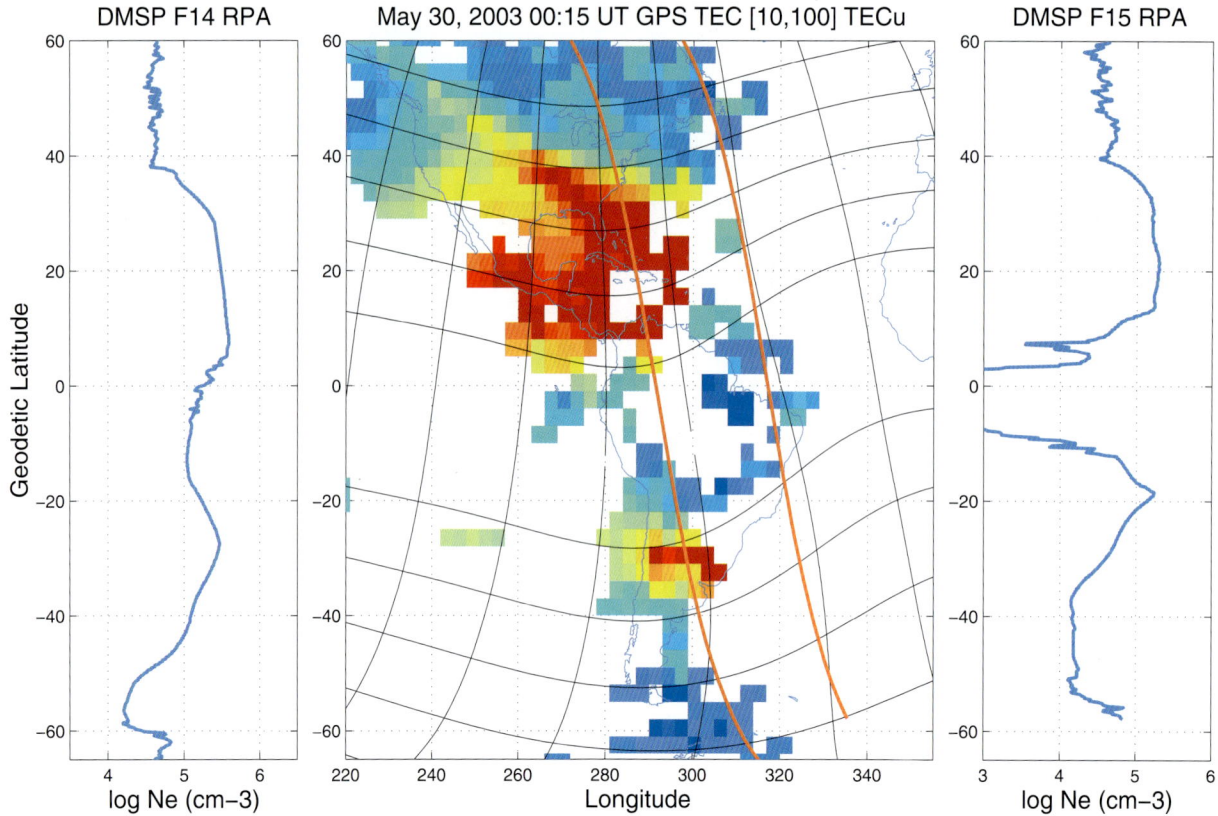

Plate 4. Simultaneous GPS TEC and DMSP in situ plasma density observations illustrate the effects of plasma-redistribution shown schematically in Plate 3. The orbital tracks for DMSP F14 (left) and F15 (right) are shown in red. A deep total content hole is formed at the magnetic equator over eastern Brazil while greatly enhanced plasma is relocated to the vicinity of the Florida coast.

It is our hypothesis that the intensity of the region of elevated TEC which forms on field lines mapping to the outer plasmasphere off the coast of Florida in these events is due to a strong enhancement of the destabilizing upward velocity ($Vz \sim |E|/|B|$) in the equatorial ionosphere at dusk at longitudes near the South Atlantic magnetic anomaly, where the strength of B is ~30% less than the global mean. Penetrating eastward electric fields near dusk (e.g. described by *Foster and Rich* [1978]) produce an unusually strong uplift of the equatorial ionosphere in the vicinity of the SAA. Our observations indicate that when the dusk terminator (associated with a further enhancement of the E-W electric fields by polarization effects) crosses the SAA as Dst is falling rapidly during a major storm onset, strong equatorial upwelling results in a large enhancement of TEC on field lines inside the plasmapause in the American sector. The plasma redistribution is more complex than would be expected if only an enhancement of the equatorial ion fountain were involved. In the northern hemisphere, plasma is relocated both poleward and westward to the Caribbean region, and is significantly enhanced poleward of the EIA crest in the vicinity of the plasmapause. *Kelley et al.* [2004] have modeled processes associated with the ionospheric redistribution on low and mid-latitude field lines in the American sector observed during the July 15, 2000 event.

For the May 29/30, 2003 event, Figure 3 presents a temporal history of the redistribution of ionospheric plasma from the equatorial region (Brazil) to higher latitudes (Florida). The temporal development of the storm is indicated by the SYM H index shown at the top of the figure. Five-min samples of the equatorial and low-latitude TEC are shown in the lower portion of the figure. Equatorial TEC over eastern Brazil drops sharply from 50 TECu to < 5 TECu beginning at ~21 UT (~19 LT over the SAA). As the SYM H index begins to decrease steeply after 22 UT, the low-latitude TEC observed over the Florida on field lines mapping into PBL increases abruptly from 50 TECu to 100 TECu and remains enhanced for > 4 hours. The vertical TEC measurements

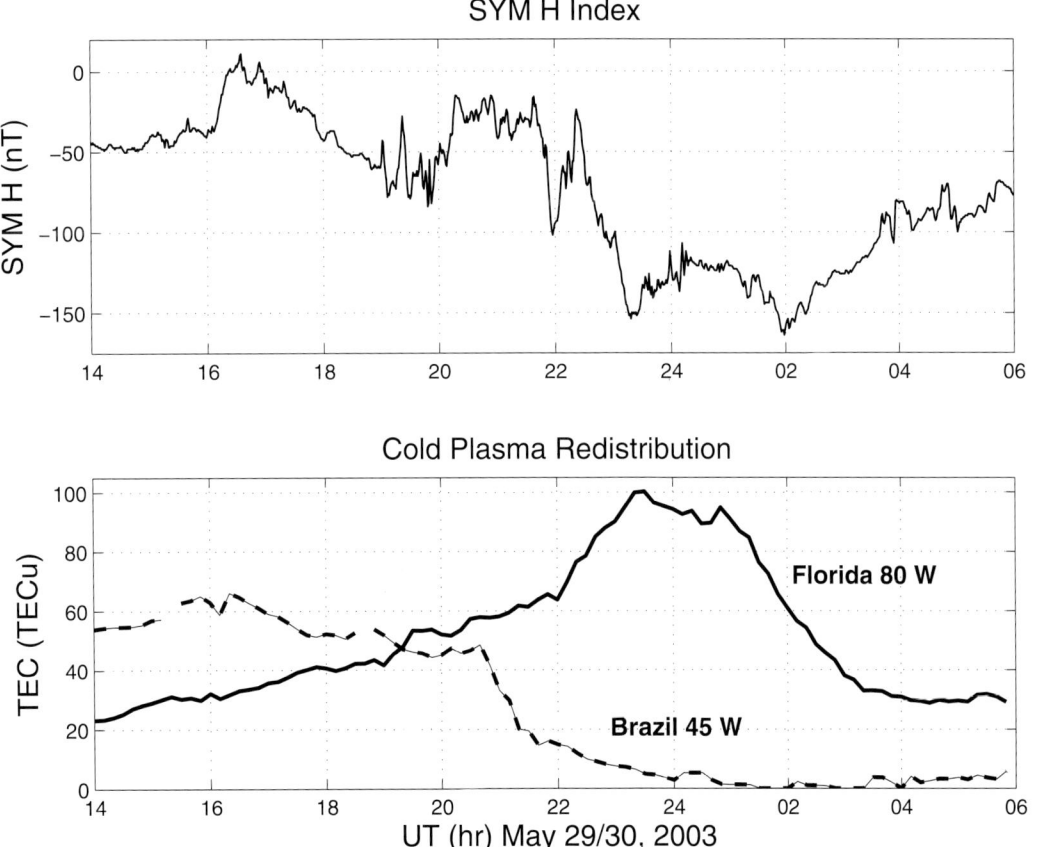

Figure 3. The redistribution of thermal plasma from the equator (Brazil) to field lines threading the outer plasmasphere (Florida) was observed during the May 2003 event. The strong enhancement in TEC off the Florida coast occurred while SYM H (a proxy for Dst; shown in the top panel) was dropping sharply (top) as the South Atlantic magnetic anomaly entered the dusk sector (21 UT = 19 LT at the SAA).

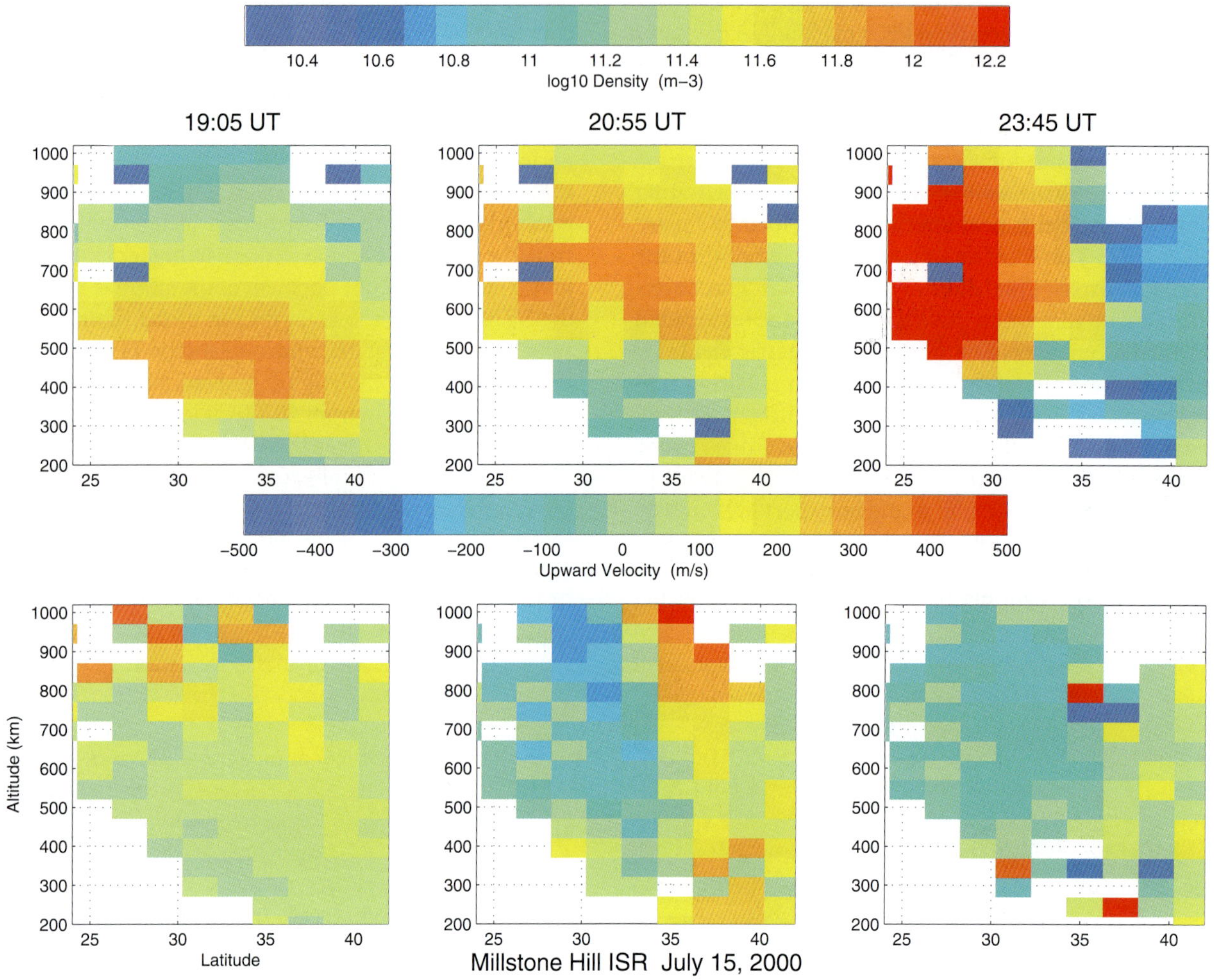

Plate 5. Elevation scans with the Millstone Hill ISR reveal the strong downwelling of plasma which accompanies the enhancement of TEC inside the plasmapause during major events. Plasma density vs. altitude and latitude is presented in the top panels, while line of sight ion velocity is presented beneath. Strong downwelling was seen at 20:55 UT as the peak height of the low-latitude F region increased to > 600 km and the F-region density increased significantly.

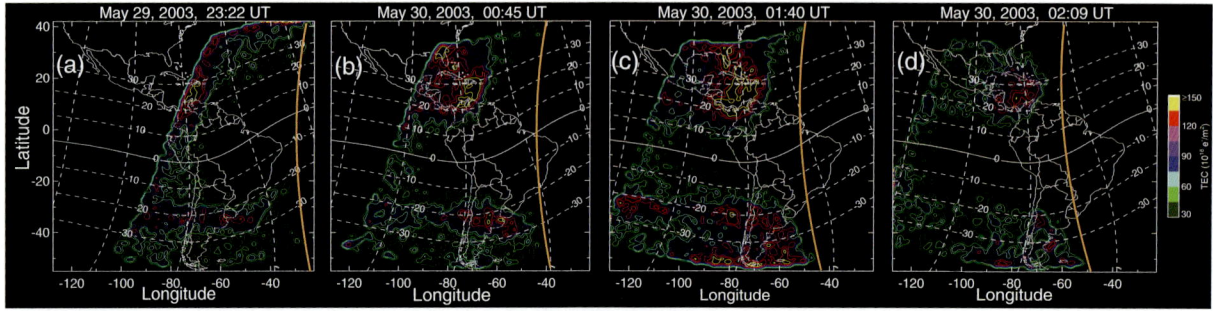

Plate 6. A montage of 4 frames of IMAGE FUV observations of ionospheric O+ emission intensity reveal the corotating enhancement which formed over the Caribbean during the May 29/30, 2003 event. Following *Immel et al.* [2004] the FUV intensity has been scaled and contoured in units of effective ionospheric TEC.

reported in this paper cannot determine the altitude distribution of the cold plasma in these regions. Incoherent scatter radar observations directly measure both the plasma altitude distribution and velocity at *F*-region heights.

Elevation scans far to the south enabled the Millstone Hill ISR to observe the plasma density enhancement and downwelling in the region of increased TEC at the base of the erosion plume during the July 15, 2000 storm. A large increase in density in the topside ionosphere near 30° geodetic latitude and 285 E longitude was seen during that event. Plate 5 presents latitude/altitude scans through the low-latitude *F* region near Florida at three times during the build up of TEC during the July 2000 event. The altitude of the low-latitude *F*-region peak, poleward of the EIA crest, increased from ~400 km to > 700 km during the event indicating the influence of an eastward electric field. Near ~21 UT, as Dst was dropping sharply, a large increase in topside density began and then increased greatly as the event progressed (c.f. 23:45 UT at latitudes <30N). The lower frames display the line-of-sight plasma velocities observed during each scan. Velocities toward the radar (poleward or down the field line) are colored blue. The large enhancement of the topside *F* region beginning near 20:55 UT is seen to be accompanied by strong (> 350 m/s) downwelling of cold plasma. Further discussion of the redistribution of low-latitude plasma during the July 15, 2000 event is presented by *Yin et al* [2004] and *Kelley et al.* [2004]. An earlier study by *Buonsanto and Foster* [1993] examined Arecibo radar observations during a similar event and noted that the high density plasma in the enhanced-TEC region moved approximately horizontally, with a strong poleward and westward component to its motion.

IMAGING THE ENHANCED EQUATORIAL ANOMALIES FROM GROUND AND SPACE

The TEC enhancements produced during the May 29/30, 2003 event have been observed from space by the IMAGE FUV instrument. The Spectrographic Imaging component of the FUV instrument [*Mende et al.*, 2000] obtains simultaneous 2D images of terrestrial FUV emissions at wavelengths of 121.8 and 135.6-nm. Important here is the 135.6- nm channel which detects O I emissions produced on the dayside, nightside and at high latitudes by several processes, including the recombination of ionospheric O^+. This process is the dominant source of nightside FUV emissions away from the auroral oval, where the intensity of the emission is expected to be proportional to the total electron content along the line of sight.

Immel et al [2005] have cross calibrated the intensity of FUV emissions with GPS TEC observations during the May 29/30, 2003 event. Plate 6, taken from that paper, presents four FUV frames looking down on the regions of enhanced sub-auroral emissions over the Americas. FUV emission intensity has been scaled to indicate effective TEC. The E-W emission band associated with the southern equatorial anomaly crest is evident. Of particular interest here is the bright patch of O+ emission over the Caribbean whose position remains fixed from frame to frame indicating that the feature approximately corotates with the Earth.

In Plate 7 we present simultaneous FUV (left) and GPS TEC (right) images of this region, at 01:35 UT. The one-to-one comparison of features and intensities seen in the FUV and TEC images is very good [*Immel et al.*, 2005]. The bright patch seen by FUV over the Caribbean corresponds to the region of TEC enhancement at the base of the SED erosion plume seen in the TEC image. Bright emissions at higher latitudes in the northern hemisphere indicate the equatorward extent of the aurora, and the narrow latitude extent of the northern EIA crest is seen faintly to the west of the TEC and FUV enhancements. Scattered sunlight near the dusk terminator contaminated the western edge of the FUV images. The ionospheric footprint of an overflight of the DMSP F-13 satellite is indicated on the TEC frame. DMSP crosses the EIA crest region near 10° N to the west of the TEC enhancement at the base of the SED plume, and then passes through the SED plume near 40° N at 110 W longitude. The DMSP observations can be used to understand better the processes leading to the cold plasma structures seen in Plate 7.

Figure 4 presents the in-situ observations of density and cross-track (westward) velocity observed by F-13. The equatorward extent of electron precipitation observed by the particle spectrometer (not shown) is indicated by vertical lines near 54° N. The extensive sub-auroral (SAPS) electric field extends equatorward from this point to ~30° N. The SED plume centered at 40° N lies entirely within the SAPS electric field, with the pronounced plume enhancement lying just equatorward of the strong SAPS peak. Some 25 min later, a similar set of observations was obtained as the DMSP F-15 satellite flew directly through the region of TEC enhancement at the base of the erosion plume.

Plate 8 presents IMAGE FUV and GPS TEC imagery of the cold plasma features seen at 02:00 UT on May 30, 2003, following the format of Plate 7. The ionospheric footprint of F-15 is indicated and in-situ observations are presented in Figure 5. The satellite sampled the region of equatorial TEC depletion at latitudes below 7° N and the TEC enhancement stretched poleward from this point until ~30° N. A strong narrow electric field enhancement (SAID) lay immediately equatorward of the precipitation boundary at 39° N, while the fringing SAPS electric field extended equatorward to

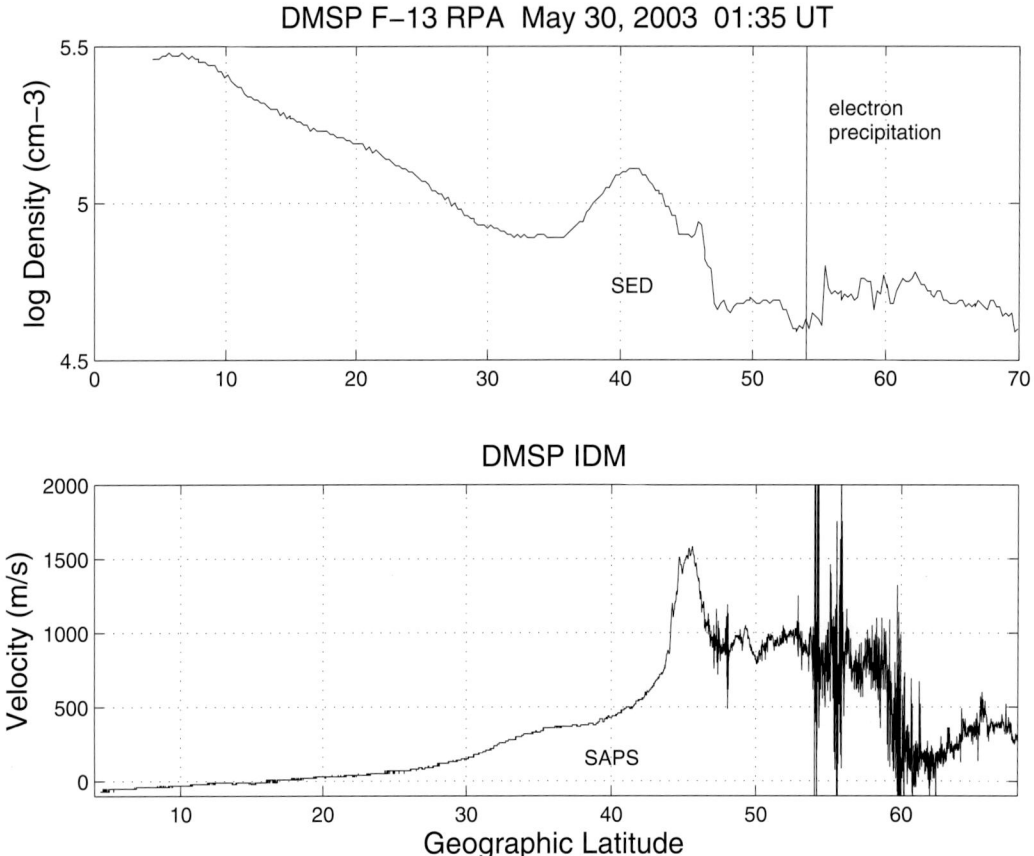

Figure 4. In-situ observations of density and cross-track (westward) velocity were observed by DMSP F-13 at the time of the FUV and TEC images shown in Plate 7. The equatorward extent of electron precipitation is indicated by vertical lines near 54° N. The SED plume lies entirely in the region of sunward convection associated with the SAPS electric field.

below ~30° N, overlapping the region of TEC erosion and plume formation. DMSP observations in this event and in prior studies [*Foster et al.* 2002; 2004] indicate that the equatorward portion of the SAPS electric field is responsible for cold plasma erosion in the dusk sector on field lines mapping to the outer plasmasphere. The eroded materials are carried sunward in the combined SAPS and convection electric fields. Ionospheric plasma redistributed from lower latitudes feeds into the erosion plume and contributes to the intensity of the plumes.

DISCUSSION

Foster et al [2004] analyzed the April 11, 2001 plasmasphere erosion event and found that at *F*-region heights a plume of storm enhanced density stretched continuously from the ionospheric projection of the dusk plasmapause to the dayside cusp. The SED plume carried a flux of $>10^{26}$ ions/s into the cusp ionosphere during the peak of the event. Assuming a dipolar magnetic field, those authors projected the low-altitude observations into the outer plasmasphere, obtaining a total sunward flux of $>10^{27}$ ions/s. High-altitude IMAGE EUV observations of the plasmasphere drainage plume provide a similar estimate of 1.5×10^{27} ions/s for the sunward flux during that event.

The solar-produced thermal plasma carried to the dayside in the erosion events constitutes a strong source of ionospheric (and plasmaspheric) ions to the acceleration processes at the ionospheric cusp and at the magnetopause. Using the Chatanika incoherent scatter radar, *Foster and Doupnik* [1984] observed the low-latitude SED plasma streaming poleward through the cusp ionosphere, and *Foster* [1989] has described the role of the SED material as a source of topside *F*-region density plumes and patches observed at polar cap latitudes. *Elphic et al.* [1997] have estimated the flux of plasmaspheric ions which are injected into the magnetotail and convected up and over the polar cap during strong disturbances to be $\sim10^{26}$ ions/s, comparable to the flux estimated from the SED plume observations. *Foster et al.* [2004] concluded that the plasmaspheric drain-

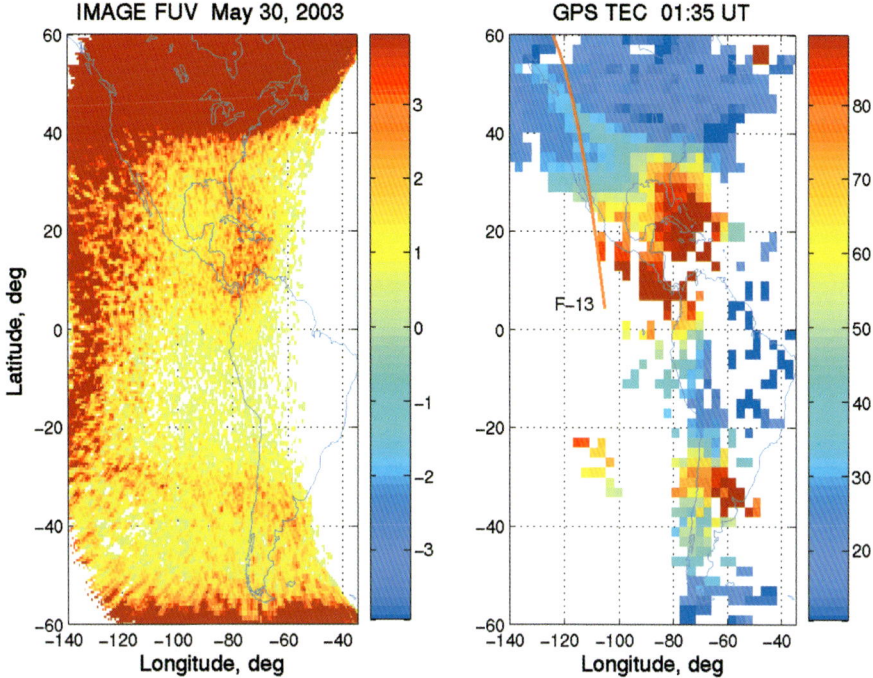

Plate 7. Simultaneous FUV and GPS TEC for 01:35 UT show a good one-to-one comparison of features and intensities. The bright patch seen by FUV over the Caribbean corresponds to the region of TEC enhancement at the base of the SED erosion plume. Bright FUV emissions at higher latitudes in the northern hemisphere indicate the equatorward extent of the aurora. An overflight of the DMSP F-13 satellite which crosses the erosion plume is indicated in red.

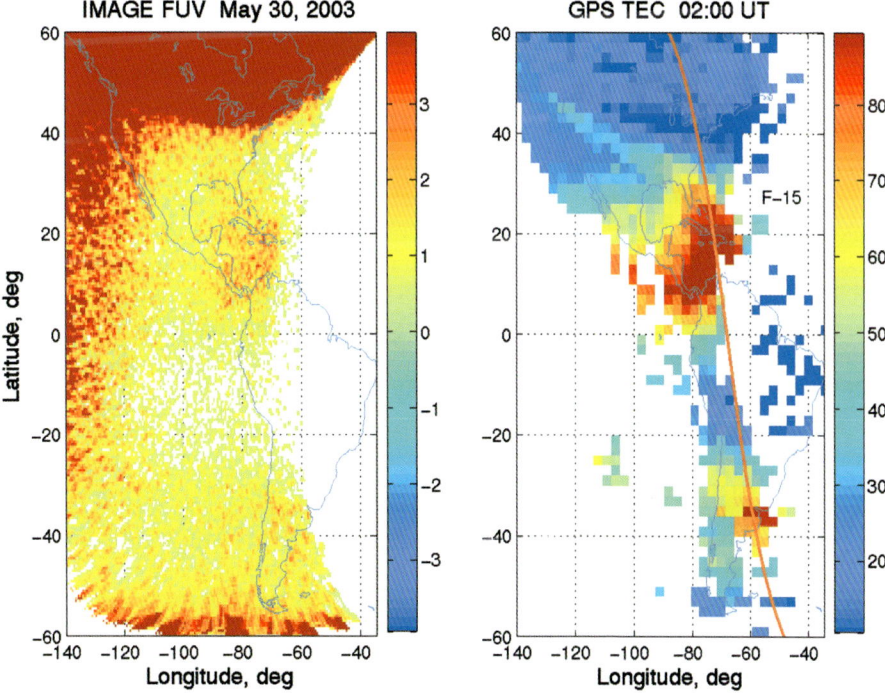

Plate 8. IMAGE FUV and GPS TEC imagery of the cold plasma features seen at 02:00 UT on May 30, 2003 are presented in the format of Plate 7. The ionospheric footprint of a DMSP F-15 overflight of the enhanced TEC region at the base of the erosion plume is shown in red.

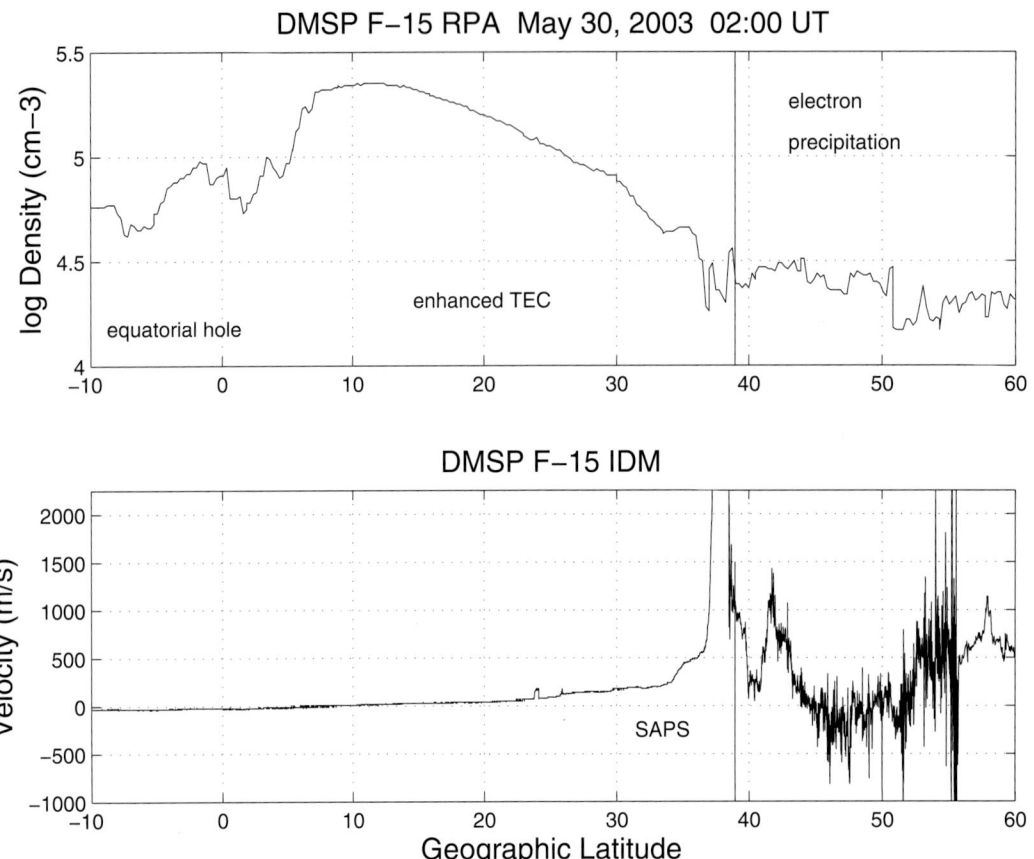

Figure 5. DMSP F-15 sampled the region of equatorial TEC depletion at latitudes below 7° N. A strong narrow electric field enhancement (SAID) lay immediately equatorward of the precipitation boundary at 39° N, while the fringing SAPS electric field extended equatorward to below ~30° N, overlapping the region of TEC erosion and plume formation.

age plumes at *F*-region and magnetospheric heights provide a sufficient source for the processes which feed and accelerate the ionospheric material into the tail. The present study indicates that the ultimate source of this thermal plasma lies on field lines which thread the inner plasmasphere and that the tongues of ionization seen over the polar caps can have very recently (2-3 hours) originated in the equatorial ionosphere.

The plasma redistribution mechanisms we have described and their enhancement at longitudes near the South Atlantic Anomaly suggest a UT dependence for the injection of ionospheric ions into the storm-time magnetosphere via such strong plasmasphere drainage plumes. A large flux of these cold ions will change the characteristics of the magnetosphere and may well alter the response of the magnetosphere to the developing magnetic storm—i.e. they may alter the geoeffectiveness of the storm drivers. We speculate that the appearance of rich concentrations of ionospheric ions in the disturbance ring current in the latter phases of great storms indicates that plasmaspheric erosion/ SED/ion injection may have preconditioned the magnetosphere before the final surges of particle injection which drive the storm-time Dst to extreme values during these events. It is possible that maximum negative Dst during the truly great storms (Dst < -300 nT) occur at similar UTs—i.e. when the SAA has recently been in the dusk/bulge sector (20 UT–00 UT). The rapidly-increasing ability to observe and characterize such events will soon make statistical tests of such hypotheses possible.

Acknowledgements. GPS and IMAGE analysis are supported by NASA SEC Guest Investigator Award (NAG5-12875) to the MIT Haystack Observatory. Radar observations and analysis at the Millstone Hill Observatory are supported by Co-operative Agreement ATM-0233230 between the National Science Foundation and the Massachusetts Institute of Technology. Work at The University of Arizona was funded by a subcontract from Southwest Research Institute, under NASA contract NAS5-96020 with SwRI. DMSP analysis is partially sponsored by the Air Force under Air Force Contract AF19628-00-C-0002.

REFERENCES

Basu, S., Sa. Basu, K. M. Groves, H. C. Yeh, F. J. Rich, P. J. Sultan, and M. J. Keskinen, Response of the equatorial ionosphere to the great magnetic storm of July 15, 2000, *Geophys. Res. Lett., 28*(18), 3577–3580, 2001.

Buonsanto, M. J., and J. C. Foster, Effects of Magnetospheric Electric Fields and Neutral Winds on the Low-Middle Latitude Ionosphere during the March 20–21, 1990 Storm, *J. Geophys. Res., 98,* 19133–19140, 1993.

Carpenter, D. L., B. L. Giles, C. R. Chappell, P. M. E., Decreau, R. R. Anderson, A. M. Persoon, A. J. Smith, Y. Corcuff, and P. Canu, Plasmaspheric dynamics in the duskside bulge region: a new look at an old topic, *J. Geophys. Res., 98,* 19243, 1993.

Carpenter, D. L., and J. Lemaire, The plasmasphere boundary layer, *Ann. Geophys., 22,* 4291, 2004.

Coster, A. J., J. Foster, and P. Erickson, Monitoring the Ionosphere with GPS: Space Weather, *GPS World, 14 (5),* 42–49, 2003.

Elphic, R. C., M. F. Thomsen, and J. E. Borovsky, The fate of the outer plasmasphere, *Geophys. Res. Lett., 24,* 365, 1997.

Foster, J. C., Storm-Time Plasma Transport at Middle and High Latitudes, *J. Geophys. Res., 98,* 1675–1689. 1993.

Foster, J. C., Plasma Transport Through the Dayside Cleft: A Source of Ionization Patches in the Polar Cap, pp 343–354, *Electromagnetic Coupling in the Polar Clefts and Caps,* P. Sandholt and A. Egeland, eds., Kluwer Acad. Pubs., Dordrecht, 1989.

Foster, J. C. and J. R. Doupnik, Plasma convection in the vicinity of the dayside cleft, *J. Geophys. Res., 89,* 9107–9113, 1984.

Foster, J. C., A. J. Coster, P. J. Erickson, J. Goldstein, and F. J. Rich, Ionospheric Signatures of Plasmaspheric Tails, *Geophys. Res. Lett., 29(13),* 10.1029/2002GL015067, 2002.

Foster, J. C., A. J. Coster, P. J. Erickson, F. J. Rich, and B. R. Sandel (2004), Stormtime observations of the flux of plasmaspheric ions to the dayside cusp/magnetopause, *Geophys. Res. Lett., 31,* L08809, doi:10.1029/ 2004GL020082, 2004.

Foster, J. C., and F. J. Rich, Prompt mid-latitude electric field effects during severe geomagnetic storms, *J. Geophys. Res., 103,* 26367–26372, 1998.

Foster, J. C., and H. B. Vo, Average Characteristics and Activity Dependence of the Subauroral Polarization Stream, *J. Geophys. Res., 107(A12),*1475, doi: 10.1029/2002JA009409, 2002.

Goldstein, J., B. R. Sandel, M. R. Hairston, and P. H. Reiff, Control of plasmaspheric dynamics by both convection and subauroral polarization stream, *Geophys. Res. Lett., 30(24),* 2243, 10.1029/2003GL018390, 2003a.

Goldstein, J., M. Spasojevic, P. H. Reiff, B. R. Sandel, W. T. Forrester, D. L. Gallagher, and B. W. Reinisch, Identifying the plasmapause in IMAGE EUV data using IMAGE RPI in situ steep density gradients, *J. Geophys. Res, 108(A4),* 1147, doi:10.1029/ 2002JA009475, 2003b.

Greenspan, M. E., C. E. Rasmussen, W. J. Burke, and M. A. Abdu, Equatorial density depletions observed at 840 km during the great magnetic storm of March 1989, *J. Geophys. Res., 96,* 13931–13942, 1991.

Immel, T. J., J. C. Foster, A. J. Coster, H. U. Frey, and S. B. Mende, Global stormtime plasma redistribution imaged from the ground and space, *Geophys. Res. Lett., 32 (3),* L03808,10.1029/ 2004GL021597, 2005.

Kelley, M. C., M. Vlassov, J. C. Foster, and A. J. Coster, A quantitative explanation for the phenomenon known as plasmaspheric tails or storm-enhanced density, *Geophys. Res. Lett., 31,* L19809, doi:10.1029/2004GL020875, 2004.

Mende, S. B., et al., Far ultraviolet imaging from the IMAGE spacecraft. 3. Spectral imaging of Lyman-alpha and OI 135.6 nm, *Space Sci. Rev., 91,* 287–318, 2000.

Sandel, B. R., R. A. King, W. T. Forrester, D. L. Gallagher, A. L. Broadfoot, and C. C. Curtis, Initial Results from the IMAGE Extreme Ultraviolet Imager, *Geophys. Res. Lett., 28,* 1439, 2001.

Tsyganenko, N. A., A Model of the Near Magnetosphere with a Dawn-Dusk Asymmetry: 1. Mathematical Structure, *J. Geophys. Res., 107(A8),* 10.1029/2001JA00219, 2002.

Yin P., C. N. Mitchell, P. S. J. Spencer, J. C. Foster, Ionospheric electron concentration imaging using GPS over the USA during the storm of July 2000, *Geophys. Res. Lett., 31,* L12806, doi:10.1029/2004GL019899, 2004.

A. J. Coster, MIT Haystack Observatory, Route 40, Westford, MA 01886, USA.

P. J. Erickson, MIT Haystack Observatory, Route 40, Westford, MA 01886, USA.

J. C. Foster, MIT Haystack Observatory, Route 40, Westford, MA 01886, USA. (jfoster@haystack.mit.edu)

T. J. Immel, Space Sciences Laboratory, University of California, Berkeley, Berkeley, CA 94720, USA.

F. J. Rich, Air Force Research Laboratory, 29 Randolph Road, Hanscom AFB, MA 01731, USA.

W. Rideout, MIT Haystack Observatory, Route 40, Westford, MA 01886, USA.

B. R. Sandel, Lunar and Planetary Laboratory, University of Arizona, 1040 East 4th Street, Room 901, Tucson, AZ 85721, USA.

Yosemite 2004—A Thirty Year Tradition

C. R. Chappell

Department of Physics and Astronomy
Vanderbilt University
Nashville, Tennessee

It has been a great privilege to be associated with the Yosemite Conference over a period of thirty years. This career-length tradition gives us a chance to see our science and ourselves through the broad brush of time. In a sense, the video recording of the very first meeting in 1974 is a time capsule that we can open and enjoy from the perspective of the twenty-first century. As such, it brings a view of how much and how little we understood about the geospace environment in the second decade after the initiation of space exploration.

The original idea for the Yosemite meeting came from discussions between Peter Banks, Andy Nagy and me in 1972. We had been working in California on topics related to the magnetosphere and the ionosphere and had become aware of the difficulty of arranging for scientists who studied these two areas to confer at meetings. We were convinced of the importance of the interdisciplinary approach to geospace—where atmospheric, ionospheric and magnetospheric physicists shared scientific progress made and taught each other about their different sub-disciplines. Today, this idea seems obvious, but in 1974 conferences that created interaction across disciplines were limited, and the national AGU meetings had so many parallel sessions that formal interdisciplinary discussions were difficult to pursue.

Peter Banks and Andy Nagy had attended meetings in Norway, in which the agenda was more relaxed with more time for interaction both within and between sessions. Since all of us had visited Yosemite and knew of its special beauty and tranquility, we saw it as a perfect place to start an interdisciplinary meeting in which the grand environment of "place" would facilitate positive and extensive interaction among all participants. We compiled a list of topics and possible speakers and gained the approval of the AGU to hold the Magnetosphere-Ionosphere Coupling Conference as one of the first Chapman Conferences.

I was anxious to capture as much of the scientific interaction as possible, so I sought a way to videotape the conference activities. In the early 70's video equipment was available, but had not reached the ease of operation that exists today. Fortunately, with the help of Dick Hoch, the Battelle Northwest Laboratory agreed to furnish the equipment and staff support to videotape the entire meeting. We have more than twenty hours of proceedings as a result, including all of the papers and the discussion. Although the quality is not what it would be today and it is in black and white, the essence of the subject and the people are there; the video is fascinating to watch.

Unfortunately, the range of topics is extensive and cannot be covered completely. It is possible, however, to provide some impressions of the meeting overall and then to identify some of the specific topics that were discussed as a touchstone with which to measure the progress in our understanding over this thirty-year period.

First, it is clear that the interdisciplinary approach was effective and successful. Bob Carovillano, who summarized the conference at the closing session, stated that scientists in the magnetospheric and ionospheric fields would forever be changed by their enhanced knowledge of each other's field. He predicted that future research in either discipline would involve knowledge of and interaction with the science and scientists of the other disciplines. The success of the interdisciplinary approach is evidenced by the long-tradition of Yosemite meetings that it spawned.

Second, the time capsule look at all of us thirty years earlier in our lives is fascinating. There is, of course, the change in appearance that comes with age. Most of the brown hair in the room then has now become grey except for Dave Evans whose hair was prematurely grey even in 1974. We were all very young in those days since space exploration itself was very young. In addition to the hair color, the amount of head hair was significantly different. The 70's were still under the influence of the turbulent 60's and the length of hair, sideburns and beards was notable. Clothing also had a distinctive 60's flavor with big ties, polyester pants with horizontal front pockets, big belts and shirts that did not have the normally expected conservative look of a serious scientist. There were few women and minorities involved in 1974—a characteristic that happily has changed over the past three decades.

Third, the science issues thirty years ago are very familiar to us today. We were on the right track then but had only the first few pieces of the puzzle to examine. Our picture was somewhat fuzzy, but it was tuned in to the correct object. Looking at the videotapes, one is struck by the similarity of the discussion then to today's discussion. We just had a lot less data and fewer models to work with in 1974. I will mention several examples of the specific discussion below.

Fourth, the magnificence of Yosemite National Park and of the Ahwahnee Lodge were most significant in the remembrance of the meeting. When one is put into a bigger-than-life place with incomparable beauty, there is a certain openness to big new ideas and to future collaborations that might not have come to be without the conversations on the bus rides up and out of the canyon to Badger Pass or the dinners in the Ahwahnee Lodge or the discussions before the gigantic fireplaces in the Great Lounge after dinner. The place itself made a difference.

Although justice cannot be done regarding the science topics that were discussed, I will try to give a flavor of what was on our minds thirty years ago, particularly as it relates to the topics of this most recent descendent meeting.

Topics could be found in one of several categories—definitions/terminology, early modeling of the ionosphere and magnetosphere, and individual ground and space-based data taken in the different regions. Ian Axford reminded us that according to the original definition, the ionosphere above 120 km was part of the magnetosphere so that discussions of ionosphere/magnetosphere coupling were somewhat misleading. This unifying definition, however, had not succeeded in bringing the separate disciplines of ionospheric and magnetospheric physics together at that time. There were also spirited discussions of other terminology as scientists from each discipline sought to clarify their fields and understand each other.

The meeting saw the beginnings of modeling efforts in both the ionosphere and the magnetosphere. The Michigan group presented several new ionospheric models and the east coast/west coast discussion of a magnetic field model with and without a tail component was active. Initial modeling of wave-particle interactions in the auroral zone and at the plasmapause/ring current interface led to a lengthy sharing of potential explanations and ideas. Dick Wolf discussed the initial version of the Rice Convection Model in which the affects of the dynamic plasma regions of the magnetosphere were connected along magnetic field lines to the Earth's ionosphere and vice versa.

Magnetospheric morphology was a hot topic in 1974. We sought, for example, to understand the dynamics of the plasmasphere and its interaction with the plasma sheet and the ring current as well as its origins in the ionosphere. Few coordinated satellite or satellite/ground-based measurements had been conducted at that time. Hence, the dynamics of the plasmasphere had been determined separately by whistlers, ionospheric satellites, ionospheric sounders, and magnetospheric satellites. A lively discussion of plasma tails versus detached plasma regions also set the stage for the innovative space-based imaging measurements that were to come almost three decades later.

Electric field measurements in the ionosphere and magnetosphere were just beginning to be made with some confidence. Extensive discussions of the relative merits of ground-based radar, barium clouds, space-borne driftmeters and electric field probes were evident at this first Yosemite meeting. In addition, Lars Block showed laboratory experimental results demonstrating the double-layer phenomenon and suggested its application to understanding auroral processes.

There were several presentations in which comprehensive multi-instrument satellite data had been analyzed with an eye toward examining the applicability of a wave-particle interaction theory. Explorer 45 data was presented by Don Williams and Jim Burch in support of Richard Thorne's theoretically predicted wave-particle interaction between the hot and cold plasma in the magnetosphere. And Keith Cole reminded everyone of the importance of Coulomb collisions as an alternate explanation for the loss of particles from the ring current.

In 1974, we all were aware that the ionosphere and magnetosphere were coupled. However, the general feeling was that the magnetosphere was a much stronger influence on the ionosphere than vice versa. It was certainly recognized that the magnetic field lines connected the two regions and that current flow between the two regions was important. The ionosphere was strongly influenced by the precipitating energetic particles from the magnetosphere, and the resulting electric field changes could in turn be mapped back up into the magnetosphere.

But how significant were the influences of the cold ionospheric plasma on the much more energetic plasmas of the plasma sheet and ring current? The possible effects of the cold plasmaspheric plasma on the energetic ring current through wave-particle interactions was being discussed, but the idea that the ionospheric plasma could be a source for the magnetospheric plasmas was not seriously considered. The Lockheed group had measured upflowing energetic ions in connection with the auroral zone, but this phenomenon was not considered as a significant source for the magnetospheric plasma at this time.

Although the polar wind was discussed at this first Yosemite meeting, it was never conceived of as a potential source for the much more energetic plasmas of the plasma sheet and ring current. That idea would come much later as polar orbiting satellites began to see the spatial breadth and magnitude of the low and medium energy ionospheric outflow. Larry Kavanagh, from NASA Headquarters, however, discussed one of those satellites, the Electrodynamics Explorer, with the interdisciplinary community at this first Yosemite meeting. His prediction of a launch between 1978 and 1980 proved to be close for Dynamics Explorer, which replaced Electrodynamics Explorer as the first ionosphere-magnetosphere coupling mission.

In summary, looking back at this first Yosemite meeting gives all of us in solar-terrestrial physics a chance to revisit our heritage and our youth. In watching the video we can view the science process at work, as the space explorers of then and now assemble the pieces of the puzzle and seek to understand the full picture of the dynamic geospace environment.

There are many colleagues who participated in that meeting who are no longer with us. Neil Brice, to whom the meeting was dedicated, was killed in an airplane crash on the way to the meeting. Bill Hanson, Chung Park, Dave Reasoner, Jim Armstrong and Lars Block, who were such tremendous contributors to our field and to that meeting, are now gone. We miss all of them now and leave the video summary as a tribute to them, our good friends and colleagues.

The significance of the Yosemite meeting series can be experienced by being part of the venue and the excitement. It has been a privilege for me to share this excitement by connecting to the group of very special people who have come together at Yosemite over the decades. Thanks to Andy Nagy and Peter Banks for their participation in the creation of the first meeting, to Dick Hoch and Battelle Northwest Laboratories for videotaping the conference, and to Jim Burch for his longstanding leadership of the continuing Yosemite series. And thanks to the Yosemite National Park and to the Ahwahnee Lodge for furnishing the magical environment which has stimulated this most successful journey of exploration.

C. R. Chappell, Vanderbilt Dyer Observatory, 1000 Oman Drive, Brentwood, Tennessee, 37027